Inverse Scattering Problems and Their Application to Nonlinear Integrable Equations

Second Edition

Inverse Scattering Problems and Their Application to Nonlinear Integrable Equations, Second Edition is devoted to inverse scattering problems (ISPs) for differential equations and their applications to nonlinear evolution equations (NLEEs). The book is suitable for anyone who has a mathematical background and interest in functional analysis, differential equations, and equations of mathematical physics. This book is intended for a wide community working with ISPs and their applications. There is an especially strong traditional community in mathematical physics.

In this monograph, the problems are presented step-by-step, and detailed proofs are given for considered problems to make the topics more accessible for students who are approaching them for the first time.

New to the Second Edition

- All new chapter dealing with the Bäcklund transformations between a common solution of both linear equations in the Lax pair and the solution of the associated IBVP for NLEEs on the half-line
- Updated references and concluding remarks

Features

- Solving the direct and ISP, then solving the associated initial value problem (IVP) or initial-boundary value problem (IBVP) for NLEEs are carried out step-by-step. The unknown boundary values are calculated with the help of the Lax (generalized) equations, then the time-dependent scattering data (SD) are expressed in terms of preassigned initial and boundary conditions. Thereby, the potential functions are recovered uniquely in terms of the given initial and calculated boundary conditions. The unique solvability of the ISP is proved, and the SD of the scattering problem is described completely. The considered ISPs are well-solved.
- The ISPs are set up appropriately for constructing the Bäckhund transformations (BTs) for solutions of associated IBVPs or IVPs for NLEEs. The procedure for finding a BT for the IBVP for NLEEs on the half-line differs from the one used for obtaining a BT for nonlinear differential equations defined in the whole space.
- The interrelations between the ISPs and the constructed BTs are established to become new powerful unified transformations (UTs) for solving IBVPs or IVPs for NLEEs, that can be used in different areas of physics and mechanics. The application of the UTs is consistent and efficiently embedded in the scheme of the associated ISP.

Pham Loi Vu, PhD is a professor and a leading Vietnamese expert in inverse problems and their applications. He has authored 50 papers published in prominent journals, including Inverse Problems (Q1), Acta Applicandae Mathematicae (Q2), Journal of Nonlinear Mathematical Physics (Q2), and others.

His previous research focused on seismic waves in seismic prospecting for petroleum-gas complexes and on determining coordinates of epicenter in near earthquakes at the Institute of Geophysics of the Vietnam Academy. He is now a researcher at VAST's Institute of Mechanics and the National Foundation for Science and Technology Development (NAFOSTED).

Monographs and Research Notes in Mathematics

Series Editors:

John A. Burns, Thomas J. Tucker, Miklos Bona, Michael Ruzhansky

About the Series

This series is designed to capture new developments and summarize what is known over the entire field of mathematics, both pure and applied. It will include a broad range of monographs and research notes on current and developing topics that will appeal to academics, graduate students, and practitioners. Interdisciplinary books appealing not only to the mathematical community, but also to engineers, physicists, and computer scientists are encouraged.

This series will maintain the highest editorial standards, publishing well-developed monographs as well as research notes on new topics that are final, but not yet refined into a formal monograph. The notes are meant to be a rapid means of publication for current material where the style of exposition reflects a developing topic.

Noncommutative Polynomial Algebras of Solvable Type and Their Modules
Basic Constructive-Computational Theory and Methods
Huishi Li

Fixed Point Results in W-Distance Spaces
Vladimir Rakočević

Analysis of a Model for Epilepsy
Application of a Max-Type Difference Equation to Mesial Temporal Lobe Epilepsy
Candace M. Kent, David M. Chan

Double Sequence Spaces and Four-Dimensional Matrices
Feyzi Başar, Medine Yeşilkayagil Savaşcı

Banach-Space Operators On C*-Probability Spaces Generated by Multi Semicircular Elements
Ilwoo Cho

Applications of Homogenization Theory to the Study of Mineralized Tissue
Robert P. Gilbert, Ana Vasilic, Sandra Klinge, Alex Panchenko, Klaus Hackl

Inverse Scattering Problems and Their Application to Nonlinear Integrable Equations, Second Edition
Pham Loi Vu

For more information about this series please visit: https://www.routledge.com/Chapman--HallCRC-Monographs-and-Research-Notes-in-Mathematics/book-series/CRCMONRESNOT

Inverse Scattering Problems and Their Application to Nonlinear Integrable Equations

Second Edition

Pham Loi Vu
Institute of Mechanics, Vietnam Academy of Science and Technology, Vietnam

CRC Press
Taylor & Francis Group
Boca Raton London New York

CRC Press is an imprint of the
Taylor & Francis Group, an **informa** business
A CHAPMAN & HALL BOOK

Second edition published 2023
by CRC Press
6000 Broken Sound Parkway NW, Suite 300, Boca Raton, FL 33487-2742

and by CRC Press
4 Park Square, Milton Park, Abingdon, Oxon, OX14 4RN

First edition published by CRC Press 2020

CRC Press is an imprint of Taylor & Francis Group, LLC

Library of Congress Cataloging-in-Publication Data

Names: Vu, Pham Loi, author.
Title: Inverse scattering problems and their application to nonlinear integrable equations / Pham Loi Vu, Institute of Mechanics and the National Foundation for Science and Technology Development (NAFOSTED), Vietnam.
Description: Second edition. | Boca Raton : C&H/CRC Press, 2023. | Series: Monographs & research notes in mathematics | Revised edition of: Inverse scattering problems and their application to nonlinear integrable equations / Pham Loi Vu. 1. 2019. | Includes bibliographical references and index.
Identifiers: LCCN 2022052684 (print) | LCCN 2022052685 (ebook) | ISBN 9781032429212 (hardback) | ISBN 9781032441160 (paperback) | ISBN 9781003370543 (ebook)
Subjects: LCSH: Inverse problems (Differential equations) | Scattering (Mathematics) | Nonlinear integral equations.
Classification: LCC QA378.5 .V87 2023 (print) | LCC QA378.5 (ebook) | DDC 515/.357--dc23/eng20230119
LC record available at https://lccn.loc.gov/2022052684
LC ebook record available at https://lccn.loc.gov/2022052685

ISBN: 978-1-032-42921-2 (hbk)
ISBN: 978-1-032-44116-0 (pbk)
ISBN: 978-1-003-37054-3 (ebk)

DOI: 10.1201/ 9781003370543

Typeset in font CMR10
by KnowledgeWorks Global Ltd.

Publisher's note: This book has been prepared from camera-ready copy provided by the authors.

Dedicated to the loving memory of
my late son Pham Vu Nam (1975-2001) and
younger brother Pham Quang Dien (1941-1999)

Contents

Acronyms

BC .... Boundary Condition

BT .... Bäcklund Transformation

BV .... Boundary Value

GLM .... Gelfand–Levitan–Marchenko

IBVP .... Initial-Boundary Value Problem

ISM .... Inverse Scattering Method

ISP .... Inverse Scattering Problem

IVP .... Initial Value Problem

KdV .... Korteweg-de Vries

mKdV .... modified Korteweg-de Vries

NLEE .... Non-linear Evolution Equation

NLS .... Non-linear Schrödinger

ODE .... Ordinary Differential Equation

R–H .... Riemann–Hilbert

SD .... Scattering Data

sG .... sine-Gordon

shG .... sinh-Gordon

SP .... Scattering Problem

Preface

The present monograph is devoted to inverse scattering problems (ISPs) for differential equations and their application to non-linear evolution equations (NLEEs). It contains most of our published papers, which are mentioned in the Introduction and at the beginning of every chapter.

In this monograph, we study the ISPs for the following differential equations:

- The systems of two first-order ordinary differential equations (ODEs) on a half-line with a non-self-adjoint potential matrix [97], and with a self-adjoint potential matrix [119] in Chapters 1–4;
- The system of n first-order ODEs and the transport equation in Chapter 5. The transport equation is a continual generalization from the system of n first-order ODEs on the whole line [124, 126];
- The Schrödinger on a half-line [76, 79, 86, 147] in Chapters 6 and 7;
- The perturbed string equation in characteristic variables on the whole line [122, 123] in Chapter 8.

The results of considered ISPs in the monograph are not only independently interesting by themselves but also are effective tools for solving initial value problems (IVPs) and initial-boundary value problems (IBVPs) for integrable NLEEs, such as the non-linear Schrödinger equation [133, 140], Korteweg-de Vries (KdV) [139, 144], modified KdV [142], KdV with dominant surface tension [146, 147], two-dimensional generalization from the KdV [148], sine-Gordon, sinh-Gordon equation [143].

The scattering data of the considered ISPs are described completely. That is to establish the necessary and sufficient conditions for given quantities to be the scattering data for the problem generated by the considered differential equation with the known initial and boundary conditions. With the help of the Lax equation or the generalized Lax equation, the application of obtained results of the ISP to solving the associated IVP or IBVP for the NLEE is carried out step by step. Namely, the NLEE can be written as the compatibility condition of two linear equations. Such NLEEs are called integrable, and the pair of linear equations is called the Lax pair [44, 69]. Then, we can derive the time-dependence of the scattering data. Hence, the time-evolution of the known functions in the time-dependent fundamental equations in the inverse problem is constructed. By virtue of the complete description of the scattering data, the fundamental equations have unique solutions. Then the solution of the NLEE is expressed through the time-dependent solution of the fundamental equations. Hence, the application of the inverse scattering method (ISM) to solving the IVPs or IBVPs for NLEEs is consistent. It is effectively embedded in the schema of the ISM.

The results of ISPs and associated IVPs, IBVPs for NLEEs in the present monograph are related with the books by Ablowitz and Clarson (1991) [1], Marchenko (1986) [86], Chadan and Sabatier (1977) [19], Fokas and Zakharov (1994) [43], Levitan (1984) [76], Nizhnik (1973) [93], Zakharov et al. (1980) [156], Beals et al. (1988) [10], Korepin et al. (1993) [62], Osborne (2010) [100], and Gladwell (1993) [49].

An important stage in the development of the ISM for solving IBVPs has witnessed a considerable number of novel concepts of interest physically and mathematically. A success in this field was achieved by Sabatier in 2000 [106], Fokas in 2000 and 2002 [36, 37, 38], Habibullin in 2002 [53], ... Due to the rate at which the field has developed, some references are missed or forgotten. We apologize for such situations whenever they occur.

The present monograph is intended for a wide community working with inverse scattering problems and their application, in particular, the traditional community in mathematical physics. This monograph is accessible to anyone having a mathematical background in functional analysis [61, 80], partial differential equations [112], equations of mathematical physics [116], and the theory of functions of a complex variable [114].

In this monograph the considered problems are solved step by step in detail. There are full proofs for solving the considered problems. By this way, the presentation in this monograph is an effort to make the material more accessible for students, who are approaching this topic for the first time.

I express my sincere gratitude to the Institute of Mathematics of Academy of Sciences of Ukraine, particularly Prof. L. P. Nizhnik for his invaluable help and guidance for my Candidate dissertation Ph.D. (1973) and the Doctoral dissertation of Science (1983). I am grateful to the Institute of Mechanics of Vietnam Academy of Science and Technology for offering me a supportive working environment throughout my time with the institute. I am very thankful to Istituto per le Applicazioni del Cacolo "Mauro Picone" in Firenze, Italy, particularly Prof. Gabriele Inglese for the valuable help during my working at the institute. I wish to thank Dr. Nguyen Huy Hoang for useful discussions on the topic. My special thanks to Tran Ngoc Trung and Truong Thi Phuong for their valuable assistance for a long time in the preparation of the manuscript to the appearance of the book. Finally, I am greatly indebted to my wife Nguyen Thi Huyen for her love and support. The text would not have been possible if it were not for the help of the mentioned people.

It is disturbing me that I can read and reread the text and still not catch all errors. My hope is that any remaining errors will not lead to undue confusion.

Pham Loi Vu
Hanoi

Preface for the second edition

The Bäcklund Transformations (BTs) for solutions of non-linear differential equations defined on the whole line or in the whole space are well known. We now contribute to constructing BTs for solutions of initial-boundary value problems (IBVPs) for non-linear evolution equations (NLEEs) defined on the half-line with the preassigned boundary condition at the origin $x = 0$. So, we add to the book's first edition a fresh chapter entitled: "The Bäcklund transformations between a common solution of both linear equations in the Lax pair and solution of the associated IBVPs for NLEEs on a half-line". In this book, the inverse scattering problems (ISPs) are well-solved and are set up appropriately for constructing BTs. The interrelation between the ISP and the constructed BT is established to become new powerful unified transformations (UTs) that can be used in different areas of physics and mechanics. The application of the obtained UTs is consistent and efficiently embedded in the scheme of the ISP (see Figure 10.11). The added chapter is based on the paper [152].

I wish to thank Mr. Tran Ngoc Trung for his important assistance for a long time in the preparation of the manuscript to the appearance of the present edition.

Pham Loi Vu
Hanoi, February 20, 2023

Author

Pham Loi Vu is a professor, a doctor of mathematics, and a leading Vietnamese expert in inverse problems and their applications. He has authored 51 papers published in prominent journals, including *Inverse Problems* (Q1), *Acta Applicandae Mathematicae* (Q2), *Journal of Nonlinear Mathematical Physics* (Q2) and others.

His previous research focused on seismic waves in seismic prospecting for petroleum-gas complexes and on determining coordinates of epicenter in near earthquakes at the Institute of Geophysics of the Vietnam Academy. He is now a researcher at VAST's Institute of Mechanics and the National Foundation for Science and Technology Development (NAFOSTED).

Introduction

An inverse problem is a usual name for any problem that requires a mathematician to ascertain the data that will determine a differential operator uniquely and a method of constructing this operator from the known data. The theory of inverse problems has been developed into some directions:

- Inverse kinematical problems of seismicity are classified as the first direction [9, 68, 67];
- Inverse problems of the Newtonian potential theory are classified as the second direction. These inverse problems are related to geophysical problems of interpretation of gravitational and magnetic data [120, 102, 105, 113];

  Various forms of inverse problems have been considered, such as inverse problems for parabolic equations, problems of determination of diffusion coefficients in the models of turbulent diffusion of contaminants [127, 128, 129, 131, 130, 132];
- The inverse scattering problems (ISPs) arising in connection with the quantum scattering theory belong to the third direction. These problems are the most interesting from the standpoint of application to the integration of non-linear evolution equations (NLEEs).

The present monograph is dedicated to the ISPs for differential equations that are associated with a class of distinguished NLEEs in mathematical physics. We study the ISPs for differential equations mentioned in the Preface and then apply the obtained results of the considered ISPs to the integration of associated integrable NLEEs by the inverse scattering method (ISM).

The discovery of the ISM was preceded by numerical simulation of the KdV. In 1964 Kruskal and Zabuskii [44], using digital simulation, concluded that solitons in the Korteweg-de Vries formalism suffer an elastic collision. This result gave an impetus to new analytical studies. Then an unending series of laws of conservation was discovered, and finally, in 1967, this progress was crowed with the discovery of the ISM [44]. This method is undoubtedly one of the most elegant discoveries of mathematical physics in the 20th century. It will also aid in advancing its applications.

We now outline the development of problems that are related to the considered problems in this monograph.

A large number of investigations in different areas of physics were devoted to the study of non-linear wave processes, for instance, various questions relating to the theory of plasma and non-linear optics. Simple "model" non-linear wave equations were constructed in the course of the development of the non-linear wave theory. In some sense, these constructed equations are universal, i.e., they may be encountered, just like the classical d'Alembert linear equations in diverse physical problems. Examples are the non-linear Schrödinger (NLS), Korteweg-de Vries (KdV), modified Korteweg-de Vries (mKdV), sine-Gordon (sG) and sinh-Gordon (shG) equations. These non-linear equations exhibit, at least in the one-dimensional case, a remarkable property. They possess hidden algebraic symmetry, as a result of which they can be "integrated" by the so-called ISM with the help of a pair of auxiliary linear operators, i.e., the non-linear equation can be written as the compatibility condition of two linear equations. The two equations constituting the Lax pair are usually designated the x part and the t (time) part. The ISM for solving the initial-value problem (IVP) for the KdV equation with decaying initial data was discovered by Gardner–Greene–Kruskal–Miura in 1967 [44]. The ISM can be thought of as a non-linear Fourier method. However, this non-linear transform is not the same for every NLEE, but it is constructed from the x part of the Lax pair. Furthermore, the direct non-linear Fourier transform of the initial data involves a linear Volterra integral equation, and the inverse non-linear Fourier transform involves a matrix Riemann–Hilbert problem. It should be emphasized that the construction of this non-linear transform is based solely on the x part of the Lax pair, and it involves the spectral analysis of the eigenvalue equation. The t part is used only to determine the evolution of the direct non-linear Fourier transform; see [34] for the early history and [43] for some developments.

The first ISP was formulated and investigated by Ambartsumyan in 1929 [8]. Uniqueness theorems in inverse problems were proved by Ambartsumyan [8], Borg in 1946 [17]. Research on quantum scattering theory inverse problem was carried out by Levison in 1949, 1953 [73, 72, 74]. Further success in inverse problem theory was achieved by Chudov in 1949 [21], Marchenko in 1952, 1955, 1972 [84, 85, 86], Krein in 1951–1958 [63, 64, 65, 66], Berezanskii in 1964, 1968 [11, 12], Gelfand and Levitan in 1951 [48], Levitan in 1973, 1984 [75, 76]. Levitan and Gasymov in 1964 [77], Faddeev in 1958, 1959, 1966 [31, 32, 33], Faddeev and Takhtajan in [34], Nizhnik in 1971, 1973, 1980, 1988 [92, 93, 94, 95], Nizhnik and Pham Loi Vu in 1974, 1982, 1983 [97, 98, 99], Pham Loi Vu in 1971–1973, 1976, 1980, 1981, 1983 [117, 118, 119, 121, 122, 123, 124, 125, 126], etc.

A decisive step was the application, first by Marchenko and then by others, of the so-called "transformation operators" to the investigation of inverse problems. These operators arose from ideas on the generalized shift operator theory, the foundations of which were laid by Delsarte in 1939 (see Levitan [75]).

The ISM is an effective tool for studying NLEEs in one space variable. It allows one to construct large classes of exact solutions, and to investigate in detail the Cauchy problems for NLEEs including the writing down explicit time asymptotics. In this way, the numerous results were achieved [1, 43, 133, 137, 135, 136, 134, 138, 139, 141, 145, 153]. In the class of non-scattering potentials, the N-soliton solutions of the NLS and KdV equations are constructed by the ISM and Hirota Direct method [133, 137, 149, 150]. A method for solving the Cauchy problems for NLEEs with space-periodic initial data was developed in the mid-1970s [25].

Further success in inverse scattering in two or three dimensions with boundary integral equations was achieved by Cakoni and Colton in the monograph Qualitative Method in Inverse Scattering Theory (2006) [18], and in Inverse Acoustic and Electromagnetic Scattering Theory by Colton and Kress (1998) [22].

There exist some essential difficulties in extending the application of the ISM to the integration of NLEEs.

An essential difficulty arises when the ISM is applied to the integration of multidimensional NLEEs. The difficulty arises from the strict restriction on coefficients of multidimensional operators, which define the Lax pair. In this issue, the generalized Lax representations of multidimensional NLEEs were derived from the generalized Lax equation, and further success was achieved in [94, 95, 96].

As mentioned above, the ISM makes possible a complete investigation of the Cauchy problems for integrable NLEEs. However, this method is difficult to transfer to IBVPs for the NLEEs with arbitrary boundary and initial conditions. Such problems appear in many applications, where it is often more natural to assume that the space variable is defined only on the part of the real axis. For example, in the context of water waves, it is usually far easier to measure the elevation of the water surface over time at a fixed position (say $x = 0$) than to determine the surface for all values of the real axis x at a specific instant of time. This leads naturally to the mathematical formulation of an IBVP on the half-line ($x > 0$) with a prescribed boundary condition at $x = 0$ and with vanishing initial conditions.

For IBVPs, an important issue, both for mathematical and physical considerations, is the study of the effect of the boundary on the asymptotic behavior of the solution. The problem of searching boundary conditions consistent with the integrability property of the NLEEs given as well as the problem of finding proper modification of analytical integration procedure for the corresponding IBVPs is undoubtedly important.

Several attempts have been made to extend the ISM to the case of more complicated IBVPs, in which Dirichlet and/or Neumann conditions are prescribed on a half-line or on a segment. This is reminiscent of the fact that, even for linear differential equations, scattering data is given in terms of both prescribed boundary values and unknown boundary values. Thus, the essential difficulty associated with these problems is that the evolution equations for scattering data, as given by the Lax equations, cannot be integrated into most of the cases, because their coefficients depend on unknown boundary data. After the series of papers by Its and Fokas in 1989, 1992 and 1996 [35, 40, 41], it becomes clear that under arbitrary boundary conditions, solving the IBVP for the above-mentioned equations had not met the same success as solving the Cauchy problem on the whole line. There exists a specific class of boundary conditions that are completely consistent with the integrability property. Under these conditions, the IBVP is effectively embedded in the ISM schema. A number of examples of such boundary conditions were discussed in [23, 24, 38, 42, 52, 54, 51, 53, 140, 142, 143, 144, 146, 147]. In [70, 71] the inverse spectral transform for the Zakharov–Shabat equation on the semi-line is reconsidered as a Hilbert problem. Then solving the inverse problem means solving a Hilbert problem with particular prescribed behavior. An important success in methodology for analysis of IBVPs for non-linear integrable partial differential equations on a half-line was achieved by Fokas in 2002 [38]. This methodology is first presented in general, and then applied to the NLS equation, sG equation and KdV equation with dominant surface tension.

This book is organized as follows. Chapter 1 presents two ISPs for the system of two first-order ODEs on a half-line:

$$-iJu_x + C(x)u = \lambda u, \quad u = (u_1(x,\lambda), u_2(x,\lambda)), \quad 0 \leq x < \infty, \tag{0.1}$$

with the boundary condition (BC) at the origin $x = 0$:

$$u_1(0,\lambda) = u_2(0,\lambda), \tag{0.2}$$

where $J = \begin{pmatrix} -1 & 0 \\ 0 & 1 \end{pmatrix}, C(x) = \begin{pmatrix} 0 & c_1(x) \\ c_2(x) & 0 \end{pmatrix}, c_1(x)$ and $c_2(x)$ are complex-valued measurable functions.

The first inverse scattering problem (ISP) (0.1)–(0.2) with non-self-adjoint matrix $C(x)$ is studied in Section 1.1 [97]. The second ISP (0.1)–(0.2) with self-adjoint matrix $C(x)$ is studied in Section 1.2 [119]. The potentials $c_1(x)$ and $c_2(x)$ of the self-adjoint scattering problem (SP) are power decreasing, while the potentials $c_1(x)$ and $c_2(x)$ of the non-self-adjoint SP are required to decrease exponentially. Owing to the self-adjoint condition of the potential matrix, the self-adjoint SP possesses a series of characters [119], which are essentially different from the corresponding characters of the non-self-adjoint SP [97]. The scattering data (SD) set s of the non-self-adjoint problem consists of the scattering function $S(\lambda)$ defined in the strip $|\mathrm{Im}\lambda| < \varepsilon_0, \varepsilon_0 > 0$, and singular numbers, standardized polynomials. The self-adjoint problem has no discrete spectrum, therefore it characterizes by the scattering function $S(\lambda)$ defined on the real axis λ. The systems of fundamental equations (fund. eqs) are derived for connecting the SD set s with the kernels of the transformation operator. The known functions in these systems are constructed from the SD set s. The systems of fund. eqs have unique solutions in terms of kernels of the transformation operator. Then the potentials $c_1(x)$ and $c_2(x)$ of the matrix

$C(x)$ are uniquely reconstructed. They are expressed through the solutions of the found solutions of the systems of fund. eqs. The schema for restoring the matrix potential $C(x)$ is deduced from the SP and ISP for the system (0.1) with the boundary condition (0.2):

$$\text{SP with potential } C(x) \text{ and BC } \leftrightarrow \text{ SD set } s \leftrightarrow \text{ Fund. Eqs} \rightarrow C(x). \quad (0.3)$$

In Chapter 2, a Cauchy problem and some IBVPs for the system of NLEEs:

$$\begin{cases} ic_{1t} = c_{1xx} - 2c_1^2 c_2, \\ -ic_{2t} = c_{2xx} - 2c_1 c_2^2. \end{cases} \quad 0 \leq x < \infty, \quad -\infty < t < \infty \quad (0.4)$$

are considered. The attractive and repulsive non-linear Schrödinger (NLS) equations are obtained from system (0.4) when $c_1 = -\overline{c}_2$ and $c_1 = \overline{c}_2$, respectively.

Chapter 3 is devoted to solving the IBVPs for the modified Korteweg-de Vries (mKdV) equations:

$$q_t + q_{xxx} + 6q^2 q_x = 0, \quad 0 \leq x < \infty, \quad -\infty < t < \infty \quad (0.5)$$

$$q_t + q_{xxx} - 6q^2 q_x = 0, \quad 0 \leq x < \infty, \quad -\infty < t < \infty \quad (0.6)$$

The Cauchy problem for the cubic non-linear evolution equation:

$$q_t + q_{xxx} - 6|q|^2 q_x = 0, \quad 0 \leq x < \infty, \quad -\infty < t < \infty. \quad (0.7)$$

is considered in this chapter.

In Chapter 4 the Dirichlet IBVPs for the sine-Gordon (sG) equation:

$$v_{xt} = \sin v, \quad 0 \leq x < \infty, \quad 0 \leq t \leq T, \quad (0.8)$$

and sinh-Gordon (shG) equation:

$$v_{xt} = \text{sh} v, \quad 0 \leq x < \infty, \quad 0 \leq t \leq T, \quad (0.9)$$

are studied.

The Cauchy problems and IBVPs for the NLEEs in Chapters 2–4 have been solved by the ISM in [133, 137, 135, 136, 134, 138, 139] and [140, 150, 142, 143]. It should be noted that the considered IBVPs for the attractive NLS, mKdV and sG equations are associated with their non-self-adjoint SPs, whereas the Cauchy problems for the repulsive NLS, cubic non-linear evolution equation are associated with their self-adjoint SPs. However, the Dirichlet IBVP for the shG equation in Chapter 4 is associated with its self-adjoint SP. Furthermore, in these chapters, the N-soliton solutions of the considered Cauchy problems are found by the ISM in the class of non-scattering potentials [133, 137, 138].

The results obtained in Chapter 1 are applied to solving the Cauchy problems and IBVPs for the NLEEs in Chapters 2–4. Indeed, by using the known initial and boundary conditions of the considered SP and t part of the Lax pair, i.e., the time-evolution of eigenfunctions of this problem, the time-dependent SD set $s(t)$ is derived from the SD set s of the considered SP. Then the time-dependence of the known functions in the systems of fund. eqs is constructed from the set $s(t)$. By the Lax compatibility condition of system (0.1) and the operator governing the time-evolution of eigenfunctions of system (0.1), the solution of the considered IBVP for NLEEs is found, that is expressed through solutions of systems of time-dependent fund. eqs. Hence, the algorithm of the main steps for finding the potentials $c_1(x,t)$ and $c_2(x,t)$ of the matrix $C(x,t)$ on the half-line is constructed. Thus, we obtain the following ISM schema in finding the solution of the considered NLEEs.

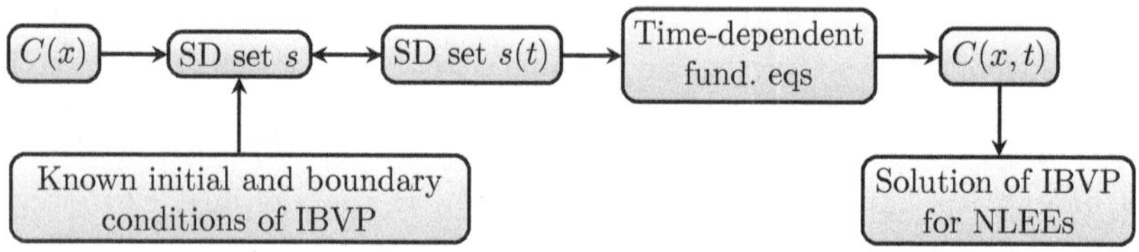

The schema for finding the solution of the KdV equation on the whole line was presented in [10].

Chapter 5 presents the Cauchy problem for the continual system of non-linear interaction waves:

$$\begin{aligned} &\frac{\partial q(x,y,\tau;t)}{\partial t}+w(y,\tau)\frac{\partial q(x,y,\tau;t)}{\partial x} \\ &+i\int_{-1}^{1}\beta(y,\tau,\tau_1)q(x,y,\tau_1;t)q(x,\tau_1,\tau;t)d\tau_1=0, \end{aligned} \tag{0.10}$$

where $x\in(-\infty,\infty)$; $y,\tau\in(-1,1)$, $w(y,\tau)=\frac{\alpha(\tau)-\alpha(y)}{y-\tau}$, $\beta(y,\tau,\tau_1)=\alpha(\tau_1)-\alpha(\tau)+\frac{(\tau-\tau_1)}{y-\tau}(\alpha(y)-\alpha(\tau))$, $\alpha(y)$ is any continuous function, $\alpha(y)\neq 0$ for any $y\in(-1,1)$. The continual system (0.10) can be represented by the Lax condition of compatibility of the operator transport equation:

$$\begin{aligned} &i\frac{\partial\varphi}{\partial x}+U\varphi=\lambda J\varphi, \\ &U\varphi(x,y;t)=\int_{-1}^{1}u(x,y,\tau;t)\varphi(x,\tau;t)d\tau,\ J\varphi(x,y;\tau)=y\varphi(x,y;\tau), \end{aligned} \tag{0.11}$$

and the operator governing the time-evolution of eigenfunctions of the transport equation (0.11).

The transport equation (0.11) is the continual generalization from the system of n first-order ordinary differential equations (ODEs). In fact, if in the transport equation the parameter y takes on a finite number of values y_j, $j=1,...,n$, and if the measure $d\tau$ is concentrated at the points y_j, $j=1,...,n$, then the transport equation (0.11) represents the system of n first-order ODEs:

$$-i\frac{d}{dx}u(x,\lambda)+C(x)u(x,\lambda)=\lambda J_n u(x,\lambda)\quad -\infty<x<\infty, \tag{0.12}$$

where λ is a complex valued spectral parameter, $u(x,\lambda) = (u_1(x,\lambda),...,u_n(x,\lambda))$, $C(x)$ is a potential matrix $C(x) = [c_{jk}(x)]$ with the zero elements along the principal diagonal, i.e., $c_{jj}(x) = 0$, and the elements $c_{jk}(x)$, $j \neq k$ are the complex-valued measurable power decreasing functions, $J_n = J_n^* = diag(a_1,...,a_n)$, a_j, $j = 1,...,n$, are real regulated numbers: $a_1 > ... > a_n$, J_n^* denotes the complex conjugate of J_n.

Solving the SP and ISP for system (0.12) and transport equation (0.11) is more complex than solving SPs and ISPs for system (0.1). Indeed, the role of the special Volterra integral representations of solutions of the transport equation (0.11) is analogous to the role of transformation operator. Both of them are significant in solving the ISP. Furthermore, the transition operator for the scattering problem for the transport equation (0.11) admits the right and left Volterra factorization and connects with an intermediate operator. The intermediate operator is constructed from the known transition operator and admits not only the right and left Volterra factorization but also the analytic factorization. Then the functions appearing in the fund. eqs in the inverse problem are constructed from the intermediate operator. By this fact, the potential in the transport equation is uniquely reconstructed through the solutions of the fund. eqs. Note that the known functions in systems of fund. eqs in the inverse problem considered in the previous chapters are constructed directly from known scattering functions, (see Remark 5.1).

Further, in Section 5.3 of this chapter, the generalized Lax equation for the transport equation is introduced. This enables us to derive the time-dependence of the transition operator. Thus, the solution of the considered Cauchy problem for the continual system (0.10) is found uniquely from the known initial condition in terms of solutions of the fund. eqs in the inverse problem.

Chapter 6 is devoted to solving the IBVP for the KdV equation with the positive dispersive coefficient on the positive quarter-plane [144]:

$$p_t - 6pp_x + p_{xxx} = 0, \quad x \geq 0, \quad t \geq 0.$$

Further, we present explicit solutions of the Cauchy problem for the KdV equation for $(x,t) \in [0,\infty) \times (-\infty,\infty)$ in the class of non-scattering potentials [139].

In Chapter 7 we study the inverse scattering problem, which is associated with the IBVP for the KdV with dominant surface tension on a half-line, i.e., for the KdV equation with the negative dispersive coefficient on a half-line [146, 147]:

$$p_t - p_{xxx} + 6pp_x = 0, \quad x > 0, \quad t > 0.$$

We prove that the solution of this IBVP can be found by solving two ISPs on two different spectral planes. The first SP is associated with the considered KdV equation with the negative dispersive coefficient. The second SP is self-adjoint. Its scattering function is expressed through entries of the given scattering matrix $s(k)$ of the first SP. The relationship between these SPs is established. The necessary and sufficient conditions for given functions to be the right- and left-reflection coefficients of the first SP, and the necessary and sufficient conditions for a given function to be the scattering function for the self-adjoint SP are established. Knowing the scattering function of the self-adjoint SP, we solve the system of fundamental equations for finding the potential self-adjoint matrix. Consequently, the unknown object entering coefficients of the system of evolution equations for the scattering matrix $s(k,t)$ is found. Furthermore, the time-dependence of entries of $s(k,t)$ is defined in every sector by expansion formulas, which are constructed from known initial and boundary conditions. Knowing the right-reflection coefficient constructed from the found entries of $s(k,t)$, we solve the time-dependent Gelfand–Levitan–Marchenko (GLM) equation. By the Lax compatibility condition for the Lax pair, the solution of the considered IBVP is found and is expressed through the found solution of this GLM equation. The asymptotic behavior at infinity of time of the found solution of this IBVP is shown. The comparison between the method used in Chapter 7 and the methods given by Fokas [38] and Habibullin [53] are written in Chapter 9.

In Chapter 8, we study the ISP for the perturbed string equation in characteristic variables on the whole axis [122, 123]. The results of this ISP are applied to integration of the two-dimensional generalization from the KdV equation. This generalization from the KdV equation is derived from the generalized Lax equation, which is generated by the associated perturbed string equation [123]. With the help of the generalized Lax equation, the two-dimensional generalization from the KdV equation is derived and the time-evolution of the scattering operator of the considered SP is established. Then the known functions in the system of fund. eqs. in inverse problem are constructed from the time-dependent kernel matrix of the known scattering operator. This enables us to solve the system of time-dependent fund. eqs. and the solution of this generalization from the KdV equation is expressed through the found solution of the system of fund. eqs. in the inverse problem [148].

In Chapter 9, first we introduce preliminary Fokas's methodology for the analysis of IBVPs for NLEEs on the half-line [38], and a Riemann–Hilbert problem [51, 53]. Further, the connections between the Hirota's method and the ISM are considered. The solutions of the KdV and NLS equations are constructed by Hirota's method and the ISM [133, 139, 149, 150].

In Chapter 10, we construct the Bäcklund transformations (BTs) from the one-to-one correspondence relations between the scattering data (SD) of scattering problems (SPs) generated by both linear equations in the Lax pair and the recovered potentials in these equations. These recovered potentials are the solutions of the IBVPs. If the solution of the SP associated with the KdV with dominant surface tension satisfies only the first equation in the Lax pair and does not satisfy the second equation in this pair, we must therefore correct this solution. More specifically, we multiply the solution of the first equation from the right by a factor function appropriately chosen such that the resulting factor function satisfies both equations in the Lax pair. The correspondence relation between the common solution of both equations is a BT, which transforms the common solution into the solution of the NLEEs with the given initial and boundary conditions, and conversely.

Hence, the solutions of NLEEs on a half-line or on the whole line considered in this monograph can be obtained by three different methods: the ISM, the Hirota's method and the BT. The considered NLEEs are most widely studied non-linear wave equations because of intrinsic mathematical properties and their wide applicability in physics, plasma physics, fluid mechanics, etc.

In this monograph, the considered problems are solved step-by-step, and detail proofs are given. With the help of the Lax equation or generalized Lax equation, the time-evolutions of the scattering data were derived and expressed in terms of known initial and boundary conditions of the SP. The ISPs are well-solved and are set up appropriately for constructing the Bäcklund transformations (BTs) for solutions of associated initial-boundary value problems (IBVPs) or initial-value problems (IVPs) for NLEEs. The interrelations between the ISPs and constructed BTs are established to become new powerful unified transformations (UTs) for solving IBVPs or IVPs for NLEEs, that can be used in different areas of physics and mechanics. The application of the UTs is consistent and efficiently embedded in the scheme of the considered ISP.

Chapter 1

Inverse Scattering Problems for Systems of First-Order ODEs on A Half-Line

In this chapter, two inverse scattering problems (ISPs) on a half-line for systems of first-order ordinary differential equations (ODEs) with non-self-adjoint and self-adjoint potential matrices are considered. The necessary and sufficient conditions for given quantities to be the scattering data for the considered problems for systems of ODEs on a half-line with the boundary condition at the origin $x = 0$ are established. The algorithms for finding the potential matrices in these systems are constructed in terms of the given scattering data.

The self-adjoint problem does not have a discrete spectrum. Therefore, it is associated mostly with a Cauchy initial value problem (IVP) for integrable non-linear equations (NLEEs), whereas the non-self-adjoint problem is associated mostly with an initial-boundary value problem (IBVP) for integrable NLEEs. The results of the considered ISPs are not only independently interesting by themselves but also are effective tools for solving the IVP and IBVP for integrable NLEEs. In Chapters 2, 3 and 4 we shall apply the results of ISPs obtained in this chapter for solving IVPs and IBVPs for integrable NLEEs, like the non-linear Shrödinger (NLS), modified Korteweg-de Vries (mKdV), cubic non-linear equation, sine-Gordon (SG) and sinh-Gordon (shG).

This chapter is based on published papers [97, 119].

We study two ISPs for a system of first-order ODEs of the form [97]:

$$-iJu_x + C(x)u = \lambda u, \quad u = (u_1(x,\lambda), u_2(x,\lambda)), \quad 0 \le x < \infty \tag{1.1}$$

with the boundary condition at the origin $x = 0$:

$$u_1(0,\lambda) = u_2(0,\lambda), \tag{1.2}$$

where $J = \begin{pmatrix} -1 & 0 \\ 0 & 1 \end{pmatrix}$, λ is a complex-valued parameter, $C(x) = \begin{pmatrix} 0 & c_1(x) \\ c_2(x) & 0 \end{pmatrix}$, $c_1(x)$ and $c_2(x)$ are complex-valued measurable functions.

In Section 1.1 the direct and ISP (1.1)–(1.2) are studied with a potential non-self-adjoint matrix $C(x)$ satisfying the estimate:

$$|c_k(x)| \le \tilde{C}e^{-\varepsilon x}, \quad \tilde{C} \text{ is a constant}, \quad \varepsilon > 0, \quad k = 1, 2. \tag{1.3}$$

DOI: 10.1201/9781003370543-1

In Section 1.2 the direct and ISP is studied with a potential self-adjoint matrix $C(x)$ satisfying the condition:

$$c_1(x) = \overline{c_2(x)}, \tag{1.4}$$

and the estimate:

$$|c_k(x)| \leq \frac{\tilde{C}}{1+x^{1+\varepsilon}}, \quad \tilde{C} \text{ is a constant}, \quad \varepsilon > 0, \quad k = 1, 2. \tag{1.5}$$

By using the simple substitution:

$$y_1(x,\lambda) = \frac{1}{2}\left(u_1(x,\lambda) + u_2(x,\lambda)\right), \quad y_2(x,\lambda) = \frac{1}{2i}\left(u_2(x,\lambda) - u_1(x,\lambda)\right),$$

we can reduce the system (1.1) to the more usual form of the so-called Dirac system of equations with mass $m = 0$:

$$By_x + \Omega(x)y(x,\lambda) = \lambda y(x,\lambda), \tag{1.6}$$

where

$$B = \begin{pmatrix} 0 & 1 \\ -1 & 0 \end{pmatrix}, \quad y(x,\lambda) = (y_1(x,\lambda) \;\; y_2(x,\lambda)), \quad \Omega(x) = \begin{pmatrix} P(x) & Q(x) \\ Q(x) & -P(x) \end{pmatrix},$$

and

$$P(x) = \frac{1}{2}\left(c_1(x) + c_2(x)\right), \quad Q(x) = \frac{i}{2}\left(c_1(x) - c_2(x)\right).$$

We mention some works that closely border on the problems to be discussed here. In [47] the ISP on a half-line was solved for a Dirac system of equations with a self-adjoint potential. The attractive non-linear Schrödinger equation on the whole line was solved in 1972 by Zakharov and Shabat [111] by reducing it to the ISP on the whole line for the Dirac system of the form (1.6). However, in [111] there was no full and detailed exposition of mathematical solutions of the inverse problem. In 1971–1972 the ISP on the whole line for the system of the form (1.6) was solved with rigorous mathematical proofs by Pham Loi Vu [117, 118].

The ISPs for the system (1.1) on a half-line with the potential self-adjoint matrix and with a potential non-self-adjoint matrix were solved in 1973 [119] and 1974 [97], respectively. In 1975 the ISP on the whole line for a non-self-adjoint Dirac operator was studied in [81].

The ISP for the Schrödinger equation:

$$-\frac{d^2}{dx^2}\phi(x,-k) + p(x)\phi(x,-k) = k^2\phi(x,-k),$$

where $p(x) = [Q(x)]^2 + \frac{d}{dx}Q(x)$, and k is a parameter that is obtained from (1.6) when $c_1(x) = -c_2(x)$, has been solved for the half-line by Ljance [78, 79], and for the whole line by Blashchak [14, 15] without the assumption that the potential is self-adjoint. The case of a non-stationary potential has been investigated in detail in [93].

1.1 The inverse scattering problem on a half-line with a potential non-self-adjoint matrix

In this section we study the direct and ISP for the system of first-order ODEs on the half-line of the form (1.1):

$$-iJu_x + C(x)u = \lambda u, \quad u = \Big(u_1(x,\lambda), u_2(x,\lambda)\Big), \quad 0 \le x < \infty \tag{1.7}$$

with the boundary condition at the origin $x = 0$:

$$u_1(0,\lambda) = u_2(0,\lambda), \tag{1.8}$$

where $J = \begin{pmatrix} -1 & 0 \\ 0 & 1 \end{pmatrix}$, λ is a complex-valued parameter, $C(x) = \begin{pmatrix} 0 & c_1(x) \\ c_2(x) & 0 \end{pmatrix}$, $C(x)$ is a potential non-self-adjoint matrix, $c_1(x)$ and $c_2(x)$ are the complex-valued measurable functions satisfying the estimate:

$$|c_k(x)| \le \tilde{C}e^{-\varepsilon x}, \quad \tilde{C} \text{ is a constant}, \quad \varepsilon > 0,\ k = 1,2 \tag{1.9}$$

and $c_1(x)$, $c_2(x) \ne 0$ for any $x \ge 0$.

The following eigenvalue problem for the one-dimensional Schrödinger equation:

$$L\psi = -\frac{d^2\psi}{dx^2} + p\,(x)\,\psi = \lambda\psi, \quad -\infty < x < \infty,$$

has been studied in a class of functions $\psi\,(x)$ bounded on the whole axis [76]. We know that the number of eigenvalues of the Schrödinger operator L may be infinite; they can accumulate in the neighborhood of the point $\lambda = 0$. This is the case for a potential, which drops slowly as $x \to \infty$, for instance, the Coulomb potential $p(x) = -\frac{\text{const}}{x}$, or $p(x) = \frac{\text{const}}{x^2}$. To eliminate potentials with an infinite number of eigenvalues, the potential $p(x)$ of the Schrödinger operator L is required to be a real-valued function satisfying the usual condition in scattering theory [76, 156]:

$$\int_{-\infty}^{\infty} |p\,(x)|\,(1 + |x|)\,dx < \infty.$$

We prove in Subsection 1.1.3 that the non-self-adjoint problem (1.7)–(1.8) with the elements of the potential matrix $C(x)$ satisfying the constraint (1.9) has only a finite of singular numbers, which are the same as non-real eigenvalues of the problem (1.7)–(1.8).

1.1.1 The representation of the solution of system (1.7)

We begin by establishing the representation of the solution of system (1.7) in terms of the kernel of the transformation operator.

Lemma 1.1. *Let* λ *be a real number, and* $u(x,\lambda) = \big(u_1(x,\lambda), u_2(x,\lambda)\big)$ *be a bounded solution of system* (1.7), *then*

$$u_1(x,\lambda) = A(\lambda)e^{-i\lambda x} + o(1),\quad u_2(x,\lambda) = B(\lambda)e^{i\lambda x} + o(1) \text{ as } x \to \infty. \tag{1.10}$$

The solution of the problem (1.7), (1.10) *is represented in the form:*

$$\begin{aligned} u_1(x,\lambda) &= A(\lambda)e^{-i\lambda x} + A(\lambda)\int_x^\infty H_{11}(x,s)e^{-i\lambda s}ds + B(\lambda)\int_x^\infty H_{12}(x,s)e^{i\lambda s}ds,\\ u_2(x,\lambda) &= B(\lambda)e^{i\lambda x} + A(\lambda)\int_x^\infty H_{21}(x,s)e^{-i\lambda s}ds + B(\lambda)\int_x^\infty H_{22}(x,s)e^{i\lambda s}ds. \end{aligned} \tag{1.11}$$

The matrix function $H(x,s) = [H_{jk}(x,s)]$ *is related to the potential matrix* $C(x)$ *by the condition:*

$$2i\begin{pmatrix} 0 & H_{12}(x,x) \\ -H_{21}(x,x) & 0 \end{pmatrix} = \begin{pmatrix} 0 & c_1(x) \\ c_2(x) & 0 \end{pmatrix}, \tag{1.12}$$

and it has the estimate:

$$|H_{jk}(x,s)| \le \tilde{C}e^{-\varepsilon\frac{x+s}{2}},\quad 0 \le x \le s,\ \varepsilon > 0,\quad j,k = 1,2, \tag{1.13}$$

where $\tilde{C}$ *is a constant.*

Proof. Let $u(x,\lambda) = \begin{pmatrix} u_1(x,\lambda) \\ u_2(x,\lambda) \end{pmatrix}$ be a bounded solution of system (1.7). Then by the uniqueness of the Cauchy problem for system (1.7), we have

$$\begin{cases} u_1(x,\lambda) = u_1(0,\lambda)e^{-i\lambda x} + i\displaystyle\int_0^x e^{-i\lambda(x-y)}c_1(y)u_2(y,\lambda)dy,\\ u_2(x,\lambda) = u_2(0,\lambda)e^{i\lambda x} - i\displaystyle\int_0^x e^{i\lambda(x-y)}c_2(y)u_1(y,\lambda)dy, \end{cases}$$

whence it follows that when λ is real, the solution $u(x,\lambda)$ of system (1.7) has the asymptotic form (1.10) as $x \to \infty$, where

$$A(\lambda) = u_1(0,\lambda) + i\int_0^\infty e^{i\lambda y}c_1(y)u_2(y,\lambda)dy,$$

$$B(\lambda) = u_2(0,\lambda) - i\int_0^\infty e^{-i\lambda y}c_2(y)u_1(y,\lambda)dy.$$

Obviously, the solution of system (1.7) with the boundary condition (1.10) is equivalent to the following system of integral equations:

$$\begin{cases} u_1(x,\lambda) = A(\lambda)e^{-i\lambda x} + \dfrac{1}{i}\displaystyle\int_x^\infty e^{-i\lambda(x-t)}c_1(t)u_2(t,\lambda)dt, \\ u_2(x,\lambda) = B(\lambda)e^{i\lambda x} - \dfrac{1}{i}\displaystyle\int_x^\infty e^{i\lambda(x-t)}c_2(t)u_1(t,\lambda)dt. \end{cases} \tag{1.14}$$

We seek a solution of this system in the form (1.11). By starting from (1.11) and (1.14), we obtain the systems of integral equations:

$$\begin{cases} H_{11}(x,s) = \dfrac{1}{i}\displaystyle\int_x^\infty H_{21}(t,t+s-x)c_1(t)dt, \\ H_{21}(x,s) = -\dfrac{1}{2i}c_2\left(\dfrac{x+s}{2}\right) - \dfrac{1}{i}\displaystyle\int_x^{\frac{x+s}{2}} H_{11}(t,x+s-t)c_2(t)dt, \end{cases} \quad 0 \le x \le s \tag{1.15}$$

$$\begin{cases} H_{22}(x,s) = -\dfrac{1}{i}\displaystyle\int_x^\infty H_{12}(t,t+s-x)c_2(t)dt, \\ H_{12}(x,s) = \dfrac{1}{2i}c_1\left(\dfrac{x+s}{2}\right) + \dfrac{1}{i}\displaystyle\int_x^{\frac{x+s}{2}} H_{22}(t,x+s-t)c_1(t)dt. \end{cases} \quad 0 \le x \le s \tag{1.16}$$

The integral equations for $H_{jk}(x,s)$ are derived from (1.15) and (1.16):

$$\begin{aligned} H_{11}(x,s) =& \frac{1}{2}\int_x^\infty c_1(\tau)c_2\left(\tau + \frac{s-x}{2}\right)d\tau + \\ & \int_x^\infty c_1(\tau)d\tau \int_\tau^{\tau+\frac{s-x}{2}} c_2(\xi)H_{11}(\xi, 2\tau+s-x-\xi)d\xi, \end{aligned} \tag{1.17}$$

$$\begin{aligned} H_{21}(x,s) =& -\frac{1}{2i}c_2\left(\frac{x+s}{2}\right) + \\ & \int_x^{\frac{x+s}{2}} c_2(\tau)d\tau \int_\tau^\infty c_1(\xi)H_{21}(\xi, \xi+x+s-2\tau)d\xi, \end{aligned} \tag{1.18}$$

$$H_{12}(x,s) = \frac{1}{2i}c_1\left(\frac{x+s}{2}\right) + \int\limits_x^{\frac{x+s}{2}} c_1(\tau)d\tau \int\limits_\tau^\infty c_2(\xi)H_{12}(\xi,\xi+x+s-2\tau)d\xi, \tag{1.19}$$

$$H_{22}(x,s) = \frac{1}{2}\int\limits_x^\infty c_1\left(\tau+\frac{s-x}{2}\right)c_2(\tau)d\tau + \int\limits_x^\infty c_2(\tau)d\tau \int\limits_\tau^{\tau+\frac{s-x}{2}} c_1(\xi)H_{22}(\xi,2\tau+s-x-\xi)d\xi. \tag{1.20}$$

The solutions of integral equations (1.17)–(1.20) can be found by the method of successive approximations. Indeed, we put

$$H_{12}(x,s) = \sum_{n=0}^\infty H_{12,n}(x,s), \tag{1.21}$$

where

$$H_{12,0}(x,s) = \frac{1}{2i}c_1\left(\frac{x+s}{2}\right),$$

$$H_{12,n}(x,s) = \int\limits_x^{\frac{x+s}{2}} c_1(\tau)d\tau \int\limits_\tau^\infty c_2(\xi)H_{12,n-1}(\xi,\xi+x+s-2\tau)d\xi,\ n=1,2,...$$

Using the above successive approximations with due regard for estimate (1.9), we construct the majorant series:

$$\sum_{n=0}^\infty \zeta_n(x,s),$$

where

$$\zeta_0(x,s) = \tilde{C}e^{-\varepsilon\frac{x+s}{2}},\ \varepsilon>0,\ \tilde{C} \text{ is a constant},$$

$$\zeta_n(x,s) = \tilde{C}^2 \int\limits_x^{\frac{x+s}{2}} e^{-\varepsilon\tau}d\tau \int\limits_\tau^\infty e^{-\varepsilon\xi}\zeta_{n-1}(\xi,\xi+x+s-2\tau)d\xi,\ n=1,2,...$$

In view of iterations $\zeta_n(x,s)$, the series (1.21) is dominated by the constructed majorant series:

$$\sum_{n=0}^\infty \zeta_n(x,s) \le \tilde{C}e^{-\varepsilon\frac{x+s}{2}}\Big\{1+\int\limits_x^\infty \tilde{C}e^{-\varepsilon\tau_1}d\tau_1 \int\limits_{\tau_1}^\infty \tilde{C}e^{-\varepsilon\xi_1}d\xi_1 + ... + \int\limits_x^\infty \tilde{C}e^{-\varepsilon\tau_1}d\tau_1 \int\limits_{\tau_1}^\infty \tilde{C}e^{-\varepsilon\xi_1}d\xi_1 ... \int\limits_{\xi_{n-1}}^\infty \tilde{C}e^{-\varepsilon\tau_n}d\tau_n \int\limits_{\tau_n}^\infty \tilde{C}e^{-\varepsilon\xi_n}d\xi_n + ...\Big\}.$$

We denote the expression in the bracket by $\Phi(x)$ and write it in the form:

$$\Phi(x) = 1 + \int_x^\infty \alpha(\tau_1)d\tau_1 + ... + \int_x^\infty \alpha(\tau_1)d\tau_1 ... \int_{\xi_{n-1}}^\infty \alpha(\tau_n)d\tau_n + ...$$

where $\alpha(\tau_n) = \tilde{C}^2 e^{-\varepsilon\tau_n} \int_{\tau_n}^\infty e^{-\varepsilon\xi_n} d\xi_n$.

It is easy to verify that $\Phi'(x) = -\alpha(x)\Phi(x)$ and $\Phi(x) = \exp\left\{ \int_x^\infty \alpha(\tau)d\tau \right\}$. Hence, the majorant series converges uniformly, whence it follows that the series (1.21) is summarized. Its sum satisfies the integral equation (1.19) and the inequality:

$$|H_{12}(x,s)| \leq \tilde{C}e^{-\varepsilon\frac{x+s}{2}} \exp\left\{ \int_x^\infty \tilde{C}e^{-\varepsilon\tau}d\tau \int_\tau^\infty \tilde{C}e^{-\varepsilon\xi}d\xi \right\}.$$

To prove the uniqueness of the solution of equation (1.19), it suffices to show that the homogeneous equation corresponding to (1.19) has only zero solution. Indeed, the solution of this homogeneous equation is dominated by the n-th term of the above majorant series. The n-th term tends to zero as $n \to \infty$.

Similarly prove that the solutions of integral equations (1.17), (1.18) and (1.20) exist and are unique in the class of bounded functions. In addition, the solutions $H_{jk}(x,s), j,k = 1,2$, of these equations satisfy estimate (1.13).

When we let $s \to x$ in the sense of the metric in $L_1[0,\infty)$, we obtain from (1.18) and (1.19) that $2iH_{12}(x,x) = c_1(x)$, and $-2iH_{21}(x,x) = c_2(x)$, that is, the relation equality (1.12) holds. From the integral equations (1.17) and (1.19), we have:

$$H_{11}(x,x) = H_{22}(x,x) = \frac{1}{2}\int_x^\infty c_1(\tau)\, c_2(\tau)\, d\tau,$$

$$\frac{d}{dx}H_{11}(x,x) = \frac{d}{dx}H_{22}(x,x) = -\frac{1}{2}c_1(x)\, c_2(x) \neq 0 \text{ for any } x \geq 0.$$

Hence, $H_{11}(x,x) = H_{22}(x,x) \neq 0$ for any $x \geq 0$.

Conversely, if the functions $H_{jk}(x,s)$, $0 \leq x \leq s$, $j,k = 1,2$, satisfy systems (1.15) and (1.16), then the formula (1.11) gives a bounded solution $u(x,\lambda)$ of system (1.14). By the Volterra property, the solution $u(x,\lambda)$ of this system is unique in the class of bounded functions. Hence, $u(x,\lambda)$ satisfies system (1.7) with the boundary condition (1.10). This proves the lemma. □

Here the matrix function $H(x,s)$ is an analog of the kernel of the transformation operator usually used in scattering problems [7, 75].

1.1.2 The Jost solutions of system (1.7)

Let $v(x,\lambda)=\begin{pmatrix} v_1(x,\lambda) \\ v_2(x,\lambda)\end{pmatrix}$ and $z(x,\lambda)=\begin{pmatrix} z_1(x,\lambda) \\ z_2(x,\lambda)\end{pmatrix}$ be any solutions of system (1.7). It is easily shown that when λ is real, the Wronskian determinant $W[v,z]=\begin{vmatrix} v_1(x,\lambda) & z_1(x,\lambda) \\ v_2(x,\lambda) & z_2(x,\lambda)\end{vmatrix}$ of the solutions $v(x,\lambda)$ and $z(x,\lambda)$ is independent of x, and the solutions are linearly independent if and only if $W[v,z]\neq 0$.

We consider the vector functions:

$$e^{(1)}(x,\lambda)=\begin{pmatrix} e_1^{(1)}(x,\lambda) \\ e_2^{(1)}(x,\lambda)\end{pmatrix}=\begin{pmatrix} e^{-i\lambda x}+e^{-i\lambda x}\int\limits_0^\infty H_{11}(x,x+t)e^{-i\lambda t}dt \\ e^{-i\lambda x}\int\limits_0^\infty H_{21}(x,x+t)e^{-i\lambda t}dt \end{pmatrix},$$

$$e^{(2)}(x,\lambda)=\begin{pmatrix} e_1^{(2)}(x,\lambda) \\ e_2^{(2)}(x,\lambda)\end{pmatrix}=\begin{pmatrix} e^{i\lambda x}\int\limits_0^\infty H_{12}(x,x+t)e^{i\lambda t}dt \\ e^{i\lambda x}+e^{i\lambda x}\int\limits_0^\infty H_{22}(x,x+t)e^{i\lambda t}dt \end{pmatrix},$$

where λ is a real number and the functions $H_{jk}(x,x+t)$, $x,t\geq 0$, $j,k=1,2$ are the kernels of the transformation operator defined by Lemma 1.1.

It is clear that $e^{(1)}(x,\lambda)$ and $e^{(2)}(x,\lambda)$ are the solutions of system (1.7). Taking into account the fact that when λ is real $W[e^{(1)},e^{(2)}]$ is independent of x, we let $x\to\infty$ and obtain that $W[e^{(1)},e^{(2)}]=1$. Hence, when λ is real, the solutions $e^{(1)}(x,\lambda)$ and $e^{(2)}(x,\lambda)$ form a fundamental system of solutions of system (1.7). These solutions are called the Jost solutions of system (1.7).

1.1.3 The scattering function $S(\lambda)$ and non-real eigenvalues

Suppose that for some real λ $u(x,\lambda)=\begin{pmatrix} u_1(x,\lambda) \\ u_2(x,\lambda)\end{pmatrix}$ is a bounded solution of system (1.7) with boundary condition (1.8). Then by Lemma 1.1, the solution $u(x,\lambda)$ is representable in the form (1.11). It follows from (1.8) and (1.11) that

$$A\left(1+\int\limits_0^\infty [H_{11}(0,t)-H_{21}(0,t)]e^{-i\lambda t}dt\right)$$
$$=B\left(1+\int\limits_0^\infty [H_{22}(0,t)-H_{12}(0,t)]e^{i\lambda t}dt\right).$$

Introduce the function:

$$S(\lambda)=\frac{N(\lambda)}{D(\lambda)},\tag{1.22}$$

where

$$N(\lambda) = e_1^{(1)}(0,\lambda) - e_2^{(1)}(0,\lambda) = 1 + \int_0^\infty [H_{11}(0,t) - H_{21}(0,t)]e^{-i\lambda t}dt,$$

$$D(\lambda) = e_2^{(2)}(0,\lambda) - e_1^{(2)}(0,\lambda) = 1 + \int_0^\infty [H_{22}(0,t) - H_{12}(0,t)]e^{i\lambda t}dt.$$

By analogy with the self-adjoint case (see formula (1.105) in Section 1.2), we call $S(\lambda)$ the scattering function for the scattering problem (1.7)–(1.8). We call $N(\lambda)$ the numerator and $D(\lambda)$ the denominator of the scattering function. Owing to estimate (1.13), the numerator $N(\lambda)$ and the denominator $D(\lambda)$ admit an analytic continuation into the lower half-plane $\mathrm{Im}\lambda < \frac{\varepsilon}{2}$, and into upper half-plane $\mathrm{Im}\lambda > -\frac{\varepsilon}{2}$, respectively. Hence, $N(\lambda)$ can have only a finite number of zeros in any closed lower half-plane $\mathrm{Im}\lambda < \frac{\varepsilon}{2}$, and $D(\lambda)$ can have only a finite number of zeros in any closed upper half-plane $\mathrm{Im}\lambda > -\frac{\varepsilon}{2}$ [114]. Therefore, there exists a positive number ε_0 such that

$$\varepsilon_0 = \min\left(\varepsilon_1, \frac{\varepsilon}{2}\right), \tag{1.23}$$

where ε_1 is the distance from the real axis to the non-real zeros of the numerator $N(\lambda)$ and denominator $D(\lambda)$.

Lemma 1.2. *Let the kernel functions $H_{jk}(x,s)$ satisfy estimate* (1.13) *and the number ε_0 be defined by* (1.23). *Then*

(1) *The function $1+\int_0^\infty [H_{11}(0,t)-H_{21}(0,t)]e^{-i\lambda t}dt$ is holomorphic for $\mathrm{Im}\lambda < \varepsilon_0$ and the function $1 + \int_0^\infty [H_{22}(0,t) - H_{12}(0,t)]e^{i\lambda t}dt$ is holomorphic for $\mathrm{Im}\lambda > -\varepsilon_0$.*
For every $\eta < \varepsilon_0$ the following asymptotic relations hold as $|\lambda| \to \infty$:

$$1 + \int_0^\infty [H_{11}(0,t) - H_{21}(0,t)]e^{-i\lambda t}dt = 1 + o(1)$$

uniformly in the half-plane $\mathrm{Im}\lambda \le \eta$,

$$1 + \int_0^\infty [H_{22}(0,t) - H_{12}(0,t)]e^{i\lambda t}dt = 1 + o(1)$$

uniformly in the half-plane $\mathrm{Im}\lambda \ge -\eta$;

(2)

$$\left.\begin{aligned} 1+\int_0^\infty [H_{11}(0,t)-H_{21}(0,t)]e^{-i\lambda t}dt \neq 0,\\ 1+\int_0^\infty [H_{22}(0,t)-H_{12}(0,t)]e^{i\lambda t}dt \neq 0, \end{aligned}\right\} \quad \text{for } 0<|\mathrm{Im}\lambda|<\varepsilon_0.$$

(3) If $\mathrm{Im}\lambda = 0$ *and* $1+\int_0^\infty [H_{22}(0,t)-H_{12}(0,t)]e^{i\lambda t}dt = 0$, *then* $1+\int_0^\infty [H_{11}(0,t)-H_{21}(0,t)]e^{-i\lambda t}dt \neq 0$.

If $\mathrm{Im}\lambda = 0$ *and* $1+\int_0^\infty [H_{11}(0,t)-H_{21}(0,t)]e^{-i\lambda t}dt = 0$, *then* $1+\int_0^\infty [H_{22}(0,t)-H_{12}(0,t)]e^{i\lambda t}dt \neq 0$.

Proof. For every $\varepsilon_0 \leq \frac{\varepsilon}{2}$ the assertion (1) follows from estimate (1.13) for the functions $H_{jk}(0,t),\quad t\geq 0,\quad j,k=1,2$. The assertion (2) is deduced from the definition (1.23) of the number ε_0. We now prove the assertion (3). We assume that the numerator and denominator of the scattering function are simultaneously zero for some $\lambda=\lambda_0$, $\mathrm{Im}\lambda_0=0$. Consider the vector function $u(x,\lambda_0)=\Big(u_1(x,\lambda_0),u_2(x,\lambda_0)\Big)$, $\mathrm{Im}\lambda_0=0$, such that

$$\begin{cases} u_1(x,\lambda_0)=e^{-i\lambda_0 x}-\dfrac{1}{i}\displaystyle\int_0^x e^{i\lambda_0(\xi-x)}c_1(\xi)u_2(\xi,\lambda_0)d\xi,\\ u_2(x,\lambda_0)=\dfrac{1}{i}\displaystyle\int_0^x e^{i\lambda_0(x-\xi)}c_2(\xi)u_1(\xi,\lambda_0)d\xi. \end{cases}$$

It is clear that $u(x,\lambda_0)$ is a bounded solution of system (1.7) and satisfies the boundary conditions:

$$\begin{aligned} &u_1(0,\lambda_0)=1,\quad u_2(0,\lambda_0)=0,\\ &\lim_{x\to\infty}u_1(x,\lambda_0)e^{i\lambda_0 x}=A(\lambda_0),\quad \lim_{x\to\infty}u_2(x,\lambda_0)e^{-i\lambda_0 x}=B(\lambda_0), \end{aligned}$$

where

$$A(\lambda_0)=1-\frac{1}{i}\int_0^\infty e^{i\lambda_0\xi}c_1(\xi)u_2(\xi,\lambda_0)d\xi,\quad B(\lambda_0)=\frac{1}{i}\int_0^\infty e^{-i\lambda_0\xi}c_2(\xi)u_1(\xi,\lambda_0)d\xi.$$

According to Lemma 1.1, the considered bounded solution $u(x,\lambda_0) = \big(u_1(x,\lambda_0),\, u_2(x,\lambda_0)\big)$ of system (1.7) is represented in the form (1.11):

$$u_1(x,\lambda_0) = A(\lambda_0)e^{-i\lambda_0 x} + A(\lambda_0)\int_x^\infty H_{11}(x,t)e^{-i\lambda_0 t}dt$$

$$+ B(\lambda_0)\int_x^\infty H_{12}(x,t)e^{i\lambda_0 t}dt,$$

$$u_2(x,\lambda_0) = B(\lambda_0)e^{i\lambda_0 x} + A(\lambda_0)\int_x^\infty H_{21}(x,t)e^{-i\lambda_0 t}dt$$

$$+ B(\lambda_0)\int_x^\infty H_{22}(x,t)e^{i\lambda_0 t}dt,$$

and satisfies the boundary conditions:

$$\lim_{x\to\infty} u_1(x,\lambda_0)e^{i\lambda_0 x} = A(\lambda_0), \quad \lim_{x\to\infty} u_2(x,\lambda_0)e^{-i\lambda_0 x} = B(\lambda_0).$$

From the assumption it follows that

$$u_1(0,\lambda_0) - u_2(0,\lambda_0) = A(\lambda_0)\left(1 + \int_0^\infty [H_{11}(0,t) - H_{21}(0,t)]e^{-i\lambda_0 t}dt\right) -$$

$$B(\lambda_0)\left(1 + \int_0^\infty [H_{22}(0,t) - H_{12}(0,t)]e^{i\lambda_0 t}dt\right) = 0,$$

but this contradicts the fact that

$$u_1(0,\lambda_0) = 1, \quad u_2(0,\lambda_0) = 0.$$

□

The following lemma about the properties of the scattering function is deduced from Lemma 1.2 and definition (1.22).

Lemma 1.3. *The scattering function* $S(\lambda)$ *defined by* (1.22) *for the scattering problem* (1.7)–(1.8) *with a potential non-self-adjoint matrix satisfying estimate* (1.9) *possesses the properties:*

(1) The functions $S(\lambda)$ *and* $S^{-1}(\lambda)$ *are meromorphic in the strip* $|\mathrm{Im}\lambda| < \varepsilon_0$*, and for every* $\eta < \varepsilon_0$*:*
$S(\lambda) = 1 + o(1)$, $S^{-1}(\lambda) = 1 + o(1)$ *as* $|\lambda| \to \infty$ *uniformly in the strip* $|\mathrm{Im}\lambda| \le \eta$*,*

(2) The functions $S(\lambda)$ *and* $S^{-1}(\lambda)$ *have no non-real poles in the strip* $|\mathrm{Im}\lambda| < \varepsilon_0$.

We call functions $S(\lambda)$ and $S^{-1}(\lambda)$ with the properties (1) and (2) functions of the type (S) in the strip $|\mathrm{Im}\lambda| < \varepsilon_0$.

Definition 1.1. *We call the zeros* λ_j^-, $j = 1, ..., \gamma_2$ *of the numerator* $N(\lambda)$ *that lie in the domain* $\mathrm{Im}\lambda < 0$, *and the zeros* λ_k^+, $k = 1, ..., \gamma_1$ *of the denominator* $D(\lambda)$ *that lie in the domain* $\mathrm{Im}\lambda > 0$, *the singular numbers of the problem* (1.7)–(1.8). *The multiplicities of these zeros are called the multiplicities of the singular numbers and denoted by* n_j *and* m_k, *respectively.*

Lemma 1.4. *The singular numbers of the problem* (1.7)–(1.8) *are the same as its non-real eigenvalues, and the multiplicity of a singular number is the same as the multiplicity of the corresponding eigenfunction.*

Proof. We only prove the first assertion; the second follows easily from the definition of the multiplicity of the eigenfunction corresponding to a complex eigenvalue. Let λ_k^+ be a singular number with $\mathrm{Im}\lambda_k^+ > 0$, then $e_1^{(2)}(0, \lambda_k^+) = e_2^{(2)}(0, \lambda_k^+)$, and so $e^{(2)}(x, \lambda_k^+)$ is a solution of the problem (1.7)–(1.8), that belongs to $L_2[0, \infty)$; that is, $e^{(2)}(x, \lambda_k^+)$ is an eigenfunction, and λ_k^+ is a non-real eigenvalue of the problem (1.7)–(1.8).

Conversely, if λ_0^+, $\mathrm{Im}\lambda_0^+ > 0$ is any eigenvalue of the problem (1.7)–(1.8), then there exists a solution $u(x, \lambda_0^+) \neq 0$ of the system (1.7) that belongs to $L_2[0, \infty)$, and $u_1(0, \lambda_0^+) = u_2(0, \lambda_0^+)$. The numbers $u_1(0, \lambda_0^+)$ and $u_2(0, \lambda_0^+)$ are non-zero for otherwise we would have $u(x, \lambda_0^+) \equiv 0$ by the uniqueness of the solution of the Cauchy problem for the system (1.7).

Since the Wronskian determinant of the solutions $u(x, \lambda_0^+)$ and $e^{(2)}(x, \lambda_0^+)$, which belong to $L_2[0, \infty)$, is independent of x, there exists a sequence $\left\{x_n\right\}$ such that

$$u_1(0, \lambda_0^+)\left\{e_2^{(2)}(0, \lambda_0^+) - e_1^{(2)}(0, \lambda_0^+)\right\} = W\left[u, e^{(2)}\right] =$$
$$\lim_{n\to\infty} \left\{u_1(x_n, \lambda_0^+)e_2^{(2)}(x_n, \lambda_0^+) - u_2(x_n, \lambda_0^+)e_1^{(2)}(x_n, \lambda_0^+)\right\} = 0,$$

that is, λ_0^+ is a singular number of the problem (1.7)–(1.8).

Analogously we can prove that if $\lambda_j^-, \mathrm{Im}\lambda_j^- < 0$, is any eigenvalue of the problem (1.7)–(1.8), then it is a singular number of this problem. □

In what follows the solution $\varphi(x, \lambda) = \Big(\varphi_1(x, \lambda), \varphi_2(x, \lambda)\Big)$ of system (1.7) with the initial condition:

$$\varphi_1(0, \lambda) = \varphi_2(0, \lambda) = 1, \tag{1.24}$$

plays an important role.

It is obvious that the solution $\varphi(x,\lambda)$ satisfies the system of integral equations:

$$\begin{cases} \varphi_1(x,\lambda) = e^{-i\lambda x} - \dfrac{1}{i}\displaystyle\int\limits_0^x e^{i\lambda(y-x)}c_1(y)\varphi_2(y,\lambda)dy, \\ \varphi_2(x,\lambda) = e^{i\lambda x} + \dfrac{1}{i}\displaystyle\int\limits_0^x e^{i\lambda(x-y)}c_2(y)\varphi_1(y,\lambda)dy. \end{cases} \tag{1.25}$$

It is easily shown that this system has a solution that can be derived by the method of successive approximations. Here the following estimate holds for every complex number λ:

$$|\varphi_k(x,\lambda)| \le e^{|\mathrm{Im}\lambda|x} \exp\left\{\tilde{C}\int\limits_0^x e^{-\varepsilon y}dy\right\}, \quad \varepsilon > 0, \quad \tilde{C} \text{ is a constant}, \quad k = 1,2.$$

By the uniqueness of the solution of the Cauchy problem (1.7), (1.24), the solution $\varphi(x,\lambda)$ is unique. Obviously, for $x \ge 0$ the vector function $\varphi(x,\lambda)$ is an entire analytic function of the parameter λ.

1.1.4 Connection between the analytic solution and Jost solutions

We discuss the connection between the analytic solution $\varphi(x,\lambda)$ and Jost solutions $e^{(1)}(x,\lambda)$, $e^{(2)}(x,\lambda)$. When we take into account the fact that the Wronskian determinant of solutions $e^{(2)}(x,\lambda)$ and $\varphi(x,\lambda)$, $\mathrm{Im}\lambda > 0$, is independent of x, we obtain

$$W\left[\varphi, e^{(2)}\right] = \begin{cases} e_2^{(2)}(0,\lambda) - e_1^{(2)}(0,\lambda) \ne 0 \text{ for } \lambda \ne \lambda_k^+, \\ e_2^{(2)}(0,\lambda) - e_1^{(2)}(0,\lambda) = 0 \text{ for } \lambda = \lambda_k^+. \end{cases} \mathrm{Im}\lambda_k^+ \ge \varepsilon_0, k = 1, ..., \gamma_1.$$

Hence, for $\mathrm{Im}\lambda > 0$ the solutions $e^{(2)}(x,\lambda)$ and $\varphi(x,\lambda)$ are linearly independent for $\lambda \ne \lambda_k^+$, $k = 1, ..., \gamma_1$, and they are linearly dependent for $\lambda = \lambda_k^+$, that is

$$\begin{pmatrix} \varphi_1(x,\lambda_k^+) \\ \varphi_2(x,\lambda_k^+) \end{pmatrix} = A_0(\lambda_k^+)\begin{pmatrix} e_1^{(2)}(x,\lambda_k^+) \\ e_2^{(2)}(x,\lambda_k^+) \end{pmatrix}, \quad \mathrm{Im}\lambda_k^+ \ge \varepsilon_0, \ j = 1, ..., \gamma_1, \tag{1.26}$$

where taking (1.24) into account, from the relation (1.26), we obtain

$$A_0(\lambda_k^+) = \frac{1}{e_2^{(2)}(0,\lambda_k^+)}.$$

Here the number $e_2^{(2)}(0,\lambda_k^+)$ is non-zero because otherwise we would have that $e_1^{(2)}(0,\lambda_k^+) = 0$, and then by virtue of the uniqueness of the solution of the

Cauchy problem for system (1.7), we obtain that $e_1^{(2)}(x,\lambda_k^+) = e_2^{(2)}(x,\lambda_k^+) \equiv 0$. This is impossible because due to the estimates (1.13) for $H_{22}(x, x+\xi)$, the function

$$e_2^{(2)}(x,\lambda_k^+) = e^{i\lambda_k^+ x}\left(1 + \int\limits_0^\infty H_{22}(x,x+\xi)e^{i\lambda_k^+\xi}d\xi\right),$$

is non-zero for large x.

Analogously one can show that in the lower half-plane $\mathrm{Im}\lambda < 0$:

$$W\left[\varphi, e^{(1)}\right] = \begin{cases} e_2^{(1)}(0,\lambda) - e_1^{(1)}(0,\lambda) \neq 0 & \text{for } \lambda \neq \lambda_j^-, \\ e_2^{(1)}(0,\lambda) - e_1^{(1)}(0,\lambda) = 0 & \text{for } \lambda = \lambda_j^-. \end{cases}$$
$$\mathrm{Im}\lambda_j^- \leq -\varepsilon_0,\ j = 1, ..., \gamma_2.$$

Hence,

$$\begin{pmatrix} \varphi_1(x,\lambda_j^-) \\ \varphi_2(x,\lambda_j^-) \end{pmatrix} = B_0(\lambda_j^-)\begin{pmatrix} e_1^{(1)}(x,\lambda_j^-) \\ e_2^{(1)}(x,\lambda_j^-) \end{pmatrix},\ \mathrm{Im}\lambda_j^- \leq -\varepsilon_0,\ j = 1, ..., \gamma_2, \tag{1.27}$$

where

$$B_0(\lambda_j^-) = \frac{1}{e_1^{(1)}(0,\lambda_j^-)},\ e_1^{(1)}(0,\lambda_j^-) \neq 0.$$

Lemma 1.5. *Let m_k be the multiplicity of the zero λ_k^+ of the denominator and n_j be the multiplicity of the zero λ_j^- of the numerator. Then the following relations hold for every $x \geq 0$:*

$$\begin{aligned}(\varphi(x,\lambda))^{(q)}_{\lambda=\lambda_k^+} &= A_q e^{(2)}(x,\lambda_k^+) + qA_{q-1}\left(e^{(2)}(x,\lambda)\right)'_{\lambda=\lambda_k^+} + ... + \\ &\frac{q(q-1)...(q-i+1)}{i!}A_{q-i}\left(e^{(2)}(x,\lambda)\right)^{(i)}_{\lambda=\lambda_k^+} + ... + \\ &A_0\left(e^{(2)}(x,\lambda)\right)^{(q)}_{\lambda=\lambda_k^+},\ q = 0, 1, ..., m_k - 1,\end{aligned} \tag{1.28}$$

$$\begin{aligned}(\varphi(x,\lambda))^{(p)}_{\lambda=\lambda_j^-} &= B_p e^{(1)}(x,\lambda_j^-) + pB_{p-1}\left(e^{(1)}(x,\lambda)\right)'_{\lambda=\lambda_j^-} + ... + \\ &\frac{p(p-1)...(p-i+1)}{i!}B_{p-i}\left(e^{(1)}(x,\lambda)\right)^{(i)}_{\lambda=\lambda_j^-} + ... + \\ &B_0\left(e^{(1)}(x,\lambda)\right)^{(p)}_{\lambda=\lambda_j^-},\ p = 0, 1, ..., n_j - 1,\end{aligned} \tag{1.29}$$

where $A_0, ..., A_q$ are numbers that depend on the zero λ_k^+ of the denominator and $B_0, ..., B_p$ are numbers that depend on the zero λ_j^- of the numerator. These numbers are defined by the formulas:

$$A_q(\lambda_k^+) = \left[\frac{d^q}{d\lambda^q}\frac{1}{e_2^{(2)}(0,\lambda)}\right]_{\lambda=\lambda_k^+}, \ q = 0, 1, ..., m_k - 1, \ k = 1, ..., \gamma_1, \quad (1.30)$$

$$B_p(\lambda_j^-) = \left[\frac{d^p}{d\lambda^p}\frac{1}{e_1^{(1)}(0,\lambda)}\right]_{\lambda=\lambda_j^-}, \ p = 0, 1, ..., n_j - 1, \ j = 1, ..., \gamma_2. \quad (1.31)$$

Proof. Let m_k be the multiplicity of the zero λ_k^+ of the denominator, then by taking into account the initial condition (1.24) and the fact that the Wronskian determinant of solutions $\varphi(x, \lambda)$ and $e^{(2)}(x, \lambda)$ does not depend on x, we obtain

$$\frac{d^q}{d\lambda^q}\begin{vmatrix} \varphi_1(x,\lambda) & e_1^{(2)}(x,\lambda) \\ \varphi_2(x,\lambda) & e_2^{(2)}(x,\lambda) \end{vmatrix}_{\lambda=\lambda_k^+} = \frac{d^q}{d\lambda^q}\left[e_2^{(2)}(0,\lambda) - e_1^{(2)}(0,\lambda)\right]_{\lambda=\lambda_k^+} = 0$$

$$\text{for } q = 0, 1, ..., m_k - 1. \quad (1.32)$$

We prove the lemma by induction. For $q = 0$, the relation (1.28) is proved by the formula (1.26). In the case $q = 1$, using (1.26), from (1.32) gives

$$\frac{d}{d\lambda}\left[\varphi_1(x,\lambda)e_2^{(2)}(x,\lambda) - \varphi_2(x,\lambda)e_1^{(2)}(x,\lambda)\right]_{\lambda=\lambda_k^+} =$$

$$\left\{(\varphi_1(x,\lambda))'_{\lambda=\lambda_k^+} - A_0(\lambda_k^+)\left(e_1^{(2)}(x,\lambda)\right)'_{\lambda=\lambda_k^+}\right\}e_2^{(2)}(x,\lambda_k^+) -$$

$$\left\{(\varphi_2(x,\lambda))'_{\lambda=\lambda_k^+} - A_0(\lambda_k^+)\left(e_2^{(2)}(x,\lambda)\right)'_{\lambda=\lambda_k^+}\right\}e_1^{(2)}(x,\lambda_k^+) = 0. \quad (1.33)$$

Upon setting:

$$\psi(x,\lambda_k^+) = (\varphi(x,\lambda))'_{\lambda=\lambda_k^+} - A_0(\lambda_k^+)\left(e^{(2)}(x,\lambda)\right)'_{\lambda=\lambda_k^+},$$

we show that the vector function $\psi(x, \lambda_k^+)$ satisfies system (1.7). Indeed, since $\varphi(x, \lambda)$ and $e^{(2)}(x, \lambda)$ are solutions of system (1.7), then

$$-iJ\frac{d}{dx}\left(\frac{\partial}{\partial\lambda}\varphi(x,\lambda)\right)_{\lambda=\lambda_k^+} + C(x)\left(\frac{\partial}{\partial\lambda}\varphi(x,\lambda)\right)_{\lambda=\lambda_k^+} =$$

$$\left[\varphi(x,\lambda) + \lambda\frac{\partial}{\partial\lambda}\varphi(x,\lambda)\right]_{\lambda=\lambda_k^+},$$

$$A_0(\lambda_k^+)iJ\frac{d}{dx}\left(\frac{\partial}{\partial\lambda}e^{(2)}(x,\lambda)\right)_{\lambda=\lambda_k^+} - A_0(\lambda_k^+)C(x)\left(\frac{\partial}{\partial\lambda}e^{(2)}(x,\lambda)\right)_{\lambda=\lambda_k^+} =$$

$$-A_0(\lambda_k^+)\left[e^{(2)}(x,\lambda) + \lambda\frac{\partial}{\partial\lambda}e^{(2)}(x,\lambda)\right]_{\lambda=\lambda_k^+}.$$

We add together the last two equalities with due regard for (1.26) and obtain:

$$-iJ\frac{d}{dx}\psi(x,\lambda_k^+)+C(x)\psi(x,\lambda_k^+)=\lambda_k^+\psi(x,\lambda_k^+),$$

i.e., $\psi(x,\lambda_k^+)$ is the solution of (1.7). Due to (1.33), the Wronskian determinant of the solutions $\psi(x,\lambda_k^+)$ and $e^{(2)}(x,\lambda_k^+)$ of (1.7) is equal to zero; therefore, there exists a number $A_1(\lambda_k^+)$ such that

$$(\varphi(x,\lambda))'_{\lambda=\lambda_k^+}-A_0(\lambda_k^+)\left(e^{(2)}(x,\lambda)\right)'_{\lambda=\lambda_k^+}=A_1(\lambda_k^+)e^{(2)}(x,\lambda_k^+).$$

From the latter equality, using (1.24) and (1.26), we obtain the formula for $A_1(\lambda_k^+)$:

$$A_1(\lambda_k^+)=\frac{d}{d\lambda}\left[\frac{1}{e_2^{(2)}(0,\lambda)}\right]_{\lambda=\lambda_k^+}.$$

Thus, the relation (1.28) is proved for $q=1$. Assume that the relation (1.28) holds for $q\le q_0\le m_k-2$. We are going to prove that it is true for $q=q_0+1$. Denote

$$\Phi(x,\lambda)=\begin{pmatrix}\Phi_1(x,\lambda)\\ \Phi_2(x,\lambda)\end{pmatrix}=\begin{pmatrix}\varphi_1(x,\lambda)\\ \varphi_2(x,\lambda)\end{pmatrix}_\lambda^{(q_0+1)}-(q_0+1)A_{q_0}\begin{pmatrix}e_1^{(2)}(x,\lambda)\\ e_2^{(2)}(x,\lambda)\end{pmatrix}_\lambda^{\prime}$$
$$-\frac{(q_0+1)q_0}{2!}A_{q_0-1}\begin{pmatrix}e_1^{(2)}(x,\lambda)\\ e_2^{(2)}(x,\lambda)\end{pmatrix}_\lambda^{\prime\prime}-\ldots-A_0\begin{pmatrix}e_1^{(2)}(x,\lambda)\\ e_2^{(2)}(x,\lambda)\end{pmatrix}_\lambda^{(q_0+1)}. \tag{1.34}$$

If $u(x,\lambda)$ is a solution of system (1.7), then $\frac{\partial^q u(x,\lambda)}{\partial\lambda^q}$ satisfies the system:

$$\left(-iJ\frac{d}{dx}+C(x)-\lambda\right)\frac{\partial^q u(x,\lambda)}{\partial\lambda^q}=q\frac{\partial^{q-1}u(x,\lambda)}{\partial\lambda^{q-1}},\ J=\begin{pmatrix}-1&0\\0&1\end{pmatrix}.$$

Apply the latter system to each term of the right-hand side of equality (1.34), then add together all terms, we have:

$$\left(-iJ\frac{d}{dx}+C(x)-\lambda\right)\Phi(x,\lambda_k^+)=$$
$$(q_0+1)\Bigg\{\begin{pmatrix}\varphi_1(x,\lambda)\\ \varphi_2(x,\lambda)\end{pmatrix}_{\lambda=\lambda_k^+}^{(q_0)}-A_{q_0}\begin{pmatrix}e_1^{(2)}(x,\lambda)\\ e_2^{(2)}(x,\lambda)\end{pmatrix}_{\lambda=\lambda_k^+}-$$
$$-q_0A_{q_0-1}\begin{pmatrix}e_1^{(2)}(x,\lambda)\\ e_2^{(2)}(x,\lambda)\end{pmatrix}_{\lambda=\lambda_k^+}^{\prime}-\ldots-A_0\begin{pmatrix}e_1^{(2)}(x,\lambda)\\ e_2^{(2)}(x,\lambda)\end{pmatrix}_{\lambda=\lambda_k^+}^{(q_0)}\Bigg\}.$$

Due to the induction hypothesis, the expression in the parentheses is equal to zero, whence it follows that $\Phi(x, \lambda_k^+)$ is a solution of system (1.7). Taking into account of (1.32) and (1.34), we see that

$$\left(\varphi_1(x,\lambda)e_2^{(2)}(x,\lambda) - \varphi_2(x,\lambda)e_1^{(2)}(x,\lambda)\right)^{(q_0+1)}_{\lambda=\lambda_k^+} = \begin{vmatrix} \Phi_1(x,\lambda_k^+) & e_1^{(2)}(x,\lambda_k^+) \\ \Phi_2(x,\lambda_k^+) & e_2^{(2)}(x,\lambda_k^+) \end{vmatrix} = 0,$$

that is, the Wronskian determinant of the solutions $\Phi(x, \lambda_k^+)$ and $e^{(2)}(x, \lambda_k^+)$ of system (1.7) is equal to zero; i.e., these solutions are linearly dependent. Then there exists a number $A_{q_0+1} = A_{q_0+1}(\lambda_k^+)$ such that

$$\Phi(x, \lambda_k^+) = A_{q_0+1}(\lambda_k^+)e^{(2)}(x, \lambda_k^+).$$

Using (1.24) and (1.34) at $x = 0$, we write the latter equality at $x = 0$:

$$A_{q_0+1}\begin{pmatrix} e_1^{(2)}(0,\lambda_k^+) \\ e_2^{(2)}(0,\lambda_k^+) \end{pmatrix} + (q_0+1)A_{q_0}\begin{pmatrix} e_1^{(2)}(0,\lambda) \\ e_2^{(2)}(0,\lambda) \end{pmatrix}'_{\lambda=\lambda_k^+} +$$

$$\frac{(q_0+1)q_0}{2!}A_{q_0-1}\begin{pmatrix} e_1^{(2)}(0,\lambda) \\ e_2^{(2)}(0,\lambda) \end{pmatrix}''_{\lambda=\lambda_k^+} + ... + A_0\begin{pmatrix} e_1^{(2)}(0,\lambda) \\ e_2^{(2)}(0,\lambda) \end{pmatrix}^{(q_0+1)}_{\lambda=\lambda_k^+} = 0.$$

where by induction the numbers $A_{q_0}, A_{q_0-1}, \dots, A_0$ are calculated by formula (1.30).

It follows from the Leibniz rule for differentiating a product of functions that

$$\left(\frac{1}{e_2^{(2)}(0,\lambda)}e_2^{(2)}(0,\lambda)\right)^{(q_0+1)}_{\lambda=\lambda_k^+} = \left(\frac{1}{e_2^{(2)}(0,\lambda)}\right)^{(q_0+1)}_{\lambda=\lambda_k^+} e_2^{(2)}(0,\lambda_k)$$

$$+ (q_0+1)\left(\frac{1}{e_2^{(2)}(0,\lambda)}\right)^{(q_0)}_{\lambda=\lambda_k^+}\left(e_2^{(2)}(0,\lambda)\right)'_{\lambda=\lambda_k^+} + \frac{(q_0+1)q_0}{2!}\left(\frac{1}{e_2^{(2)}(0,\lambda)}\right)^{(q_0-1)}_{\lambda=\lambda_k^+}$$

$$\times\, e_2^{(2)}(0,\lambda)''_{\lambda=\lambda_k^+} + ... + \frac{1}{e_2^{(2)}(0,\lambda_k^+)}\left(e_2^{(2)}(0,\lambda)\right)^{(q_0+1)}_{\lambda=\lambda_k^+} = 0.$$

By comparison of every term of the latter equality with the corresponding term of the previous equality, we obtain formula (1.30) for calculating the number A_{q_0+1}. Thus, the relations (1.28) are proved. The relations (1.29) with the numbers $B_0, ..., B_p$ calculated by formula (1.31) are proved analogously. □

We continue our study of the connection between the analytic solution $\varphi(x, \lambda)$ and Jost solutions $e^{(1)}(x, \lambda)$ and $e^{(2)}(x, \lambda)$ of system (1.7). Since the Jost solutions form a fundamental system of solutions of system (1.7), then the analytic solution $\varphi(x, \lambda)$ is a linear combination of Jost solutions:

$$\varphi(x,\lambda) = a_1(\lambda)e^{(1)}(x,\lambda) + a_2(\lambda)e^{(2)}(x,\lambda) \quad \text{for real } \lambda, \tag{1.35}$$

where taking the initial condition (1.24) into account, from (1.35) we calculate

$$a_1(\lambda) = e_2^{(2)}(0,\lambda) - e_1^{(2)}(0,\lambda), \;\; a_2(\lambda) = e_1^{(1)}(0,\lambda) - e_2^{(1)}(0,\lambda).$$

If λ is not a real pole of the scattering function $S(\lambda)$, then from (1.35) we have

$$u(x,\lambda) = \begin{pmatrix} u_1(x,\lambda) \\ u_2(x,\lambda) \end{pmatrix} = \frac{1}{e_2^{(2)}(0,\lambda) - e_1^{(2)}(0,\lambda)} \begin{pmatrix} \varphi_1(x,\lambda) \\ \varphi_2(x,\lambda) \end{pmatrix} = e^{(1)}(x,\lambda) + S(\lambda)e^{(2)}(x,\lambda).$$

If λ is not a real pole of $S^{-1}(\lambda)$, then

$$z(x,\lambda) = \begin{pmatrix} z_1(x,\lambda) \\ z_2(x,\lambda) \end{pmatrix} = \frac{1}{e_1^{(1)}(0,\lambda) - e_2^{(1)}(0,\lambda)} \begin{pmatrix} \varphi_1(x,\lambda) \\ \varphi_2(x,\lambda) \end{pmatrix} = S^{-1}(\lambda)e^{(1)}(x,\lambda) + e^{(2)}(x,\lambda).$$

The considered vector functions $u(x,\lambda)$ and $z(x,\lambda)$ are the solutions of system (1.7). By Lemma 1.1 these solutions are representable in the form:

$$u_1(x,\lambda) = e^{-i\lambda x} + \int_x^\infty H_{11}(x,t)e^{-i\lambda t}dt + S(\lambda)\int_x^\infty H_{12}(x,t)e^{i\lambda t}dt,$$

$$u_2(x,\lambda) = S(\lambda)e^{i\lambda x} + \int_x^\infty H_{21}(x,t)e^{-i\lambda t}dt + S(\lambda)\int_x^\infty H_{22}(x,t)e^{i\lambda t}dt,$$

$$z_1(x,\lambda) = S^{-1}(\lambda)e^{-i\lambda x} + S^{-1}(\lambda)\int_x^\infty H_{11}(x,t)e^{-i\lambda t}dt + \int_x^\infty H_{12}(x,t)e^{i\lambda t}dt,$$

$$z_2(x,\lambda) = e^{i\lambda x} + S^{-1}(\lambda)\int_x^\infty H_{21}(x,t)e^{-i\lambda t}dt + \int_x^\infty H_{22}(x,t)e^{i\lambda t}dt.$$

The following lemma holds.

Lemma 1.6. *The vector function* $\begin{pmatrix} u_1(x,\lambda)e^{i\lambda x} - 1 \\ u_2(x,\lambda)e^{i\lambda x} - e^{2i\lambda x} \end{pmatrix}$ *can be extended analytically into the upper half-plane* $\mathrm{Im}\lambda > 0$ *apart from the singular numbers* $\lambda = \lambda_k^+,\ \mathrm{Im}\lambda_k^+ > 0,\ k = 1, ..., \gamma_1,$ *and* $\begin{pmatrix} z_1(x,\lambda)e^{-i\lambda x} - e^{-2i\lambda x} \\ z_2(x,\lambda)e^{-i\lambda x} - 1 \end{pmatrix}$ *can be extended analytically into the lower half-plane* $\mathrm{Im}\lambda < 0$ *apart from the singular numbers* $\lambda = \lambda_j^-,\ \mathrm{Im}\lambda_j^- < 0,\ j = 1, ..., \gamma_2,$ *here*

$$\begin{aligned} &\lim_{\lambda\to\infty} \begin{pmatrix} u_1(x,\lambda)e^{i\lambda x} - 1 \\ u_2(x,\lambda)e^{i\lambda x} - e^{2i\lambda x} \end{pmatrix} = 0,\ \mathrm{Im}\lambda \geq 0, \\ &\lim_{\lambda\to\infty} \begin{pmatrix} z_1(x,\lambda)e^{-i\lambda x} - e^{-2i\lambda x} \\ z_2(x,\lambda)e^{-i\lambda x} - 1 \end{pmatrix} = 0,\ \mathrm{Im}\lambda \leq 0. \end{aligned} \tag{1.36}$$

Proof. The analytic properties of these vector functions follow from their definition and the analyticity of $\varphi(x,\lambda)$ in the whole plane of λ. We prove the assertions (1.36) and put

$$\psi_1(x,\lambda)=\varphi_1(x,\lambda)e^{i\lambda x}-1,\ \psi_2(x,\lambda)=\varphi_2(x,\lambda)e^{i\lambda x}-e^{2i\lambda x},\ \mathrm{Im}\lambda\geq 0.$$

It follows from the system (1.25) that the vector function $\psi(x,\lambda)=\begin{pmatrix}\psi_1(x,\lambda)\\ \psi_2(x,\lambda)\end{pmatrix}$ satisfies the system of integral equations:

$$\begin{cases}\psi_1(x,\lambda)=h_1(x,\lambda)-\dfrac{1}{i}\displaystyle\int\limits_0^x c_1(y)\psi_2(y,\lambda)dy, \\ \psi_2(x,\lambda)=h_2(x,\lambda)+\dfrac{1}{i}\displaystyle\int\limits_0^x e^{2i\lambda(x-y)}c_2(y)\psi_1(y,\lambda)dy,\end{cases}\qquad \mathrm{Im}\lambda\geq 0,$$

where $h_1(x,\lambda)=-\dfrac{1}{i}\displaystyle\int\limits_0^x e^{2i\lambda y}c_1(y)dy,\ h_2(x,\lambda)=\dfrac{1}{i}\displaystyle\int\limits_0^x e^{2i\lambda(x-y)}c_2(y)dy.$

Owing to estimate (1.9), this system has a unique bounded solution that depends analytically on λ for $\mathrm{Im}\lambda\geq 0$. It can be obtained by the method of successive approximations. In addition, the obtained solution $\psi(x,\lambda)$ satisfies in the space $C[0,\infty)$ the estimate:

$$|\psi_k(x,\lambda)|\leq ||h(x,\lambda)||\exp\left\{\tilde{C}\int\limits_0^x e^{-\varepsilon y}dy\right\},\ k=1,2,$$

where $||h(x,\lambda)||=\max\left(\max\limits_x|h_1(x,\lambda)|,\max\limits_x|h_2(x,\lambda)|\right)$, $\tilde{C}$ is constant.

Due to estimate (1.9): $\lim_{\lambda\to\infty}||h(x,\lambda)||=0$ for $\mathrm{Im}\lambda\geq 0$. Then from the estimate for $|h_k(x,\lambda)|$ we get:

$$\lim_{\lambda\to\infty}\psi_k(x,\lambda)=0\ \text{ for }\mathrm{Im}\lambda\geq 0,\ k=1,2.$$

Due to the first assertion of Lemma 1.2, we have

$$\lim_{\lambda\to\infty}\left\{e_2^{(2)}(0,\lambda)-e_1^{(2)}(0,\lambda)\right\}=$$

$$\lim_{\lambda\to\infty}\left\{1+\int\limits_0^\infty\left[H_{22}(0,t)-H_{12}(0,t)\right]e^{i\lambda t}dt\right\}=1,\ \mathrm{Im}\lambda\geq 0.$$

The first assertion of (1.36) is deduced from the above estimates:

$$\lim_{\lambda\to\infty}\begin{pmatrix} u_1(x,\lambda)e^{i\lambda x}-1 \\ u_2(x,\lambda)e^{i\lambda x}-e^{2i\lambda x}\end{pmatrix} =$$

$$\lim_{\lambda\to\infty}\left\{\frac{1}{e_2^{(2)}(0,\lambda)-e_1^{(2)}(0,\lambda)}\begin{pmatrix} \varphi_1(x,\lambda)e^{i\lambda x}-1 \\ \varphi_2(x,\lambda)e^{i\lambda x}-e^{2i\lambda x}\end{pmatrix}+\right.$$

$$\left.\frac{1}{e_2^{(2)}(0,\lambda)-e_1^{(2)}(0,\lambda)}\begin{pmatrix} 1 \\ e^{2i\lambda x}\end{pmatrix}-\begin{pmatrix} 1 \\ e^{2i\lambda x}\end{pmatrix}\right\}=0,\ \mathrm{Im}\lambda\geq 0.$$

The second assertion of (1.36) is proved analogously. □

1.1.5 The scattering data

The following functions constructed from the given scattering function will play an important role in what follows:

$$f_S(x)=\frac{1}{2\pi}\int\limits_{-\infty+i\eta}^{\infty+i\eta}[S(\lambda)-1]e^{-i\lambda x}d\lambda,$$
$$g_S(x)=\frac{1}{2\pi}\int\limits_{-\infty-i\eta}^{\infty-i\eta}[S^{-1}(\lambda)-1]e^{-i\lambda x}d\lambda, \tag{1.37}$$

where $S(\lambda)$ is a function of type (S) in the strip $|\mathrm{Im}\lambda|<\varepsilon_0$, and η is some number such that $0<\eta<\varepsilon_0$. Since $S(\lambda)-1$ and $S^{-1}(\lambda)-1$ are analytic in the strip $0<|\mathrm{Im}\lambda|<\varepsilon_0$, the value of these integrals is independent of η. In the sense of generalized functions [116], $f_S(x)$ and $g_S(x)$ can be considered as the inverse Fourier transformations of $S(\lambda)-1$ and $S^{-1}(\lambda)-1$, respectively. The functions (1.37) characterize the problem (1.7)–(1.8) on the continuous spectrum. The following functions characterize this problem on the point spectrum:

$$f_k^+(-x)=\frac{1}{2\pi}\int\limits_{c_k^+}\frac{1}{e_2^{(2)}(0,\lambda)-e_1^{(2)}(0,\lambda)}\Big[A_0+A_1(\lambda-\lambda_k^+)+...+$$
$$\frac{A_{m_k-1}}{(m_k-1)!}(\lambda-\lambda_k^+)^{m_k-1}\Big]e^{i\lambda x}d\lambda,\ k=1,...,\gamma_1, \tag{1.38}$$

$$g_j^-(x)=\frac{1}{2\pi}\int\limits_{c_j^-}\frac{1}{e_1^{(1)}(0,\lambda)-e_2^{(1)}(0,\lambda)}\Big[B_0+B_1(\lambda-\lambda_j^-)+...+$$
$$\frac{B_{n_j-1}}{(n_j-1)!}(\lambda-\lambda_j^-)^{n_j-1}\Big]e^{-i\lambda x}d\lambda,\ j=1,...,\gamma_2. \tag{1.39}$$

where c_k^+ and c_j^- stand for circles with a sufficiently small radius that avoid the singular points λ_k^+ and λ_j^-, respectively. The numbers $A_q = A_q(\lambda_k^+)$, $q = 0, 1, ..., m_k - 1$ and $B_p = B_p(\lambda_j^-)$, $p = 0, 1, ..., n_j - 1$ are calculated by formulas (1.30) and (1.31), respectively.

By using the definition of a residue [114], from (1.38) and (1.39), we can verify that

$$f_k^+(-x) = p_k(x)e^{i\lambda_k^+ x} \quad k = 1, ..., \gamma_1, \quad g_j^-(x) = \tilde{q}_j(x)e^{-i\lambda_j^- x}, \; j = 1, ..., \gamma_2, \tag{1.40}$$

where $p_k(x)$ and $\tilde{q}_j(x)$ are polynomials of degree $m_k - 1$ and of $n_j - 1$, respectively.

We call the functions $p_1(x), ..., p_{\gamma_1}(x)$ and $\tilde{q}_1(x), ..., \tilde{q}_{\gamma_2}(x)$ the standardized polynomials [97] of the problem (1.7)–(1.8). When the potential matrix $C(x)$ is self-adjoint, all the standardized polynomials are equal to the zero polynomial [119].

Definition 1.2. *The scattering function $S(\lambda)$ of the type (S), the non-real singular numbers $\lambda_1^+, ..., \lambda_{\gamma_1}^+; \lambda_j^-, ..., \lambda_{\gamma_2}^-$ and the standardized polynomials $p_1(x), ..., p_{\gamma_1}(x)$, $\tilde{q}_1(x), ..., \tilde{q}_{\gamma_2}(x)$ are called the scattering data of the problem* (1.7)–(1.8).

We turn to the inverse scattering problem, which is to determine the potential matrix from the scattering data, and also to describe the scattering data, that is, to establish necessary and sufficient conditions for given quantities to be scattering data for the system (1.7) on a half-line with the boundary condition (1.8). We shall see that, in exactly the same way as in the self-adjoint case [119], the scattering data uniquely determine the problem (1.7)–(1.8). To recover this problem from the scattering data, we first obtain systems of equations connecting the scattering data with the kernels of the transformation operator. We call these systems the systems of fundamental equations.

1.1.6 Derivation of systems of fundamental equations

We construct the following functions from the scattering data given by (1.37)–(1.40):

$$f(-x) = f_S(-x) - \sum_{k=1}^{\gamma_1} f_k^+(-x), \quad g(x) = g_S(x) - \sum_{j=1}^{\gamma_2} g_j^-(x), \quad x \geq 0. \tag{1.41}$$

The following lemma about the derivation of systems of fundamental equations holds.

Lemma 1.7. *For every $x \geq 0,\ \xi > 0$:*

$$H_{11}(x, x+\xi) + \int_0^\infty H_{12}(x, x+t) f(-2x-\xi-t)dt = 0, \tag{1.42}$$

$$g(2x+\xi) + H_{12}(x, x+\xi) + \int_0^\infty H_{11}(x, x+t) g(2x+\xi+t)dt = 0, \tag{1.43}$$

$$H_{22}(x, x+\xi) + \int_0^\infty H_{21}(x, x+t) g(2x+\xi+t)dt = 0, \tag{1.44}$$

$$f(-2x-\xi) + H_{21}(x, x+\xi) + \int_0^\infty H_{22}(x, x+t) f(-2x-\xi-t)dt = 0. \tag{1.45}$$

Proof. When we apply Cauchy's Theorem to the vector function $\begin{pmatrix} u_1(x,\lambda)e^{i\lambda x} - 1 \\ u_2(x,\lambda)e^{i\lambda x} - e^{2i\lambda x} \end{pmatrix}$ for $\mathrm{Im}\lambda \geq \eta$, from Lemma 1.6 we obtain for $\xi > 0$:

$$\sum_{k=1}^{\gamma_1} \int_{c_k^+} \begin{pmatrix} u_1(x,\lambda)e^{i\lambda x} - 1 \\ u_2(x,\lambda)e^{i\lambda x} - e^{2i\lambda x} \end{pmatrix} e^{i\lambda\xi} d\lambda =$$

$$\int_{-\infty+i\eta}^{\infty+i\eta} \begin{pmatrix} \int_x^\infty H_{11}(x,t)e^{-i\lambda(t-x)}dt + S(\lambda)\int_x^\infty H_{12}(x,t)e^{i\lambda(t+x)}dt \\ (S(\lambda)-1)e^{2i\lambda x} + \int_x^\infty H_{21}(x,t)e^{-i\lambda(t-x)}dt + S(\lambda)\int_x^\infty H_{22}(x,t)e^{i\lambda(t+x)}dt \end{pmatrix}$$
$$\times\, e^{i\lambda\xi} d\lambda.$$

On the right-hand side we use the theorem about the Fourier transform of a convolution of functions, bearing in mind that $S(\lambda)$ is a function of the type (S) in the strip $0 < |\mathrm{Im}\lambda| < \eta$, we find that

$$\begin{pmatrix} H_{11}(x, x+\xi) + \int_0^\infty H_{12}(x, x+t) f_S(-2x-\xi-t)dt \\ f_S(-2x-\xi) + H_{21}(x, x+\xi) + \int_0^\infty H_{22}(x, x+t) f_S(-2x-\xi-t)dt \end{pmatrix} =$$

$$\sum_{k=1}^{\gamma_1} \frac{1}{2\pi} \int_{c_k^+} \frac{1}{e_2^{(2)}(0,\lambda) - e_1^{(2)}(0,\lambda)} \begin{pmatrix} \varphi_1(x,\lambda) \\ \varphi_2(x,\lambda) \end{pmatrix} e^{i\lambda(x+\xi)} d\lambda =$$

$$\sum_{k=1}^{\gamma_1} i\mathrm{Res}_{\lambda=\lambda_k^+} \frac{1}{e_2^{(2)}(0,\lambda) - e_1^{(2)}(0,\lambda)} \begin{pmatrix} \varphi_1(x,\lambda) \\ \varphi_2(x,\lambda) \end{pmatrix} e^{i\lambda(x+\xi)} d\lambda,\ \xi > 0. \tag{1.46}$$

By Lemma 1.5 and the definition of a residue, after some uncomplicated transformations, we obtain

$$\mathrm{Res}_{\lambda=\lambda_k^+}\frac{1}{e_2^{(2)}(0,\lambda)-e_1^{(2)}(0,\lambda)}\varphi(x,\lambda)e^{i\lambda(x+\xi)} =$$

$$\frac{1}{2\pi i}\int\limits_{c_k^+}\frac{e^{i\lambda(x+\xi)}}{e_2^{(2)}(0,\lambda)-e_1^{(2)}(0,\lambda)}\Big[A_0+A_1(\lambda-\lambda_k^+)+...+$$

$$\frac{A_{m_k-1}}{(m_k-1)!}(\lambda-\lambda_k^+)^{m_k-1}\Big]e^{(2)}(x,\lambda)d\lambda =$$

$$\frac{1}{i}\begin{pmatrix} \int\limits_0^\infty H_{12}(x,x+t)f_k^+(-2x-\xi-t)dt \\ f_k^+(-2x-\xi)+\int\limits_0^\infty H_{22}(x,x+t)f_k^+(-2x-\xi-t)dt \end{pmatrix}, \quad \xi>0. \qquad (1.47)$$

When we take the result (1.47) into account, from (1.46) with due regard for (1.41), we derive equations (1.42) and (1.45).

By using the second assertions of Lemmas 1.5 and 1.6, we analogously obtain:

$$\begin{pmatrix} g_S^-(2x+\xi)+H_{12}(x,x+\xi)+\int\limits_0^\infty H_{11}(x,x+t)g_S^-(2x+\xi+t)dt \\ H_{22}(x,x+\xi)+\int\limits_0^\infty H_{21}(x,x+t)g_S^-(2x+\xi+t)dt \end{pmatrix} =$$

$$\sum_{j=1}^{\gamma_2}\frac{1}{2\pi}\int\limits_{c_j^-}\frac{1}{e_1^{(1)}(0,\lambda)-e_2^{(1)}(0,\lambda)}\begin{pmatrix}\varphi_1(x,\lambda)\\ \varphi_2(x,\lambda)\end{pmatrix}e^{-i\lambda(x+\xi)}d\lambda =$$

$$\sum_{j=1}^{\gamma_2}\begin{pmatrix} g_j^-(2x+\xi)+\int\limits_0^\infty H_{11}(x,x+t)g_j^-(2x+\xi+t)dt \\ \int\limits_0^\infty H_{21}(x,x+t)g_j^-(2x+\xi+t)dt \end{pmatrix}, \quad \xi>0. \qquad (1.48)$$

Then with due regard for (1.41), the equations (1.43) and (1.44) are derived from (1.48).

Using (1.24) and the definition of the standardized polynomials, from (1.47) and (1.48) one can prove the following lemma. □

Lemma 1.8. *The following compatibility relations between the scattering data of the problem (1.7)–(1.8) hold:*

$$p_k(\xi)+\int\limits_0^\infty p_k(\xi+t)e^{i\lambda_k^+t}\big[H_{22}(0,t)-H_{12}(0,t)\big]dt=0,\ \xi>0,\ k=1,...,\gamma_1,$$

$$\tilde q_j(\xi)+\int\limits_0^\infty \tilde q_j(\xi+t)e^{-i\lambda_j^-t}\big[H_{11}(0,t)-H_{21}(0,t)\big]dt=0,\ \xi>0,\ j=1,...,\gamma_2.$$

1.1.7 The estimates for the functions $f(-x)$ and $g(x)$

To prove the unique solvability of systems of fundamental equations, the functions $f(-x)$ and $g(x)$ in these equations need to be estimated.

Lemma 1.9. *Let the kernel functions* $H_{jk}(x, x+\xi)$, $x \geq 0, \xi > 0$, $j, k = 1, 2$ *satisfy the inequalities* (1.13) *and let the scattering function* (1.22) *be the function of the type* (S) *in the strip* $|\mathrm{Im}\lambda| < \varepsilon_0$, *where the number* ε_0 *is defined by* (1.23). *Then the following estimates for* $f(-x)$ *and* $g(x)$ *are valid:*

$$|f(-x)| \leq \tilde{C}(\eta)e^{-\frac{\varepsilon}{4}x}, \quad |g(x)| \leq \tilde{C}(\eta)e^{-\frac{\varepsilon}{4}x}, \tag{1.49}$$

where $\tilde{C} = \tilde{C}(\eta)$ *is a constant depending on* $\eta:\ 0 < \eta < \frac{1}{12}\varepsilon_0$.

Proof. Consider the numerator $N(\lambda)$ and denominator $D(\lambda)$ of the factorization (1.22):

$$\begin{aligned} N(\lambda) &= 1 + \int_0^\infty [H_{11}(0,t) - H_{21}(0,t)]\, e^{-i\lambda t} dt, \quad \mathrm{Im}\lambda = -\eta, \\ & \qquad\qquad\qquad\qquad\qquad\qquad\qquad\qquad 0 < \eta < \frac{1}{12}\varepsilon_0. \\ D(\lambda) &= 1 + \int_0^\infty [H_{22}(0,t) - H_{12}(0,t)]\, e^{i\lambda t} dt, \quad \mathrm{Im}\lambda = \eta, \end{aligned} \tag{1.50}$$

Owing to estimate (1.13) and Lemma 1.2, the functions (1.50) are continuous and different from zero and satisfy the asymptotic relations:

$$\lim_{|\lambda|\to\infty} N(\lambda) = 1, \ \ \mathrm{Im}\lambda = -\eta; \quad \lim_{|\lambda|\to\infty} D(\lambda) = 1, \ \ \mathrm{Im}\lambda = \eta,$$

whence it follows that there exist positive numbers m and M such that

$$m \leq |N(\lambda)| \leq M, \ \ \mathrm{Im}\lambda = -\eta; \ \ m \leq |D(\lambda)| \leq M, \ \ \mathrm{Im}\lambda = \eta.$$

Using the last inequalities, from the factorization (1.22), we obtain that the denominators of functions $S(\sigma + i\eta) - 1$ and $S^{-1}(\sigma - i\eta) - 1$ are bounded from below for $\mathrm{Im}\lambda = \eta$, $0 < \eta < \frac{1}{12}\varepsilon_0$. Then

$$\begin{aligned} |S(\sigma + i\eta) - 1| \leq \frac{1}{m}\Bigg\{ \Bigg| &\int_0^\infty \big[H_{11}(0,t) - H_{21}(0,t)\big] e^{\eta t} e^{-i\sigma t} dt - \\ &\int_0^\infty \big[H_{22}(0,t) - H_{12}(0,t)\big] e^{-\eta t} e^{i\sigma t} dt \Bigg| \Bigg\}, \end{aligned}$$

$$\begin{aligned} \big|S^{-1}(\sigma - i\eta) - 1\big| \leq \frac{1}{m}\Bigg\{ \Bigg| &\int_0^\infty \big[H_{22}(0,t) - H_{12}(0,t)\big] e^{\eta t} e^{i\sigma t} dt - \\ &\int_0^\infty \big[H_{11}(0,t) - H_{21}(0,t)\big] e^{-\eta t} e^{-i\sigma t} dt \Bigg| \Bigg\}, \end{aligned}$$

whence from the estimates (1.13) of $H_{jk}(0,t),\quad t \geq 0,\quad j,k = 1,2$ and the definition (1.23) of the number ε_0, it follows that the functions $S(\sigma + i\eta)$ and $S^{-1}(\sigma - i\eta)$ are bounded and absolutely integrable with respect to σ. Hence,

$$\int_{-\infty}^{\infty} |S(\sigma + i\eta) - 1|^2 d\sigma < \infty, \quad \int_{-\infty}^{\infty} \left|S^{-1}(\sigma - i\eta) - 1\right|^2 d\sigma < \infty,$$

and the formulas (1.37) can be written in the form:

$$f_S(t)e^{-\eta t} = \frac{1}{2\pi} \int_{-\infty}^{\infty} [S(\sigma + i\eta) - 1]e^{-i\sigma t} d\sigma,$$

$$g_S(t)e^{\eta t} = \frac{1}{2\pi} \int_{-\infty}^{\infty} \left[S^{-1}(\sigma - i\eta) - 1\right] e^{-i\sigma t} d\sigma.$$

Then by Parseval's equality, we have

$$\begin{aligned} \int_{-\infty}^{\infty} \left|f_S(t)e^{-\eta t}\right|^2 dt &= \frac{1}{2\pi} \int_{-\infty}^{\infty} |S(\sigma + i\eta) - 1|^2 d\sigma < \infty, \\ \int_{-\infty}^{\infty} \left|g_S(t)e^{\eta t}\right|^2 dt &= \frac{1}{2\pi} \int_{-\infty}^{\infty} \left|S^{-1}(\sigma - i\eta) - 1\right|^2 d\sigma < \infty. \end{aligned} \tag{1.51}$$

Because $\text{Im}\lambda_k^+ > \varepsilon_0,\ k = 1, ..., \gamma_1,\ \text{Im}\lambda_j^- < -\varepsilon_0,\ j = 1, ..., \gamma_2$, and $0 < \eta < \frac{\varepsilon_0}{12}$, we have $f_k^+(-x)e^{-\eta x} = p_k(x)e^{i\lambda_k^+ x}e^{-\eta x} \in L_2[0,\infty)$ and $g_j^-(x)e^{\eta x} = \tilde{q}_j(x)e^{-i\lambda_j^- x}e^{\eta x} \in L_2[0,\infty)$. Due to this fact and (1.51) with due regard for (1.41), we have:

$$f(-x)e^{-\eta x} \in L_2(0,\infty), \quad g(x)e^{\eta x} \in L_2(0,\infty).$$

By using this result, estimate (1.13) and the Cauchy-Bunyakovskii inequality, from the equation (1.45), we obtain the estimate (1.49) for $f(-x)$:

$$\begin{aligned} &|f(-x)| \leq \left|H_{21}\left(\frac{x}{3}, \frac{x}{3} + \frac{x}{3}\right)\right| + \int_0^{\infty} \left|H_{22}\left(\frac{x}{3}, \frac{x}{3} + t\right)\right| \left|f(-x-t)\right| dt \leq \\ &\tilde{C}_1 e^{-\frac{1}{2}\varepsilon x} + \tilde{C}_1 e^{-\frac{1}{3}\varepsilon x} \int_0^{\infty} e^{-\frac{1}{2}\varepsilon t} e^{\eta(x+t)} \left|e^{-\eta(x+t)} f(-x-t)\right| dt \leq \\ &\tilde{C}_1 e^{-\frac{1}{4}\varepsilon x} \left\{1 + \left[\int_0^{\infty} \left|e^{-(\frac{\varepsilon}{2}-\eta)t}\right|^2 dt\right]^{\frac{1}{2}} \left[\int_0^{\infty} \left|e^{-\eta(x+t)} f(-x-t)\right|^2 dt\right]^{\frac{1}{2}}\right\} \leq \\ &\tilde{C}(\eta)e^{-\frac{1}{4}\varepsilon x}, \end{aligned}$$

where $0 < \eta < \frac{\varepsilon_0}{12}$, ε_0 is defined by (1.23) and $\tilde{C}_1, \tilde{C}$ are constants depending on η.

In a manner analogous to the previous one, we get the estimate for $g(x)$ from the equation (1.43). □

1.1.8 The unique solvability of systems of fundamental equations

The systems of fundamental equations (1.42)–(1.43) and (1.44)–(1.45) can be written in the matrix form:

$$\begin{pmatrix} H_{11}(x,t) & H_{21}(x,t) \\ H_{12}(x,t) & H_{22}(x,t) \end{pmatrix} + \begin{pmatrix} 0 & f(-x-t) \\ g(x+t) & 0 \end{pmatrix} +$$
$$\int_x^\infty \begin{pmatrix} 0 & f(-t-s) \\ g(t+s) & 0 \end{pmatrix} \begin{pmatrix} H_{11}(x,s) & H_{21}(x,s) \\ H_{12}(x,s) & H_{22}(x,s) \end{pmatrix} ds = 0, \quad 0 \le x < t. \tag{1.52}$$

Using estimate (1.49) and the Fredholm's Theorems [61, 80, 101], we can verify that the matrix equation (1.52) is the Fredholm equation of the second kind in the space $L_1[0,\infty)$.

Lemma 1.10. *For every fixed $x \ge 0$ the matrix equation* (1.52) *has a unique solution in the space $L_1[0,\infty)$.*

Proof. To prove the uniqueness of the solution of (1.52), it is sufficient to show that the system of homogeneous equations:

$$\begin{pmatrix} h_1(t) \\ h_2(t) \end{pmatrix} = -\int_x^\infty \begin{pmatrix} 0 & f(-t-\tau) \\ g(t+\tau) & 0 \end{pmatrix} \begin{pmatrix} h_1(\tau) \\ h_2(\tau) \end{pmatrix} d\tau, \quad 0 \le x < t, \tag{1.53}$$

has only the zero solution in the space of continuous functions determined in $[0,\infty)$. Let $\big(h_1(t), h_2(t)\big)$ be a solution of equation (1.53). We consider the vector function $(z_1(t), z_2(t))$, which is a solution of the equation:

$$\begin{pmatrix} h_1(t) \\ h_2(t) \end{pmatrix} = \begin{pmatrix} z_1(t) \\ z_2(t) \end{pmatrix} + \int_x^t \begin{pmatrix} H_{11}(\xi,t) & H_{21}(\xi,t) \\ H_{12}(\xi,t) & H_{22}(\xi,t) \end{pmatrix} \begin{pmatrix} z_1(\xi) \\ z_2(\xi) \end{pmatrix} d\xi, \quad 0 \le x < t, \tag{1.54}$$

where $H_{jk}(\xi,t)$, $j,k = 1,2$, are the kernels of the transformation operators.

In view of the Volterra character of equation (1.54), the functions $z_1(t)$ and $z_2(t)$ are uniquely determined by $h_1(t)$ and $h_2(t)$.

By substituting (1.54) into (1.53), we obtain the equation for $z_1(t)$ and $z_2(t)$ for $0 \leq x < t$:

$$\begin{pmatrix} z_1(t) \\ z_2(t) \end{pmatrix} + \int_x^t \begin{pmatrix} H_{11}(\xi,t) & H_{21}(\xi,t) \\ H_{12}(\xi,t) & H_{22}(\xi,t) \end{pmatrix} \begin{pmatrix} z_1(\xi) \\ z_2(\xi) \end{pmatrix} d\xi =$$

$$-\int_x^\infty \begin{pmatrix} 0 & f(-t-\tau) \\ g(t+\tau) & 0 \end{pmatrix} \begin{pmatrix} z_1(\tau) \\ z_2(\tau) \end{pmatrix} d\tau -$$

$$\left(\int_x^t + \int_t^\infty\right) d\xi \int_\xi^\infty \begin{pmatrix} 0 & f(-t-\tau) \\ g(t+\tau) & 0 \end{pmatrix} \begin{pmatrix} H_{11}(\xi,\tau) & H_{21}(\xi,\tau) \\ H_{12}(\xi,\tau) & H_{22}(\xi,\tau) \end{pmatrix} d\tau \begin{pmatrix} z_1(\xi) \\ z_2(\xi) \end{pmatrix}.$$

In view of the matrix equation (1.52):

$$\int_\xi^\infty \begin{pmatrix} 0 & f(-t-\tau) \\ g(t+\tau) & 0 \end{pmatrix} \begin{pmatrix} H_{11}(\xi,\tau) & H_{21}(\xi,\tau) \\ H_{12}(\xi,\tau) & H_{22}(\xi,\tau) \end{pmatrix} d\tau =$$

$$-\begin{pmatrix} H_{11}(\xi,t) & H_{21}(\xi,t) \\ H_{12}(\xi,t) & H_{22}(\xi,t) \end{pmatrix} - \begin{pmatrix} 0 & f(-t-\xi) \\ g(t+\xi) & 0 \end{pmatrix}, \quad t > \xi.$$

Due to the latest matrix equation, the previous equation is reduced to the equation:

$$\begin{pmatrix} z_1(t) \\ z_2(t) \end{pmatrix} = \int_t^\infty \begin{pmatrix} H_{11}(\xi,t) & H_{21}(\xi,t) \\ H_{12}(\xi,t) & H_{22}(\xi,t) \end{pmatrix} \begin{pmatrix} z_1(\xi) \\ z_2(\xi) \end{pmatrix} d\xi.$$

Taking (1.52) into account again, from the latter equation we get the homogeneous integral equation for $(z_1(t),\ z_2(t))$ for $t > x$:

$$\begin{pmatrix} z_1(t) \\ z_2(t) \end{pmatrix} = -\int_t^\infty d\xi \Bigg[\begin{pmatrix} 0 & f(-t-\xi) \\ g(t+\xi) & 0 \end{pmatrix}$$

$$+ \int_\xi^\infty \begin{pmatrix} 0 & f(-t-\tau) \\ g(t+\tau) & 0 \end{pmatrix} \times \begin{pmatrix} H_{11}(\xi,\tau) & H_{21}(\xi,\tau) \\ H_{12}(\xi,\tau) & H_{22}(\xi,\tau) \end{pmatrix} d\tau \Bigg] \begin{pmatrix} z_1(\xi) \\ z_2(\xi) \end{pmatrix},$$

where the kernels $H_{ij}(\xi,\tau), 0 \leq \xi \leq \tau,\ i,j = 1,2$ and functions $f(-t-\xi), g(t+\xi)$ satisfy estimates (1.13) and (1.49), respectively.

The kernels of this equation have the estimate of the form $\tilde{C}(\eta)e^{-\frac{\varepsilon}{4}(\xi+t)}$, where $\tilde{C}(\eta)$ is a constant depending on $\eta:\ 0 < \eta < \frac{1}{12}\varepsilon_0$. Thus, $\begin{pmatrix} z_1(t) \\ z_2(t) \end{pmatrix}$ is a solution of the homogeneous Volterra integral equation with the exponentially decreasing kernel. Therefore, $z(t) = z_2(t) \equiv 0$ for $t > x$, and by (1.54) $h_1(t) = h_2(t) \equiv 0$ for $t > x$.

Since the matrix equation (1.52) is the Fredholm integral equation of the second kind in the space $L_1[0,\infty)$, then the unique solvability of this equation is deduced from the uniqueness of the solution in this space [61, 80, 116]. □

1.1.9 The description of the scattering data

We now turn to the main theorem regarding the complete description of the scattering data, that is the establishment of necessary and sufficient conditions for given quantities to be the scattering data for the system (1.7) on a half-line with boundary condition (1.8).

Theorem 1.1. *Suppose that we are given:*

(a) A function $\gamma(\rho)$;

(b) Numbers ρ_k^+, $\mathrm{Im}\rho_k^+ > \varepsilon_0$, $k = 1, ..., \alpha_1$ *and* ρ_j^-, $\mathrm{Im}\rho_j^- < -\varepsilon_0$, $j = 1, ..., \alpha_2$, *(ε_0 is some positive number);*

(c) Polynomials $P_k(x)$, $k = 1, ..., \alpha_1$, *and* $Q_j(x)$, $j = 1, ..., \alpha_2$.

For the set of quantities (a)–(c) to form the scattering data for a system of the form (1.7) on a half-line with boundary condition (1.8) and an exponentially decreasing potential, it is necessary and sufficient that the given quantities satisfy the conditions:

(1) $\gamma(\rho)$ *and* $\gamma^{-1}(\rho)$ *can be extended analytically into the strip* $0 < |\mathrm{Im}\rho| < \varepsilon_0$, *and in the strip* $|\mathrm{Im}\rho| < \varepsilon_0$ *they are functions of the type (S);*

(2) $\gamma(\rho)$ *has the factorization:*

$$\gamma(\rho) = \frac{1 + \int\limits_0^\infty K_-(t)e^{-i\rho t}dt}{1 + \int\limits_0^\infty K_+(t)e^{i\rho t}dt}, \quad 0 \le |\mathrm{Im}\rho| < \varepsilon_0, \tag{1.55}$$

where $K_\pm(t) \in L_1[0,\infty)$;

(3) The following compatibility relations hold:

$$P_k(\xi) + \int\limits_0^\infty e^{i\rho_k^+ t} P_k(\xi + t)K_+(t)dt = 0, \ k = 1, ..., \alpha_1, \ \xi > 0, \tag{1.56}$$

$$Q_j(\xi) + \int\limits_0^\infty e^{-i\rho_j^- t} Q_j(\xi + t)K_-(t)dt = 0, \ j = 1, ..., \alpha_2, \ \xi > 0; \tag{1.57}$$

(4) The homogeneous system of equations:

$$\begin{cases} h_1(t) = -\int\limits_x^\infty F(-t-\tau)h_2(\tau)d\tau, \\ h_2(t) = -\int\limits_x^\infty G(t+\tau)h_1(\tau)d\tau, \end{cases} \qquad 0 \le x < t$$

has in $L_1[0,\infty)$ only the trivial solution, here

$$F(-x) = F_\gamma(-x) - \sum_{k=1}^{\alpha_1} P_k(x)e^{i\rho_k^+ x}, \quad G(x) = G_\gamma(x) - \sum_{j=1}^{\alpha_2} Q_j(x)e^{-i\rho_j^- x}, \tag{1.58}$$

and

$$\begin{aligned} F_\gamma(x) &= \frac{1}{2\pi}\int\limits_{-\infty+i\eta}^{\infty+i\eta} [\gamma(\rho)-1]e^{-i\rho x}d\rho, \\ G_\gamma(x) &= \frac{1}{2\pi}\int\limits_{-\infty-i\eta}^{\infty-i\eta} [\gamma^{-1}(\rho)-1]e^{-i\rho x}d\rho, \end{aligned} \qquad 0 < \eta < \varepsilon_0. \tag{1.59}$$

(5) $F(-x)$ and $G(x)$ satisfy the inequalities:

$$|F(-x)| \le \tilde{C}e^{-\frac{1}{2}\varepsilon_0 x}, \ \ |G(x)| \le \tilde{C}e^{-\frac{1}{2}\varepsilon_0 x}, \ \tilde{C} \text{ is a constant.} \tag{1.60}$$

The necessity of these conditions is given by Lemmas 1.2, 1.3 and 1.8–1.10. We divide the proof of the sufficiency into several lemmas.

By formulas (1.58) and (1.59), we construct the functions $F(-x)$ and $G(x)$. It follows from the conditions (4) and (5) that the systems of fundamental equations:

$$\begin{aligned} &G(x+s) + \int\limits_x^\infty K_{11}(x,t)G(s+t)dt = -K_{12}(x,s), \\ &\int\limits_x^\infty K_{12}(x,t)F(-s-t)dt = -K_{11}(x,s), \end{aligned} \qquad 0 \le x < s, \tag{1.61}$$

$$\begin{aligned} &F(-x-s) + \int\limits_x^\infty K_{22}(x,t)F(-s-t)dt = -K_{21}(x,s), \\ &\int\limits_x^\infty K_{21}(x,t)G(s+t)dt = -K_{22}(x,s), \end{aligned} \tag{1.62}$$

are Fredholm systems, and in $L_1[0,\infty)$ they have the unique solutions $K_{jk}(x,s)$, $0 \le x < \infty$, $j,k = 1,2$.

We are going to study the properties of the solutions of these systems in greater detail.

Lemma 1.11. *Suppose that for every $x \ge 0$ the functions $F(-x)$ and $G(x)$ satisfy the inequalities* (1.60). *Then the solutions of the systems* (1.61) *and* (1.62) *satisfy the following estimate:*

$$|K_{jk}(x,s)| \le \tilde{C}e^{-\frac{1}{2}\varepsilon_0(x+s)},\ 0 \le x < s,\ \tilde{C} \text{ is a constant},\ j,k = 1,2. \quad (1.63)$$

Proof. It is easily proved that for sufficiently large x, the systems (1.61) and (1.62) can be solved by the method of successive approximations, and the obtained solutions satisfy estimate (1.63). Then, for sufficiently large x we have $\int_0^\infty |K_{jk}(x,t)|dt < \tilde{C}$, where $\tilde{C}$ is a constant, $j,k = 1,2$.

By noting that in $L_1[0,\infty)$ the systems (1.61) and (1.62) depend continuously on x as parameter, we find that $\int_0^\infty |K_{jk}(x,t)|dt$, $j,k = 1,2$, is continuous with respect to x, and so it is uniformly bounded on $0 \le x < \infty$. Then using estimates (1.60), from (1.61) and (1.62), we have the estimate (1.63).

From (1.61) and (1.62) we also find that

$$K_{12}(x,s) \to K_{12}(x,x),\quad K_{21}(x,s) \to K_{21}(x,x) \text{ as } s \to x,$$

in the sense of the metric in $L_1[0,\infty)$. □

Lemma 1.12. *The solution of systems of fundamental equations* (1.61) *and* (1.62) *with the conditions*

$$K_{jj}(x,x+\xi) \to 0\ (x \to \infty),\ \ j = 1,2, \quad (1.64)$$

satisfy systems of integral equations of the form (1.15) *and* (1.16), *where*

$$c_1(x) = 2iK_{12}(x,x),\ \ c_2(x) = -2iK_{21}(x,x). \quad (1.65)$$

Proof. We shall first prove that the solutions of the systems (1.61), (1.62) satisfy, in the sense of generalized functions [116], the respective differential equations:

$$i(\frac{\partial}{\partial x} + \frac{\partial}{\partial s})K_{11}(x,s) = -c_1(x)K_{21}(x,s),\quad 0 \le x < \infty, \quad (1.66)$$

$$i(\frac{\partial}{\partial x} - \frac{\partial}{\partial s})K_{12}(x,s) = -c_1(x)K_{22}(x,s),\quad 0 \le x < \infty, \quad (1.67)$$

$$i(\frac{\partial}{\partial x} - \frac{\partial}{\partial s})K_{21}(x,s) = c_2(x)K_{11}(x,s),\quad 0 \le x < \infty, \quad (1.68)$$

$$i(\frac{\partial}{\partial x} + \frac{\partial}{\partial s})K_{22}(x,s) = c_2(x)K_{12}(x,s),\quad 0 \le x < \infty. \quad (1.69)$$

We define the operators $\Delta_{x,\delta}$ and $\Delta_{s,\delta}$ by

$$\left.\begin{aligned} \Delta_{x,\delta}\Phi(x,s) &= \frac{\Phi(x+\delta,s)-\Phi(x,s)}{\delta}, \\ \Delta_{s,\delta}\Phi(x,s) &= \frac{\Phi(x,s+\delta)-\Phi(x,s)}{\delta}, \end{aligned}\right. \quad \delta > 0$$

where $\Phi(x,s)$ is a continuously differentiable function.

Applying the operators $\Delta_{x,\delta}$, $\Delta_{s,\delta}$ to the first equation of the system (1.61), we obtain

$$\begin{aligned} -i\Delta_{x,\delta}K_{12}(x,s) =& i\Delta_{x,\delta}G(x+s) - iK_{11}(x,x)G(x+s) + \\ & \int_{x+s}^{\infty} i\Delta_{x,\delta}K_{11}(x,\tau-s)G(\tau)d\tau + \alpha_1(x,s,\delta), \\ i\Delta_{s,\delta}K_{12}(x,s) =& - i\Delta_{s,\delta}G(x+s) + iK_{11}(x,x)G(x+s) + \\ & \int_{x+s}^{\infty} i\Delta_{s,\delta}K_{11}(x,\tau-s)G(\tau)d\tau + \alpha_2(x,s,\delta), \end{aligned}$$

where $\int_x^{\infty} |\alpha_j(x,s,\delta)|ds \to 0$ for $\delta \to 0$, $j = 1,2$.

Adding together the last two equations, we get:

$$\begin{aligned} -i(\Delta_{x,\delta} - \Delta_{s,\delta})K_{12}(x,s) = & \int_x^{\infty} i(\Delta_{x,\delta} + \Delta_{s,\delta})K_{11}(x,t)G(s+t)dt + \\ & \alpha_1(x,s,\delta) + \alpha_2(x,s,\delta), \end{aligned} \tag{1.70}$$

where $\int_x^{\infty} |\alpha_k(x,s,\delta)|ds \to 0$ as $\delta \to 0$, $k = 1,2$.

Analogously applying the operators $\Delta_{x,\delta}$, $\Delta_{s,\delta}$ to the second equation of the system (1.61), then adding together the resulting equations, yields:

$$\begin{aligned} & i(\Delta_{x,\delta} + \Delta_{s,\delta})K_{11}(x,s) = 2iK_{12}(x,x)F(-x-s) - \\ & \int_x^{\infty} i(\Delta_{x,\delta} - \Delta_{s,\delta})K_{12}(x,t)F(-s-t)dt + \alpha_3(x,s,\delta) + \alpha_4(x,s,\delta), \end{aligned} \tag{1.71}$$

where $\int_x^{\infty} |\alpha_j(x,s,\delta)|ds \to 0$ as $\delta \to 0$, $j = 3,4$.

Multiplying both sides of the first and the second identities of the system (1.62) by $c_1(x) = 2iK_{12}(x,x)$, gives

$$-c_1(x)K_{21}(x,s) = c_1(x)F(-x-s) + \int_x^\infty c_1(x)K_{22}(x,t)F(-s-t)dt, \quad (1.72)$$

$$-c_1(x)K_{22}(x,s) = \int_x^\infty c_1(x)K_{21}(x,t)G(s+t)dt. \quad (1.73)$$

Subtracting (1.72) from (1.71), and (1.73) from (1.70), we obtain the system of equations:

$$\begin{aligned}
&i(\Delta_{x,\delta}+\Delta_{s,\delta})K_{11}(x,s)+c_1(x)K_{21}(x,s) = \\
&-\int_x^\infty [i(\Delta_{x,\delta}-\Delta_{s,\delta})K_{12}(x,t)+c_1(x)K_{22}(x,t)]F(-s-t)dt+ \\
&\qquad\qquad \alpha_3(x,s,\delta)+\alpha_4(x,s,\delta), \qquad (1.74)\\
&-i(\Delta_{x,\delta}-\Delta_{s,\delta})K_{12}(x,s)-c_1(x)K_{22}(x,s) = \\
&\int_x^\infty [i(\Delta_{x,\delta}+\Delta_{s,\delta})K_{11}(x,t)+c_1(x)K_{21}(x,t)]G(s+t)dt+ \\
&\qquad\qquad \alpha_1(x,s,\delta)+\alpha_2(x,s,\delta),
\end{aligned}$$

where $\int_x^\infty |\alpha_j(x,s,\delta)|ds \to 0$ as $\delta \to 0,\ j=1,2,3,4.$

Due to the uniqueness of the solution of systems of fundamental equations (1.61) and (1.62) and their Fredholm properties, we obtain the continuous dependence of the solution on the free term of these systems. Therefore, it follows from (1.74) that there exist the following limits:

$$\begin{aligned}
\lim_{\delta\to 0}\,[i(\Delta_{x,\delta}+\Delta_{s,\delta})K_{11}(x,s)] &= -c_1(x)K_{21}(x,s),\\
\lim_{\delta\to 0}\,[i(\Delta_{x,\delta}-\Delta_{s,\delta})K_{12}(x,s)] &= -c_1(x)K_{22}(x,s).
\end{aligned} \quad (1.75)$$

We have the following limits for a continuously differentiable function $\Phi(x,s)$:

$$\begin{aligned}
\lim_{\delta\to 0}\,[i(\Delta_{x,\delta}+\Delta_{s,\delta})\Phi(x,s)] &= i\left(\frac{\partial}{\partial x}+\frac{\partial}{\partial s}\right)\Phi(x,s),\\
\lim_{\delta\to 0}\,[i(\Delta_{x,\delta}-\Delta_{s,\delta})\Phi(x,s)] &= i\left(\frac{\partial}{\partial x}-\frac{\partial}{\partial s}\right)\Phi(x,s).
\end{aligned}$$

Owing to these limits, from (1.75) we obtain the fulfillment of differential equations (1.66) and (1.67) in the sense of generalized functions.

Analogously, we can prove the fulfillment of differential equations (1.68) and (1.69) in this sense.

By taking the conditions (1.64) into account, from (1.66)–(1.69) with due regard for (1.65), we obtain systems of integral equations of the form (1.15) and (1.16). The lemma is proved. □

We now give a direct proof of the sufficiency of the conditions enumerated in the main Theorem 1.1.

We consider the system of the form (1.7):

$$-iJu_x + C(x)u = \lambda u, \quad u = (u_1(x,\lambda), u_2(x,\lambda)), \quad 0 \le x < \infty \tag{1.76}$$

with the boundary condition:

$$u_1(0,\lambda) = u_2(0,\lambda), \tag{1.77}$$

where $C(x) = \begin{pmatrix} 0 & c_1(x) \\ c_2(x) & 0 \end{pmatrix}, 0 \le x < \infty$, and $c_1(x) = 2iK_{12}(x,x)$ and $c_2(x) = -2iK_{21}(x,x)$, here the functions $K_{12}(x,s)$ and $K_{21}(x,s)$, $0 \le x < s$ are solutions of the systems (1.61) and (1.62) respectively.

Owing to inequality (1.63) the constructed potentials in (1.76) satisfy the estimate:

$$|c_k(x)| \le \tilde{C}e^{-\varepsilon_0 x}, \ x \ge 0, \ \tilde{C} \text{ is a positive number}, \ k = 1,2,$$

that is, the scattering problem (1.76)–(1.77) belongs to the class of problems discussed in this section. The scattering data set for the scattering problem (1.76)–(1.77) is denoted by s:

$$s = \Big\{ S(\lambda) \text{ for } |\text{Im}\lambda| < \varepsilon_0, \ \lambda_k^+, \ \text{Im}\lambda_k^+ > 0, \ k = 1, ..., \gamma_1; \\ \lambda_j^-, \ \text{Im}\lambda_j^- < 0, \ j = 1, ..., \gamma_2; \ p_1(x), ..., p_{\gamma_1}(x); \ \tilde{q}_1(x), ..., \tilde{q}_{\gamma_2}(x) \Big\}. \tag{1.78}$$

Let $H_{jk}(x,s)$, $0 \le x < \infty$, $j,k = 1,2$, be the kernels of the transformation operator for the scattering problem (1.76)–(1.77), then they satisfy the systems of integral equations (1.15) and (1.16).

By Lemma 1.12 the solutions $K_{jk}(x,s)$, $0 \le x < s$, $j,k = 1,2$, of the systems (1.61), (1.62) satisfy the same systems of integral equations of the form (1.15), (1.16), and because the solutions of these systems are unique, we obtain

$$H_{jk}(x,s) = K_{jk}(x,s), \ 0 \le x < s, \ j,k = 1,2. \tag{1.79}$$

From the factorization (1.55), we have

$$F_\gamma(-\xi) + \int_0^\infty F_\gamma(-\xi - t)K_+(t)dt = K_-(\xi), \qquad \xi > 0, \tag{1.80}$$

$$G_\gamma(\xi) + \int_0^\infty G_\gamma(\xi + t)K_-(t)dt = K_+(\xi), \qquad \xi > 0, \tag{1.81}$$

where $F_\gamma(x)$ and $G_\gamma(x)$, $x \geq 0$, are determined by (1.59).

Adding equality (1.56) to (1.80) and (1.57) to (1.81), we obtain the system of the form:

$$F(-\xi) + \int_0^\infty F(-\xi - t)K_+(t)dt = K_-(\xi), \qquad \xi > 0,$$

$$G(\xi) + \int_0^\infty G(\xi + t)K_-(t)dt = K_+(\xi), \qquad \xi > 0,$$

where $F(-x)$ and $G(x)$, $x \geq 0$, are given by (1.58).

On the other hand, from (1.61) and (1.62), we obtain a system of equations of the same form:

$$F(-\xi) + \int_0^\infty F(-\xi - t)[K_{22}(0,t) - K_{12}(0,t)]dt = K_{11}(0,\xi) - K_{21}(0,\xi), \xi > 0,$$

$$G(\xi) + \int_0^\infty G(\xi + t)[K_{11}(0,t) - K_{21}(0,t)]dt = K_{22}(0,\xi) - K_{12}(0,\xi), \xi > 0,$$

where $F(-\xi)$ and $G(\xi)$ are known, and $K_{11}(0,\xi) - K_{21}(0,\xi)$ and $K_{22}(0,\xi) - K_{12}(0,\xi)$, $\xi > 0$ are unknown.

Hence by the condition (4) of the theorem:

$$K_+(\xi) = K_{22}(0,\xi) - K_{12}(0,\xi), \;\; K_-(\xi) = K_{11}(0,\xi) - K_{21}(0,\xi), \; \xi > 0.$$

By taking (1.79) into account, we have

$$K_+(\xi) = H_{22}(0,\xi) - H_{12}(0,\xi), \;\; K_-(\xi) = H_{11}(0,\xi) - H_{21}(0,\xi), \; \xi > 0.$$

Then by the factorizations (1.22) and (1.55), we obtain the equality: $\gamma(\lambda) = S(\lambda)$. From the last equality and from (1.59), (1.37), we have

$$F_\gamma(-x) = f_S(-x), \;\; G_\gamma(x) = g_S(x), \; x \geq 0. \tag{1.82}$$

With due regard for (1.79), from (1.61), (1.43) and (1.62), (1.45), we obtain

$$G(x+s)-g(x+s)+\int_x^\infty K_{11}(x,t)\,[G(s+t)-g(s+t)]\,dt=0,\quad 0\le x<s,$$

$$F(-x-s)-f(-x-s)+\int_x^\infty K_{22}(x,t)\,[F(-s-t)-f(-s-t)]\,dt=0,$$

$$0\le x<s.$$

When we use (1.41), (1.58) and (1.82), from the last equations we have

$$\sum_{j=1}^{\gamma_2}\tilde{q}_j(x+s)e^{-i\lambda_j^-(x+s)}-\sum_{j=1}^{\alpha_2}Q_j(x+s)e^{-i\rho_j^-(x+s)}+$$

$$\int_x^\infty K_{11}(x,t)\left\{\sum_{j=1}^{\gamma_2}\tilde{q}_j(s+t)e^{-i\lambda_j^-(s+t)}-\sum_{j=1}^{\alpha_2}Q_j(s+t)e^{-i\rho_j^-(s+t)}\right\}dt=0,$$

$$\sum_{k=1}^{\gamma_1}p_k(x+s)e^{i\lambda_k^+(x+s)}-\sum_{k=1}^{\alpha_1}P_k(x+s)e^{i\rho_k^+(x+s)}+$$

$$\int_x^\infty K_{22}(x,t)\left\{\sum_{k=1}^{\gamma_1}p_k(s+t)e^{i\lambda_k^+(s+t)}-\sum_{k=1}^{\alpha_1}P_k(s+t)e^{i\rho_k^+(s+t)}\right\}dt=0.$$

Since $|K_{jj}(x,t)|\le\tilde{C}e^{-\frac{1}{2}\varepsilon_0(x+t)}$, $0\le x<t$, $j=1,2$, it is obvious that for sufficiently large x, we have

$$\sum_{k=1}^{\alpha_1}P_k(x)e^{i\rho_k^+x}=\sum_{k=1}^{\gamma_1}p_k(x)e^{i\lambda_k^+x},\quad \sum_{j=1}^{\alpha_2}Q_j(x)e^{-i\rho_j^-x}=\sum_{j=1}^{\gamma_2}\tilde{q}_j(x)e^{-i\lambda_j^-x}.$$

Hence it follows that

$$\alpha_1=\gamma_1,\ \ \alpha_2=\gamma_2,\ \ \rho_k^+=\lambda_k^+,\ \ P_k(x)=p_k(x),\ k=1,...,\gamma_1,$$
$$\rho_j^-=\lambda_j^-,\ \ Q_j(x)=\tilde{q}_j(x),\ j=1,...,\gamma_2.$$

Thus, we have proved that when the conditions (1)–(5) of the main theorem hold, the given quantities of the set $\tilde{s}$:

$$\tilde{s}=\Big\{\gamma(\rho)\text{ for }|\mathrm{Im}\rho|<\varepsilon_0;\ \ \rho_k^+,\ \mathrm{Im}\rho_k^+>0,\ k=1,...,\alpha_1,$$
$$\rho_j^-,\ \mathrm{Im}\rho_j^-<0,\ j=1,...,\alpha_2\ \ P_1(x),...,P_{\alpha_1}(x);\ \ Q_1(x),...,Q_{\alpha_2}(x)\Big\}\tag{1.83}$$

coincide with the quantities, respectively, of the scattering data set s given by (1.78) for the system of equations (1.76) on a half-line with the boundary condition (1.77) and the constructed potentials (1.65). □

One should note that in the proof of the Theorem 1.1, we have constructed an algorithm for finding the potentials $c_1(x)$, $c_2(x)$ in terms of the known data set (1.83). This consists of constructing the functions $F(-x)$, $G(x)$ by the formulas (1.58) from the given data (1.83), solving the systems (1.61), (1.62) and recovering the system (1.7), that is the potential functions $c_1(x)$ and $c_2(x)$ from the formulas (1.65).

1.2 The inverse scattering problem on a half-line with a potential self-adjoint matrix

We consider the scattering problem for the system of first-order ODEs on the half-line:

$$-iJu_x + C(x)u = \lambda u,\ u = (u_1(x,\lambda), u_2(x,\lambda)),\ 0 \leq x < \infty, \tag{1.84}$$

with the boundary condition at the origin $x = 0$:

$$u_1(0,\lambda) = u_2(0,\lambda), \tag{1.85}$$

where $J = \begin{pmatrix} -1 & 0 \\ 0 & 1 \end{pmatrix}$, λ is a complex-valued parameter, $C(x) = \begin{pmatrix} 0 & c_1(x) \\ c_2(x) & 0 \end{pmatrix}$, $C(x)$ is a potential self-adjoint matrix:

$$c_1(x) = \overline{c_2(x)}, \tag{1.86}$$

where $c_1(x)$ and $c_2(x)$ are the complex-valued measurable functions satisfying the estimate:

$$|c_k(x)| \leq \frac{\tilde{C}}{1 + x^{1+\varepsilon}},\ \tilde{C} \text{ is a positive number},\ \varepsilon > 0,\ k = 1,2. \tag{1.87}$$

In this section we shall prove that under condition (1.86) the self-adjoint problem (1.84)–(1.86) possesses a series of specific characters which are essentially different from the corresponding characters of the non-self-adjoint problem (1.7)–(1.8) considered in the previous Section. Note that the potentials of this self-adjoint problem are power decreasing, while the potentials of the non-self-adjoint problem (1.7)–(1.8) are required to decrease exponentially.

1.2.1 The unique solvability of the self-adjoint problem

Lemma 1.13. *Let λ be a real number, then for any number $A(\lambda)$ there exists a unique bounded solution $(u_1(x,\lambda), u_2(x,\lambda))$ of the problem (1.84)–(1.85)–(1.86) such that:*

$$u_1(x,\lambda) = A(\lambda)e^{-i\lambda x} + o(1) \ \ (x \to \infty). \tag{1.88}$$

In addition there exists a unique number $B(\lambda)$ defined from the given $A(\lambda)$ such that:

$$u_2(x,\lambda) = B(\lambda)e^{i\lambda x} + o(1) \ \ (x \to \infty). \tag{1.89}$$

Furthermore, the solution of this problem is presented in the form:

$$\begin{aligned} u_1(x,\lambda) &= A(\lambda)e^{-i\lambda x} + A(\lambda)\int_x^\infty H_{11}(x,s)e^{-i\lambda s}ds + B(\lambda)\int_x^\infty H_{12}(x,s)e^{i\lambda s}ds, \\ u_2(x,\lambda) &= B(\lambda)e^{i\lambda x} + A(\lambda)\int_x^\infty H_{21}(x,s)e^{-i\lambda s}ds + B(\lambda)\int_x^\infty H_{22}(x,s)e^{i\lambda s}ds. \end{aligned} \tag{1.90}$$

The matrix function $H(x,s) = [H_{jk}(x,s)]$ relates to the potential matrix $C(x)$ by the relation:

$$2i\begin{pmatrix} 0 & H_{12}(x,x) \\ -H_{21}(x,x) & 0 \end{pmatrix} = \begin{pmatrix} 0 & c_1(x) \\ c_2(x) & 0 \end{pmatrix}, \tag{1.91}$$

and satisfies the estimate:

$$|H_{jk}(x,s)| \le \frac{\tilde{C}}{1+(x+s)^{1+\varepsilon}}, \quad 0 \le x \le s, \ \ j,k = 1,2, \tag{1.92}$$

where $\tilde{C}$ is a constant, $\varepsilon > 0$.

Proof. To prove the uniqueness of the solution of the problem (1.84)–(1.85)–(1.86), it is sufficient to show that the number $B(\lambda)$ is uniquely determined from the given number $A(\lambda)$. Indeed, assume that for $\lambda = \lambda_0$, $A(\lambda_0) = 0$, i.e., $u_1(x,\lambda_0) = o(1)$ $(x \to \infty)$, and there exists a number $B(\lambda_0)$ different from zero such that

$$u_2(x,\lambda_0) = B(\lambda_0)e^{i\lambda_0 x} + o(1) \ \ (x \to \infty).$$

By using the system (1.84) and the self-adjoint condition (1.86), we verify that

$$\frac{d}{dx}\left\{i\overline{u_1(x,\lambda_0)}u_1(x,\lambda_0) - i\overline{u_2(x,\lambda_0)}u_2(x,\lambda_0)\right\} = 0,$$

i.e., the quantity $\overline{iu_1(x,\lambda_0)}u_1(x,\lambda_0) - \overline{iu_2(x,\lambda_0)}u_2(x,\lambda_0)$ does not depend on x. Therefore, by the assumption we have:

$$\overline{iu_1(x,\lambda_0)}u_1(x,\lambda_0) - \overline{iu_2(x,\lambda_0)}u_2(x,\lambda_0) =$$
$$\lim_{x\to\infty}\left\{\overline{iu_1(x,\lambda_0)}u_1(x,\lambda_0) - \overline{iu_2(x,\lambda_0)}u_2(x,\lambda_0)\right\} = -i|B(\lambda_0)|^2,$$

but on the other hand by the boundary condition (1.85):

$$\overline{iu_1(x,\lambda_0)}u_1(x,\lambda_0) - \overline{iu_2(x,\lambda_0)}u_2(x,\lambda_0) =$$
$$\overline{iu_1(0,\lambda_0)}u_1(0,\lambda_0) - \overline{iu_2(0,\lambda_0)}u_2(0,\lambda_0) = 0,$$

whence it follows that $B(\lambda_0) = 0$ and $u_1(x,\lambda_0) = u_2(x,\lambda_0) \equiv 0$. This contradicts the assumption that $B(\lambda_0) \neq 0$. Hence, the solution of this problem is unique.

Consider the system of integral equations:

$$\begin{cases} u_1(x,\lambda) = A(\lambda)e^{-i\lambda x} + \dfrac{1}{i}\displaystyle\int_x^\infty e^{-i\lambda(x-\xi)}c_1(\xi)u_2(\xi,\lambda)d\xi, \\ u_2(x,\lambda) = B(\lambda)e^{i\lambda x} - \dfrac{1}{i}\displaystyle\int_x^\infty e^{i\lambda(x-\xi)}c_2(\xi)u_1(\xi,\lambda)d\xi, \end{cases} \tag{1.93}$$

which is equivalent to system (1.84) with boundary conditions (1.88) and (1.89). In the class of bounded functions the system (1.93) is a system of Volterra integral equations, therefore there exists a bounded solution of system (1.93) for any $A(\lambda)$ and $B(\lambda)$.

We seek the solution of (1.93) in the form (1.90). By starting from (1.90) and (1.93), we derive the systems of integral equations of the form (1.15), (1.16) for the kernel functions $H_{jk}(x,s)$, $0 \le x \le s$, $j,k = 1,2$. These kernels satisfy the integral equations of the same form (1.17)–(1.20) respectively, but in which the potentials obey the self-adjoint condition: $c_1(x) = \overline{c_2(x)}$. Owing to this self-adjoint condition, from (1.17)–(1.20) it follows that

$$H_{11}(x,s) = \overline{H_{22}(x,s)},\;\; H_{12}(x,s) = \overline{H_{21}(x,s)},\;\; 0 \le x \le s. \tag{1.94}$$

By an argument analogous to that used for the proof of Lemma 1.1, using estimate (1.87), we prove that the solutions $H_{jk}(x,s), 0 \le x \le s, j,k = 1,2$ of equations (1.17)–(1.20) with potentials satisfying condition (1.86) exist and are unique in the class of bounded functions. In addition the functions $H_{jk}(x,s)$ satisfy the estimate (1.92) and relate to the potential matrix $C(x)$ by the relation matrix (1.91).

Conversely, if the functions $H_{jk}(x,s), 0 \le x \le s, j,k = 1,2$ satisfy equations of the form (1.17)–(1.20), then the formula (1.90) gives a bounded solution of system (1.93). By the Volterra property of system (1.93) the solution

$u(x,\lambda)$ of this system is unique in the class of bounded functions. Hence, $u(x,\lambda)$ satisfies system (1.84) with boundary conditions (1.88) and (1.89).

It remains to satisfy the boundary condition (1.85) by determining a number $B(\lambda)$ from the given number $A(\lambda)$. Substituting (1.90) for $x = 0$ into (1.85), gives

$$B(\lambda)\left\{1+\int_0^\infty [H_{22}(0,s)-H_{12}(0,s)]e^{i\lambda s}ds\right\} = A(\lambda)\left\{1+\int_0^\infty [H_{11}(0,s)-H_{21}(0,s)]e^{-i\lambda s}ds\right\}. \tag{1.95}$$

We prove that for any real number λ

$$1+\int_0^\infty [H_{22}(0,s)-H_{12}(0,s)]^{i\lambda s}ds \neq 0. \tag{1.96}$$

Indeed, assume that for some real number λ_0, we have

$$1+\int_0^\infty [H_{22}(0,s)-H_{12}(0,s)]e^{i\lambda_0 s}ds = 0.$$

Consider the vector-function

$$u(x,\lambda_0) = \begin{pmatrix} u_1(x,\lambda_0) \\ u_2(x,\lambda_0) \end{pmatrix} = \begin{pmatrix} \int_x^\infty H_{12}(x,\xi)e^{i\lambda_0\xi} \\ e^{i\lambda_0 x} + \int_x^\infty H_{22}(x,\xi)e^{i\lambda_0\xi} \end{pmatrix}, \tag{1.97}$$

where $H_{12}(x,\xi)$ and $H_{22}(x,\xi), 0 \le x \le \xi$ are determined by Lemma 1.13.

This vector-function $u(x,\lambda_0)$ is a solution of system (1.84) with the potentials satisfying conditions (1.86), (1.87) and with the boundary condition (1.85). Due to estimate (1.92), the solution (1.97) of the self-adjoint problem (1.84)–(1.85) satisfies the asymptotic condition as $x \to \infty$:

$$u_1(x,\lambda_0) = o(1)\ \ (x\to\infty),\ \ u_2(x,\lambda_0) = e^{i\lambda_0 x} + o(1)\ \ (x\to\infty),$$

which conflicts with the uniqueness of the solution of the self-adjoint problem (1.84)–(1.85). Hence, the assertion (1.96) is proved and the number $B(\lambda)$ is determined uniquely by equality (1.95) from the given number $A(\lambda)$. This completes the proof of Lemma 1.13. □

Analogously, one can prove that for a real number λ and any given number $B(\lambda)$ there exists a unique bounded solution $(u_1(x,\lambda), u_2(x,\lambda))$ of the self-adjoint problem (1.84)–(1.85) such that

$$u_1(x,\lambda) = A(\lambda)e^{-i\lambda x} + o(1)\ \ (x\to\infty),\ \ u_2(x,\lambda) = B(\lambda)e^{i\lambda x} + o(1)\ \ (x\to\infty),$$

where the number $A(\lambda)$ is determined uniquely from the given number $B(\lambda)$. Furthermore, the solution $u(x,\lambda)$ has the representation (1.90) and

$$1+\int_0^\infty [H_{11}(0,s)-H_{21}(0,s)]e^{-i\lambda s}ds \neq 0 \text{ for any real number } \lambda. \qquad (1.98)$$

Thus, the unique solvability of the self-adjoint problem (1.84)–(1.85) is proved. The function matrix $H(x,s)$ is an analog of the kernel of the transformation operator usually used in the scattering problems [7, 75].

1.2.2 The Jost solutions of system (1.84)

Lemma 1.14. *There exist the bounded vector-solutions* $\mathrm{e}^{(1)}(x,\lambda) = \left(\mathrm{e}_1^{(1)}(x,\lambda), \mathrm{e}_2^{(1)}(x,\lambda)\right)$, $Im\rho \le 0$ *and* $\mathrm{e}^{(2)}(x,\lambda) = \left(\mathrm{e}_1^{(2)}(x,\lambda), \mathrm{e}_2^{(2)}(x,\lambda)\right)$, $Im\rho \ge 0$ *of system* (1.84) *with the potential matrix satisfying the self-adjoint condition* (1.86), *estimate* (1.87) *and with the boundary condition at infinity:*

$$\begin{aligned} \mathrm{e}_1^{(1)}(x,\lambda)e^{i\lambda x} &= 1+o(1), \\ \mathrm{e}_2^{(1)}(x,\lambda)e^{-i\lambda x} &= o(1), \end{aligned} \qquad Im\lambda \le 0 \;\; (x\to\infty) \qquad (1.99)$$

and

$$\begin{aligned} \mathrm{e}_1^{(2)}(x,\lambda)e^{i\lambda x} &= o(1), \\ \mathrm{e}_2^{(2)}(x,\lambda)e^{-i\lambda x} &= 1+o(1), \end{aligned} \qquad Im\lambda \ge 0 \;\; (x\to\infty) \qquad (1.100)$$

Furthermore, these vector-solutions have the representations:

$$\begin{aligned} \mathrm{e}_1^{(1)}(x,\lambda) &= e^{-i\lambda x} + e^{-i\lambda x}\int_0^\infty H_{11}(x,x+\xi)e^{-i\lambda\xi}d\xi, \\ \mathrm{e}_2^{(1)}(x,\lambda) &= e^{-i\lambda x}\int_0^\infty H_{21}(x,x+\xi)e^{-i\lambda\xi}d\xi, \end{aligned} \qquad Im\lambda \le 0 \qquad (1.101)$$

and

$$\begin{aligned} \mathrm{e}_1^{(2)}(x,\lambda) &= e^{i\lambda x}\int_0^\infty H_{12}(x,x+\xi)e^{i\lambda\xi}d\xi, \\ \mathrm{e}_2^{(2)}(x,\lambda) &= e^{i\lambda x} + e^{i\lambda x}\int_0^\infty H_{22}(x,x+\xi)e^{i\lambda\xi}d\xi, \end{aligned} \qquad Im\lambda \ge 0. \qquad (1.102)$$

Proof. Consider the system of integral equations:

$$\begin{cases} e_1^{(1)}(x,\lambda) = e^{-i\lambda x} + \dfrac{1}{i}\displaystyle\int_x^\infty e^{-i\lambda(x-\xi)}c_1(\xi)e_2^{(1)}(\xi,\lambda)d\xi, \\ e_2^{(1)}(x,\lambda) = -\dfrac{1}{i}\displaystyle\int_x^\infty e^{i\lambda(x-\xi)}c_2(\xi)e_1^{(1)}(\xi,\lambda)d\xi, \end{cases} \quad \mathrm{Im}\lambda \leq 0, \quad (1.103)$$

which is equivalent to system (1.84) with boundary condition (1.99). In the class of bounded functions the system (1.103) is a system of Volterra equations, therefore there exists its bounded solution. By starting from (1.101) and (1.103), we obtain the system of the form (1.15) for $(H_{11}(x,x+\xi), H_{21}(x,x+\xi))$. Thus, the solution $e^{(1)}(x,\lambda)$ of system (1.103) has the representation (1.101).

In the same way as the previous one, we can prove that there exists a bounded vector-solution $e^{(2)}(x,\lambda)$ of system (1.84) with boundary condition (1.100), which has the representation (1.102).

Consider the Wronskian determinant of the solutions $e^{(1)}(x,\lambda)$ and $e^{(2)}(x,\lambda)$ for real λ:

$$W[e^{(1)},e^{(2)}] = \begin{vmatrix} e_1^{(1)}(x,\lambda) & e_1^{(2)}(x,\lambda) \\ e_2^{(1)}(x,\lambda) & e_2^{(2)}(x,\lambda) \end{vmatrix} = e_1^{(1)}(x,\lambda)e_2^{(2)}(x,\lambda) - e_1^{(2)}(x,\lambda)e_2^{(1)}(x,\lambda).$$

It is easily verified that this determinant does not depend on x, therefore in view of estimate (1.92):

$$W[e^{(1)},e^{(2)}] = \lim_{x\to\infty}\Bigg\{1 + \int_0^\infty H_{22}(x,x+\xi)e^{i\lambda\xi}d\xi + \int_0^\infty H_{11}(x,x+\xi)e^{-i\lambda\xi}d\xi +$$
$$\int_0^\infty H_{11}(x,x+\xi)e^{-i\lambda\xi}d\xi \times \int_0^\infty H_{22}(x,x+\xi)e^{i\lambda\xi}d\xi -$$
$$\int_0^\infty H_{21}(x,x+\xi)e^{-i\lambda\xi}d\xi \times \int_0^\infty H_{12}(x,x+\xi)e^{i\lambda\xi}d\xi\Bigg\} = 1. \quad (1.104)$$

Due to (1.104) for any real λ the solutions $e^{(1)}(x,\lambda)$ and $e^{(2)}(x,\lambda)$ form a fundamental system of solutions of system (1.84). □

1.2.3 The scattering function and its properties

Definition 1.3. *According to Lemma 1.13, any given number $A(\lambda)$ corresponds to a unique bounded solution $u(x,\lambda) = \big(u_1(x,\lambda), u_2(x,\lambda)\big)$ of the* *problem* (1.84)–(1.85)–(1.86) *such that*

$$u_1(x,\lambda) = A(\lambda)e^{-i\lambda x} + o(1) \ \ (x \to \infty),$$
$$u_2(x,\lambda) = B(\lambda)e^{i\lambda x} + o(1) \ \ (x \to \infty),$$

where the number $B(\lambda)$ *is determined uniquely from the given number* $A(\lambda)$ *by equality* (1.95).

The one-to-one correspondence between the numbers $A(\lambda)$ *and* $B(\lambda)$ *determines the function* $S(\lambda) : S(\lambda)A(\lambda) = B(\lambda)$. *The function* $S(\lambda)$ *is called the scattering function.*

From (1.85) and (1.90) it follows that

$$A(\lambda) + A(\lambda)\int_0^\infty H_{11}(0,s)e^{-i\lambda s}ds + B(\lambda)\int_0^\infty H_{12}(0,s)e^{i\lambda s}ds =$$
$$B(\lambda) + A(\lambda)\int_0^\infty H_{21}(0,s)e^{-i\lambda s}ds + B(\lambda)\int_0^\infty H_{22}(0,s)e^{i\lambda s}ds.$$

Hence, according to Definition 1.3, the scattering function $S(\lambda)$ admits its factorization:

$$S(\lambda) = \frac{1 + \int_0^\infty [H_{11}(0,s) - H_{21}(0,s)]e^{-i\lambda s}ds}{1 + \int_0^\infty [H_{22}(0,s) - H_{12}(0,s)]e^{i\lambda s}ds}, \tag{1.105}$$

where owing to (1.96) and (1.98) for any real λ the numerator and the denominator of the factorization (1.105) are different from zero.

In view of (1.94) the scattering function (1.105) possesses the property:

$$\overline{S(\lambda)} = S^{-1}(\lambda).$$

Analogously, we define for any $a \geq 0$ the scattering function for system (1.84) on a half-line $a \leq x < \infty$. Let $v(x,\lambda,a) = (v_1(x,\lambda,a), v_2(x,\lambda,a))$, $0 \leq a \leq x < \infty$, be a vector solution of system (1.84) with the boundary condition at $x = a$:

$$v_1(a,\lambda,a) = v_2(a,\lambda,a),$$

and the boundary condition at infinity:

$$\begin{aligned} v_1(x,\lambda,a) &= A_a(\lambda)e^{-i\lambda(x-a)} + o(1), \\ v_2(x,\lambda,a) &= B_a(\lambda)e^{i\lambda(x-a)} + o(1), \end{aligned} \qquad (x \to \infty) \ 0 \leq a \leq x < \infty.$$

The one-to-one correspondence between the numbers $A_a(\lambda)$ and $B_a(\lambda)$ determines the scattering function $S(a,\lambda)$, $-\infty < \lambda < \infty$ for system (1.84) on the half-line $a \le x < \infty$: $S(a,\lambda)A_a(\lambda) = B_a(\lambda)$. Furthermore, $S(a,\lambda)$ admits the factorization:

$$S(a,\lambda) = \frac{1 + \int\limits_a^\infty [H_{11}(a,s) - H_{21}(a,s)]e^{-i\lambda(s-a)}ds}{1 + \int\limits_a^\infty [H_{22}(a,s) - H_{12}(a,s)]e^{i\lambda(s-a)}ds} =$$

$$\frac{1 + \int\limits_0^\infty [H_{11}(a,a+\xi) - H_{21}(a,a+\xi)]e^{-i\lambda\xi}d\xi}{1 + \int\limits_0^\infty [H_{22}(a,a+\xi) - H_{12}(a,a+\xi)]e^{i\lambda\xi}d\xi}, \tag{1.106}$$

and $\overline{S(a,\lambda)} = S^{-1}(a,\lambda)$.

We prove that for any $a \ge 0$ and λ, $\text{Im}\lambda < 0$

$$1 + \int\limits_0^\infty [H_{11}(a,a+\xi) - H_{21}(a,a+\xi)]e^{-i\lambda\xi}d\xi \ne 0.$$

Assume that for some λ_0, $\text{Im}\lambda_0 < 0$, we have:

$$1 + \int\limits_0^\infty [H_{11}(a,a+\xi) - H_{21}(a,a+\xi)]e^{-i\lambda_0\xi}d\xi = 0.$$

Then the functions

$$u_1(x,\lambda_0) = e^{-i\lambda_0(x-a)}\left(1 + \int\limits_0^\infty H_{11}(x,x+\xi)e^{-i\lambda_0\xi}d\xi\right),$$

$$u_2(x,\lambda_0) = e^{-i\lambda_0(x-a)}\int\limits_0^\infty H_{21}(x,x+\xi)e^{-i\lambda_0\xi}d\xi$$

satisfy the boundary condition: $u_1(a,\lambda_0) = u_2(a,\lambda_0)$ and the estimate:

$$|u_j(x,\lambda_0)| \le \frac{\tilde{C}}{1+x^{1+\varepsilon}}, \quad 0 \le a \le x, \quad j = 1,2, \quad \tilde{C} \text{ is a positive number,}$$

i.e., $u(x,\lambda_0) = (u_1(x,\lambda_0), u_2(x,\lambda_0)) \in L_2[a,\infty)$. Furthermore, $u_1(x,\lambda_0) \ne 0$ for sufficient large x. Thus, $u(x,\lambda_0)$ is the eigenfunction of the problem for system (1.84) on the half-line $a \le x < \infty$ with the boundary condition: $u_1(a,\lambda_0) = u_2(a,\lambda_0)$, and the non-real number λ_0 is the eigenvalue. This contradicts the self-adjoint property of the considered problem.

Analogously, one can prove that for any $a \ge 0$ and $\lambda, \text{Im}\lambda > 0$

$$1 + \int\limits_0^\infty [H_{22}(a,a+\xi) - H_{12}(a,a+\xi)]e^{i\lambda\xi}d\xi \ne 0.$$

This result together with (1.96) and (1.98) shows that the numerator and denominator of factorization (1.106) are different from zero for $Im\lambda \leq 0$ and $Im\lambda \geq 0$, respectively. Thus, the following lemma is proved. □

Lemma 1.15. *Every system of first-order ODEs of the form* (1.84) *on a half-line* $a \leq x < \infty,\ \ a \geq 0$ *with a potential matrix satisfying the self-adjoint condition* (1.86) *and estimate* (1.87) *corresponds to the scattering function* $S(a, \lambda)$ *admitting the factorization:*

$$S(a,\lambda) = \frac{1 + \int\limits_0^\infty \left[\overline{H_{22}(a,a+\xi)} - \overline{H_{12}(a,a+\xi)}\right] e^{-i\lambda\xi} d\xi}{1 + \int\limits_0^\infty [H_{22}(a,a+\xi) - H_{12}(a,a+\xi)] e^{i\lambda\xi} d\xi},$$

where the function $1 + \int\limits_0^\infty [H_{22}(a,a+\xi) - H_{12}(a,a+\xi)]e^{i\lambda\xi}$ *is analytical and different from zero in the upper half-plane* $Im\lambda \geq 0$.

Putting

$$\begin{aligned} H_+(x,\xi) &= H_{22}(x,x+\xi) - H_{12}(x,x+\xi), \\ H_-(x,\xi) &= H_{11}(x,x+\xi) - H_{21}(x,x+\xi), \end{aligned} \qquad 0 \leq x, \xi < \infty,$$

then according to Lemma 1.15 there exist absolutely integrable with respect to ξ functions $K_+(x, \xi)$ and $K_-(x, \xi)$ such that

$$\left(1 + \int\limits_0^\infty H_+(x,\xi) e^{i\lambda\xi} d\xi\right)^{-1} = 1 + \int\limits_0^\infty K_+(x,\xi) e^{i\lambda\xi} d\xi,$$

$$\left(1 + \int\limits_0^\infty H_-(x,\xi) e^{-i\lambda\xi} d\xi\right)^{-1} = 1 + \int\limits_0^\infty K_-(x,\xi) e^{-i\lambda\xi} d\xi.$$

Using the above equalities, from the factorization (1.106), we obtain

$$S(x,\lambda) = 1 + F(x,\lambda) = 1 + \int\limits_{-\infty}^{\infty} f(x,\xi) e^{i\lambda\xi} d\xi,$$

$$S^{-1}(x,\lambda) = 1 + G(x,\lambda) = 1 + \int\limits_{-\infty}^{\infty} g(x,\xi) e^{i\lambda\xi} d\xi,$$

where

$$f(x,\xi) = H_-(x,-\xi) + K_+(x,\xi) + \int_0^\infty K_+(x,t)H_-(x,t-\xi)dt, \qquad (1.107)$$

$$g(x,\xi) = H_+(x,\xi) + K_-(x,-\xi) + \int_{-\infty}^0 K_-(x,-t)H_+(x,\xi-t)dt. \qquad (1.108)$$

By using estimates (1.92) for $H_{jk}(x, x+\xi)$ and the absolute integrability with respect to ξ of the functions $K_\pm(x,\xi)$, we can verify that for any $x \geq 0$ the functions $f(x,\xi)$ and $g(x,\xi)$ are absolutely integrable with respect to ξ.

The following special case is deduced from (1.105), (1.106), (1.107), (1.108):

$$S(0,\lambda) = S(\lambda) = 1 + \int_{-\infty}^\infty f(\xi)e^{i\lambda\xi}d\xi,$$

$$S^{-1}(0,\lambda) = S^{-1}(\lambda) = 1 + \int_{-\infty}^\infty g(\xi)e^{i\lambda\xi}d\xi,$$

where $f(\xi) = f(0,\xi)$, $g(\xi) = g(0,\xi)$.

Since $\overline{S(\lambda)} = S^{-1}(\lambda)$, then

$$\overline{f(-\xi)} = g(\xi). \qquad (1.109)$$

Thus, the following lemma is proved. □

Lemma 1.16. *For any $x \geq 0$ the functions $S(x,\lambda) - 1$ and $S^{-1}(x,\lambda) - 1$ are the Fourier transformations of the absolutely integrable with respect to ξ functions $f(x,\xi)$ and $g(x,\xi)$ defined by* (1.107) *and* (1.108), *respectively.*

1.2.4 The relation between the functions $f(x,\xi)$, $g(x,\xi)$ and $f(\xi)$, $g(\xi)$

Denote by $\varphi(x,\lambda) = (\varphi_1(x,\lambda), \varphi_2(x,\lambda))$ the solution of system (1.84) with the initial condition:

$$\varphi_1(0,\lambda) = \varphi_2(0,\lambda) = 1. \qquad (1.110)$$

It is easy to verify that this solution satisfies the system of integral equations:

$$\begin{cases} \varphi_1(x,\lambda) = e^{-i\lambda x} - \dfrac{1}{i}\displaystyle\int_0^x e^{i\lambda(y-x)}c_1(y)\varphi_2(y,\lambda)dy, \\ \varphi_2(x,\lambda) = e^{i\lambda x} + \dfrac{1}{i}\displaystyle\int_0^x e^{i\lambda(x-y)}c_2(y)\varphi_1(y,\lambda)dy, \end{cases} \qquad (1.111)$$

where the potentials $c_1(x)$ and $c_2(x)$ satisfy the self-adjoint condition (1.86) and the estimate (1.87).

When we apply the method of successive approximations to the system (1.111), we obtain that the solution of this system exists. Owing to the uniqueness of the solution of the Cauchy problem (1.84), (1.110), the obtained solution is unique. It is clear that for any $x \geq 0$ the vector-function $\varphi(x, \lambda)$ is an entire analytic function of the parameter λ. Since the Jost solutions $e^{(1)}(x, \lambda)$ and $e^{(2)}(x, \lambda)$ form a fundamental system of solutions of system (1.84), then

$$\varphi(x, \lambda) = A_1(\lambda)e^{(1)}(x, \lambda) + A_2(\lambda)e^{(2)}(x, \lambda), \tag{1.112}$$

where the numbers $A_1(\lambda)$ and $A_2(\lambda)$ are determined by the initial condition (1.110). In fact, using (1.110), we derive the following system from (1.112):

$$A_1(\lambda)e_1^{(1)}(0, \lambda) + A_2(\lambda)e_1^{(2)}(0, \lambda) = 1,$$
$$A_1(\lambda)e_2^{(1)}(0, \lambda) + A_2(\lambda)e_2^{(2)}(0, \lambda) = 1.$$

Using (1.104) and Lemma 1.15, from the latter system we find

$$A_1(\lambda) = e_2^{(2)}(0, \lambda) - e_1^{(2)}(0, \lambda) =$$
$$1 + \int_0^\infty [H_{22}(0, \xi) - H_{12}(0, \xi)]e^{i\lambda\xi}d\xi \neq 0 \ \text{ for } \operatorname{Im}\lambda \geq 0,$$
$$A_2(\lambda) = e_1^{(1)}(0, \lambda) - e_2^{(1)}(0, \lambda) =$$
$$1 + \int_0^\infty [H_{11}(0, \xi) - H_{21}(0, \xi)]e^{-i\lambda\xi}d\xi \neq 0 \ \text{ for } \operatorname{Im}\lambda \leq 0.$$

Taking the calculated numbers $A_1(\lambda)$ and $A_2(\lambda)$ into account, we rewrite the relation (1.112) between solutions $\varphi(x, \lambda)$ and $e^{(1)}(x, \lambda), e^{(2)}(x, \lambda)$ of system (1.84) as follows:

$$\frac{1}{e_2^{(2)}(0, \lambda) - e_1^{(2)}(0, \lambda)}\varphi(x, \lambda) = e^{(1)}(x, \lambda) + S(\lambda)e^{(2)}(x, \lambda) =$$
$$\begin{pmatrix} e^{-i\lambda x} + \int_x^\infty H_{11}(x, t)e^{-i\lambda t} + S(\lambda)\int_x^\infty H_{12}(x, t)e^{i\lambda t}dt \\ S(\lambda)e^{i\lambda x} + \int_x^\infty H_{21}(x, t)e^{-i\lambda t}dt + S(\lambda)\int_x^\infty H_{22}(x, t)e^{i\lambda t}dt \end{pmatrix}. \tag{1.113}$$

We return to the question raised about the relationship between the functions $f(x, \xi)$, $g(x, \xi)$ and $f(\xi)$, $g(\xi)$.

Lemma 1.17. *For any $x \geq 0$ the functions $f(x, \xi)$ and $g(x, \xi)$ are closely related to the functions $f(\xi)$ and $g(\xi)$, respectively by the equalities:*

$$f(x, \xi) = f(\xi - 2x) \ \textit{for } \xi < 0, \ \ g(x, \xi) = g(\xi + 2x) \ \textit{for } \xi > 0. \tag{1.114}$$

Proof. If in (1.90) $A(\lambda) = 1$, then $B(\lambda) = S(\lambda)$, then in this case the solution (1.90) has the representation:

$$u_1(x,\lambda) = e^{-i\lambda x} + \int_x^\infty H_{11}(x,t)e^{-i\lambda t}dt + S(\lambda)\int_x^\infty H_{12}(x,t)e^{i\lambda t},$$

$$u_2(x,\lambda) = S(\lambda)e^{i\lambda x} + \int_x^\infty H_{21}(x,t)e^{-i\lambda t}dt + S(\lambda)\int_x^\infty H_{22}(x,t)e^{i\lambda t},$$

whence it follows that

$$u_1(x,\lambda)e^{i\lambda x} - 1 = \int_x^\infty H_{11}(x,t)e^{-i\lambda(t-x)}dt + S(\lambda)\int_x^\infty H_{12}(x,t)e^{i\lambda(t+x)}dt,$$

$$\begin{aligned} u_2(x,\lambda)e^{i\lambda x} - e^{2i\lambda x} =& (S(\lambda)-1)e^{2i\lambda x} + \int_x^\infty H_{21}(x,t)e^{-i\lambda(t-x)}dt + \\ & S(\lambda)\int_x^\infty H_{22}(x,t)e^{i\lambda(t+x)}dt. \end{aligned}$$

By comparison of the solution $u(x,\lambda) = (u_1(x,\lambda), u_2(x,\lambda))$ with the vector function (1.113), owing to the uniqueness of the solution of problem (1.84)–(1.85), (1.88), we have:

$$u(x,\lambda) = \frac{1}{e_2^{(2)}(0,\lambda) - e_1^{(2)}(0,\lambda)}\varphi(x,\lambda).$$

Since the vector-function $\varphi(x,\lambda)$ is an entire analytic function of λ and the function $e_2^{(2)}(0,\lambda) - e_1^{(2)}(0,\lambda)$ is analytic and different from zero in the upper half-plane $\mathrm{Im}\lambda \geq 0$, then the vector-function

$$\begin{pmatrix} u_1(x,\lambda)e^{i\lambda x} - 1 \\ u_2(x,\lambda)e^{i\lambda x} - e^{2i\lambda x} \end{pmatrix} = \frac{1}{e_2^{(2)}(0,\lambda) - e_1^{(2)}(0,\lambda)} \begin{pmatrix} \varphi_1(x,\lambda)e^{i\lambda x} \\ \varphi_2(x,\lambda)e^{i\lambda x} \end{pmatrix} - \begin{pmatrix} 1 \\ e^{2i\lambda x} \end{pmatrix}, \tag{1.115}$$

is analytic in the upper half-plane $\mathrm{Im}\lambda \geq 0$ and tends uniformly to zero as $|\lambda| \to \infty, \mathrm{Im}\lambda > 0$. The vector-function in the right-hand side of (1.115) does not have singular numbers on the real axis. Therefore, by an application of the Jordan's Lemma [114], from (1.115) we obtain for any $s > 0$:

$$\int_{-\infty}^{\infty} e^{i\lambda s} d\lambda \int_{x}^{\infty} H_{11}(x,t)e^{-i\lambda(t-x)}dt + \int_{-\infty}^{\infty} S(\lambda)e^{i\lambda s} d\lambda \int_{x}^{\infty} H_{12}(x,t)e^{i\lambda(x+t)}dt = 0$$

$$\int_{-\infty}^{\infty} F(\lambda)e^{i\lambda(2x+s)}d\lambda + \int_{-\infty}^{\infty} e^{i\lambda s} d\lambda \int_{x}^{\infty} H_{21}(x,t)e^{-i\lambda(t-x)}dt +$$

$$\int_{-\infty}^{\infty} S(\lambda)e^{i\lambda s} d\lambda \int_{x}^{\infty} H_{22}(x,t)e^{i\lambda(t+x)}dt = 0,$$

where $F(\lambda) = S(\lambda) - 1 = \int\limits_{-\infty}^{\infty} f(\xi)e^{i\lambda\xi}d\xi$.

Applying the theorem about the convolution Fourier transform of integrable functions to the left-hand sides of the last equalities, we find that [61]:

$$H_{11}(x, x+s) + \int_{0}^{\infty} H_{12}(x, x+\tau)f(-2x - s - \tau)d\tau = 0 \quad s > 0, \tag{1.116}$$

$$f(-2x - s) + H_{21}(x, x+s) + \int_{0}^{\infty} H_{22}(x, x+\tau)f(-2x - s - \tau)d\tau = 0 \quad s > 0, \tag{1.117}$$

where $f(x) = \frac{1}{2\pi} \int\limits_{-\infty}^{\infty} F(\lambda)e^{-i\lambda x}d\lambda$.

Due to (1.94) and (1.109), the equalities (1.116) and (1.117) are rewritten as follows

$$H_{22}(x, x+s) + \int_{0}^{\infty} H_{21}(x, x+\tau)g(2x + s + \tau)d\tau = 0, s > 0, \tag{1.118}$$

$$g(2x + s) + H_{12}(x, x+s) + \int_{0}^{\infty} H_{11}(x, x+\tau)g(2x + s + \tau)d\tau = 0, s > 0. \tag{1.119}$$

Making use of estimates (1.92) and the absolute integrability of functions $f(x)$ and $g(x)$, by starting from (1.117) and (1.119), we obtain the estimates:

$$|f(-x)| \leq \frac{\tilde{C}}{1 + x^{1+\varepsilon}}, |g(x)| \leq \frac{\tilde{C}}{1 + x^{1+\varepsilon}}, \varepsilon > 0, \tilde{C} \text{ is a positive number.} \tag{1.120}$$

Subtracting (1.116) from (1.117), we have

$$f(-2x-s)+H_{21}(x,x+s)-H_{11}(x,x+s)+$$
$$\int_0^\infty [H_{22}(x,x+\tau)-H_{12}(x,x+\tau)]\,f(-2x-s-\tau)d\tau=0 \text{ for } s>0.$$

Putting

$$r(x,s)=\begin{cases} f(-2x-s)+H_{21}(x,x+s)-H_{11}(x,x+s)+ \\ \int_0^\infty [H_{22}(x,x+\tau)-H_{12}(x,x+\tau)]f(-2x-s-\tau)d\tau \text{ for } s<0 \\ \\ 0 \text{ for } s>0, \end{cases}$$

then for any s, we obtain

$$\int_{-\infty}^{\infty} f(-2x-s)e^{-i\lambda s}ds+\int_0^\infty [H_{21}(x,x+s)-H_{11}(x,x+s)]e^{-i\lambda s}ds+$$
$$\int_{-\infty}^{\infty} f(-2x-s)e^{-i\lambda s}ds\int_0^\infty [H_{22}(x,x+\tau)-H_{12}(x,x+\tau)]e^{i\lambda\tau}=$$
$$\int_{-\infty}^{0} e^{-i\lambda s}r(x,s)ds,$$

or

$$\left\{1+\int_0^\infty [H_{22}(x,x+\tau)-H_{12}(x,x+\tau)]e^{i\lambda\tau}d\tau\right\}F_x(\lambda)=$$
$$R_+(x,\lambda)+\int_0^\infty [H_{11}(x,x+\tau)-H_{21}(x,x+\tau)]e^{-i\lambda\tau}d\tau, \tag{1.121}$$

where $F_x(\lambda)=\int\limits_{-\infty}^{\infty} f(-2x-s)e^{-i\lambda s}ds$, $R_+(x,\lambda)=\int\limits_{\infty}^{0} r(x,s)e^{-i\lambda s}ds$.

Owing to estimates (1.120) for $f(-x)$ and (1.92) for $H_{jk}(x,x+s), x,s\geq 0, j,k=1,2$, the function $r(x,s)$ is square-integrable with respect to s. Then obviously the function $R_+(x,\lambda)$ admits a bounded analytic continuation into the upper half-plane $\mathrm{Im}\lambda>0$. On the other hand, from the factorization

(1.106) and Lemma 1.16 we deduce

$$\left\{1+\int_0^\infty [H_{22}(x,x+\tau)-H_{12}(x,x+\tau)]e^{i\lambda\tau}d\tau\right\}F(x,\lambda)=$$
$$\int_0^\infty [H_{11}(x,x+\tau)-H_{21}(x,x+\tau)]e^{-i\lambda\tau}d\tau-$$
$$\int_0^\infty [H_{22}(x,x+\tau)-H_{12}(x,x+\tau)]e^{i\lambda\tau}d\tau. \tag{1.122}$$

Subtracting (1.122) from (1.121), yields

$$\left\{1+\int_0^\infty [H_{22}(x,\xi+\tau)-H_{12}(x,x+\tau)]e^{i\lambda\tau}d\tau\right\}\{F_x(\lambda)-F(x,\lambda)\}=$$
$$R_+(x,\lambda)+\int_0^\infty [H_{22}(x,x+\tau)-H_{12}(x,x+\tau)]e^{i\lambda\tau}d\tau,$$

whence it follows that

$$F_x(\lambda)-F(x,\lambda)=\frac{R_+(x,\lambda)}{1+\int_0^\infty [H_{22}(x,x+\tau)-H_{12}(x,x+\tau)]e^{i\lambda\tau}d\tau}+$$
$$\frac{\int_0^\infty [H_{22}(x,x+\tau)-H_{12}(x,x+\tau)]e^{i\lambda\tau}}{1+\int_0^\infty [H_{22}(x,x+\tau)-H_{12}(x,x+\tau)]e^{i\lambda\tau}d\tau}.$$

The function $F_x(\lambda)-F(x,\lambda)$ admits a bounded analytic continuation into the upper half-plane $\mathrm{Im}\lambda>0$ and tends uniformly to zero as $|\lambda|\to\infty$, then by an application of the Jordan's Lemma, the Fourier transformation of this function vanishes on the right half-axis of s [114]:

$$\int_{-\infty}^\infty e^{i\lambda s}[F_x(\lambda)-F(x,\lambda)]d\lambda=\int_{-\infty}^\infty d\lambda\int_{-\infty}^\infty [f(-2x-\xi)-f(x,-\xi)]\,e^{-i\lambda(\xi-s)}d\xi$$
$$=f(-2x-s)-f(x,-s)=0 \text{ for } s>0,$$

whence it follows that

$$f(x,\xi)=f(\xi-2x) \text{ for } \xi<0.$$

Using the factorization (1.106) and equalities (1.118), (1.119), analogously one can prove the second assertion of (1.114). □

We turn to the inverse scattering problem, which is to determine the potential matrix from the scattering function and also to describe the scattering function of the self-adjoint problem.

1.2.5 The inverse scattering problem

With due regard for Lemma 1.17, from (1.117) and (1.118), we obtain the system of fundamental equations connecting the scattering function with the kernel of the transformation operator:

$$\begin{aligned} &H_{22}(x,x+\xi)+\int_0^\infty H_{21}(x,x+\tau)g(x,\xi+\tau)d\tau=0, \\ &f(x,\xi)+H_{21}(x,x+\xi)+\int_0^\infty H_{22}(x,x+\tau)f(x,\xi+\tau)d\tau=0, \end{aligned} \qquad \xi>0.$$

Consider the solvability of the system of the form:

$$\begin{aligned} &h_1(x,\xi)+N(x,x+\xi)+\int_0^\infty M(x,x+\tau)g(x,\xi+\tau)d\tau=0, \\ &h_2(x,\xi)+M(x,x+\xi)+\int_0^\infty N(x,x+\tau)f(x,\xi+\tau)d\tau=0, \end{aligned} \qquad \xi>0, \tag{1.123}$$

where $f(x,\xi)$ and $g(x,\xi)$ are the given functions satisfying relations (1.114) and estimates (1.120), the functions $h_1(x,\xi)$, $h_2(x,\xi)$ are arbitrarily integrable with respect to ξ, and $N(x,x+\xi)$, $M(x,x+\xi)$ are unknown functions of ξ for every $x\in[0,\infty)$.

Lemma 1.18. *For every fixed $x\geq 0$ the system* (1.123) *has a unique solution in the space $L_1[0,\infty)$.*

Proof. To prove the uniqueness of the solution of (1.123), it is sufficient to show that the system of homogeneous equations:

$$\begin{aligned} &b_2(\xi)+\int_0^\infty b_1(\tau)g_x(\xi+\tau)d\tau=0, \\ &b_1(\xi)+\int_0^\infty b_2(\tau)f_x(-\xi-\tau)d\tau=0, \end{aligned} \qquad \xi>0,$$

has only the zero solution in the space $L_1[0,\infty)$.

The last system can be rewritten in the form:

$$\begin{aligned} b_2(\xi) + \int_{\xi}^{\infty} b_1(\eta-\xi) g_x(\eta) d\eta = 0, \\ b_1(\xi) + \int_{\xi}^{\infty} b_2(\eta-\xi) f_x(-\eta) d\eta = 0, \end{aligned} \qquad \xi > 0. \tag{1.124}$$

Eliminating $b_1(\eta-\xi)$ from the system (1.124), we have:

$$b_2(\xi) - \int_0^{\infty} b_2(z)dz \int_{\xi}^{\infty} g_x(\eta) f_x(\xi - z - \eta) d\eta = 0, \quad \xi > 0. \tag{1.125}$$

According to the definitions of $f(x,\xi)$ and $g(x,\xi)$, we have

$$-\int_{\xi}^{\infty} f_x(\xi - z - \eta) g_x(\eta) d\eta = f_x(\xi - z) + g_x(\xi - z) + \int_{-\infty}^{\xi} f_x(\xi - z - \eta) g_x(\eta) d\eta.$$

Using the last equality, we rewrite equation (1.125):

$$b_2(\xi) + \int_0^{\infty} b_2(z) f_x(\xi - z) dz + \int_0^{\infty} b_2(z) g_x(\xi - z) dz +$$
$$\int_0^{\infty} dz \int_0^{\infty} b_2(\eta) f_x(z - \eta) g_x(\xi - z) d\eta = 0 \text{ for } \xi > 0.$$

Putting

$$w(\xi) = b_2(\xi) + \int_0^{\infty} b_2(z) f_x(\xi - z) dz, \quad \xi > 0, \tag{1.126}$$

then the last equation can be written in the form:

$$w(\xi) + \int_0^{\infty} w(z) g_x(\xi - z) dz = 0, \quad \xi > 0.$$

Instead of $g_x(\xi - z)$ in this equation, we substitute its expression (1.108) and get:

$$w(\xi) + \int_0^{\infty} w(z) H_+(x, \xi - z) dz + \int_0^{\infty} w(\tau) K_-(x, \tau - \xi) d\tau +$$
$$\int_0^{\infty} w(z) dz \int_{\xi}^{\infty} K_-(x, \tau - \xi) H_+(x, \tau - z) d\tau = 0, \ \xi > 0.$$

Putting

$$P(\xi) = w(\xi) + \int_0^\infty w(z)H_+(x, \xi - z)dz, \tag{1.127}$$

then the last equation is presented in the form:

$$P(\xi) + \int_\xi^\infty P(\tau)K_-(x, \tau - \xi)d\tau = 0, \quad \xi > 0. \tag{1.128}$$

Since the function $1 + \int_0^\infty H_-(x,\xi)e^{-i\lambda\xi}d\xi$ does not have zeros in the lower half-plane $\mathrm{Im}\lambda \leq 0$, and $\lim_{|\lambda|\to\infty}\left\{1 + \int_0^\infty H_-(x,\xi)e^{-i\lambda\xi}d\xi\right\} = 1$, then there exists the function:

$$1 + \int_0^\infty K_-(\xi)e^{-i\lambda\xi}d\xi = \left(1 + \int_0^\infty H_-(x,\xi)e^{-i\lambda\xi}d\xi\right)^{-1}, \quad \mathrm{Im}\lambda \leq 0,$$

which is different from zero and admits a bounded analytic continuation into the lower half-plane $\mathrm{Im}\lambda \leq 0$. Therefore by Krein's Theorem [66] the equation (1.128) has in the space $L_1[0,\infty)$ only the zero solution $P(\xi) \equiv 0$. Hence, from (1.127) we obtain the Volterra integral equation:

$$w(\xi) + \int_0^\xi w(z)H_+(x, \xi - z)dz = 0, \quad \xi > 0,$$

which, as known, has only the zero solution $w(\xi) \equiv 0 \;\; \xi > 0$ in $L_1[0,\infty)$. Then from (1.126) we have the equation:

$$b_2(\xi) + \int_0^\infty b_2(z)f_x(\xi - z)dz = 0, \quad \xi > 0.$$

Instead of $f_x(\xi - z)$ in the last equation, we substitute its expression (1.107) and obtain:

$$b_2(\xi) + \int_0^\infty b_2(z)K_+(x, \xi - z)dz + \int_0^\infty b_2(\tau)H_-(x, \tau - \xi)d\tau$$
$$+ \int_0^\infty b_2(z)dz \int_z^\infty K_+(x, \tau - z)H_-(x, \tau - \xi)d\tau = 0, \quad \xi > 0.$$

Putting

$$\tilde{w}(\xi) = b_2(\xi) + \int_0^\infty b_2(z)K_+(x, \xi - z)dz, \quad \xi > 0,$$

from the above equality we get the equation:

$$\tilde{w}(\xi) + \int_\xi^\infty \tilde{w}(\tau)H_-(x, \tau - \xi)d\tau = 0, \quad \xi > 0.$$

Since the function $1 + \int_0^\infty H_-(x,\xi)e^{-i\lambda\xi}d\xi$ $\mathrm{Im}\lambda \leq 0$ is different from zero and admits a bounded analytic continuation into the lower half-plane $\mathrm{Im}\lambda \leq 0$, then by an argument analogous to the case of equation (1.128), we prove that the last equation has only the zero solution $\tilde{w}(\xi) \equiv 0$ for $\xi > 0$ in $L_1[0, \infty)$. Hence, we obtain the Volterra integral equation for $b_2(\xi)$:

$$b_2(\xi) + \int_0^\xi b_2(z)K_+(x, \xi - z)dz = 0, \quad \xi > 0,$$

which has only the zero solution $b_2(\xi) \equiv 0$. Then from system (1.124) it follows that $b_1(\xi) \equiv 0$.

By using estimates (1.120), we can show that in the space $L_1[0, \infty)$ the system (1.123) is a Fredholm system of the second kind. Therefore, the unique solvability of this system is deduced from the uniqueness of the solution of the system [61, 80, 101]. □

1.2.6 The complete description of the scattering function

We turn to the main theorem regarding the complete description of the scattering function, that is the establishment of necessary and sufficient conditions for a given function to be the scattering function for system (1.84) on a half-line with a potential self-adjoint matrix and with boundary condition (1.85).

Theorem 1.2. *For a given function $\gamma(\lambda)$ to be the scattering function for a system of the form* (1.84) *on a half-line with a potential self-adjoint matrix $C(x)$ and with boundary condition* (1.85), *it is necessary and sufficient that for any $x \geq 0$ there exists a function $\gamma(x, \lambda)$ such that $\gamma(\lambda) = \gamma(0, \lambda)$ and*

(1) The function $\gamma(\lambda, x)$ admits the following factorization:

$$\gamma(x, \lambda) = \left(1 + \int_0^\infty \overline{T(x, x+t)e^{-i\lambda t}}dt\right)\left(1 + \int_0^\infty T(x, x+t)e^{i\lambda t}dt\right)^{-1}, \tag{1.129}$$

where for every fixed $x \geq 0$ *the function* $T(x, x+t)$ *is absolutely integrable with respect to* t*, and the function* $1 + \int_0^\infty T(x, x+t)e^{i\lambda t}dt$ *does not have zeros in the upper half-plane* $Im\lambda \geq 0$*;*

(2) For any $x \geq 0$ *the functions* $\gamma(x, \lambda) - 1$ *and* $\gamma^{-1}(x, \lambda) - 1$ *are the inverse Fourier transformations of absolutely integrable with respect to* ξ *functions* $f(x, \xi)$ *and* $g(x, \xi)$*:*

$$\gamma(x, \lambda) - 1 = \int_{-\infty}^{\infty} f(x, \xi)e^{i\lambda\xi}d\xi, \quad \gamma^{-1}(x, \lambda) - 1 = \int_{-\infty}^{\infty} g(x, \xi)e^{i\lambda\xi}d\xi;$$

(3) For any $x \geq 0$ *the functions* $f(x, \xi)$ *and* $g(x, \xi)$ *are closely related to* $f(\xi)$ *and* $g(\xi)$ *by the equalities:*

$$f(x, \xi) = f(\xi - 2x) \text{ for } \xi < 0, \quad g(x, \xi) = g(\xi + 2x) \text{ for } \xi > 0, \tag{1.130}$$

where

$$f(x, \xi) = \frac{1}{2\pi} \int_{-\infty}^{\infty} [\gamma(x, \lambda) - 1]e^{-i\lambda\xi}d\lambda,$$

$$g(x, \xi) = \frac{1}{2\pi} \int_{-\infty}^{\infty} [\gamma^{-1}(x, \lambda) - 1]e^{-i\lambda\xi}d\lambda, f(\xi) = f(0, \xi), \quad g(\xi) = g(0, \xi);$$

(4) For any $x \geq 0$ *the functions* $f(-x)$ *and* $g(x)$ *satisfy the inequality:*

$$|f(-x)| \leq \frac{\tilde{C}}{1 + (x)^{1+\varepsilon}}, \quad |g(x)| \leq \frac{\tilde{C}}{1 + (x)^{1+\varepsilon}}, \tag{1.131}$$

where $\tilde{C}$ *is a constant,* $\varepsilon > 0$*.*

The necessity of these conditions is given by Lemmas 1.15–1.18. We divide the proof of sufficiency into several lemmas. First, we prove the following lemma.

Lemma 1.19. *If we are given the factorization* (1.129)*, relation equalities* (1.130) *and estimates* (1.131)*, then the function* $T(x, x+\xi)$ *in* (1.129) *satisfies the estimate:*

$$|T(x, x+\xi)| \leq \frac{\tilde{C}}{1 + (2x + \xi)^{1+\varepsilon}}, \quad x, \xi \geq 0 \tag{1.132}$$

where $\tilde{C}$ *is a constant,* $\varepsilon > 0$*.*

Proof. From the factorization (1.129), we obtain

$$\left(1+\int_0^\infty \overline{T(x,x+\xi)}e^{-i\lambda\xi}d\xi\right) =$$

$$\left(1+\int_{-\infty}^\infty f(x,\xi)e^{i\lambda\xi}d\xi\right)\left(1+\int_0^\infty T(x,x+\xi)e^{i\lambda\xi}d\xi\right) =$$

$$1+\int_{-\infty}^\infty f(x,-\xi)e^{-i\lambda\xi}d\xi+\int_{-\infty}^0 T(x,x-\xi)e^{-i\lambda\xi}d\xi+$$

$$\int_{-\infty}^\infty e^{-i\lambda\xi}d\xi\int_0^\infty T(x,x+t)f(x,-\xi-t)dt. \tag{1.133}$$

Since $\overline{\gamma(\lambda)}=\gamma^{-1}(\lambda)$, then $\overline{f(-\xi)}=g(\xi)$. Taking this fact and (1.130) into account, from (1.133) we have

$$\begin{aligned} &f(-\xi-2x)+\int_0^\infty T(x,x+t)f(-2x-\xi-t)dt=\overline{T(x,x+\xi)}, \\ & && \xi>0. \\ &g(\xi+2x)+\int_0^\infty \overline{T(x,x+t)}g(2x+\xi+t)dt=T(x,x+\xi), \end{aligned} \tag{1.134}$$

By using estimate (1.131), we prove that for sufficiently large x the system (1.134) is solved by the method of successive approximations. In addition, the obtained solution satisfies the inequality (1.132). Then for sufficiency large x:

$$\int_0^\infty |T(x,x+t)|dt<\tilde{C},\ \ \tilde{C} \text{ is a constant.}$$

Since the system (1.134) in the space $L_1[0,\infty)$ depends continuously on x as a parameter, then the function $\int_0^\infty |T(x,x+t)|dt$ is continuous with respect to x, and so it is bounded uniformly on all half-axes: $0\le x<\infty$. Then using estimate (1.131), from (1.134) we obtain the estimate (1.132). □

Consider the system of equations:

$$\begin{aligned} &g(x,\xi)+\int_0^\infty K_{11}(x,x+t)g(x,\xi+t)dt=-K_{12}(x,x+\xi), \\ & && \xi>0, \\ &-\int_0^\infty K_{12}(x,x+t)f(x,\xi+t)dt=K_{11}(x,x+\xi), \end{aligned} \tag{1.135}$$

where the functions $f(x,\xi)$, $g(x,\xi)$ are known and $K_{11}(x,x+\xi)$, $K_{12}(x,x+\xi)$ are unknown.

The system of equations (1.135) has a unique solution in $L_1[0,\infty)$. Indeed, this assertion is deduced from the proof of Lemma 1.18, because in the proof of this lemma we used only a factorization of the form (1.129) and estimates of the type (1.131).

In view of relation equalities (1.130), the system (1.135) can be written in the form:

$$\begin{aligned} &g(2x+\xi)+\int_0^\infty K_{11}(x,x+t)g(2x+\xi+t)dt=-K_{12}(x,x+\xi), \\ &\qquad\qquad\qquad\qquad\qquad\qquad\qquad\qquad\qquad\qquad\qquad \xi>0. \\ &-\int_0^\infty K_{12}(x,x+t)f(-2x-\xi-t)dt=K_{11}(x,x+\xi), \end{aligned} \tag{1.136}$$

Since $\overline{g(x,\xi)}=f(x,-\xi)$, then from (1.136) we derive the system:

$$\begin{aligned} &f(-2x-\xi)+\int_0^\infty K_{22}(x,x+t)f(-2x-\xi-t)dt=-K_{21}(x,x+\xi), \\ &\qquad\qquad\qquad\qquad\qquad\qquad\qquad\qquad\qquad\qquad\qquad \xi>0, \\ &-\int_0^\infty K_{21}(x,x+t)g(2x+\xi+t)dt=K_{22}(x,x+\xi), \end{aligned} \tag{1.137}$$

where $K_{22}(x,x+\xi)=\overline{K_{11}(x,x+\xi)}$, $K_{21}(x,x+\xi)=\overline{K_{12}(x,x+\xi)}$.

The systems (1.136) and (1.137) connecting the operator $\gamma(x,\lambda)$ with the functions $K_{jk}(x,x+\xi)$, $x\geq 0$, $\xi>0$, $j,k=1,2$, are called the systems of fundamental equations. Owing to estimate (1.131) and Lemma 1.18, these systems are Fredholm systems and have a unique solution in $L_1[0,\infty)$. We show that the function $T(x,x+\xi)$ from the factorization (1.129) is related to the functions $K_{22}(x,x+\xi)$ and $K_{12}(x,x+\xi)$ by the equality:

$$T(x,x+\xi)=K_{22}(x,x+\xi)-K_{12}(x,x+\xi),\quad \xi>0. \tag{1.138}$$

Indeed, from (1.136) and (1.137), we obtain the system of equations:

$$\begin{aligned} &f(-2x-\xi)+\int_0^\infty [K_{22}(x,x+t)-K_{12}(x,x+t)]f(-2x-\xi-t)dt= \\ &\qquad \overline{K_{22}(x,x+\xi)}-\overline{K_{12}(x,x+\xi)},\xi>0, \\ &g(2x+\xi)+\int_0^\infty [\overline{K_{22}(x,x+t)}-\overline{K_{12}(x,x+t)}]g(2x+\xi+t)dt= \\ &\qquad K_{22}(x,x+\xi)-K_{12}(x,x+\xi),\xi>0. \end{aligned} \tag{1.139}$$

By Lemma 1.18 the system (1.139) has a unique solution in the space $L_1[0,\infty)$. On the other hand from the factorization (1.129), we obtain the system of equations of the same form (1.139) with the same known functions $\overline{f(-2x-\xi)}$, $\overline{g(2x+\xi)}$ and unknown functions $T(x,x+\xi)$, $\overline{T(x,x+\xi)}$. Owing to the unique solvability of system (1.139), we obtain the equality (1.138).

In the same way used in the proof of Lemma 1.19, we prove the following lemma.

Lemma 1.20. *Let for any $x \geq 0$ the functions $f(-x)$ and $g(x)$ satisfy the inequalities* (1.131)*, then the solutions of systems of fundamental equations* (1.136) *and* (1.137) *satisfy the estimate:*

$$|K_{jk}(x,x+\xi)| \leq \frac{\tilde{C}}{1+(2x+\xi)^{1+\varepsilon}}, \quad x,\xi \geq 0, \quad j,k=1,2 \tag{1.140}$$

where $\tilde{C}$ is a constant, $\varepsilon > 0$.

By argument analogous to that used for the proof of Lemma 1.12 in the previous section, one can prove the following lemma.

Lemma 1.21. *The solutions of systems of fundamental equations* (1.136) *and* (1.137) *with the conditions:*

$$K_{jj}(x,x+\xi) \to 0 \;\; (x \to \infty), \;\; j=1,2, \tag{1.141}$$

satisfy systems of integral equations of the form (1.15), (1.16)*, where*

$$c_1(x) = 2iK_{12}(x,x), \;\; c_2(x) = -2iK_{21}(x,x). \tag{1.142}$$

We are now in a position to give a direct proof of the sufficiency of the conditions in the main Theorem 1.2. Let $\gamma(\lambda)$ be a given function possessing the conditions (1)–(4) enumerated in this theorem. Finding

$$f(\xi) = \frac{1}{2\pi}\int_{-\infty}^{\infty}[\gamma(\lambda)-1]e^{-i\lambda\xi}d\lambda,$$

$$g(\xi) = \frac{1}{2\pi}\int_{-\infty}^{\infty}[\gamma^{-1}(\lambda)-1]e^{-i\lambda\xi}d\lambda,$$

we construct systems (1.136), (1.137). Solving these systems, we obtain the solutions $K_{jk}(x,s)$, $0 \leq x \leq s$, $j,k=1,2$. According to Lemmas 1.20 and 1.21, the obtained solutions satisfy estimate (1.140) and the systems of integral equations of the form (1.15) and (1.16) for the kernels of the transformation operator of the problem generated by a system of the form (1.84) with the potentials constructed from the solutions of systems (1.136), (1.137):

$$c_1(x) = 2iK_{12}(x,x) \text{ and } c_2(x) = -2iK_{21}(x,x) = -2i\overline{K_{12}(x,x)}. \tag{1.143}$$

In view of the uniqueness of the solution of systems (1.15) and (1.16), we obtain that the functions $K_{jk}(x,s), 0 \le x \le s, j,k = 1,2$ are the kernels of the transformation operator of the problem generated by the constructed system (1.84) with the self-adjoint potentials (1.143). Then from (1.106), (1.129) and (1.138) it follows that

$$\gamma(\lambda) = S(\lambda),$$

where $S(\lambda)$ is the scattering function for the system (1.84) with the constructed self-adjoint potentials (1.143). Thus, the theorem is proved. $\square$

One should note that in the proof of the main theorem an algorithm for finding the potentials in terms of the known operator $\gamma(\lambda)$ is constructed. This algorithm consists of solving system (1.136) or (1.137) and recovering the system (1.84), that is the potentials $c_1(x)$ and $c_2(x)$ from the formulas (1.143).

Remark 1.1. The following example shows that the role of the self-adjoint condition of the potential matrix is essential. Indeed, let the potentials in the system (1.84) be given by

$$c_1(x) \neq 0, \quad c_2(x) = 0. \tag{1.144}$$

In the case (1.144), the solution $(u_1(x,\lambda), u_2(x,\lambda))$ of the problem (1.84), (1.88) is equivalent to the following system of integral equations (1.93):

$$u_1(x,\lambda) = A(\lambda)e^{-i\lambda x} + \frac{1}{i}B(\lambda)\int_x^\infty e^{-i\lambda(x-2t)}c_1(t)dt,$$

$$u_2(x,\lambda) = B(\lambda)e^{i\lambda x}.$$

If the solution $(u_1(x,\lambda), u_2(x,\lambda))$ of the above system of integral equations satisfies the boundary condition (1.85), then the number $B(\lambda)$ is defined from the given number $A(\lambda)$ by the formula: $B(\lambda) = \left(1 - \frac{1}{i}\int_0^\infty e^{2i\lambda t}c_1(t)dt\right)^{-1} A(\lambda)$. Hence, if the equation

$$1 - \frac{1}{i}\int_0^\infty e^{2i\lambda t}c_1(t)dt = 0,$$

has a real root, then there is no one-to-one correspondence between $A(\lambda)$ and $B(\lambda)$. Therefore, in the case (1.144) the assertion of Lemma 1.13 is not valid.

Remark 1.2. The potentials of the self-adjoint problem (1.84)–(1.85), (1.88) satisfy estimate (1.87); i.e., they are power decreasing functions as $x \to \infty$, whereas the potentials of the non-self-adjoint problem (1.7)–(1.8) satisfy estimate (1.9); i.e., they are exponentially decreasing functions as $x \to \infty$.

The self-adjoint condition and estimate (1.87) for the potentials ensure the unique solvability of the considered direct and inverse scattering problem.

Chapter 2

Some Problems for A System of Non-Linear Evolution Equations on A Half-Line

In this chapter we apply the results of the inverse scattering problems (ISPs) obtained in Chapter 1 to solving an initial value problem (IVP) and some initial-boundary value problems (IBVPs) for the system of non-linear evolution equations (NLEEs):

$$\begin{cases} ic_{1t} = c_{1xx} - 2c_1^2 c_2, \\ -ic_{2t} = c_{2xx} - 2c_1 c_2^2. \end{cases} \qquad 0 \le x < \infty, \quad -\infty < t < \infty$$

The attractive and repulsive non-linear Schrödinger (NLS) equations are obtained from this system when $c_1 = -\overline{c_2}$ and $c_1 = \overline{c_2}$, respectively. The IBVPs for the considered system of NLEEs and the attractive NLS equation are solved by applying the results of the ISP for a system of first-order ordinary differential equations (ODEs) on a half-line with a potential non-self-adjoint matrix. This ISP is studied in Section 1.1 of Chapter 1.

The Cauchy IVP for the repulsive NLS equation is solved by applying the results of the ISP for the system of first-order ODEs on a half-line with a potential self-adjoint matrix. This ISP is studied in Section 1.2 of Chapter 1. The chapter is based on the published papers [133, 135, 137, 140].

We consider the initial-boundary value problem (IBVP) for the system of non-linear evolution equations (NLEEs):

$$\begin{cases} ic_{1t} = c_{1xx} - 2c_1^2 c_2, \\ -ic_{2t} = c_{2xx} - 2c_1 c_2^2. \end{cases} \qquad 0 \le x < \infty \tag{2.1}$$

The system (2.1) can be written as the Lax compatibility condition of two linear eigenvalue equations, i.e., of the x part and the t part of the Lax pair. The x part is the eigenvalue problem generated by the system of first-order ODEs of the form:

$$-iJu_x + Cu = \lambda u, \ u = (u_1(x,\lambda;t), u_2(x,\lambda;t)), \ 0 \le x < \infty, \ -\infty < t < \infty \tag{2.2}$$

DOI: 10.1201/9781003370543-2

with a potential non-self-adjoint matrix $C = C(x,t) = \begin{pmatrix} 0 & c_1(x,t) \\ c_2(x,t) & 0 \end{pmatrix}$ and with the boundary condition at the origin $x = 0$:

$$u_1(0,\lambda;t) = u_2(0,\lambda;t), \quad -\infty < t < \infty, \tag{2.3}$$

where $J = \begin{pmatrix} -1 & 0 \\ 0 & 1 \end{pmatrix}$, λ is a parameter.

The potential entries $c_1(x,t)$, $c_2(x,t)$ of the matrix C are complex-valued measurable functions satisfying the estimate:

$$|c_k(x,t)| \le \tilde{C}e^{-\varepsilon x} \text{ for all } t \in (-\infty,\infty), \tag{2.4}$$

$\tilde{C}$ is a constant, $\varepsilon > 0$, $k = 1,2$.

The t part of the Lax pair is defined by the equation governing the time-dependence of eigenfunctions of the eigenvalue problem (2.2)–(2.3). The solution of the IBVP for system (2.1) is the time-dependent potential matrix C in system (2.2). Hence, the problem of solving the IBVP for the system (2.1) is reduced to deriving the time-dependence of the scattering data and to finding the time-dependent potential matrix C from the known initial and boundary conditions in terms of the solutions of the systems of fundamental equations in the inverse problem. Since the time-dependent scattering data depend on the unknown boundary values (BVs) at the origin $x = 0$ of the Jost solutions of system (2.2), then deriving the evolution equations for the unknown BVs is a crucial step in solving the IBVPs for the system of NLEEs (2.1) and for the non-linear attractive Schrödinger (NLS) equation. The evolution equations for the unknown BVs are derived in Section 2.1, then the solutions of these derived equations; i.e., the unknown BVs are found in terms of the known initial and boundary conditions of the considered IBVP. Thus, the time-dependent scattering data are found and we can solve the systems of fundamental equations considered in Section 1.1 of Chapter 1 for finding the potential matrix $C(x,t)$.

In Section 2.2 we find exact solutions of the considered NLEEs in the class of non-scattering potentials [137, 135]. Further, in Section 2.3 we apply the results of the ISP for the system of first-order ODEs on a half-line with a potential self-adjoint matrix to solving the Cauchy initial-value (IVP) for the repulsive NLS equation [133, 135].

In [13] by using the Bäcklund transformation, Bikbaev and Tarasov proved that the IBVP on a half-line for the attractive NLS equations can be reduced to the well-known Cauchy problem for the same equation on the line. Hence, the IBVP for the attractive NLS equation given in [13] is just a special case of the IBVP for the more general system of NLEEs (2.1) studied in Section 2.1 of this chapter [140].

2.1 The IBVP for the system of NLEEs

2.1.1 The Lax compatibility condition

We consider the IBVP for the system of NLEEs [140]:

$$\begin{aligned} ic_{1t} &= c_{1xx} - 2c_1^2 c_2, \\ -ic_{2t} &= c_{2xx} - 2c_1 c_2^2. \end{aligned} \quad 0 \leq x < \infty, \quad -\infty < t < \infty \tag{2.5}$$

with the initial and boundary conditions:

$$c_j(x,0) = c_j(x), \quad 0 \leq x < \infty, \tag{2.6}$$

$$c_j(0,t) = f_j(t), \quad j = 1,2, \tag{2.7}$$

where $c_j(x)$ possesses the second continuous derivative and satisfies estimate (2.4); $f_j(t)$ possesses the first continuous derivative and $f_j(t) \in L_1(-\infty,\infty), j = 1,2$.

In addition, the initial conditions (2.6) are supposed to be consistent with boundary conditions (2.7) at the corner point $(0,0)$ by the equalities:

$$\begin{aligned} c_j(0) &= f_j(0), \quad j = 1,2\,, \\ if_{1t}(0) &= c_{1xx}(0) - 2c_1^2(0)c_2(0), \\ -if_{2t}(0) &= c_{2xx}(0) - 2c_1(0)c_2^2(0). \end{aligned}$$

The IBVP (2.5)–(2.6)–(2.7) is associated with the ISP for the system of first-order ODEs (2.2) with the boundary condition at the origin (2.3) and with the potential non-self-adjoint matrix satisfying estimate (2.4). This ISP is studied in Section 1.1 of Chapter 1. We apply the results of this ISP to solving the considered IBVP (2.5)–(2.6)–(2.7).

The associated problem (2.2)–(2.3) can be formulated in the operator form:

$$u_x = Bu, B = i\lambda J + Q, u = \left(u_1(x,\lambda;t), u_2(x,\lambda;t)\right), 0 \leq x < \infty, -\infty < t < \infty, \tag{2.8}$$

with the boundary condition:

$$u_1(0,\lambda;t) = u_2(0,\lambda;t), \tag{2.9}$$

where

$$J = \begin{pmatrix} -1 & 0 \\ 0 & 1 \end{pmatrix}, Q(x,t) = \begin{pmatrix} 0 & q_1(x,t) \\ q_2(x,t) & 0 \end{pmatrix} = \begin{pmatrix} 0 & ic_1(x,t) \\ -ic_2(x,t) & 0 \end{pmatrix},$$

and the functions $q_1(x,t)$ and $q_2(x,t)$ satisfy the estimate for any $t \in (-\infty,\infty)$:

$$|q_j(x,t)| \leq \tilde{C}e^{-\varepsilon x}, \ \tilde{C} \text{ is a constant}, \ \varepsilon > 0, \ j = 1,2. \tag{2.10}$$

The time-dependence of the solution of system (2.8) is given by

$$u_t = \tilde{V}u, \tag{2.11}$$

where

$$\tilde{V} = \tilde{V}(x,\lambda,t) = \begin{pmatrix} 2i\lambda^2 + iq_1q_2 & -2\lambda q_1 - iq_{1x} \\ -2\lambda q_2 + iq_{2x} & -2i\lambda^2 - iq_1q_2 \end{pmatrix}. \tag{2.12}$$

Let $L(t)$ denote the operator defined in the space $L_2[0,\infty)$ by the system (2.2) and the boundary condition (2.3). The potential matrix C in system (2.2) is called isospectral if the spectrum of $L(t)$ is invariant with t, i.e., $\lambda_t = 0$. Denote by P the class of continuous functions $q(x,t)$ satisfying estimate (2.10) and possessing the second partial continuous derivative with respect to x and the first partial continuous derivative with respect to t.

We suppose that the potentials $q_1(x,t)$ and $q_2(x,t)$ in the considered equation (2.8) are isospectral and belong to the class P.

Since $c_1(x,t) = -iq_1(x,t)$ and $c_2(x,t) = iq_2(x,t)$, then the functions $q_1(x,t)$ and $q_2(x,t)$ satisfy the same system of the form (2.5):

$$\begin{aligned} iq_{1t} &= q_{1xx} - 2q_1^2q_2\,, \\ -iq_{2t} &= q_{2xx} - 2q_1q_2^2. \end{aligned} \qquad 0 \le x < \infty, \quad -\infty < t < \infty.$$

Lemma 2.1. *Let the potentials q_1 and q_2 in system (2.8) be isospectral and be continuous functions belonging to the class P, then the eigenvalue equations (2.8) and (2.11) are compatible, i.e., $u_{xt} = u_{tx}$, and the potentials q_1 and q_2 vary with time as the solution of system (2.5). Conversely, if q_1 and q_2 satisfy system (2.5), then $\lambda_t = 0$ and the eigenvalue equations (2.8) and (2.11) constituting the Lax pair are compatible.*

Proof. Let $\lambda_t = 0$, then differentiating (2.8) with respect to t and (2.11) with respect to x gives

$$u_{xt} = i\lambda J\tilde{V}u + B_t u + Q\tilde{V}u, \tag{2.13}$$

$$u_{tx} = \tilde{V}_x u + i\lambda\tilde{V}Ju + \tilde{V}Qu. \tag{2.14}$$

By the supposition that the functions $c_{jt}(x,t)$ and $c_{jxx}(x,t)$, $j = 1,2$ are continuous, therefore the functions $u_{xt}(x,t)$ and $u_{tx}(x,t)$ are also continuous, and the eigenvalue equations (2.8) and (2.11) are compatible, i.e., $u_{xt} = u_{tx}$. From the last equality we obtain the Lax compatibility condition for systems (2.8) and (2.11):

$$B_t - \tilde{V}_x + B\tilde{V} - \tilde{V}B = 0. \tag{2.15}$$

It is easy to verify that the matrix equation (2.15) is equivalent to the system (2.5). Thus, the system of NLEEs (2.5) is represented by the Lax compatibility condition of the eigenvalue equation (2.8) and the equation (2.11) governing the time-evolution of eigenfunctions of (2.8).

Conversely, if the potentials q_1 and q_2 satisfy the system (2.5), then $\lambda_t = 0$ and the eigenvalue equations (2.8) and (2.11) are compatible. Indeed, let $\lambda = \lambda_0$ be a point of the spectrum of $L(t)$ for $t = t_0$ and let $u(x, \lambda_0; t_0)$ be a corresponding eigenfunction. We define the function $r(x, \lambda_0; t)$ by

$$r(x, \lambda_0; t) = \frac{\partial}{\partial x} u(x, \lambda_0; t) - Bu(x, \lambda_0; t). \tag{2.16}$$

Differentiating (2.16) with respect to t, and then taking (2.11) into account gives

$$\frac{\partial r}{\partial t} = \frac{\partial \tilde{V}}{\partial x} u + \tilde{V} \frac{\partial u}{\partial x} - \frac{\partial B}{\partial t} u - B\tilde{V}u,$$

with the initial condition: $r(x, \lambda_0; t_0) = 0$.

Using (2.16) to get rid of $\frac{\partial u}{\partial x}$ in the last equality, we obtain

$$\frac{\partial r}{\partial t} - \tilde{V}r = (\tilde{V}B - B\tilde{V} + \tilde{V}_x - B_t)u,$$

where the matrices B and $\tilde{V}$ satisfy the matrix equation (2.15).

Since for each x the elements of $\tilde{V}$ are continuous functions of t, then the problem for equation (2.11) is uniquely solvable on the whole time axis [26]. Owing to this fact and (2.15) the problem for the homogeneous equation:

$$\frac{\partial r}{\partial t} - \tilde{V}r = 0,$$

with the initial condition: $r(x, \lambda_0; t_0) = 0$, has only the trivial solution $r(x, \lambda_0; t) = 0$ for all $t \in (-\infty, \infty)$. Hence, $\lambda = \lambda_0$ belongs to the spectrum for all values of time t and $u(x, \lambda_0; t)$ is the corresponding eigenfunction. Thus, the spectrum of the spectrum of the operator $L(t)$ is invariant with t, i.e., $\lambda_t = 0$. Using the invariance with t of the operator $L(t)$, from (2.8) and (2.11) we find (2.13) and (2.14), respectively. Since the potentials q_1 and q_2 satisfy the system (2.5), that is equivalent to the matrix equation (2.15), whence it follows that $u_{xt} = u_{tx}$; i.e., the equations (2.8) and (2.11) are compatible. This completes the proof. □

2.1.2 The time-dependence of the scattering function

According to (1.35), the time-dependent solution of system (2.8) with the initial condition:

$$\varphi_1(0, \lambda; t) = \varphi_2(0, \lambda; t) = 1 \text{ for any } t \in (-\infty, \infty), \tag{2.17}$$

is presented in the form:

$$\varphi(x,\lambda;t) = D(\lambda;t)e^{(1)}(x,\lambda;t) + N(\lambda;t)e^{(2)}(x,\lambda;t), \tag{2.18}$$

where $e^{(1)}(x,\lambda;t)$ and $e^{(2)}(x,\lambda;t)$ are the Jost solutions of the problem for system (2.8) with the boundary condition at infinity:

$$e^{(1)}(x,\lambda;t) = \begin{pmatrix} e^{-i\lambda x} + o(1) \\ o(1) \end{pmatrix}, \quad e^{(2)}(x,\lambda;t) = \begin{pmatrix} o(1) \\ e^{i\lambda x} + o(1) \end{pmatrix} \text{ as } x \to \infty, \tag{2.19}$$

$$N(\lambda;t) = e_1^{(1)}(0,\lambda;t) - e_2^{(1)}(0,\lambda;t), \quad D(\lambda;t) = e_2^{(2)}(0,\lambda;t) - e_1^{(2)}(0,\lambda;t),$$
$$N(\lambda;0) = N(\lambda), \quad D(\lambda;0) = D(\lambda).$$

We call

$$S(\lambda;t) = \frac{N(\lambda;t)}{D(\lambda;t)} \text{ for } |\text{Im}\lambda| < \varepsilon_0, \ -\infty < t < \infty, \tag{2.20}$$

the scattering function for the scattering problem (2.8)–(2.9), where $S(\lambda;0) = S(\lambda)$.

From (2.19) we see that the asymptotic behavior of the Jost solutions $e^{(1)}(x,\lambda;t)$ and $e^{(2)}(x,\lambda;t)$ at infinity of t does not depend on time, i.e.,

$$\lim_{x\to\infty} e_{jt}^{(1)}(x,\lambda;t) = 0, \quad \lim_{x\to\infty} e_{jt}^{(2)}(x,\lambda;t) = 0, \ j = 1,2. \tag{2.21}$$

The vector functions $D(\lambda;t)e^{(1)}(x,\lambda;t)$ and $N(\lambda;t)e^{(2)}(x,\lambda;t)$ are solutions of the system (2.8). The time-dependence of these solutions obeys evolutionary equation (2.11):

$$[D(\lambda;t)e^{(1)}(x,\lambda;t)]_t = \tilde{V}[D(\lambda;t)e^{(1)}(x,\lambda;t)], \tag{2.22}$$

$$[N(\lambda;t)e^{(2)}(x,\lambda;t)]_t = \tilde{V}[N(\lambda;t)e^{(2)}(x,\lambda;t)]. \tag{2.23}$$

In (2.22) and (2.23) by letting x tend to infinity and using (2.10), (2.12) and (2.21), we obtain

$$D_t(\lambda;t) = -2i\lambda^2 D(\lambda;t), \quad N_t(\lambda;t) = 2i\lambda^2 N(\lambda;t). \tag{2.24}$$

The following lemma is deduced from (2.24), Lemma 1.3 and the invariance with time of the spectrum of the operator $L(t)$.

Lemma 2.2. *Let the isospectral potentials q_1 and q_2 be continuous functions belonging to the class P and satisfying estimate (2.10). Then the scattering function $S(\lambda;t)$ for the scattering problem (2.8)–(2.9) associated with the IBVP (2.5)–(2.6)–(2.7) possesses the properties:*

1. *$S(\lambda;t)$ and $S^{-1}(\lambda;t)$ are functions of type (S) in the strip $|\text{Im}\lambda| < \varepsilon_0$ and are meromorphic in the strip $0 < |\text{Im}\lambda| < \varepsilon_0$, where the positive number ε_0 is determined by (1.23);*

2. *The scattering function* $S(\lambda;t)$ *admits the factorization* (2.20). *The numerator* $N(\lambda;t)$ *and the denominator* $D(\lambda;t)$ *of* (2.20) *satisfy the evolutionary equations:*

$$\begin{aligned} N(\lambda;t) &= N(\lambda)e^{2i\lambda^2 t}, \ \mathrm{Im}\lambda \le \varepsilon_0; \\ D(\lambda;t) &= D(\lambda)e^{-2i\lambda^2 t}, \ \mathrm{Im}\lambda \ge -\varepsilon_0. \end{aligned} \tag{2.25}$$

3. $S(\lambda;t)$ *obeys the evolutionary equation:*

$$S(\lambda;t) = S(\lambda)e^{4i\lambda^2 t} \ \textit{for} \ |\mathrm{Im}\lambda| < \varepsilon_0. \tag{2.26}$$

The scattering function $S(\lambda;t)$ gives the complete characterization of the operator $L(t)$ on the continuous spectrum.

Remark 2.1. Due to (1.38)–(1.40) the time-dependence of the standardized polynomials is determined by the unknown BVs $e^{(1)}(\lambda;t)$ and $e^{(2)}(\lambda;t)$ evaluated at $x=0$ of the Jost solutions $e^{(1)}(x,\lambda;t)$ and $e^{(2)}(x,\lambda;t)$ of the problem (2.8), (2.19). Therefore, the main difficulty in solving the IBVP (2.5)–(2.6)–(2.7) lies in the evaluation of the unknown BVs in terms of the given initial and boundary conditions (2.6) and (2.7). We deal with this difficulty in the next two subsections.

2.1.3 Evaluation of unknown BVs

In order to find the unknown BVs evaluated at $x=0$ of the Jost solutions, we initially derive evolutionary equations for the Jost solutions $e^{(1)}(x,\lambda;t)$ and $e^{(2)}(x,\lambda;t)$ of the problem (2.8), (2.19). Using (2.22), (2.23) and (2.24), we get the equalities:

$$\begin{aligned} D(\lambda;t)e_t^{(1)}(x,\lambda;t) &= [D(\lambda;t)e^{(1)}(x,\lambda;t)]_t - D_t(\lambda;t)e^{(1)}(x,\lambda;t) = \\ &\tilde{V}[D(\lambda;t)e^{(1)}(x,\lambda;t)] + 2i\lambda^2 D(\lambda;t)e^{(1)}(x,\lambda;t), \\ N(\lambda;t)e_t^{(2)}(x,\lambda;t) &= [N(\lambda;t)e^{(2)}(x,\lambda;t)]_t - N_t(\lambda;t)e^{(2)}(x,\lambda;t) = \\ &\tilde{V}[N(\lambda;t)e^{(2)}(x,\lambda;t)] - 2i\lambda^2 N(\lambda;t)e^{(2)}(x,\lambda;t). \end{aligned}$$

Instead of $\tilde{V}$ in these equalities, we substitute matrix (2.12) and derive the evolutionary equations:

$$\begin{aligned} e_{1t}^{(1)}(x,\lambda;t) =& -iq_1(x,t)q_2(x,t)e_1^{(1)}(x,\lambda;t) + \\ & [2\lambda q_1(x,t) + iq_{1x}(x,t)]e_2^{(1)}(x,\lambda;t), \end{aligned} \tag{2.27}$$

$$\begin{aligned} e_{2t}^{(1)}(x,\lambda;t) =& [2\lambda q_2(x,t) - iq_{2x}(x,t)]e_1^{(1)}(x,\lambda;t) + \\ & [4i\lambda^2 + iq_1(x,t)q_2(x,t)]e_2^{(1)}(x,\lambda;t), \end{aligned} \tag{2.28}$$

$$e_{1t}^{(2)}(x,\lambda;t) = -[4i\lambda^2 + iq_1(x,t)q_2(x,t)]e_1^{(2)}(x,\lambda;t)+ \\ [2\lambda q_1(x,t) + iq_{1x}(x,t)]e_2^{(2)}(x,\lambda;t), \qquad (2.29)$$

$$e_{2t}^{(2)}(x,\lambda;t) = [2\lambda q_2(x,t) - iq_{2x}(x,t)]e_1^{(2)}(x,\lambda;t)+ \\ iq_1(x,t)q_2(x,t)e_2^{(2)}(x,\lambda;t). \qquad (2.30)$$

Owing to estimate (2.10) and conditions (2.19), the right-hand sides of (2.27)–(2.30) tend to zero as $x \to \infty$ uniformly in t, i.e., the derived evolutionary equations (2.27)–(2.30) obey the required asymptotic condition (2.21) at infinity.

The solution (2.18) of the problem (2.8), (2.17) evolves according to the evolution equation (2.11):

$$\varphi_t(x,\lambda;t) = \tilde{V}[\varphi(x,y;t)]. \qquad (2.31)$$

From the initial condition (2.17) at $x = 0$ it follows that

$$\varphi_{1t}(0,\lambda;t) = \varphi_{2t}(0,\lambda;t) = 0. \qquad (2.32)$$

Using conditions (2.17), (2.32) and equations (2.27)–(2.30) at $x = 0$, from evolutionary equation (2.31) evaluated at $x = 0$, we obtain the important relations between the boundary values of functions $q_j(x,t)$ and their first x-derivatives $q_{jx}(x,t)$ at $x = 0$:

$$-2i\lambda^2 - iq_1(0,t)q_2(0,t) + 2\lambda q_1(0,t) + iq_{1x}(0,t) = 0, \qquad (2.33)$$

$$2i\lambda^2 + iq_1(0,t)q_2(0,t) + 2\lambda q_2(0,t) - iq_{2x}(0,t) = 0. \qquad (2.34)$$

Making use of relations (2.33) and (2.34) to get rid of $q_{jx}(0,t)$, $j = 1,2$ in derived equations (2.27)–(2.30) at $x = 0$, we obtain the systems of first-order ODEs for the unknown BVs of the Jost solutions $e^{(1)}(x,\lambda;t)$ and $e^{(2)}(x,\lambda;t)$ of system (2.8) evaluated at $x = 0$:

$$e_{1t}^{(1)}(0,\lambda;t) = -iq_1(0,t)q_2(0,t)e_1^{(1)}(0,\lambda;t)+ \\ [2i\lambda^2 + iq_1(0,t)q_2(0,t)]e_2^{(1)}(0,\lambda;t), \qquad (2.35)$$

$$e_{2t}^{(1)}(0,\lambda;t) = [4i\lambda^2 + iq_1(0,t)q_2(0,t)]e_2^{(1)}(0,\lambda;t)- \\ [2i\lambda^2 + iq_1(0,t)q_2(0,t)]e_1^{(1)}(0,\lambda;t), \qquad (2.36)$$

$$e_{1t}^{(2)}(0,\lambda;t) = -[4i\lambda^2 + iq_1(0,t)q_2(0,t)]e_1^{(2)}(0,\lambda;t)+ \\ [2i\lambda^2 + iq_1(0,t)q_2(0,t)]e_2^{(2)}(0,\lambda;t), \qquad (2.37)$$

$$e_{2t}^{(2)}(0,\lambda;t) = iq_1(0,t)q_2(0,t)e_2^{(2)}(0,\lambda;t) - [2i\lambda^2 + iq_1(0,t)q_2(0,t)]e_1^{(2)}(0,\lambda;t). \tag{2.38}$$

Using (2.25), we reduce evolution equations (2.35)–(2.38) for the unknown BVs to the form:

$$e_{nt}^{(1)}(0,\lambda;t) = 2i\lambda^2 e_n^{(1)}(0,\lambda;t) - iN(\lambda)[2\lambda^2 + q_1(0,t)q_2(0,t)]e^{2i\lambda^2 t},\ n=1,2, \tag{2.39}$$

$$e_{nt}^{(2)}(0,\lambda;t) = -2i\lambda^2 e_n^{(2)}(0,\lambda;t) + iD(\lambda)[2\lambda^2 + q_1(0,t)q_2(0,t)]e^{-2i\lambda^2 t},\ n=1,2. \tag{2.40}$$

The solutions of first-order ODEs (2.39) and (2.40) are found by the method of variation constants:

$$e_n^{(1)}(0,\lambda;t) = \left\{ e_n^{(1)}(0,\lambda;0) - iN(\lambda)\left[2\lambda^2 t + \int_0^t q_1(0,\tau)q_2(0,\tau)d\tau\right]\right\} e^{2i\lambda^2 t},\quad n=1,2, \tag{2.41}$$

$$e_n^{(2)}(0,\lambda;t) = \left\{ e_n^{(2)}(0,\lambda;0) + iD(\lambda)\left[2\lambda^2 t + \int_0^t q_1(0,\tau)q_2(0,\tau)d\tau\right]\right\} e^{-2i\lambda^2 t},\quad n=1,2. \tag{2.42}$$

In the right-hand sides of equalities (2.41) and (2.42) the second terms are the particular solutions of equations (2.39) and (2.40), while the first terms are the solutions of the corresponding homogeneous equations. Thus, the unknown BVs $e^{(1)}(0,\lambda;t)$ and $e^{(2)}(0,\lambda;t)$ are found by formulas (2.41) and (2.42) in terms of quantities: $e^{(1)}(0,\lambda;0)$, $e^{(2)}(0,\lambda;0)$, $N(\lambda)$, $D(\lambda)$ and $q_1(0,t)q_2(0,t)$, which are evaluated from known initial and boundary conditions (2.6) and (2.7).

Since $D(\lambda_k^+) = 0$, $\mathrm{Im}\lambda_k^+ \geq \varepsilon_o, k = 1, ..., \gamma_1$, and $N(\lambda_j^-) = 0$, $\mathrm{Im}\lambda_j^- \leq -\varepsilon_o, j = 1, ..., \gamma_2$, then from (2.41) and (2.42) it follows that the BVs at $x = 0$ and at non-real singular numbers of the Jost solutions of problem (2.8), (2.19) are calculated by the formulas:

$$e_n^{(2)}(0,\lambda_k^+;t) = e_n^{(2)}(0,\lambda_k^+;0)e^{-2i(\lambda_k^+)^2 t} \neq 0,\ k = 1,...,\gamma_1,\ \ n=1,2 \tag{2.43}$$

$$e_n^{(1)}(0,\lambda_j^-;t) = e_n^{(1)}(0,\lambda_j^-;0)e^{2i(\lambda_j^-)^2 t} \neq 0,\ j = 1,...,\gamma_2,\ \ n=1,2. \tag{2.44}$$

Hence, the calculated BVs (2.43) and (2.44) of the Jost solutions $e^{(1)}(x,\lambda;t)$ and $e^{(2)}(x,\lambda;t)$ at $x = 0$ and at singular numbers do not depend on the boundary conditions (2.7).

2.1.4 The time-dependence of the scattering data

Now knowing the BVs $e_2^{(2)}(0,\lambda;t)$ and $e_1^{(1)}(0,\lambda;t)$ defined by formulas (2.42) and (2.41), we use formulas (1.30) and (1.31) for calculating the time-dependent coefficients:

$$A_q(\lambda_k^+;t) = \left[\frac{d^q}{d\lambda^q}\frac{1}{e_2^{(2)}(0,\lambda;t)}\right]_{\lambda=\lambda_k^+}, \; q=0,1,...,m_k-1, \; k=1,...,\gamma_1, \tag{2.45}$$

$$B_p(\lambda_j^-;t) = \left[\frac{d^p}{d\lambda^p}\frac{1}{e_1^{(1)}(0,\lambda;t)}\right]_{\lambda=\lambda_j^-}, \; p=0,1,...,n_j-1, \; j=1,...,\gamma_2, \tag{2.46}$$

Substituting (2.45) and (2.46) into (1.38) and (1.39) respectively, we obtain the time-dependent standardized polynomials (1.40):

$$\begin{aligned} f_k^+(-x;t) &= p_k(x;t)e^{i\lambda_k^+ x}, \; k=1,...,\gamma_1, \\ g_j^-(x;t) &= \tilde{q}_j(x;t)e^{-i\lambda_j^- x}, \; j=1,...,\gamma_2. \end{aligned} \tag{2.47}$$

The functions (2.47) characterize the scattering problem (2.8)–(2.9) on the discrete spectrum. The scattering function (2.20) possessing the properties enumerated in Lemma 2.2 characterizes the problem on the continuous spectrum of this problem. Hence, according to formulas (1.41), we obtain the following formulas for the time-dependent scattering data of the scattering problem (2.8)–(2.9) on the complete spectrum:

$$f(-x;t) = f_S(-x;t) - \sum_{k=1}^{\gamma_1} f_k^+(-x;t), \;\; g(x;t) = g_S(x;t) - \sum_{j=1}^{\gamma_2} g_j^-(x;t), \tag{2.48}$$

where

$$\begin{aligned} f_S(x;t) &= \frac{1}{2\pi}\int_{-\infty+i\eta}^{\infty+i\eta}\left[S(\lambda)e^{4i\lambda^2 t}-1\right]e^{-i\lambda x}d\lambda, \\ g_S(x;t) &= \frac{1}{2\pi}\int_{-\infty-i\eta}^{\infty-i\eta}\left[S^{-1}(\lambda)e^{-4i\lambda^2 t}-1\right]e^{-i\lambda x}d\lambda. \end{aligned} \tag{2.49}$$

Thus, by Lemma 2.2 and formulas (2.41)–(2.49) the difficulty that was mentioned in Remark 2.1, is overcome. By the above discussion the following lemma is proved.

Lemma 2.3. *(1) The unknown BVs $e^{(1)}(0,\lambda;t)$ and $e^{(2)}(0,\lambda;t)$ of the Jost solutions of the problem* (2.8), (2.19) *obey evolutionary equations* (2.39) *and* (2.40), *respectively. The solutions of these equations, i.e., the BVs* *are found by explicit formulas* (2.41) *and* (2.42) *in terms of the quantities which are calculated from the known initial and boundary conditions* (2.6) *and* (2.7)*;*

(2) *The time-dependencies of the scattering data of the scattering problem* (2.8)–(2.9) *on the discrete spectrum and complete spectrum are determined by expressions* (2.45)–(2.46)–(2.47) *and* (2.48)–(2.49)*, respectively.*

2.1.5 The solution of the IBVP for the system of NLEEs (2.5)

We consider the IBVP (2.5)–(2.6)–(2.7) in the class of isospectral potentials, which are continuous functions belonging to the class P and satisfying estimate (2.10). In according with Definition 1.2 in Chapter 1, we construct a data set s given by (1.78) from known initial condition (2.6). These quantities satisfy conditions (1)–(5) enumerated in the Theorem 1.1 in Chapter 1. Then in accordance with Lemmas 2.2 and 2.3 a time-dependent set $s(t)$ is constructed from the set s and known initial and boundary conditions (2.6) and (2.7):

$$s(t) = \Big\{S(\lambda)e^{4i\lambda^2 t} \text{ for } |\mathrm{Im}\lambda| < \varepsilon_0;\ \lambda_k^+,\ \mathrm{Im}\lambda_k^+ \geq \varepsilon_0,\ p_k(x;t),\ k = 1, ..., \gamma_1;$$
$$\lambda_j^-,\ \mathrm{Im}\lambda_j^- \leq -\varepsilon_0,\ \tilde{q}_j(x;t),\ j = 1, ..., \gamma_2\Big\}, \tag{2.50}$$

where the positive number ε_0 is determined by (1.23), and for every $t \in (-\infty, \infty)$ the data set $s(t)$ satisfies conditions (1)–(5) enumerated in Theorem 1.1.

The Theorem 1.1 and Lemmas 2.2 and 2.3 reduce the problem of solving the IBVP (2.5)–(2.6)–(2.7) to that of solving the systems of fundamental equations of the form (1.42)–(1.43) and (1.44)–(1.45) for every $x \geq 0,\ \xi > 0$:

$$H_{11}(x, x+\xi; t) + \int_0^\infty H_{12}(x, x+\tau; t) f(-2x-\xi-\tau; t)d\tau = 0,$$

$$g(2x+\xi; t) + H_{12}(x, x+\xi; t) + \int_0^\infty H_{11}(x, x+\tau; t) g(2x+\xi+\tau; t)d\tau = 0, \tag{2.51}$$

$$f(-2x-\xi; t) + H_{21}(x, x+\xi; t) + \int_0^\infty H_{22}(x, x+\tau; t) f(-2x-\xi-\tau; t)d\tau = 0,$$

$$H_{22}(x, x+\xi; t) + \int_0^\infty H_{21}(x, x+\tau; t) g(2x+\xi+\tau; t)d\tau = 0, \tag{2.52}$$

where both x and t enter systems (2.51) and (2.52) merely as parameters, $f(-x;t)$ and $g(x;t)$ are known functions constructed by formulas (2.48) and (2.49) from given data set (2.50), $H_{jk}(x,x+\xi;t)$, $j,k=1,2$ are unknown functions of ξ for every $(x,t) \in [0,\infty) \times (-\infty,\infty)$.

The potentials q_1 and q_2 in system (2.8) belong to the class P, therefore from integral equations of the form (1.17)–(1.20) with the potentials q_1 and q_2 belonging to the class P, it follows that the solutions $H_{ij}(x,s;t)$, $0 \le x \le s$, $i,j=1,2$ of these integral equations also belong to this class. Further, using systems (2.51) and (2.52), we can verify that the functions $f(-2x-\xi;t)$ and $g(2x+\xi;t)$ in these systems possess properties, which are analogous with the properties of $H_{ij}(x,s;t), 0 \le x \le s, i,j=1,2$; i.e., these functions also belong to the class P of functions possessing the second partial continuous derivative with respect to x and the first partial continuous derivative with respect to t. Owing to the fifth condition of Theorem 1.1, the constructed functions $f(-x;t)$ and $g(x;t)$ satisfy the estimate:

$$|f(-x;t)| \le \tilde{C}\exp\left\{-\frac{1}{4}\varepsilon x\right\}, \quad |g(x;t)| \le \tilde{C}\exp\left\{-\frac{1}{4}\varepsilon x\right\}, \tag{2.53}$$

for any $t \in (-\infty,\infty)$ where $\tilde{C}$ is a constant, $\varepsilon > 0$.

Taking into account that the functions $f(-x;t)$ and $g(x;t)$ belong to the class P and satisfy estimate (2.53), and arguing as in the proof of Theorem 1.1 and Lemmas 2.2 and 2.3, we easily prove the following assertions:

(I). The systems (2.51) and (2.52) are Fredholm systems of integral equations in $L_1[0,\infty)$ and have in $L_1[0,\infty)$ for every $(x,t) \in [0,\infty) \times (-\infty,\infty)$ unique solutions: $(H_{11}(x,x+\xi;t), H_{12}(x,x+\xi;t))$ and $(H_{21}(x,x+\xi;t), H_{22}(x,x+\xi;t))$, respectively. These found solutions satisfy estimate of the type (2.10).

(II). The vector functions

$$\left(e_1^{(1)}(x,\lambda;t),\ e_2^{(1)}(x,\lambda;t)\right) = \left(e^{-i\lambda x} + e^{-i\lambda x}\int_0^\infty H_{11}(x,x+\tau;t)e^{-i\lambda\tau}d\tau, e^{-i\lambda x}\int_0^\infty H_{21}(x,x+\tau;t)e^{-i\lambda\tau}d\tau\right), \tag{2.54}$$

$$\left(e_1^{(2)}(x,\lambda;t),\ e_2^{(2)}(x,\lambda;t)\right) = \left(e^{i\lambda x}\int_0^\infty H_{12}(x,x+\tau;t)e^{i\lambda\tau}d\tau,\ e^{i\lambda x} + e^{i\lambda x}\int_0^\infty H_{22}(x,x+\tau;t)e^{i\lambda\tau}d\tau\right), \tag{2.55}$$

are Jost solutions of a system of the form (2.8) with the reconstructed potential matrix:

$$2i\begin{pmatrix} 0 & H_{12}(x,x;t) \\ -H_{21}(x,x;t) & 0 \end{pmatrix} = \begin{pmatrix} 0 & c_1(x,t) \\ c_2(x,t) & 0 \end{pmatrix},$$
$$(x,t)\in[0,\infty)\times(-\infty,\infty), \quad (2.56)$$

where the constructed potentials $c_1(x,t)$ and $c_2(x,t)$ are continuous functions belonging to the class P and satisfying estimate of the type (2.10);

(III). The quantities of the data set (2.50) coincide with the respective quantities of the scattering data set of the scattering problem for the system of the form (2.8) with the constructed potential matrix (2.56) and with the boundary condition (2.9). In fact, according to (2.19) and (2.20), the BVs of the Jost solutions (2.54) and (2.55) evaluated at $x = 0$ are related to the scattering function $S(\lambda;t)$ by the identity:

$$e_1^{(1)}(0,\lambda;t) - e_2^{(1)}(0,\lambda;t) - [e_2^{(2)}(0,\lambda;t) - e_1^{(2)}(0,\lambda;t)]S(\lambda;t) = 0$$
$$\text{for any } t\in(-\infty,\infty) \text{ and } |\mathrm{Im}\lambda| < \varepsilon_0,\ \varepsilon_0 > 0. \quad (2.57)$$

Owing to the Definition 1.1 and equations (2.25), the BVs of the Jost solutions at $x = 0$ and at non-real singular numbers, the numerator and denominatior of $S(\lambda;t)$ satisfy the following identities:

$$\left[\frac{d^p}{d\lambda^p}\left(e_1^{(1)}(0,\lambda;t) - e_2^{(1)}(0,\lambda;t)\right)\right]_{\lambda=\lambda_j^-} =$$
$$\left[\frac{d^p}{d\lambda^p}\left(e_1^{(1)}(0,\lambda;0) - e_2^{(1)}(0,\lambda;0)\right)e^{2i\lambda^2 t}\right]_{\lambda=\lambda_j^-} = 0,$$
$$\mathrm{Im}\lambda_j^- \le -\varepsilon_0,\ p = 0,1,...,n_j-1,\ j = 1,...,\gamma_2,$$
$$(2.58)$$
$$\left[\frac{d^q}{d\lambda^q}\left(e_2^{(2)}(0,\lambda;t) - e_1^{(2)}(0,\lambda;t)\right)\right]_{\lambda=\lambda_k^+} =$$
$$\left[\frac{d^q}{d\lambda^q}\left(e_2^{(2)}(0,\lambda;0) - e_1^{(2)}(0,\lambda;0)\right)e^{-2i\lambda^2 t}\right]_{\lambda=\lambda_k^+} = 0,$$
$$\mathrm{Im}\lambda_k^+ \ge \varepsilon_0,\ q = 0,1,...,m_k-1,\ k = 1,...,\gamma_1,$$

where m_k is the multiplicity of the zero λ_k^+ of the denominator and n_j is the multiplicity of the zero λ_j^- of the numerator of the scattering function $S(\lambda;t)$.

The identity (2.57) characterizes the problem (2.8)–(2.9) with the constructed potential matrix (2.56) on the continuous spectrum, while the identities (2.58) characterize this problem on the discrete spectrum;

(IV). A correspondence between the scattering data set (2.50) and the reconstructed potential matrix (2.56) is one-to-one, and the potential matrices $Q(x,t)$ and $\tilde{V}(x,t)$ in equations (2.8) and (2.11) are reconstructed uniquely from the potential matrix (2.56). This matrix is recovered in terms of solutions of systems (2.51) and (2.52). Due to condition (2.53) and the properties of the functions $f(-x;t)$ and $g(x;t)$, the solutions of the systems (2.51) and (2.52) are found uniquely from the given data set (2.50). By Lemmas 1.11 and 1.12 and properties of known functions $f(-2x-\xi;t)$ and $g(2x+\xi;t)$, the obtained solutions of systems (2.51) and (2.52) satisfy an estimate of the type (1.63). Therefore these solutions belong to the class P and satisfy systems of integral equations of the form (1.17)–(1.20) with the reconstructed potential matrix (2.56). Owing to the unique solvability of these integral equations, the solutions of the integral equations coincide with the solutions of systems (2.51) and (2.52). Hence, the obtained solutions have the first partial continuous derivative with respect to t and the second partial continuous derivative with respect to x. Due to this fact, the eigenvalue equations (2.8) and (2.11) with the matrices B and $\tilde{V}$ constructed from the matrix (2.56) are compatible. Then by Lemma 2.1 the constructed potentials (2.56) are isospectral and are the solution of the IBVP (2.5)–(2.6)–(2.7).

Thus, the IBVP (2.5)–(2.6)–(2.7) is solved.

Theorem 2.1. *The IBVP* (2.5)–(2.6)–(2.7) *can be solved by the ISM in the class of isospectral potentials which are continuous functions belonging to the class P and satisfying the estimate* (2.10). *Namely,*

(1) The quantities of the data set (2.50) *constructed from known initial and boundary conditions* (2.6) *and* (2.7) *are the scattering data of the associated problem* (2.8)–(2.9) *with reconstructed isospectral potentials* (2.56). *The time-dependence of this set is determined by Lemmas 2.2 and 2.3;*

(2) Every scattering data set of the type (2.50) *corresponds to a unique solution of the IBVP* (2.5)–(2.6)–(2.7). *The solution of this IBVP is found by formula* (2.56) *and is expressible in terms of the solutions of systems of fundamental equations* (2.51) *and* (2.52), *i.e., of the kernels of the transformation operator. The solutions of these systems are constructed from the data set* (2.50).

2.1.6 The IBVP for the attractive NLS equation

We now turn our attention to a case of special interest of the system (2.5):

$$c_1(x,t) = -\overline{c_2(x,t)} = c(x,t). \tag{2.59}$$

Under reduction (2.59), the system (2.5) reduces to the attractive NLS equation:

$$ic_t(x,t) = c_{xx}(x,t) + 2|c(x,t)|^2 c(x,t), \quad (x,t) \in [0,\infty) \times (-\infty,\infty). \tag{2.60}$$

We consider the IBVP for the attractive NLS equation (2.60) with the given initial and boundary conditions:

$$c(x,0) = c(x), \quad 0 \le x < \infty, \tag{2.61}$$
$$c(0,t) = f(t), \quad -\infty < t < \infty, \tag{2.62}$$

where $c(x)$ possesses the continuous derivative of the second-order and satisfies estimate (2.10), and $f(t)$ possesses the continuous derivative of the first-order. In addition, the initial value (2.61) is supposed to be consistent with boundary condition (2.62) at the corner point $(0,0)$:

$$\begin{aligned} &c(0) = f(0), \\ &if_t(0) = c_{xx}(0) + 2\left|c(0)\right|^2 c(0). \end{aligned}$$

The considered IBVP (2.60)–(2.61)–(2.62) is associated with the scattering problem (2.8)–(2.9) with the isospectral potential matrix:

$$Q(x,t) = \begin{pmatrix} 0 & ic(x,t) \\ -i\,\overline{c(x,t)} & 0 \end{pmatrix}, \tag{2.63}$$

where the potential $c(x,t)$ is a continuous function belonging to the class P and satisfying estimate of the type (2.10).

We prove that under reduction (2.59) the solution of the IBVP for the attractive NLS equation (2.60) can be obtained from the solution of the IBVP for system (2.5)–(2.6)–(2.7). Indeed, by Theorem 2.1 the solution of the IBVP (2.5)–(2.6)–(2.7) is constructed from the solution of the systems of fundamental equations (2.51) and (2.52). By Lemma 1.12 the solutions of these systems satisfy the integral equations of the form (1.17)–(1.20) with the reconstructed isospectral potentials (2.56) for the kernels $H_{jk}(x, x+\xi; t)$, $x \ge 0$, $\xi > 0$, $j,k = 1,2$ of the transformation operator of the scattering problem (2.8)–(2.9). Under reduction (2.59), from equations (1.17)–(1.20) we obtain the following relations for any $t \in (-\infty,\infty)$:

$$\begin{aligned} H_{11}(x,x+\xi;t) &= \overline{H_{22}(x,x+\xi;t)}, \\ H_{12}(x,x;t) &= -\overline{H_{21}(x,x;t)}, \end{aligned} \qquad x \ge 0,\ \xi > 0. \tag{2.64}$$

The problem of solving the IBVP (2.60)–(2.61)–(2.62) is reduced to that of solving the system of fundamental equations of the form (2.51), in which the known functions $f(-x;t)$ and $g(x;t)$ belong to the class P and satisfy estimate (2.53). In system (2.51) these functions are known functions constructed from the data set (2.50), which is calculated from given initial and boundary conditions (2.61) and (2.62). The solution of system (2.51) satisfies the integral equations of the form (1.17) and (1.19) with the potential constructed from the solution of the system of the form (2.51) by formulas (1.12) and (2.64):

$$c(x,t) = 2iH_{12}(x,x;t) = -2i\,\overline{H_{21}(x,x;t)}. \tag{2.65}$$

Owing to the unique solvability of integral equations (1.17) and (1.19) the found potential (2.65) belongs to the class P and satisfies estimate (2.10). By Lemma 2.1 the found potential (2.65) is the solution of the IBVP (2.60)–(2.61)–(2.62).

Thus, the following theorem is a corollary of Theorem 2.1.

Theorem 2.2. *Under condition* (2.59), *the IBVP for the attractive NLS equation* (2.60) *with known initial and boundary conditions* (2.61) *and* (2.62) *is a special case of the IBVP* (2.5)–(2.6)–(2.7). *The IBVP* (2.60)–(2.61)–(2.62) *can be solved by the ISM in the class of isospectral potentials which are continuous functions belonging to the class P and satisfying estimate (2.10). Namely, every scattering data set of the type (2.50) of the associated scattering problem generated by systems* (2.8) *and* (2.9) *with the reconstructed potential* (2.65) *corresponds to a unique solution of the IBVP* (2.60)–(2.61)–(2.62). *This solution is presented by the formula* (2.65) *in terms of the solution of the system of fundamental equations* (2.51). *The known functions* $f(-2x-\xi;t)$ *and* $g(2x+\xi;t)$ *in system (2.51) are found from the data set (2.50). This set is constructed from the known initial and boundary conditions* (2.61) *and* (2.62).

Remark 2.2. The potentials $c_1(x,t)$ and $c_2(x,t)$ in the system of NLEEs (2.1) depend on an additional parameter t (time) and satisfy constraint (2.4). The initial point is $t = 0$, therefore, it is more natural to confine the t-values belonging to $(-\infty,\infty)$ to values $t \in [0,\infty)$. Due to this fact, the IBVPs (2.5)–(2.6)–(2.7) and (2.60)–(2.61)–(2.62) can be solved for the region $(x,t) \in [0,\infty) \times [0,\infty)$.

There exists a remarkable class of potentials, for which the ISP can be solved exactly. These potentials are non-scattering potentials, for which the scattering function $S(\lambda;t)$ is identically equal to the identity for all λ in the strip $|\mathrm{Im}\lambda| < \varepsilon_0$, where ε_0 is a positive number. In this case the functions (2.48) and (2.49) are reduced to the form [133]:

$$\begin{gathered} S(\lambda;t) \equiv 1 \ \text{ for } \ |\mathrm{Im}\lambda| < \varepsilon_0,\ \varepsilon_0 > 0,\ f_S(-x;t) = 0,\ \ g_S(x;t) = 0, \\ f(-x;t) = -\sum_{k=1}^{\gamma_1} f_k^+(-x;t),\ \ g(x;t) = -\sum_{j=1}^{\gamma_2} g_j^-(x;t). \end{gathered} \tag{2.66}$$

Our definition of non-scattering potentials is similar to the definition of reflectionless potentials for which the reflection coefficient is identically zero [156].

2.2 Exact solutions of the system of NLEEs

2.2.1 Exact solutions of fundamental equations

In this subsection we use formulas (2.66) with t merely as a parameter, so we omit it for convenience. Due to (2.66) the non-scattering data of the scattering problem (2.2)–(2.3) with the non-scattering potential matrix are completely determined by the discrete spectrum and can be formulated in the form:

$$s = \Big\{ S(\lambda) \equiv 1 \text{ for } |\mathrm{Im}\lambda| < \varepsilon_0;\ \lambda_k^+ \geq \varepsilon_0,\ p_k(x),\ k = 1, ..., \gamma_1; \\ \lambda_j^-,\ \mathrm{Im}\lambda_j^- \leq -\varepsilon_0,\ \varepsilon_0 > 0,\ \tilde{q}_j(x),\ j = 1, ..., \gamma_2 \Big\} \tag{2.67}$$

Under conditions (2.66), the systems of fundamental equations (2.51) and (2.52) can be solved exactly. They readily yield to solutions in explicit form, because they are systems of equations with degenerated kernels. The solutions $H_{jk}(x,u),\ 0 \leq x \leq u,\ j,k = 1,2$ are to be sought in the form:

$$H_{n1}(x,u) = \sum_{k=1}^{\gamma_1} H_{n1,k}(x) e^{i\lambda_k^+ u},\quad H_{n2}(x,u) = \sum_{j=1}^{\gamma_2} H_{n2,j}(x) e^{-i\lambda_j^- u},\quad n = 1,2. \tag{2.68}$$

Hence, in the class of non-scattering potentials the systems (2.51) and (2.52) are systems of linear algebraic equations for $H_{n_1,k}(x)$ and $H_{n_2,j}(x)$:

$$H_{11,k}(x) - \sum_{j=1}^{\gamma_2} H_{12,j}(x) \int_x^\infty p_k(u+\tau) e^{i(\lambda_k^+ - \lambda_j^-)\tau} d\tau = 0, \\ k = 1, ..., \gamma_1, 0 \leq x < \infty, \tag{2.69}$$

$$-\tilde{q}_j(x+u) e^{-i\lambda_j^- x} + H_{12,j}(x) - \sum_{k=1}^{\gamma_1} H_{11,k}(x) \int_x^\infty \tilde{q}_j(u+\tau) e^{i(\lambda_k^+ - \lambda_j^-)\tau} d\tau = 0, \\ j = 1, ..., \gamma_2, \tag{2.70}$$

$$H_{22,j}(x) - \sum_{k=1}^{\gamma_1} H_{21,k}(x) \int_x^\infty \tilde{q}_j(u+\tau) e^{i(\lambda_k^+ - \lambda_j^-)\tau} d\tau = 0,\ j = 1, ..., \gamma_2, \\ 0 \leq x < \infty, \tag{2.71}$$

$$-p_k(x+u)e^{i\lambda_k^+ x} + H_{21,k}(x) - \sum_{j=1}^{\gamma_2} H_{22,j}(x)\int_x^{\infty} p_k(u+\tau)e^{i(\lambda_k^+ - \lambda_j^-)\tau}d\tau = 0,$$

$$k = 1, ..., \gamma_1. \quad (2.72)$$

For simplicity we suppose that the given non-scattering data set (2.67) consists of only one pair of discrete spectra λ^+, $\mathrm{Im}\lambda^+ \geq \varepsilon_0$ and λ^-, $\mathrm{Im}\lambda^- \leq -\varepsilon_0$, $\varepsilon_0 > 0$, which are simple poles. We use formulas (1.38) and (1.39) for calculating the functions $f^+(-x)$ and $g^-(x)$ for this case [114]:

$$f^+(-x) = i\frac{e^{i\lambda^+ x}}{D_\lambda(\lambda^+)e_2^{(2)}(0,\lambda^+)}, \quad g^-(x) = i\frac{e^{-i\lambda^- x}}{N_\lambda(\lambda^-)e_1^{(1)}(0,\lambda^-)}, \quad 0 \leq x < \infty, \tag{2.73}$$

where $D(\lambda^+;0) = D(\lambda^+) = e_2^{(2)}(0,\lambda^+) - e_1^{(2)}(0,\lambda^+) = 0, D_\lambda(\lambda^+) \neq 0$, $N(\lambda^-;0) = N(\lambda^-) = e_1^{(1)}(0,\lambda^-) - e_2^{(1)}(0,\lambda^-) = 0, N_\lambda(\lambda^-) \neq 0$, due to (1.26), (1.27) $e_1^{(1)}(0,\lambda^-) \neq 0$, $e_2^{(2)}(0,\lambda^+) \neq 0$.

Hence, in this case the standardized polynomials are reduced to the standardized multipliers: $i\left(D_\lambda(\lambda^+)e_2^{(2)}(0,\lambda^+)\right)^{-1}$ and $i\left(N_\lambda(\lambda^-)e_1^{(1)}(0,\lambda^-)\right)^{-1}$.

Using (2.73), we write systems (2.69)–(2.70) and (2.71)–(2.72) for this case:

$$H_{11}(x) + \frac{e^{i(\lambda^+ - \lambda^-)x}}{(\lambda^+ - \lambda^-)e_2^{(2)}(0,\lambda^+)D_\lambda(\lambda^+)}H_{12}(x) = 0, \tag{2.74}$$

$$H_{12}(x) + \frac{e^{i(\lambda^+ - \lambda^-)x}}{(\lambda^+ - \lambda^-)e_1^{(1)}(0,\lambda^-)N_\lambda(\lambda^-)}H_{11}(x) = i\frac{e^{-i\lambda^- x}}{N_\lambda(\lambda^-)e_1^{(1)}(0,\lambda^-)}, \tag{2.75}$$

$$H_{22}(x) + \frac{e^{i(\lambda^+ - \lambda^-)x}}{(\lambda^+ - \lambda^-)e_1^{(1)}(0,\lambda^-)N_\lambda(\lambda^-)}H_{21}(x) = 0, \tag{2.76}$$

$$H_{21}(x) + \frac{e^{i(\lambda^+ - \lambda^-)x}}{(\lambda^+ - \lambda^-)e_2^{(2)}(0,\lambda^+)D_\lambda(\lambda^+)}H_{22}(x) = i\frac{e^{i\lambda^+ x}}{D_\lambda(\lambda^+)e_2^{(2)}(0,\lambda^+)}, \tag{2.77}$$

with $0 \leq x < \infty$.

Since the potential matrix $C(x)$ is determined by the formula (1.12), we calculate $H_{12}(x)$ and $H_{21}(x)$ from the systems (2.74)–(2.75) and (2.76)–(2.77), and then with the help of the formula (2.68) we find the non-scattering potentials

$c_1(x)$, $c_2(x)$, $0 \le x < \infty$:

$$c_1(x) = 2iH_{12}(x,x) = 2iH_{12}(x)e^{-i\lambda^- x} =$$
$$-2\frac{(\lambda^+ - \lambda^-)^2 e_2^{(2)}(0,\lambda^+)D_\lambda(\lambda^+)e^{-2i\lambda^- x}}{(\lambda^+ - \lambda^-)^2 e_1^{(1)}(0,\lambda^-)e_2^{(2)}(0,\lambda^+)N_\lambda(\lambda^-)D_\lambda(\lambda^+) - e^{2i(\lambda^+ - \lambda^-)x}}, \quad (2.78)$$

$$c_2(x) = -2iH_{21}(x,x) = -2iH_{21}(x)e^{i\lambda^+ x} =$$
$$2\frac{(\lambda^+ - \lambda^-)^2 e_1^{(1)}(0,\lambda^-)N_\lambda(\lambda^-)e^{2i\lambda^+ x}}{(\lambda^+ - \lambda^-)^2 e_1^{(1)}(0,\lambda^-)e_2^{(2)}(0,\lambda^+)N_\lambda(\lambda^-)D_\lambda(\lambda^+) - e^{2i(\lambda^+ - \lambda^-)x}}. \quad (2.79)$$

2.2.2 The time-dependence of standardized multipliers and an exact solution of system (2.5)

The time-evolution of non-scattering potentials (2.78), (2.79) is obtained with the help of the substitution:

$$e_1^{(1)}(0,\lambda^-) \to e_1^{(1)}(0,\lambda^-;t), \quad e_2^{(2)}(0,\lambda^+) \to e_2^{(2)}(0,\lambda^+;t),$$
$$N_\lambda(\lambda^-) \to N_\lambda(\lambda^-;t), \quad D_\lambda(\lambda^+) \to D_\lambda(\lambda^+;t). \quad (2.80)$$

In order to derive the time-dependence of the non-scattering potentials (2.78), (2.79), we need to derive the time-dependence of the standardized multipliers: $i\left(e_1^{(1)}(0,\lambda^-;t)N_\lambda(\lambda^-;t)\right)^{-1}$ and $i\left(e_2^{(2)}(0,\lambda^+;t)D_\lambda(\lambda^+;t)\right)^{-1}$.
Suppose that for any time t the function $D(\lambda;t)$ has only one simple zero λ^+, $\mathrm{Im}\lambda^+ \ge \varepsilon_0$ and the function $N(\lambda;t)$ has only one simple zero λ^-, $\mathrm{Im}\lambda^- \le -\varepsilon_0$, $\varepsilon_0 > 0$, then by Lemma 1.4, these simple zeros are simple non-real eigenvalues of the problem (2.2)–(2.3) and:

$$e_1^{(1)}(0,\lambda^-;t) = e_2^{(1)}(0,\lambda^-;t), \quad e_1^{(2)}(0,\lambda^+;t) = e_2^{(2)}(0,\lambda^+;t). \quad (2.81)$$

In this considered case the following quantities:

$$S(\lambda;t) \equiv 1 \text{ for } |\mathrm{Im}\lambda| < \epsilon_0, \ \epsilon_0 > 0,$$
$$\lambda^+, \ \mathrm{Im}\lambda^+ \ge \varepsilon_0, \ \lambda^-, \ \mathrm{Im}\lambda^- \le -\varepsilon_0,$$
$$i\left(e_1^{(1)}(0,\lambda^-;t)N_\lambda(\lambda^-;t)\right)^{-1}, \ i\left(e_2^{(2)}(0,\lambda^+;t)D_\lambda(\lambda^+;t)\right)^{-1}$$

are the non-scattering data for recovering the non-scattering potentials $c_1(x,t)$ and $c_2(x,t)$.

Taking (2.81) into account, from the evolution equation (2.11), we obtain the time-evolution of $e_1^{(1)}(0,\lambda^-;t)$ and of $e_2^{(2)}(0,\lambda^+;t)$:

$$e_1^{(1)}(0,\lambda^-;t) = e_1^{(1)}(0,\lambda^-;0)\exp\left\{V(\lambda^-;t)\right\}, \quad (2.82)$$
$$e_2^{(2)}(0,\lambda^+;t) = e_2^{(2)}(0,\lambda^+;0)\exp\left\{V(\lambda^+;t)\right\}, \quad (2.83)$$

where the formula (2.12) and condition (2.81) imply the following equalities:

$$V(\lambda^-;t) = 2i(\lambda^-)^2 t - 2\lambda^- \int_0^t q_1(0,\tau)d\tau + i\int_0^t q_1(0,\tau)q_2(0,\tau)d\tau - i\int_0^t q_{1x}(0,\tau)d\tau = -2i(\lambda^-)^2 t - 2\lambda^- \int_0^t q_2(0,\tau)d\tau - i\int_0^t q_1(0,\tau)q_2(0,\tau)d\tau + i\int_0^t q_{2x}(0,\tau)d\tau, \tag{2.84}$$

$$V(\lambda^+;t) = 2i(\lambda^+)^2 t - 2\lambda^+ \int_0^t q_1(0,\tau)d\tau + i\int_0^t q_1(0,\tau)q_2(0,\tau)d\tau - i\int_0^t q_{1x}(0,\tau)d\tau = -2i(\lambda^+)^2 t - 2\lambda^+ \int_0^t q_2(0,\tau)d\tau - i\int_0^t q_1(0,\tau)q_2(0,\tau)d\tau + i\int_0^t q_{2x}(0,\tau)d\tau. \tag{2.85}$$

The equalities (2.84) and (2.85) are consistent conditions of the matrix $\tilde{V} = \tilde{V}(x,\lambda;t)$ at the origin $x = 0$ and the singular numbers of the problem (2.2)–(2.3).

After substitution (2.80) into (2.78) and (2.79), we obtain the time-dependent potentials:

$$c_1(x,t) = \frac{2}{(\lambda^+ - \lambda^-)^{-2}\alpha - \beta^{-1}}, \quad c_2(x,t) = \frac{2}{\alpha^{-1} - (\lambda^+ - \lambda^-)^{-2}\beta}, \qquad 0 \le x < \infty \tag{2.86}$$

where

$$\alpha = \left(e_{20}^{(2)} D_\lambda\right)^{-1} e^{2i\lambda^+ x - V^+}, \quad \beta = \left(e_{10}^{(1)} N_\lambda\right)^{-1} e^{-2i\lambda^- x - V^-},$$
$$e_{20}^{(2)} = e_2^{(2)}(0,\lambda^+;0) \ne 0, \quad e_{10}^{(1)} = e_1^{(1)}(0,\lambda^-;0) \ne 0, \quad D_\lambda = D_\lambda(\lambda^+;t) \ne 0, \tag{2.87}$$
$$N_\lambda = N_\lambda(\lambda^-;t) \ne 0, \quad V^+ = V^+(\lambda^+;t) \ \mathrm{Im}\lambda^+ \ge \varepsilon_0,$$
$$V^- = V^-(\lambda^-;t) \ \mathrm{Im}\lambda^- \le -\varepsilon_0, \quad \varepsilon_0 > 0.$$

Now we find the time-dependence of the functions $D_\lambda(\lambda;t)$ and $N_\lambda(\lambda;t)$ at the singular numbers. After substitution of (2.86) into the system of NLEEs (2.5), we obtain the following relations:

$$[(\lambda^+-\lambda^-)^{-2}\alpha-\beta^{-1}]\left\{(\lambda^+-\lambda^-)^{-2}\alpha\left(4(\lambda^+)^2+i\dot{V}^++i\frac{\partial}{\partial t}\ln e_{20}^{(2)}D_\lambda\right)\right.$$
$$\left.+\beta^{-1}\left(i\dot{V}^-+i\frac{\partial}{\partial t}\ln e_{10}^{1)}N_\lambda-4(\lambda^-)^2\right)\right\}=0, \qquad (2.88)$$

$$[(\lambda^+-\lambda^-)^{-2}\beta-\alpha^{-1}]\left\{(\lambda^+-\lambda^-)^{-2}\beta\left(4(\lambda^-)^2-i\dot{V}^--i\frac{\partial}{\partial t}\ln e_{10}^{(1)}N_\lambda\right)\right.$$
$$\left.-\alpha^{-1}\left(4(\lambda^+)^2+i\dot{V}^++i\frac{\partial}{\partial t}\ln e_{20}^{(2)}D_\lambda\right)\right\}=0, \qquad (2.89)$$

where $\dot{V}^+$ and $\dot{V}^-$ are the derivatives of $V^+(\lambda^+;t)$ and of $V^-(\lambda^-;t)$ with respect to t, respectively.

In view of (2.84), (2.85) and (2.87), we have for any t:

$$(\lambda^+-\lambda^-)^{-2}\alpha\beta=(\lambda^+-\lambda^-)^{-2}\left(e_{10}^{(1)}N_\lambda\right)^{-1}\left(e_{20}^{(2)}D_\lambda\right)^{-1}\times$$
$$e^{2i(\lambda^+-\lambda^-)x}e^{-V(\lambda^+;t)-V(\lambda^-;t)}\neq 1,$$

whence it follows that for any t:

$$(\lambda^+-\lambda^-)^{-2}\alpha\neq\beta^{-1}\ \text{ or }\ (\lambda^+-\lambda^-)^{-2}\beta\neq\alpha^{-1}. \qquad (2.90)$$

Due to (2.90), from (2.88) and (2.89) it follows that

$$(\lambda^+-\lambda^-)^{-2}\alpha\beta\left\{4(\lambda^+)^2+i\dot{V}^++i\frac{\partial}{\partial t}\ln e_{20}^{(2)}D_\lambda\right\}=$$
$$4(\lambda^-)^2-i\dot{V}^--i\frac{\partial}{\partial t}\ln e_{10}^{(1)}N_\lambda, \qquad (2.91)$$

$$(\lambda^+-\lambda^-)^{-2}\alpha\beta\left\{4(\lambda^-)^2-i\dot{V}^--i\frac{\partial}{\partial t}\ln e_{10}^{(1)}N_\lambda\right\}=$$
$$4(\lambda^+)^2+i\dot{V}^++i\frac{\partial}{\partial t}\ln e_{20}^{(2)}D_\lambda. \qquad (2.92)$$

Substituting the left-hand side of (2.92)(and of (2.91), respectively) in the left-hand side of (2.91) (and of (2.92), respectively), we obtain

$$(\lambda^+ - \lambda^-)^{-4}\alpha^2\beta^2 \left\{4(\lambda^-)^2 - i\dot{V}^- - i\frac{\partial}{\partial t}\ln e_{10}^{(1)} N_\lambda\right\} =$$
$$4(\lambda^-)^2 - i\dot{V}^- - i\frac{\partial}{\partial t}\ln e_{10}^{(1)} N_\lambda, \tag{2.93}$$
$$(\lambda^+ - \lambda^-)^{-4}\alpha^2\beta^2 \left\{4(\lambda^+)^2 + i\dot{V}^+ + i\frac{\partial}{\partial t}\ln e_{20}^{(2)} D_\lambda\right\} =$$
$$4(\lambda^+)^2 + i\dot{V}^+ + i\frac{\partial}{\partial t}\ln e_{20}^{(2)} D_\lambda. \tag{2.94}$$

Owing to (2.90), the equalities (2.93), (2.94) are fulfilled if and only if

$$-i\frac{\partial}{\partial t}\ln e_{10}^{(1)} N_\lambda = i\dot{V}^- - 4(\lambda^-)^2, \quad i\frac{\partial}{\partial t}\ln e_{20}^{(2)} D_\lambda = -4(\lambda^+)^2 - i\dot{V}^+,$$

whence it follows that $N_\lambda(\lambda^-;t)$ and $D_\lambda(\lambda^+;t)$ satisfy the evolution equation:

$$N_\lambda(\lambda^-;t) = N_\lambda(\lambda^-;0)e^{-4i(\lambda^-)^2 t - V(\lambda^-;t)}, \tag{2.95}$$
$$D_\lambda(\lambda^+;t) = D_\lambda(\lambda^+;0)e^{4i(\lambda^+)^2 t - V(\lambda^+;t)}. \tag{2.96}$$

From (2.82), (2.83), (2.95) and (2.96) it follows that the time-dependence of the standardized multipliers is defined by the formulas:

$$i\left(e_1^{(1)}(0,\lambda^-;t)N_\lambda(\lambda^-;t)\right)^{-1} = i\left(e_1^{(1)}(0,\lambda^-;0)N_\lambda(\lambda^-;0)\right)^{-1} e^{4i(\lambda^-)^2 t}, \tag{2.97}$$
$$i\left(e_2^{(2)}(0,\lambda^+;t)D_\lambda(\lambda^+;t)\right)^{-1} = i\left(e_2^{(2)}(0,\lambda^+;0)D_\lambda(\lambda^+;0)\right)^{-1} e^{-4i(\lambda^+)^2 t}. \tag{2.98}$$

If the vector solution $(\varphi_1(x,\lambda;t), \varphi_2(x,\lambda;t))$ of the problem (2.8), (2.17) satisfies the condition:

$$\frac{d}{dt}\left(\lim_{x\to\infty}\varphi_1(x,\lambda;t)e^{i\lambda x}\right) = \frac{d}{dt}\left(\lim_{x\to\infty}\varphi_2(x,\lambda;t)e^{-i\lambda x}\right)$$
$$\text{for } |\mathrm{Im}\lambda| < \varepsilon_0 \text{ and any } t \in (-\infty,\infty), \tag{2.99}$$

then using (2.18) and (2.19), from (2.99) we have the equality:

$$\frac{d}{dt}D(\lambda;t) = \frac{d}{dt}N(\lambda;t) \text{ for } |\mathrm{Im}\lambda| < \varepsilon_0 \text{ and any } t \in (-\infty,\infty). \tag{2.100}$$

Since the potentials are non-scattering, for which $S(\lambda;0) = S(\lambda) = 1$, i.e., $N(\lambda) = D(\lambda)$ for $|\mathrm{Im}\lambda| < \varepsilon_0$, then from (2.100) it follows that $D(\lambda;t) = N(\lambda;t)$ for $|Im\lambda| < \varepsilon_0$ and any $t \in (-\infty,\infty)$, i.e.,

$$S(\lambda;t) \equiv 1 \text{ for } |\mathrm{Im}\lambda| < \varepsilon_0, \ \varepsilon_0 > 0 \text{ and all } t \in (-\infty,\infty). \tag{2.101}$$

This means that if the condition (2.99) is satisfied, then the non-scattering property (2.101) of the scattering function $S(\lambda;t)$ is conserved in time when these potentials change according to the system of NLEEs (2.5).

Hence, the exact non-scattering potentials (2.78), (2.79) are the solution of the system of NLEEs (2.5) if and only if the condition (2.99) is satisfied and the time-dependence of the standardized multipliers obey the evolution equations (2.97) and (2.98). The time-dependent non-scattering potentials are derived from (2.78) and (2.79) with standardized multipliers (2.97) and (2.98) are:

$$c_1(x,t) = -2\frac{(\lambda^+-\lambda^-)^2 e_{20}^{(2)} D_{\lambda_0} e^{-2i\lambda^- x+4i(\lambda^+)^2 t}}{(\lambda^+-\lambda^-)^2 e_{10}^{(1)} e_{20}^{(2)} N_{\lambda_0} D_{\lambda_0} e^{4i[(\lambda^+)^2-(\lambda^-)^2]t} - e^{2i(\lambda^+-\lambda^-)x}}, \tag{2.102}$$

$$c_2(x,t) = 2\frac{(\lambda^+-\lambda^-)^2 e_{10}^{(1)} N_{\lambda_0} e^{2i\lambda^+ x-4i(\lambda^-)^2 t}}{(\lambda^+-\lambda^-)^2 e_{10}^{(1)} e_{20}^{(2)} N_{\lambda_0} D_{\lambda_0} e^{4i[(\lambda^+)^2-(\lambda^-)^2]t} - e^{2i(\lambda^+-\lambda^-)x}}, \tag{2.103}$$

where $e_{20}^{(2)} = e_2^{(2)}(0,\lambda^+;0) \neq 0, e_{10}^{(1)} = e_1^{(1)}(0,\lambda^-;0) \neq 0, N_{\lambda_0} = N_\lambda(\lambda^-;0) \neq 0, D_{\lambda_0} = D_{\lambda_0}(\lambda^+;0) \neq 0$.

Thus, we have solved in the class of non-scattering potentials the Cauchy initial-value problem for the system (2.5) with the initial condition (2.6):

$$c_j(x,t)|_{t=0} = c_j(x), \quad j=1,2, \tag{2.104}$$

where $c_1(x)$ and $c_2(x)$ are continuous functions belonging to the class P and satisfying estimate (2.4).

The solution (2.102), (2.103) of the Cauchy problem (2.5), (2.104) is found from the given non-scattering data consisting of two simple poles at λ^+, $Im\lambda^+ > \varepsilon_0$ and λ^-, $Im\lambda^- \leq -\varepsilon_0$, $\varepsilon_0 > 0$, and two standardized multipliers (2.97) and (2.98). Thus, the following theorem is proved.

Theorem 2.3. *In the class of non-scattering potentials belonging to the class P and satisfying estimate* (2.4), *the Cauchy initial-value problem for the system of NLEEs* (2.5) *with known initial condition* (2.104) *is solved exactly by the ISM. If the non-scattering set* (2.67) *has only two simple poles at λ^+, $Im\lambda^+ \geq \varepsilon_0$ and λ^-, $Im\lambda^- \leq -\varepsilon_0$, $\varepsilon_0 > 0$ and two standardized multipliers defined by* (2.73), *then the non-scattering potentials are constructed by formulas* (2.78) *and* (2.79) *in terms of solutions of systems* (2.74)–(2.75) *and* (2.76)–(2.77). *These constructed potentials are solution of the system of NLEEs* (2.5) *if and only if the condition (2.99) is satisfied and the time-dependence of the standardized multipliers obeys time evolution equations* (2.97) *and* (2.98). *Then the potentials* (2.102) *and* (2.103) *are an exact solution of the Cauchy problem* (2.5) *and* (2.104).

2.2.3 An exact solution of the attractive NLS equation

We consider the Cauchy problem for the attractive NLS equation (2.60) with known initial condition (2.61) in the class of non-scattering potentials satisfying the estimate (2.4). The considered Cauchy problem is associated with the scattering problem (2.2)–(2.3) with a potential matrix of the form (2.63) in this class. The non-scattering data set (2.67) for this case is constructed from known initial condition (2.61) and can be formulated in the form:

$$s = \Big\{ S(\lambda;t) \equiv 1 \text{ for } |\mathrm{Im}\lambda| < \varepsilon_0, \ \lambda_k^+, \ \mathrm{Im}\lambda_k^+ \geq \varepsilon_0, \ \varepsilon_0 > 0, \ \lambda_k^- = \overline{\lambda_k^+},$$

$$p_k(x), \ \overline{p_k(x)}, \ k = 1, ..., N \Big\}. \tag{2.105}$$

The non-scattering data set (2.105) consists of N pairs of discrete spectra, therefore the solution of the system of fundamental equations for $H_{11}(x,u)$, $H_{12}(x,u)$, $0 \leq x \leq u$ is to be sought in the form:

$$H_{11}(x,u) = \sum_{k=1}^{N} H_{11,k}(x) e^{i\lambda_k^+ u}, \quad H_{12}(x,u) = \sum_{k=1}^{N} H_{12,k}(x) e^{-i\lambda_k^- u}, \tag{2.106}$$

where $\lambda_k^- = \overline{\lambda_k^+}$, $\mathrm{Im}\lambda_k^+ \geq \varepsilon_0, \ \varepsilon_0 > 0, \ k = 1, ..., N$.

Under the condition (2.59) the kernels of the transformation operator satisfy the relations (2.64). Using (2.64), from (2.69) and (2.72), we formulate the system of linear algebraic equations for $H_{11,k}(x)$, $H_{12,k}(x)$:

$$H_{11,k}(x) - \sum_{j=1}^{N} H_{12,j}(x) \int_x^{\infty} p_k(u+\tau) e^{i(\lambda_k^+ - \lambda_j^-)\tau} d\tau = 0,$$

$$\overline{p_k(x+u)} e^{-i\lambda_k^- x} + H_{12,k}(x) + \sum_{j=1}^{N} H_{11,j}(x) \int_x^{\infty} \overline{p_k(u+\tau)} e^{i(\lambda_k^+ - \lambda_j^-)\tau} d\tau = 0,$$

$$k = 1, ..., N, \qquad 0 \leq x < \infty. \tag{2.107}$$

Calculating the solution of system (2.107), then with the help of (2.106) and (2.65) we find the exact non-scattering potential (2.59).

For simplicity we suppose that the associated scattering problem (2.2)–(2.3) with the potentials satisfying the condition (2.59) has only one pair of simple poles at $\lambda^+, \lambda^- = \overline{\lambda^+}, \mathrm{Im}\lambda^+ \geq \varepsilon_0, \varepsilon_0 > 0$. In this case the non-scattering data set (2.105) constructed from the initial condition (2.61) is formulated in the form:

$$s = \Big\{ S(\lambda;t) \equiv 1 \text{ for } |\mathrm{Im}\lambda| < \varepsilon_0; \ \lambda^+, \ \mathrm{Im}\lambda^+ \geq \varepsilon_0, \ \varepsilon_0 > 0, \ \lambda^- = \overline{\lambda^+},$$

$$i(e_2^{(2)}(0,\lambda^+) D_\lambda(\lambda^+))^{-1}, \ -i\overline{(e_2^{(2)}(0,\lambda^+) D_\lambda(\lambda^+))^{-1}} \Big\} \tag{2.108}$$

Due to (2.64) and (2.108), the equations (2.74) and (2.77) are written in the form:

$$H_{11}(x) + \frac{e^{i(\lambda^+-\lambda^-)x}}{(\lambda^+-\lambda^-)e_2^{(2)}(0,\lambda^+)D_\lambda(\lambda^+)}H_{12}(x) = 0,$$
$$H_{12}(x) + \frac{e^{i(\lambda^+-\lambda^-)x}}{(\lambda^+-\lambda^-)\overline{e_2^{(2)}(0,\lambda^+)}\,\overline{D_\lambda(\lambda^+)}}H_{11}(x) = i\frac{e^{-i\lambda^- x}}{\overline{D_\lambda(\lambda^+)}\,\overline{e_2^{(2)}(0,\lambda^+)}},$$
$$0 \le x < \infty. \tag{2.109}$$

With the help of the substitution (2.98), from the system (2.109) we obtain the system of linear algebraic equations for $H_{11}(x;t)$ and $H_{12}(x;t)$:

$$H_{11}(x;t) + \frac{e^{i(\lambda^+-\lambda^-)x}}{(\lambda^+-\lambda^-)e_{20}^{(2)}D_{\lambda_0}\exp\{4i(\lambda^+)^2t\}}H_{12}(x;t) = 0,$$
$$\tag{2.110}$$
$$H_{12}(x;t) + \frac{e^{i(\lambda^+-\lambda^-)x}}{(\lambda^+-\lambda^-)\overline{e_{20}^{(2)}}\,\overline{D_{\lambda_0}}\exp\{-4i(\lambda^-)^2t\}}H_{11}(x;t)$$
$$= i\frac{e^{-i\lambda^- x}}{\overline{e_{20}^{(2)}}\,\overline{D_{\lambda_0}}\exp\{-4i(\lambda^-)^2t\}}$$

where $\overline{\lambda^+} = \lambda^-$, $\mathrm{Im}\lambda^+ \ge \varepsilon_0$, $\varepsilon_0 > 0$, $e_{20}^{(2)} = e_2^{(2)}(0,\lambda^+;0) \ne 0$, $D_{\lambda_0} = D_\lambda(\lambda^+;0) \ne 0$, $H_{12}(x;0) = H_{12}(x)$, $H_{11}(x;0) = H_{11}(x)$.

Calculating $H_{12}(x;t)$ from system (2.110), then with the help of (2.106) we find the exact solution of the Cauchy problem for equation (2.60) from the given non-scattering data set (2.108):

$$c(x,t) = 2iH_{12}(x,x;t) =$$
$$-\frac{8(\mathrm{Im}\lambda^+)^2e_{20}^{(2)}D_{\lambda_0}\exp\{4i(\lambda^+)^2t\}\exp\{-2i\lambda^- x\}}{4(\mathrm{Im}\lambda^+)^2|e_{20}^{(2)}D_{\lambda_0}|^2\exp\{-16\mathrm{Re}\lambda^+\mathrm{Im}\lambda^+ t\} + \exp\{-4\mathrm{Im}\lambda^+ x\}} \tag{2.111}$$

Theorem 2.4. *In the class of non-scattering potentials satisfying estimate* (2.4)*, the Cauchy IVP for the attractive NLS equation* (2.60) *with known initial condition* (2.61) *is solved by the ISM. If the non-scattering data of the associated scattering problem is given by the set* (2.108) *and if the condition* (2.99) *is satisfied, then the non-scattering potential is constructed from this set by formula* (2.111)*. This constructed potential is expressed in terms of the solution of system* (2.110) *and is an exact solution of this Cauchy problem.*

2.3 The Cauchy IVP problem for the repulsive NLS equation

Under the reduction:

$$c_1(x,t) = \overline{c_2(x,t)} = c(x,t), \;\; 0 \le x < \infty, \;\; -\infty < t < \infty, \tag{2.112}$$

the system (2.5) reduces to the repulsive NLS equation:

$$ic_t(x,t) = c_{xx}(x,t) - 2|c(x,t)|^2 c(x,t), \;\; 0 \le x < \infty, \;\; -\infty < t < \infty. \tag{2.113}$$

We consider the Cauchy IVP problem for the repulsive NLS equation (2.113) with the known initial condition:

$$c(x,0) = c(x) = \overline{c(x)}, \;\; 0 \le x < \infty, \tag{2.114}$$

where $c(x)$ is a continuous function possessing the continuous derivative of the second-order and satisfying estimate of the type (1.87):

$$|c(x,t)| \le \frac{\tilde{C}}{1+x^{1+\varepsilon}} \text{ for any } t \in (-\infty,\infty), \;\; \tilde{C} \text{ is a constant, } \varepsilon \ge 0. \tag{2.115}$$

The Cauchy problem (2.113)–(2.114) is associated with the self-adjoint scattering problem for the system (2.2) with boundary condition (2.3) and the potential self-adjoint matrix:

$$C = \begin{pmatrix} 0 & c(x,t) \\ \overline{c(x,t)} & 0 \end{pmatrix}, \tag{2.116}$$

where $c(x,t)$ is a continuous function satisfying estimate (2.115) and belonging to the class P.

Under the reduction (2.112) and estimate (2.115), the scattering problem (2.2)–(2.3) is self-adjoint and has no discrete spectrum. By Lemma 1.13 the integral equations of the form (1.17)–(1.20) with the potential belonging to the class P and satisfying the condition (2.112) and estimate (2.115) have unique solutions $H_{ij}(x,s;t)$, $0 \le x \le s$, $i,j = 1,2$. The obtained solutions of these integral equations are the kernels of the transformation operator defined by the associated self-adjoint problem (2.2)–(2.3). The kernels belong to the class P and satisfy the equalities:

$$H_{11}(x,s;t) = \overline{H_{22}(x,s;t)}, \;\; H_{12}(x,s;t) = \overline{H_{21}(x,s;t)}, \;\; 0 \le x \le s,$$

where t enters these equalities merely as parameter.

We apply results of the self-adjoint problem considered in Section 1.2 of Chapter 1 to solving the Cauchy problem (2.113)–(2.114). In fact, the correspondence between the numbers $A(\lambda)$ and $B(\lambda)$ established by Lemma 1.13 is one-to-one. Therefore the scattering function $S(\lambda)$ is uniquely defined by the relation: $S(\lambda)A(\lambda) = B(\lambda)$. The scattering function $S(a, \lambda)$ for the self-adjoint problem on the half-line $a \le x < \infty, \quad a \ge 0$, is determined analogously: $S(a, \lambda)A_a(\lambda) = B_a(\lambda)$.

Theorem 1.2 about the description of the scattering function $S(\lambda)$ of the scattering problem (2.2)–(2.3) with the potential self-adjoint matrix is valid, i.e., for any $x \ge 0$ there exists a function $S(x, \lambda)$ such that $S(\lambda) = S(0, \lambda)$ and this function satisfies the conditions 1, 2, 3 and 4 enumerated in Theorem 1.2.

Moreover, by Lemma 2.2 the scattering function $S(x, \lambda; t)$ obeys the evolutionary equation for every $x \ge 0$:

$$S(x, \lambda; t) = S(x, \lambda)e^{4i\lambda^2 t}, \quad S^{-1}(x, \lambda; t) = S^{-1}(x, \lambda)e^{-4i\lambda^2 t} = \overline{S(x, \lambda)}e^{-4i\lambda^2 t}$$
$$\text{for } |\mathrm{Im}\lambda| < \varepsilon_0, \quad -\infty < t < \infty. \tag{2.117}$$

Thus, Theorem 1.2 and Lemma 2.2 reduce the problem of solving the Cauchy problem (2.113)–(2.114) to that of solving the system of fundamental equations (1.116) and (1.119). Indeed, the functions $f(x, -\xi; t)$ and $g(x, \xi; t)$ appeared in system (1.116), (1.119) are constructed from the given function $S(x, \lambda)$ by the formulas:

$$f(x, -\xi; t) = \frac{1}{2\pi} \int_{-\infty}^{\infty} [S(x, \lambda; t) - 1] \, e^{i\lambda\xi} d\lambda$$
$$= \frac{1}{2\pi} \int_{-\infty}^{\infty} \left[S(x, \lambda)e^{4i\lambda^2 t} - 1\right] e^{i\lambda\xi} d\lambda,$$
$$g(x, \xi; t) = \frac{1}{2\pi} \int_{-\infty}^{\infty} [S^{-1}(x, \lambda; t) - 1] \, e^{-i\lambda\xi} d\lambda$$
$$= \frac{1}{2\pi} \int_{-\infty}^{\infty} \left[S^{-1}(\lambda)e^{-4i\lambda^2 t} - 1\right] e^{-i\lambda\xi} d\lambda,$$
$$f(x, -\xi; t) = \overline{g(x, \xi; t)} \text{ for any } x \ge 0,$$

where $S(x, \lambda; t)$ is the scattering function of the associated self-adjoint problem (2.2)–(2.3) for every $x \ge 0$ and $t \in (-\infty, \infty)$. It possesses the conditions 1–4 enumerated in Theorem 1.2.

By the condition 4 of Theorem 1.2 the constructed functions $f(x; t), \ g(x; t)$ satisfy the estimate for any $x \ge 0$ and $t \in (-\infty, \infty)$:

$$|f(-x; t)| \le \frac{\tilde{C}}{1 + (x)^{1+\varepsilon}}, \quad |g(x; t)| \le \frac{\tilde{C}}{1 + (x)^{1+\varepsilon}}, \quad \tilde{C} \text{ is a constant, } \varepsilon > 0.$$

Using system (1.116), (1.119), we verify that the functions $f(-x;t)$ and $g(x;t)$ possess the properties which are analogous with the properties of the kernels functions $H_{11}(x,s;t)$ and $H_{12}(x,s;t)$, $0 \geq x \geq s$, i.e., they possess the first partial continuous derivative with respect to t and the second partial continuous derivative with respect to x.

Owing to the above estimate for the functions $f(-x;t)$ and $g(x;t)$ and the properties of these functions, by Lemmas 1.18, 1.20 and 1.21 the system of fundamental equations (1.116) and (1.119) has for every $(x,t) \in [0,\infty) \times (-\infty,\infty)$ a unique solution $(H_{11}(x,s;t),\ H_{12}(x,s;t))$, $0 \leq x \leq s$. The found solution of this system is related to the reconstructed potential matrix by the formula:

$$C = \begin{pmatrix} 0 & c(x,t) \\ \overline{c(x,t)} & 0 \end{pmatrix} = \begin{pmatrix} 0 & 2iH_{12}(x,x;t) \\ -2i\,\overline{H_{12}(x,x;t)} & 0 \end{pmatrix},$$
$$0 \leq x < \infty, \quad -\infty < t < \infty, \qquad (2.118)$$

and satisfies the estimate:

$$|H_{12}(x,s;t)| \leq \frac{\tilde{C}}{1+(x+s)^{1+\varepsilon}} \quad \text{for any } t \in (-\infty,\infty),$$
$$0 \leq x \leq s, \quad \varepsilon > 0, \quad \tilde{C} \text{ is a constant.}$$

The found solution of system (1.116), (1.119) belongs to the class P and satisfies integral equations of the form (1.17) and (1.19), respectively with the constructed potential self-adjoint (2.118). Owing to the uniqueness of the solution of equations (1.17) and (1.19), the potential self-adjoint matrix (2.118) is uniquely recovered from the given function $S(\lambda)$ satisfying the conditions 1–4 enumerated in the Theorem 1.2. Hence, the eigenvalue equations (2.8) and (2.11) with the reconstructed potential matrix (2.118) are compatible, i.e., $u_{xt} = u_{tx}$. Then by Lemma 2.1 the potential $c(x,t)$ defined by formula (2.118) is the solution of the repulsive NLS equation (2.113).

Thus, the Cauchy problem (2.113)–(2.114) is solved.

Theorem 2.5. *Under the reduction* (2.112) *and estimate* (2.115)*, the associated scattering problem* (2.2)–(2.3) *is self-adjoint and has no discrete spectrum. The scattering function $S(\lambda)$ of this self-adjoint scattering problem possesses the properties 1–4 enumerated in the Theorem 1.2. The time-dependence of the time-dependent scattering function $S(x,\lambda;t)$ is defined by formulas* (2.117)*. A correspondence between the scattering function $S(x,\lambda;t)$ and the reconstructed potential matrix* (2.118) *is one-to-one. Therefore, the potential self-adjoint matrix in system* (2.2) *is uniquely recovered from the scattering function $S(x,\lambda)$ by the formula* (2.118)*. Thus, the Cauchy IVP for the repulsive NLS equation* (2.113) *with the known initial condition* (2.114) *has a unique solution $c(x,t)$ defined by* (2.118) *in the class of isospectral potentials, which are continuous functions satisfying the estimate of the type* (2.115)*, and possessing the first partial continuous derivative with respect to t and the second partial continuous derivative with respect to x.*

Chapter 3

Some Problems for Cubic Non-Linear Evolution Equations on A Half-Line

In this chapter we shall apply results of the inverse scattering problem (ISP) for the system of first-order ordinary differential equations (ODEs), which is similar to system (1.7) considered in Chapter 1 for solving the initial-boundary value problems (IBVPs) for the modified Korteweg-de Vries (mKdV) equations:

$$q_t + q_{xxx} + 6q^2 q_x = 0, \qquad 0 \leq x < \infty, \ -\infty < t < \infty \tag{3.1}$$

$$q_t + q_{xxx} - 6q^2 q_x = 0, \qquad 0 \leq x < \infty, \ -\infty < t < \infty \tag{3.2}$$

and for solving the initial value problem (IVP) for the cubic non-linear evolution equation:

$$q_t + q_{xxx} - 6|q|^2 q_x = 0, \ \ 0 \leq x < \infty, \qquad -\infty < t < \infty. \tag{3.3}$$

The chapter is based on the published papers [97, 119, 138, 142].

The main difficulty in the study of the IBVPs for the mKdV equations (3.1) and (3.2) is given by the presence of unknown boundary values (BVs) at the origin $x = 0$ of the Jost solutions of the associated system of first-order ODEs. In Section 3.2 this difficulty is overcome by deriving evolution equations for the unknown BVs. The solutions of the derived equations, i.e., the unknown boundary values (BVs) are found by explicit formulas in terms of the known initial and boundary conditions of the considered IBVPs. Then we prove that every scattering data set of the associated scattering problem corresponds a unique solution of the IBVP for the mKdV equations on the half-line. Consequently, the IBVPs for the mKdV equations (3.1) and (3.2) are solved. Further, the relations between the solutions of these IBVPs for the mKdV equations and the solutions of the KdV equation on the half-line are established. In Section 3.3 the exact solutions of the considered mKdV equations are found in the class of non-scattering potentials. The Cauchy IVP problem for the cubic non-linear evolution equation (3.3) is studied in Section 3.4.

DOI: 10.1201/9781003370543-3

3.1 The direct and inverse scattering problem

The IBVPs for the mKdV equations (3.1) and (3.2) are associated with the scattering problem for the system of first-order ODEs of the form:

$$-iJ_1u_x + Cu = \lambda u, \quad u = (u_1(x,\lambda), u_2(x,\lambda)), \quad 0 \le x < \infty \tag{3.4}$$

with the boundary condition at the origin:

$$u_1(0,\lambda) = u_2(0,\lambda), \tag{3.5}$$

where $J_1 = \begin{pmatrix} 1 & 0 \\ 0 & -1 \end{pmatrix}$, λ is a complex-valued parameter, $C = C(x) = \begin{pmatrix} 0 & c_1(x) \\ c_2(x) & 0 \end{pmatrix}$, $C(x)$ is a non-self-adjoint potential matrix, $c_1(x)$ and $c_2(x)$ are complex-valued measurable functions satisfying the estimate:

$$|c_k(x)| \le \tilde{C}e^{-\varepsilon x}, \ \tilde{C} \text{ is a constant}, \ \varepsilon > 0, \ k = 1, 2. \tag{3.6}$$

The direct and inverse scattering problem for the system of the form:

$$-iJu_x + Cu = \lambda u, \quad J = \begin{pmatrix} -1 & 0 \\ 0 & 1 \end{pmatrix},$$

with the boundary condition (3.5) is investigated in detail in Section 1.1 of Chapter 1 [97]. Although only the minor changes in system (1.7), as $J_1 = -J$, were made, there is a difference between the formulas of the problem (3.4)–(3.5) and the corresponding formulas of the problem considered in Section 1.1 of Chapter 1. Therefore, we present the necessary results of the direct and inverse scattering problem (3.4)–(3.5) without proving.

3.1.1 The representation of the solution of system (3.4)

Lemma 3.1. *Let λ be a real number, and $u(x,\lambda) = \big(u_1(x,\lambda), u_2(x,\lambda)\big)$ be a bounded solution of the system* (3.4), *then*

$$u_1(x,\lambda) = A(\lambda)e^{i\lambda x} + o(1), \quad u_2(x,\lambda) = B(\lambda)e^{-i\lambda x} + o(1) \text{ as } x \to \infty. \tag{3.7}$$

The solution of the problem (3.4), (3.7) *is represented in the form:*

$$u_1(x,\lambda) = A(\lambda)e^{i\lambda x} + A(\lambda)\int_x^\infty H_{11}(x,s)e^{i\lambda s}ds + B(\lambda)\int_x^\infty H_{12}(x,s)e^{-i\lambda s}ds,$$

$$u_2(x,\lambda) = B(\lambda)e^{-i\lambda x} + A(\lambda)\int_x^\infty H_{21}(x,s)e^{i\lambda s}ds + B(\lambda)\int_x^\infty H_{22}(x,s)e^{-i\lambda s}ds. \tag{3.8}$$

The matrix function $H(x,s) = [H_{jk}(x,s)]$ *is related to the potential matrix* $C(x)$ *by*

$$2i\begin{pmatrix} 0 & -H_{12}(x,x) \\ H_{21}(x,x) & 0 \end{pmatrix} = \begin{pmatrix} 0 & c_1(x) \\ c_2(x) & 0 \end{pmatrix}, \tag{3.9}$$

and it has the estimate:

$$|H_{jk}(x,s)| \le \tilde{C}e^{-\varepsilon\frac{x+s}{2}},\ 0 \le x \le s,\ j,k = 1,2, \tag{3.10}$$

where $\tilde{C}$ *is a constant.*

By starting from (3.8) and the system of integral equations for $u(x,\lambda)$, we derive the following systems of integral equations for the kernel functions $H_{jk}(x,s)$:

$$\begin{cases} H_{11}(x,s) = -\dfrac{1}{i}\displaystyle\int_x^\infty c_1(\tau)H_{21}(\tau,\tau+s-x)d\tau, \\ H_{21}(x,s) = \dfrac{1}{2i}c_2\left(\dfrac{x+s}{2}\right) + \dfrac{1}{i}\displaystyle\int_x^{\frac{x+s}{2}} c_2(\tau)H_{11}(\tau,x+s-\tau)d\tau, \end{cases} \tag{3.11}$$

$$\begin{cases} H_{22}(x,s) = \dfrac{1}{i}\displaystyle\int_x^\infty c_2(\tau)H_{12}(\tau,\tau+s-x)d\tau, \\ H_{12}(x,s) = -\dfrac{1}{2i}c_1\left(\dfrac{x+s}{2}\right) - \dfrac{1}{i}\displaystyle\int_x^{\frac{x+s}{2}} c_1(\tau)H_{22}(\tau,x+s-\tau)d\tau. \end{cases} \tag{3.12}$$

The integral equations for the kernels $H_{jk}(x,s)$ of the transformation operator are

$$H_{11}(x,s) = \frac{1}{2}\int_x^\infty c_1(\tau)c_2\left(\tau+\frac{s-x}{2}\right)d\tau + \int_x^\infty c_1(\tau)d\tau \int_\tau^{\tau+\frac{s-x}{2}} c_2(\xi)H_{11}(\xi,2\tau+s-x-\xi)d\xi, \tag{3.13}$$

$$H_{21}(x,s) = \frac{1}{2i}c_2\left(\frac{x+s}{2}\right) + \int_x^{\frac{x+s}{2}} c_2(\tau)d\tau \int_\tau^\infty c_1(\xi)H_{21}(\xi,\xi+x+s-2\tau)d\xi, \tag{3.14}$$

$$H_{12}(x,s) = -\frac{1}{2i}c_1\left(\frac{x+s}{2}\right) + \int_x^{\frac{x+s}{2}} c_1(\tau)d\tau \int_\tau^\infty c_2(\xi)H_{12}(\xi,\xi+x+s-2\tau)d\xi, \tag{3.15}$$

$$H_{22}(x,s) = \frac{1}{2}\int_x^\infty c_1\left(\tau+\frac{s-x}{2}\right)c_2(\tau)d\tau + \int_x^\infty c_2(\tau)d\tau \int_\tau^{\tau+\frac{s-x}{2}} c_1(\xi)H_{22}(\xi, 2\tau+s-x-\xi)d\xi. \tag{3.16}$$

3.1.2 The Jost solutions of system (3.4)

Consider the systems of integral equations:

$$\begin{cases} e_1^{(1)}(x,\lambda) = e^{i\lambda x} - \dfrac{1}{i}\displaystyle\int_x^\infty c_1(\xi)e_2^{(1)}(\xi,\lambda)e^{i\lambda(x-\xi)}d\xi, \\ e_2^{(1)}(x,\lambda) = \dfrac{1}{i}\displaystyle\int_x^\infty c_2(\xi)e_1^{(1)}(\xi,\lambda)e^{-i\lambda(x-\xi)}d\xi. \end{cases} \tag{3.17}$$

$$\begin{cases} e_1^{(2)}(x,\lambda) = -\dfrac{1}{i}\displaystyle\int_x^\infty c_1(\xi)e_2^{(2)}(\xi,\lambda)e^{i\lambda(x-\xi)}d\xi, \\ e_2^{(2)}(x,\lambda) = e^{-i\lambda x} + \dfrac{1}{i}\displaystyle\int_x^\infty c_2(\xi)e_1^{(2)}(\xi,\lambda)e^{-i\lambda(x-\xi)}d\xi. \end{cases} \tag{3.18}$$

Owing to estimate (3.6), the systems (3.17) and (3.18) satisfy the asymptotic conditions as $x \to \infty$:

$$e_1^{(1)}(x,\lambda) = e^{i\lambda x} + o(1), \quad e_2^{(1)}(x,\lambda) = o(1), \tag{3.19}$$

$$e_1^{(2)}(x,\lambda) = o(1), \quad e_2^{(2)}(x,\lambda) = e^{-i\lambda x} + o(1). \tag{3.20}$$

It is easy to verify that the systems (3.17) and (3.18) are equivalent to the system of ODEs (3.4) with the boundary conditions (3.19) and (3.20), respectively. By Lemma 3.1 the solutions of systems (3.17) and (3.18) are represented in the form:

$$e^{(1)}(x,\lambda) = \begin{pmatrix} e_1^{(1)}(x,\lambda) \\ e_2^{(1)}(x,\lambda) \end{pmatrix} = \begin{pmatrix} e^{i\lambda x} + e^{i\lambda x}\int_0^\infty H_{11}(x,x+t)e^{i\lambda t}dt \\ e^{i\lambda x}\int_0^\infty H_{21}(x,x+t)e^{i\lambda t}dt \end{pmatrix}, \tag{3.21}$$

$$e^{(2)}(x,\lambda) = \begin{pmatrix} e_1^{(2)}(x,\lambda) \\ e_2^{(2)}(x,\lambda) \end{pmatrix} = \begin{pmatrix} e^{-i\lambda x}\int_0^\infty H_{12}(x,x+t)e^{-i\lambda t}dt \\ e^{-i\lambda x} + e^{-i\lambda x}\int_0^\infty H_{22}(x,x+t)e^{-i\lambda t}dt \end{pmatrix}. \tag{3.22}$$

By taking into account the fact that the Wronskian determinant $W[e^{(1)}, e^{(2)}]$ of solutions $e^{(1)}(x, \lambda)$ and $e^{(2)}(x, \lambda)$ is independent of x, when we let $x \to \infty$, we obtain that for real values of λ $W[e^{(1)}, e^{(2)}] = 1$. Hence, when λ is real, $e^{(1)}(x, \lambda)$ and $e^{(2)}(x, \lambda)$ form a fundamental system of solutions of system (3.4) and are called the Jost solutions of this system.

3.1.3 The scattering function $S(\lambda)$ and non-real eigenvalues

Suppose that for some real λ $u(x, \lambda) = \begin{pmatrix} u_1(x, \lambda) \\ u_2(x, \lambda) \end{pmatrix}$ is a bounded solution of system (3.4) with boundary condition (3.5). Then by Lemma 3.1 the solution $u(x, \lambda)$ is represented in the form (3.8). From (3.5) and (3.8) we have

$$A(\lambda)\left(1 + \int_0^\infty [H_{11}(0,t) - H_{21}(0,t)]e^{i\lambda t}dt\right) =$$

$$B(\lambda)\left(1 + \int_0^\infty [H_{22}(0,t) - H_{12}(0,t)]e^{-i\lambda t}dt\right).$$

Introduce the function:

$$S(\lambda) = \frac{N(\lambda)}{D(\lambda)}, \tag{3.23}$$

where

$$N(\lambda) = e_1^{(1)}(0, \lambda) - e_2^{(1)}(0, \lambda) = 1 + \int_0^\infty [H_{11}(0,t) - H_{21}(0,t)]e^{i\lambda t}dt,$$

$$D(\lambda) = e_2^{(2)}(0, \lambda) - e_1^{(2)}(0, \lambda) = 1 + \int_0^\infty [H_{22}(0,t) - H_{12}(0,t)]e^{-i\lambda t}dt.$$

We call $S(\lambda)$ the scattering function for the scattering problem (3.4)–(3.5). We call $N(\lambda)$ the numerator and $D(\lambda)$ the denominator of the scattering function. Owing to estimate (3.10), the numerator $N(\lambda)$ has a bounded analytic continuation into the upper half-plane $\mathrm{Im}\lambda > -\frac{\varepsilon}{2}$, and the denominator $D(\lambda)$ has a bounded analytic continuation into the lower half-plane $\mathrm{Im}\lambda < \frac{\varepsilon}{2}$. Hence, $N(\lambda)$ can have only a finite number of zeros in any closed upper half-plane $\mathrm{Im}\lambda > -\frac{\varepsilon}{2}$, and $D(\lambda)$ can have only a finite number of zeros in any closed lower half-plane $\mathrm{Im}\lambda < \frac{\varepsilon}{2}$. Therefore, there exists a positive number ε_0 such that

$$\varepsilon_0 = \min\left(\varepsilon_1, \frac{\varepsilon}{2}\right), \tag{3.24}$$

where ε_1 is the distance from the real axis to the non-real zeros of the numerator $N(\lambda)$ and denominator $D(\lambda)$.

Lemma 3.2. *Let the kernels functions $H_{jk}(x,s)$ satisfy estimate (3.10) and the number ε_0 be defined by (3.24). Then*

1. *The function* $1+\int_0^\infty [H_{11}(0,t)-H_{21}(0,t)]e^{i\lambda t}dt$ *is holomorphic for* $\mathrm{Im}\lambda > -\varepsilon_0$*, and the function* $1+\int_0^\infty [H_{22}(0,t)-H_{12}(0,t)]e^{-i\lambda t}$ *is holomorphic for* $\mathrm{Im}\lambda < \varepsilon_0$.
   For every $\eta < \varepsilon_0$ *the following asymptotic relations hold as* $|\lambda| \to \infty$*:*

$$1+\int_0^\infty [H_{11}(0,t)-H_{21}(0,t)]e^{i\lambda t}dt = 1+o(1)$$

   uniformly in the half-plane $\mathrm{Im}\lambda \geq -\eta$,

$$1+\int_0^\infty [H_{22}(0,t)-H_{12}(0,t)]e^{-i\lambda t}dt = 1+o(1)$$

   uniformly in the half-plane $\mathrm{Im}\lambda \leq \eta$;

2.
$$\left.\begin{aligned} &1+\int_0^\infty [H_{11}(0,t)-H_{21}(0,t)]e^{i\lambda t}dt \neq 0,\\ &1+\int_0^\infty [H_{22}(0,t)-H_{12}(0,t)]e^{-i\lambda t}dt \neq 0; \end{aligned}\right. \qquad 0<|\mathrm{Im}\lambda|<\varepsilon_0$$

3. *If* $\mathrm{Im}\lambda = 0$ *and* $1+\int_0^\infty [H_{11}(0,t)-H_{21}(0,t)]e^{i\lambda t}dt = 0$*, then* $1+\int_0^\infty [H_{22}(0,t)-H_{12}(0,t)]e^{-i\lambda t}dt \neq 0.$
   If $\mathrm{Im}\lambda = 0$ *and* $1+\int_0^\infty [H_{22}(0,t)-H_{12}(0,t)]e^{-i\lambda t}dt = 0$*, then* $1+\int_0^\infty [H_{11}(0,t)-H_{21}(0,t)]e^{i\lambda t}dt \neq 0.$

Lemma 3.3. *The scattering function $S(\lambda)$ defined by (3.23) for the scattering problem (3.4)–(3.5) with a non-self-adjoint potential matrix satisfying estimate (3.6) possesses the properties:*

1. *The functions* $S(\lambda)$ *and* $S^{-1}(\lambda)$ *are meromorphic in the strip* $|\mathrm{Im}\lambda| < \varepsilon_0$*, and for every* $\eta < \varepsilon_0$*;*
   $S(\lambda) = 1+o(1)\ |\lambda| \to \infty,\ \ S^{-1}(\lambda) = 1+o(1)\ |\lambda| \to \infty$ *uniformly in the strip* $|\mathrm{Im}\lambda| \leq \eta$*;*

2. *The functions* $S(\lambda)$ *and* $S^{-1}(\lambda)$ *have no non-real poles in the strip* $|\mathrm{Im}\lambda| < \varepsilon_0$.

We call functions $S(\lambda)$ and $S^{-1}(\lambda)$ with the properties 1 and 2 functions of the type (S) in the strip $|\mathrm{Im}\lambda| < \varepsilon_0$.

Definition 3.1. *We call the zeros* λ_j^-, $j = 1, ..., \gamma_2$ *of the denominator* $D(\lambda)$ *that lie in the domain* $\mathrm{Im}\lambda < 0$, *and the zeros* λ_k^+, $k = 1, ..., \gamma_1$ *of the numerator* $N(\lambda)$ *that lie in the domain* $\mathrm{Im}\lambda > 0$ *the singular numbers of the problem* (3.4)–(3.5). *The multiplicity of these zeros is called the multiplicity of the singular numbers, and we denote them by* n_j *and* m_k, *respectively.*

Lemma 3.4. *The singular numbers of the problem* (3.4)–(3.5) *are the same as its non-real eigenvalues, and the multiplicity of a singular number is the same as the multiplicity of the corresponding eigenfunction.*

In what follows the solution $\varphi(x,\lambda) = \big(\varphi_1(x,\lambda), \varphi_2(x,\lambda)\big)$ of system (3.4) with the initial condition at the origin $x = 0$:

$$\varphi_1(0,\lambda) = \varphi_2(0,\lambda) = 1, \tag{3.25}$$

plays an important role.

It is obvious that the solution $\varphi(x,\lambda)$ satisfies the system of integral equations:

$$\begin{cases} \varphi_1(x,\lambda) = e^{i\lambda x} + \dfrac{1}{i}\displaystyle\int_0^x e^{i\lambda(x-y)} c_1(y)\varphi_2(y,\lambda)dy, \\ \varphi_2(x,\lambda) = e^{-i\lambda x} - \dfrac{1}{i}\displaystyle\int_0^x e^{i\lambda(y-x)} c_2(y)\varphi_1(y,\lambda)dy. \end{cases} \tag{3.26}$$

The system (3.26) has a solution that can be derived by the method of successive approximations. By the uniqueness of the solution of the Cauchy problem (3.4), (3.25), the solution $\varphi(x,\lambda)$ is unique. Obviously, for $x \geq 0$ the vector function $\varphi(x,\lambda)$ is an entire analytic function of the parameter λ.

3.1.4 Connection between the analytic solution and Jost solutions

We now discuss the connection between the analytic solution $\varphi(x,\lambda)$ and Jost solutions $e^{(1)}(x,\lambda)$, $e^{(2)}(x,\lambda)$. When we take into account the fact that the Wronskian determinant of solutions $e^{(1)}(x,\lambda)$ and $\varphi(x,\lambda)$, $\mathrm{Im}\lambda > 0$, is independent of x, we obtain

$$W\left[\varphi, e^{(1)}\right] = \begin{cases} e_2^{(1)}(x,\lambda) - e_1^{(1)}(x,\lambda) \neq 0 \text{ ; for } \lambda \neq \lambda_k^+, \\ e_2^{(1)}(x,\lambda) - e_1^{(1)}(x,\lambda) = 0 \text{ for } \lambda = \lambda_k^+. \end{cases} \quad \mathrm{Im}\lambda_k^+ \geq \varepsilon_0, k = 1, ..., \gamma_1.$$

Hence, for $\mathrm{Im}\lambda > 0$ the solutions $e^{(1)}(x,\lambda)$ and $\varphi(x,\lambda)$ are linearly independent for $\lambda \neq \lambda_k^+$, $k = 1, ..., \gamma_1$, and they are linearly dependent for $\lambda = \lambda_k^+$, that is

$$\begin{pmatrix} \varphi_1(x,\lambda_k^+) \\ \varphi_2(x,\lambda_k^+) \end{pmatrix} = A_0(\lambda_k^+) \begin{pmatrix} e_1^{(1)}(x,\lambda_k^+) \\ e_2^{(1)}(x,\lambda_k^+) \end{pmatrix}, \quad \mathrm{Im}\lambda_k^+ \geq \varepsilon_0, \ j = 1, ..., \gamma_1, \tag{3.27}$$

where taking (3.25) into account, from the relation (3.27), we obtain

$$A_0(\lambda_k^+) = \frac{1}{e_1^{(1)}(0,\lambda_k^+)},$$

where the number $e_2^{(2)}(0,\lambda_k^+)$ is non-zero, because otherwise we would have $e_1^{(2)}(0,\lambda_k^+) = 0$, and then by virtue of the uniqueness of the solution of the Cauchy problem (3.4), (3.25), we obtain that $e_1^{(1)}(x,\lambda_k^+) = e_2^{(1)}(x,\lambda_k^+) \equiv 0$. This is impossible, because due to the estimate (3.10) the function

$$e_1^{(1)}(x,\lambda_k^+) = e^{i\lambda_k^+ x}\left(1 + \int_0^\infty H_{22}(x,x+\xi)e^{i\lambda_k^+\xi}d\xi\right),$$

is non-zero for large x.

Analogously one can show that in the lower half-plane $\mathrm{Im}\lambda < 0$:

$$W\left[\varphi, e^{(2)}\right] = \begin{cases} e_2^{(2)}(0,\lambda) - e_1^{(2)}(0,\lambda) \neq 0 & \text{for } \lambda \neq \lambda_j^-, \\ e_2^{(2)}(0,\lambda) - e_1^{(2)}(0,\lambda) = 0 & \text{for } \lambda = \lambda_j^-. \end{cases}$$

$$\mathrm{Im}\lambda_j^- \leq -\varepsilon_0, \ j = 1, ..., \gamma_2.$$

Hence,

$$\begin{pmatrix} \varphi_1(x,\lambda_j^-) \\ \varphi_2(x,\lambda_j^-) \end{pmatrix} = B_0(\lambda_j^-) \begin{pmatrix} e_1^{(2)}(x,\lambda_j^-) \\ e_2^{(2)}(x,\lambda_j^-) \end{pmatrix}, \quad \mathrm{Im}\lambda_j^- \leq -\varepsilon_0, \ j = 1, ..., \gamma_2, \tag{3.28}$$

where

$$B_0(\lambda_j^-) = \frac{1}{e_2^{(2)}(0,\lambda_j^-)}, \quad e_2^{(2)}(0,\lambda_j^-) \neq 0.$$

Lemma 3.5. *Let m_k be the multiplicity of the zero λ_k^+ of the numerator, and n_j be the multiplicity of the zero λ_j^- of the denominator. Then the following* *relations hold for every $x \geq 0$:*

$$(\varphi(x,\lambda))^{(q)}_{\lambda=\lambda_k^+} = A_q e^{(1)}(x,\lambda_k^+) + qA_{q-1}\left(e^{(1)}(x,\lambda)\right)'_{\lambda=\lambda_k^+} + ... + \frac{q(q-1)...(q-i+1)}{i!} A_{q-i}\left(e^{(1)}(x,\lambda)\right)^{(i)}_{\lambda=\lambda_k^+} + ... + A_0\left(e^{(1)}(x,\lambda)\right)^{(q)}_{\lambda=\lambda_k^+}, \quad q = 0,1,...,m_k-1, \tag{3.29}$$

$$(\varphi(x,\lambda))^{(p)}_{\lambda=\lambda_j^-} = B_p e^{(2)}(x,\lambda_j^-) + pB_{p-1}\left(e^{(2)}(x,\lambda)\right)'_{\lambda=\lambda_j^-} + ... + \frac{p(p-1)...(p-i+1)}{i!} B_{p-i}\left(e^{(2)}(x,\lambda)\right)^{(i)}_{\lambda=\lambda_j^-} + ... + B_0\left(e^{(2)}(x,\lambda)\right)^{(p)}_{\lambda=\lambda_j^-}, \quad p = 0,1,...,n_j-1, \tag{3.30}$$

where $A_0, ..., A_q$ are numbers that depend on the zero λ_k^+ of the numerator, and $B_0, ..., B_p$ are numbers that depend on the zero λ_j^- of the denominator. These numbers are defined by the formulas:

$$A_q(\lambda_k^+) = \left[\frac{d^q}{d\lambda^q}\frac{1}{e_1^{(1)}(0,\lambda)}\right]_{\lambda=\lambda_k^+}, \quad q = 0,1,...,m_k-1, \quad k = 1,...,\gamma_1, \tag{3.31}$$

$$B_p(\lambda_j^-) = \left[\frac{d^p}{d\lambda^p}\frac{1}{e_2^{(2)}(0,\lambda)}\right]_{\lambda=\lambda_j^-}, \quad p = 0,1,...,n_j-1, \quad j = 1,...,\gamma_2. \tag{3.32}$$

We continue our study of the connection between the analytic solution $\varphi(x,\lambda)$ and Jost solutions $e^{(1)}(x,\lambda)$ and $e^{(2)}(x,\lambda)$ of system (3.4). Since the Jost solutions form a fundamental system of solutions of system (3.4), then the analytic solution $\varphi(x,\lambda)$ is a linear combination of Jost solutions:

$$\varphi(x,\lambda) = a_1(\lambda)e^{(1)}(x,\lambda) + a_2(\lambda)e^{(2)}(x,\lambda) \text{ for real } \lambda,$$

where taking initial condition (3.25) into account, we calculate

$$a_1(\lambda) = e_2^{(2)}(0,\lambda) - e_1^{(2)}(0,\lambda), \quad a_2(\lambda) = e_1^{(1)}(0,\lambda) - e_2^{(1)}(0,\lambda).$$

If λ is not a real pole of the scattering function $S(\lambda)$, then we have

$$u(x,\lambda) = \begin{pmatrix} u_1(x,\lambda) \\ u_2(x,\lambda) \end{pmatrix} = \frac{1}{e_2^{(2)}(0,\lambda) - e_1^{(2)}(0,\lambda)} \begin{pmatrix} \varphi_1(x,\lambda) \\ \varphi_2(x,\lambda) \end{pmatrix} = e^{(1)}(x,\lambda) + S(\lambda)e^{(2)}(x,\lambda).$$

If λ is not a real pole of $S^{-1}(\lambda)$, then

$$z(x,\lambda)=\begin{pmatrix} z_1(x,\lambda)\\ z_2(x,\lambda)\end{pmatrix}=\frac{1}{e_1^{(1)}(0,\lambda)-e_2^{(1)}(0,\lambda)}\begin{pmatrix} \varphi_1(x,\lambda)\\ \varphi_2(x,\lambda)\end{pmatrix}=$$
$$S^{-1}(\lambda)e^{(1)}(x,\lambda)+e^{(2)}(x,\lambda).$$

The considered vector functions $u(x,\lambda)$ and $z(x,\lambda)$ are the solutions of system (3.4). By Lemma 3.1 these solutions are representable in the form:

$$u_1(x,\lambda)=e^{i\lambda x}+\int_x^\infty H_{11}(x,t)e^{i\lambda t}dt+S(\lambda)\int_x^\infty H_{12}(x,t)e^{-i\lambda t}dt,$$

$$u_2(x,\lambda)=S(\lambda)e^{-i\lambda x}+\int_x^\infty H_{21}(x,t)e^{i\lambda t}dt+S(\lambda)\int_x^\infty H_{22}(x,t)e^{-i\lambda t}dt,$$

$$z_1(x,\lambda)=S^{-1}(\lambda)e^{i\lambda x}+S^{-1}(\lambda)\int_x^\infty H_{11}(x,t)e^{i\lambda t}dt+\int_x^\infty H_{12}(x,t)e^{-i\lambda t}dt,$$

$$z_2(x,\lambda)=e^{-i\lambda x}+S^{-1}(\lambda)\int_x^\infty H_{21}(x,t)e^{i\lambda t}dt+\int_x^\infty H_{22}(x,t)e^{-i\lambda t}dt.$$

Lemma 3.6. *The vector function* $\begin{pmatrix} u_1(x,\lambda)e^{-i\lambda x}-1\\ u_2(x,\lambda)e^{-i\lambda x}-e^{-2i\lambda x}\end{pmatrix}$ *can be extended analytically into the lower half-plane* $\mathrm{Im}\lambda<0$ *apart from the singular numbers* $\lambda=\lambda_j^-$, $\mathrm{Im}\lambda_j^-<0$, $j=1,...,\gamma_2$, *and* $\begin{pmatrix} z_1(x,\lambda)e^{i\lambda x}-e^{2i\lambda x}\\ z_2(x,\lambda)e^{i\lambda x}-1\end{pmatrix}$ *can be extended analytically into the upper half-plane* $\mathrm{Im}\lambda>0$ *apart from the singular numbers* $\lambda=\lambda_k^+$, $\mathrm{Im}\lambda_k^+>0$, $k=1,...,\gamma_1$, *here*

$$\lim_{\lambda\to\infty}\begin{pmatrix} u_1(x,\lambda)e^{-i\lambda x}-1\\ u_2(x,\lambda)e^{i\lambda x}-e^{-2i\lambda x}\end{pmatrix}=0,\ \mathrm{Im}\lambda\le 0,$$

$$\lim_{\lambda\to\infty}\begin{pmatrix} z_1(x,\lambda)e^{i\lambda x}-e^{2i\lambda x}\\ z_2(x,\lambda)e^{i\lambda x}-1\end{pmatrix}=0,\ \mathrm{Im}\lambda\ge 0.$$

3.1.5 The scattering data

The following functions will play an important role in what follows:

$$\begin{aligned} f_S(x)&=\frac{1}{2\pi}\int_{-\infty-i\eta}^{\infty-i\eta}[S(\lambda)-1]e^{-i\lambda x}d\lambda,\\ g_S(x)&=\frac{1}{2\pi}\int_{-\infty+i\eta}^{\infty+i\eta}[S^{-1}(\lambda)-1]e^{-i\lambda x}d\lambda, \end{aligned}\tag{3.33}$$

where $S(\lambda)$ is a function of type (S) in the strip $|\text{Im}\lambda| < \varepsilon_0$, and η is some number such that $0 < \eta < \varepsilon_0$. Since $S(\lambda) - 1$ and $S^{-1}(\lambda) - 1$ are analytic in the strip $0 < |\text{Im}\lambda| < \varepsilon_0$, the value of these integrals is independent of η. In the sense of generalized functions [116], $f_S(x)$ and $g_S(x)$ can be considered as the inverse Fourier transformations of $S(\lambda)-1$ and $S^{-1}(\lambda)-1$, respectively. The functions (3.33) characterize the problem (3.4)–(3.5) on the continuous spectrum. We put:

$$g_k^+(-x) = \frac{1}{2\pi}\int\limits_{c_k^+} \frac{1}{e_1^{(1)}(0,\lambda) - e_2^{(1)}(0,\lambda)}\Bigg[A_0 + A_1(\lambda - \lambda_k^+) + ... + \frac{A_{m_k-1}}{(m_k-1)!}(\lambda - \lambda_k^+)^{m_k-1}\Bigg]e^{i\lambda x}d\lambda, \; k = 1, ..., \gamma_1, \tag{3.34}$$

$$f_j^-(x) = \frac{1}{2\pi}\int\limits_{c_j^-} \frac{1}{e_2^{(2)}(0,\lambda) - e_1^{(2)}(0,\lambda)}\Bigg[B_0 + B_1(\lambda - \lambda_j^-) + ... + \frac{B_{n_j-1}}{(n_j-1)!}(\lambda - \lambda_j^-)^{n_j-1}\Bigg]e^{-i\lambda x}d\lambda, \; j = 1, ..., \gamma_2. \tag{3.35}$$

where c_k^+ and c_j^- stand for circles with a sufficiently small radius that avoid the singular points λ_k^+ and λ_j^-, respectively. The numbers $A_q = A_q(\lambda_k^+)$, $q = 0, 1, ..., m_k - 1$ and $B_p = B_p(\lambda_j^-)$, $p = 0, 1, ..., n_j - 1$ are defined by formulas (3.31) and (3.32), respectively.

The functions $g_k^+(-x), k = 1, ..., \gamma_1$ and $f_j^-(x), j = 1, ..., \gamma_2$ characterize the problem (3.4)–(3.5) on the point spectrum. Using the definition of a residue [114], from (3.34) and (3.35) we can verify that

$$g_k^+(-x) = p_k(x)e^{i\lambda_k^+ x} \; k = 1, ..., \gamma_1, \;\; f_j^-(x) = \tilde{q}_j(x)e^{-i\lambda_j^- x}, \; j = 1, ..., \gamma_2, \tag{3.36}$$

where $p_k(x)$ and $\tilde{q}_j(x)$ are polynomials of degree m_k-1 and n_j-1, respectively.

We call the functions $p_1(x), ..., p_{\gamma_1}(x)$ and $\tilde{q}_1(x), ..., \tilde{q}_{\gamma_2}(x)$ the standardized polynomials of the problem (3.4)–(3.5). When the potential matrix $C(x)$ is self-adjoint, all the standardized polynomials are equal to the zero polynomial [119].

Definition 3.2. *The scattering function $S(\lambda)$ of the type (S), the non-real singular numbers $\lambda_1^+, ..., \lambda_{\gamma_1}^+$; $\lambda_j^-, ..., \lambda_{\gamma_2}^-$ and the standardized polynomials $p_1(x), ..., p_{\gamma_1}(x)$; $\tilde{q}_1(x), ..., \tilde{q}_{\gamma_2}(x)$ are called the scattering data of the problem* (3.4)–(3.5).

We turn to the inverse scattering problem, which is to determine the potential matrix from the scattering data, and also to describe the scattering data, that is, to establish necessary and sufficient conditions for given quantities to be scattering data for the system (3.4) on a half-line with boundary condition (3.5). We shall see that the scattering data uniquely determine the problem (3.4)–(3.5). To recover this problem from the scattering data, we first obtain systems of equations connecting the scattering data with the kernels of the transformation operator. We call these the systems of fundamental equations.

3.1.6 The systems of fundamental equations

We construct the following functions from the scattering data given by (3.33)–(3.36):

$$g(-x) = g_S(-x) - \sum_{k=1}^{\gamma_1} g_k^+(-x), \quad f(x) = f_S(x) - \sum_{j=1}^{\gamma_2} f_j^-(x), \ x \geq 0. \quad (3.37)$$

By using Lemmas 3.1, 3.2 and 3.5 and applying the Cauchy's theorem to the vector functions considered in Lemma 3.6, we derive the following systems of fundamental equations.

Lemma 3.7. *For every $x \geq 0,\ \xi > 0$:*

$$\begin{cases} H_{11}(x, x+\xi) + \int\limits_0^\infty H_{12}(x, x+t) f(2x+\xi+t)dt = 0, \\ g(-2x-\xi) + H_{12}(x, x+\xi) + \int\limits_0^\infty H_{11}(x, x+t) g(-2x-\xi-t)dt = 0, \end{cases} \quad (3.38)$$

$$\begin{cases} H_{22}(x, x+\xi) + \int\limits_0^\infty H_{21}(x, x+t) g(-2x-\xi-t)dt = 0, \\ f(2x+\xi) + H_{21}(x, x+\xi) + \int\limits_0^\infty H_{22}(x, x+t) f(2x+\xi+t)dt = 0. \end{cases} \quad (3.39)$$

Lemma 3.8. *The following compatibility relations between the scattering data hold:*

$$p_k(\xi) + \int\limits_0^\infty p_k(\xi+t) e^{i\lambda_k^+ t}\left[H_{11}(0,t) - H_{21}(0,t)\right]dt = 0, \ \xi > 0, \ k = 1, ..., \gamma_1,$$

$$\tilde{q}_j(\xi) + \int\limits_0^\infty \tilde{q}_j(\xi+t) e^{-i\lambda_j^- t}\left[H_{22}(0,t) - H_{12}(0,t)\right]dt = 0, \ \xi > 0, \ j = 1, ..., \gamma_2.$$

Lemma 3.9. *Let the kernel functions $H_{jk}(x, x+\xi),\ x \geq 0, \xi > 0,\ j, k = 1, 2$ satisfy the inequalities* (3.10) *and let the scattering function* (3.23) *be the* *function of the type (S) in the strip* $|\text{Im}\lambda| < \varepsilon_0$, *where the number* ε_0 *is defined by* (3.24). *Then the following estimates for* $f(x)$ *and* $g(-x)$ *are valid:*

$$|f(x)| \leq \tilde{C}(\eta)e^{-\frac{\varepsilon}{4}x}, \quad |g(-x)| \leq \tilde{C}(\eta)e^{-\frac{\varepsilon}{4}x}, \tag{3.40}$$

where $\tilde{C}(\eta)$ *is a constant depending on* η: $0 < \eta < \frac{1}{12}\varepsilon_0$.

By estimate (3.40) one can verify that the systems (3.38) and (3.39) are the Fredholm integral matrix equations of the second kind in the space $L_1[0,\infty)$.

Lemma 3.10. *For every fixed* $x \geq 0$ *the systems of fundamental equations* (3.38) *and* (3.39) *have a unique solution in the space* $L_1[0,\infty)$.

3.1.7 The complete description of the scattering data

We turn to the main theorem regarding the complete description of the scattering data, that is, the establishment of necessary and sufficient conditions for given quantities to be the scattering data for the system (3.4) on a half-line with boundary condition (3.5).

Theorem 3.1. *Suppose that we are given:*

(a) A function $\gamma(\rho)$;

(b) Numbers ρ_k^+, $\text{Im}\rho_k^+ > \varepsilon_0$, $k = 1, ..., \alpha_1$ *and* ρ_j^-, $\text{Im}\rho_j^- < -\varepsilon_0$, $j = 1, ..., \alpha_2$;

(c) Polynomials $P_k(x)$, $k = 1, ..., \alpha_1$, *and* $Q_j(x)$, $j = 1, ..., \alpha_2$, *(*ε_0 *is some positive number).*

For the set of quantities (a)–(c) to form the scattering data for a system of the form (3.4) *on a half-line with boundary condition* (3.5) *and an exponentially decreasing potential, it is necessary and sufficient that the given quantities satisfy the conditions:*

(1) $\gamma(\rho)$ *and* $\gamma^{-1}(\rho)$ *can be extended analytically into the strip* $0 < |\text{Im}\rho| < \varepsilon_0$, *and in the strip* $|\text{Im}\rho| < \varepsilon_0$ *they are functions of the type* (S);

(2) $\gamma(\rho)$ *has the factorization:*

$$\gamma(\rho) = \frac{1 + \int\limits_0^\infty K_+(t)e^{i\rho t}dt}{1 + \int\limits_0^\infty K_-(t)e^{-i\rho t}dt}, \quad 0 \leq |\text{Im}\rho| < \varepsilon_0,$$

where $K_\pm(t) \in L_1[0,\infty)$;

(3) The following compatibility relations hold:

$$P_k(\xi) + \int_0^\infty e^{i\rho_k^+ t} P_k(\xi + t) K_+(t) dt = 0, \ k = 1, ..., \alpha_1, \ \xi > 0,$$

$$Q_j(\xi) + \int_0^\infty e^{-i\rho_j^- t} Q_j(\xi + t) K_-(t) dt = 0, \ j = 1, ..., \alpha_2, \ \xi > 0;$$

(4) The homogeneous system of equations:

$$\begin{aligned} h_1(t) &= -\int_x^\infty F(t+\tau) h_2(\tau) d\tau, \\ h_2(t) &= -\int_x^\infty G(-t-\tau) h_1(\tau) d\tau, \end{aligned} \qquad 0 \le x < t$$

has in $L_1[0,\infty)$ only the trivial solution, here

$$F(x) = F_\gamma(x) - \sum_{j=1}^{\alpha_2} Q_j(x) e^{-i\rho_j^- x}, \ \ G(-x) = G_\gamma(-x) - \sum_{k=1}^{\alpha_1} P_k(x) e^{i\rho_j^+ x},$$

and

$$F_\gamma(x) = \frac{1}{2\pi} \int_{-\infty - i\eta}^{\infty - i\eta} [\gamma(\rho) - 1] e^{-i\rho x} d\rho,$$

$$G_\gamma(x) = \frac{1}{2\pi} \int_{-\infty + i\eta}^{\infty + i\eta} [\gamma^{-1}(\rho) - 1] e^{-i\rho x} d\rho,$$

(5) $F(x)$ and $G(-x)$ satisfy the inequalities:

$$|F(x)| \le \tilde{C} e^{-\frac{1}{2}\varepsilon_0 x}, \ \ |G(-x)| \le \tilde{C} e^{-\frac{1}{2}\varepsilon_0 x}, \ \tilde{C} \text{ is a constant.}$$

3.2 The IBVPs for the mKdV equations

We apply the results of the previous Section to considering the system of cubic non-linear evolution equations (NLEEs) [142]:

$$\begin{aligned} q_{1t} + q_{1xxx} + 6q_1q_2q_{1x} &= 0, \\ q_{2t} + q_{2xxx} + 6q_1q_2q_{2x} &= 0. \end{aligned} \qquad 0 \le x < \infty, \ \ -\infty < t < \infty \tag{3.41}$$

3.2.1 The Lax compatibility condition

Let the potentials in system (3.4) depend on an additional parameter t(time). We consider the time-dependent system (3.4) of the operator form:

$$u_x = Bu, \quad B = i\lambda J_1 + iQ, \qquad 0 \le x < \infty, \quad -\infty < t < \infty, \tag{3.42}$$

where $J_1 = \begin{pmatrix} 1 & 0 \\ 0 & -1 \end{pmatrix}$, $Q(x,t) = \begin{pmatrix} 0 & -c_1(x,t) \\ c_2(x,t) & 0 \end{pmatrix} = \begin{pmatrix} 0 & q_1(x,t) \\ q_2(x,t) & 0 \end{pmatrix}$.
The x part of the Lax pair associated with system (3.41) is the linear eigenvalue problem for system (3.42) with the boundary condition at the origin $x = 0$:

$$u_1(0,\lambda;t) = u_2(0,\lambda;t), \tag{3.43}$$

and potential functions $q_1(x,t)$, $q_2(x,t)$ satisfying the estimate for any $t \in (-\infty,\infty)$:

$$|q_j(x,t)| \le \tilde{C}e^{-\varepsilon x}, \ \tilde{C} \text{ is a constant}, \varepsilon > 0, \ j = 1,2. \tag{3.44}$$

The t part of the Lax pair, i.e., the time-dependence of the solution of the problem (3.42)–(3.43) is given by:

$$u_t = \tilde{V}u, \tag{3.45}$$

where

$$\tilde{V} = \tilde{V}(x,t) = 4i\lambda^3 J_1 + 4i\lambda^2 Q + 2i\lambda \begin{pmatrix} -q_1q_2 & -iq_{1x} \\ iq_{2x} & q_1q_2 \end{pmatrix} + \begin{pmatrix} q_1q_{2x} - q_{1x}q_2 & -2iq_1^2q_2 - iq_{1xx} \\ -2iq_1q_2^2 - iq_{2xx} & -q_1q_{2x} + q_{1x}q_2 \end{pmatrix}. \tag{3.46}$$

Denote by P the class of continuous functions $q(x;t)$ satisfying estimate (3.44) and possessing the first partial continuous derivatives with respect to t and the third partial continuous derivatives with respect to x. Here we suppose that the elements $q_1(x,t)$ and $q_2(x,t)$ of the matrix $Q(x,t)$ are isospectral potentials which are continuous functions belonging to the class P. By the supposition the mixed partial derivatives $u_{xt}(x,t)$ and $u_{tx}(x,t)$ are also continuous, and hence, the equations (3.42) and (3.45) are compatible, i.e., $u_{xt} = u_{tx}$. Then from the last equality we obtain the Lax compatibility condition for systems (3.42) and (3.45):

$$B_t - \tilde{V}_x + B\tilde{V} - \tilde{V}B = 0.$$

This Lax compatibility condition is equivalent to system (3.41) [138]. In the same way as in the proof of Lemma 2.1 we can prove the following lemma.

Lemma 3.11. *Let the potentials q_1 and q_2 in system* (3.42) *be isospectral and be continuous functions belonging to the class P, then the eigenvalue equations* (3.42) *and* (3.45) *are compatible, i.e., $u_{xt} = u_{tx}$ and the potentials q_1 and q_2 vary with time as the solution of the system* (3.41). *Conversely, if q_1 and q_2 satisfy the system* (3.41), *then $\lambda_t = 0$ and the eigenvalue equations* (3.42) *and* (3.45) *are compatible.*

3.2.2 The time-dependence of the scattering function

The time-dependent solution of system (3.26) with the initial condition:

$$\varphi_1(0,\lambda;t) = \varphi_2(0,\lambda;t) = 1 \text{ for any } t \in (-\infty,\infty), \tag{3.47}$$

is represented in the form:

$$\varphi(x,\lambda;t) = D(\lambda;t)e^{(1)}(x,\lambda;t) + N(\lambda;t)e^{(2)}(x,\lambda;t), \tag{3.48}$$

where $N(\lambda;t) = e_1^{(1)}(0,\lambda;t) - e_2^{(1)}(0,\lambda;t)$, $D(\lambda;t) = e_2^{(2)}(0,\lambda;t) - e_1^{(2)}(0,\lambda;t)$, $N(\lambda;0) = N(\lambda)$, $D(\lambda;0) = D(\lambda)$.

The functions $e^{(1)}(x,\lambda;t)$ and $e^{(2)}(x,\lambda;t)$ are the Jost solutions of the problem for system (3.42) with the boundary condition at infinity:

$$e^{(1)}(x,\lambda;t) = \begin{pmatrix} e^{i\lambda x} + o(1) \\ o(1) \end{pmatrix}, \quad e^{(2)}(x,\lambda;t) = \begin{pmatrix} o(1) \\ e^{-i\lambda x} + o(1) \end{pmatrix} \text{ as } x \to \infty. \tag{3.49}$$

We call

$$S(\lambda;t) = \frac{N(\lambda;t)}{D(\lambda;t)} \text{ for } |\mathrm{Im}\lambda| < \varepsilon_0, \quad -\infty < t < \infty, \tag{3.50}$$

the scattering function for the problem (3.42)–(3.43), where $S(\lambda;0) = S(\lambda)$.

From (3.49) it follows that the asymptotic behavior of the Jost solutions $e^{(1)}(x,\lambda;t)$ and $e^{(2)}(x,\lambda;t)$ as $x \to \infty$ does not depend on time, i.e.,

$$\lim_{x\to\infty} e_{jt}^{(1)}(x,\lambda;t) = 0, \quad \lim_{x\to\infty} e_{jt}^{(2)}(x,\lambda;t) = 0, \; j = 1,2. \tag{3.51}$$

The vector solutions $D(\lambda;t)e^{(1)}(x,\lambda;t)$ and $N(\lambda;t)e^{(2)}(x,\lambda;t)$ of system (3.42) obey evolutionary equation (3.45):

$$[D(\lambda;t)e^{(1)}(x,\lambda;t)]_t = \tilde{V}[D(\lambda;t)e^{(1)}(x,\lambda;t)], \tag{3.52}$$

$$[N(\lambda;t)e^{(2)}(x,\lambda;t)]_t = \tilde{V}[N(\lambda;t)e^{(2)}(x,\lambda;t)]. \tag{3.53}$$

In (3.52) and (3.53) by letting x tend to infinity and using (3.44), (3.46) and (3.51), we obtain

$$D_t(\lambda;t) = 4i\lambda^3 D(\lambda;t), \quad N_t(\lambda;t) = -4i\lambda^3 N(\lambda;t). \tag{3.54}$$

The following lemma is deduced from (3.54), Lemma 3.3 and the invariance with time of the spectrum of the problem (3.42)–(3.43):

Lemma 3.12. *Let the isospectral potentials q_1 and q_2 be continuous functions belonging to the class P and satisfying estimate (3.44). Then the scattering function $S(\lambda;t)$ for the scattering problem (3.42)–(3.43) associated with the system of cubic NLEEs (3.41) possesses the properties:*

(1) $S(\lambda;t)$ *and* $S^{-1}(\lambda;t)$ *are functions of the type* (S) *in the strip* $|\text{Im}\lambda| < \varepsilon_0$ *and are meromorphic in the strip* $|\text{Im}\lambda| < \varepsilon_0$, *where the positive number* ε_0 *is determined by* (3.24)*;*

(2) The scattering function $S(\lambda;t)$ *admits the factorization* (3.50)*. The numerator* $N(\lambda;t)$ *and the denominator* $D(\lambda;t)$ *of* (3.50) *satisfy the evolutionary equations:*

$$N(\lambda;t) = N(\lambda)e^{-4i\lambda^3 t}, \ \text{Im}\lambda \geq -\varepsilon_0, \tag{3.55}$$

$$D(\lambda;t) = D(\lambda)e^{4i\lambda^3 t}, \ \text{Im}\lambda \leq \varepsilon_0; \tag{3.56}$$

(3) $S(\lambda;t)$ *obeys the evolutionary equation:*

$$S(\lambda;t) = S(\lambda)e^{-8i\lambda^3 t} \ \ |\text{Im}\lambda| < \varepsilon_0. \tag{3.57}$$

The scattering function $S(\lambda;t)$ gives the complete characterization of the problem (3.42)–(3.43) on the continuous spectrum.

We now adopt the procedure for a special type of the relationship between $q_1(x,t)$ and $q_2(x,t)$, a case of special interest, i.e.,

$$q_1(x,t) = q_2(x,t) \ \text{ or } \ q_1(x,t) = -q_2(x,t). \tag{3.58}$$

Under the reduction (3.58), the system (3.41) reduces to the mKdV equations on the half-line $0 \leq x < \infty$:

$$q_t + q_{xxx} + 6q^2 q_x = 0 \ \ (q_1 = q_2), \tag{3.59}$$

$$q_t + q_{xxx} - 6q^2 q_x = 0 \ \ (q_1 = -q_2). \tag{3.60}$$

We are given the initial and boundary conditions of equations (3.59) and (3.60):

$$q(x,0) = q(x), \ \ 0 \leq x < \infty, \tag{3.61}$$

$$q(0,t) = b(t), \ \ -\infty < t < \infty, \tag{3.62}$$

where the function $q(x)$ has the continuous derivative of the third-order and satisfies the estimate (3.44). The function $b(t)$ has the continuous derivative of the first-order and $b(t) \in L_1(-\infty,\infty)$.

We denote the system (3.42) with the potential matrix $Q(x,t)$ satisfying the condition (3.58) by (3.42)*. Hence, the x part associated with the mKdV equation (3.59) or (3.60) is the problem (3.42)*, (3.43). We consider the IBVPs (3.59), (3.61), (3.62) and (3.60), (3.61), (3.62) in the class of isospectral potentials, which are continuous functions belonging to class P and satisfying estimate (3.44).

Remark 3.1. Due to (3.34), (3.35) and (3.36) the time-dependence of the standardized polynomials is determined by the unknown BVs evaluated at

$x = 0$ of the Jost solutions $e^{(1)}(x, \lambda; t)$ and $e^{(2)}(x, \lambda; t)$ of the problem (3.42)*, (3.49). Therefore, the main difficulty in solving the IBVPs (3.59), (3.61), (3.62) and (3.60), (3.61), (3.62) lies in finding the unknown BVs that can be expressed in terms of known initial and boundary conditions (3.61) and (3.62). We deal with this difficulty in the next two subsections.

3.2.3 Evaluation of unknown BVs

We first initially derive evolutionary equations for the Jost solutions $e^{(1)}(x, \lambda; t)$ and $e^{(2)}(x, \lambda; t)$ of the problem (3.42), (3.49). Taking (3.54) into account, from (3.52) and (3.53) we obtain the equalities:

$$D(\lambda; t)e_t^{(1)}(x, \lambda; t) = [D(\lambda; t)e^{(1)}(x, \lambda; t)]_t - D_t(\lambda; t)e^{(1)}(x, \lambda; t) =$$
$$\tilde{V}[D(\lambda; t)e^{(1)}(x, \lambda; t)] - 4i\lambda^3 D(\lambda; t)e^{(1)}(x, \lambda; t),$$
$$N(\lambda; t)e_t^{(2)}(x, \lambda; t) = [N(\lambda; t)e^{(2)}(x, \lambda; t)]_t - N_t(\lambda; t)e^{(2)}(x, \lambda; t) =$$
$$\tilde{V}[N(\lambda; t)e^{(2)}(x, \lambda; t)] + 4i\lambda^3 N(\lambda; t)e^{(2)}(x, \lambda; t).$$

Instead of $\tilde{V}$ in these equalities, we substitute the matrix (3.46) and derive the evolutionary equations:

$$e_{1t}^{(1)}(x, \lambda; t) = [-2i\lambda q_1(x, t)q_2(x, t) + q_1(x, t)q_{2x}(x, t) - q_{1x}(x, t)q_2(x, t)]e_1^{(1)}(x, \lambda; t)$$
$$+ [4i\lambda^2 q_1(x, t) + 2\lambda q_{1x}(x, t) - iq_{1xx}(x, t) - 2iq_1^2(x, t)q_2(x, t)]e_2^{(1)}(x, \lambda; t), \tag{3.63}$$

$$e_{2t}^{(1)}(x, \lambda; t) = [4i\lambda^2 q_2(x, t) - 2\lambda q_{2x}(x, t) - 2iq_1(x, t)q_2^2(x, t) - iq_{2xx}(x, t)]e_1^{(1)}(x, \lambda; t)$$
$$+ [2i\lambda q_1(x, t)q_2(x, t) - 8i\lambda^3 + q_{1x}(x, t)q_2(x, t) - q_1(x, t)q_{2x}(x, t)]e_2^{(1)}(x, \lambda; t), \tag{3.64}$$

$$e_{1t}^{(2)}(x,\lambda;t) = [8i\lambda^3 - 2i\lambda q_1(x,t)q_2(x,t) + q_1(x,t)q_{2x}(x,t) - q_{1x}(x,t)q_2(x,t)]e_1^{(2)}(x,\lambda;t)$$
$$+ [4i\lambda^2 q_1(x, t) + 2\lambda q_{1x}(x, t) - iq_{1xx}(x, t) - 2iq_1^2(x, t)q_2(x, t)]e_2^{(2)}(x, \lambda; t), \tag{3.65}$$

$$e_{2t}^{(2)}(x, \lambda; t) = [4i\lambda^2 q_2(x,t) - 2\lambda q_{2x}(x,t) - 2iq_1(x, t)q_2^2(x,t) - iq_{2xx}(x,t)]e_1^{(2)}(x,\lambda;t)$$
$$+ [2i\lambda q_1(x, t)q_2(x, t) + q_{1x}(x, t)q_2(x, t) - q_1(x, t)q_{2x}(x, t)]e_2^{(2)}(x, \lambda; t). \tag{3.66}$$

Owing to estimate (3.44) and boundary conditions (3.49), the right-hand sides of (3.63)–(3.66) tend to zero as $x \to \infty$ uniformly in t. Hence, the derived evolution equations (3.63)–(3.66) obey the required condition (3.51) as $x \to \infty$.

The time-dependence of solution (3.48) of the problem (3.42)*, (3.47) is defined by (3.45) and (3.46):

$$\varphi_t(x, \lambda; t) = \tilde{V}[\varphi(x, \lambda; t)]. \tag{3.67}$$

It follows from initial condition (3.47) at $x = 0$ that

$$\varphi_{1t}(0,\lambda;t) = \varphi_{2t}(0,\lambda;t) = 0. \tag{3.68}$$

Using (3.47), (3.58) and (3.68), we obtain from equation (3.67) at $x = 0$ the important relations between the boundary values of functions $q_1(x,t)$, $q_2(x,t)$ and their first and second x-derivatives evaluated at $x = 0$:

$$-4i\lambda^3 - 4i\lambda^2 q_1(0,t) + 2i\lambda q_1(0,t)q_2(0,t) + 2iq_1^2(0,t)q_2(0,t) + iq_{1xx}(0,t) - 2\lambda q_{1x}(0,t) = q_1(0,t)q_{2x}(0,t) - q_{1x}(0,t)q_2(0,t) = 0.$$

$$-4i\lambda^3 + 4i\lambda^2 q_2(0,t) + 2i\lambda q_1(0,t)q_2(0,t) - 2iq_1(0,t)q_2^2(0,t) - iq_{2xx}(0,t) - 2\lambda q_{2x}(0,t) = q_1(0,t)q_{2x}(0,t) - q_{1x}(0,t)q_2(0,t) = 0.$$

It follows from the last equalities that

$$q_1(0,t)q_{2x}(0,t) = q_{1x}(0,t)q_2(0,t),$$

$$-4i\lambda^3 + 2i\lambda q_1(0,t)q_2(0,t) = 4i\lambda^2 q_1(0,t) - 2iq_1^2(0,t)q_2(0,t) - iq_{1xx}(0,t) + 2\lambda q_{1x}, \tag{3.69}$$

$$4i\lambda^3 - 2i\lambda q_1(0,t)q_2(0,t) = 4i\lambda^2 q_2(0,t) - 2iq_1(0,t)q_2^2(0,t) - iq_{2xx}(0,t) - 2\lambda q_{2x}. \tag{3.70}$$

We use conditions (3.58) and relations (3.69), (3.70) to get rid of $q_{jx}(0,t)$ and $q_{jxx}(0,t)$, $j = 1,2$ in the derived systems (3.63)–(3.66) at $x = 0$. As a result, the systems (3.63)–(3.66) become the systems of first-order ODEs for the unknown BVs $e^{(1)}(0,\lambda;t)$ and $e^{(2)}(0,\lambda;t)$ of the Jost solutions of system (3.42)* evaluated at $x = 0$:

$$e_{1t}^{(1)}(0,\lambda;t) = -2i\lambda q_1(0,t)q_2(0,t)e_1^{(1)}(0,\lambda;t) + [2i\lambda q_1(0,t)q_2(0,t) - 4i\lambda^3]e_2^{(1)}(0,\lambda;t), \tag{3.71}$$

$$e_{2t}^{(1)}(0,\lambda;t) = [4i\lambda^3 - 2i\lambda q_1(0,t)q_2(0,t)]e_1^{(1)}(0,\lambda;t) + [2i\lambda q_1(0,t)q_2(0,t) - 8i\lambda^3]e_2^{(1)}(0,\lambda;t), \tag{3.72}$$

$$e_{1t}^{(2)}(0,\lambda;t) = [8i\lambda^3 - 2i\lambda q_1(0,t)q_2(0,t)]e_1^{(2)}(0,\lambda;t) + [2i\lambda q_1(0,t)q_2(0,t) - 4i\lambda^3]e_2^{(2)}(0,\lambda;t), \tag{3.73}$$

$$e_{2t}^{(2)}(0,\lambda;t) = [4i\lambda^3 - 2i\lambda q_1(0,t)q_2(0,t)]e_1^{(2)}(0,\lambda;t) + 2i\lambda q_1(0,t)q_2(0,t)e_2^{(2)}(0,\lambda;t). \tag{3.74}$$

Owing to (3.55) and (3.56) the evolution equations (3.71)–(3.74) for the unknown BVs of the Jost solutions at $x = 0$ are reduced to the form:

$$e_{nt}^{(1)}(0,\lambda;t) = -4i\lambda^3 e_n^{(1)}(0,\lambda;t) + iN(\lambda)[4\lambda^3 - 2\lambda q_1(0,t)q_2(0,t)]e^{-4i\lambda^3 t}, n = 1,2 \tag{3.75}$$

$$e_{nt}^{(2)}(0,\lambda;t) = 4i\lambda^3 e_n^{(2)}(0,\lambda;t) - iD(\lambda)[4\lambda^3 - 2\lambda q_1(0,t)q_2(0,t)]e^{4i\lambda^3 t}, n = 1,2. \tag{3.76}$$

The solutions of first-order ODEs (3.75) and (3.76) are found by the method of variation constants:

$$e_n^{(1)}(0,\lambda;t)=\left\{e_n^{(1)}(0,\lambda;0)+iN(\lambda)[4\lambda^3t-2\lambda\int_0^t q_1(0,\tau)q_2(0,\tau)d\tau]\right\}e^{-4i\lambda^3t}, n=1,2, \tag{3.77}$$

$$e_n^{(2)}(0,\lambda;t)=\left\{e_n^{(2)}(0,\lambda;0)-iD(\lambda)[4\lambda^3t-2\lambda\int_0^t q_1(0,\tau)q_2(0,\tau)d\tau]\right\}e^{4i\lambda^3t}, n=1,2. \tag{3.78}$$

In the right-hand side of equalities (3.77) and (3.78) the second terms of the sums are the particular solutions of equations (3.75) and (3.76), while the first terms are the solutions of the corresponding homogeneous equations. Thus, the unknown BVs $e^{(1)}(0,\lambda;t)$ and $e^{(2)}(0,\lambda;t)$ are found by the formulas (3.77) and (3.78) in terms of the quantities: $e^{(1)}(0,\lambda;0)$, $e^{(2)}(0,\lambda;0)$, $N(\lambda)$, $D(\lambda)$ and $q_1(0,t)q_2(0,t)$, which are calculated from the known initial and boundary conditions (3.61) and (3.62).

Since $N(\lambda_k^+)=0, \mathrm{Im}\lambda_k^+\geq\varepsilon_0$, $k=1,...,\gamma_1$, and $D(\lambda_j^-)=0$, $\mathrm{Im}\lambda_j^-\leq-\varepsilon_0$, $j=1,...,\gamma_2$, then from (3.77) and (3.78) it follows that the BVs at $x=0$ and non-real singular numbers of the Jost solutions of the problem (3.42)*, (3.49) are calculated by the formulas:

$$e_n^{(1)}(0,\lambda_k^+;t)=e_n^{(1)}(0,\lambda_k^+;0)e^{-4i(\lambda_k^+)^3t}\neq0,\quad n=1,2;\quad k=1,...,\gamma_1, \tag{3.79}$$

$$e_n^{(2)}(0,\lambda_j^-;t)=e_n^{(2)}(0,\lambda_j^-;0)e^{4i(\lambda_j^-)^3t}\neq0,\quad n=1,2;\quad j=1,...,\gamma_2, \tag{3.80}$$

Hence, the calculated BVs (3.77) and (3.78) of solutions $e^{(1)}(x,\lambda;t)$ and $e^{(2)}(x,\lambda;t)$ of system (3.42)* evaluated at $x=0$ and at the singular numbers do not depend on the boundary condition (3.62).

3.2.4 The time-dependence of the scattering data

Knowing the BVs $e_1^{(1)}(0,\lambda;t)$ and $e_2^{(2)}(0,\lambda;t)$ defined by formulas (3.77) and (3.78), we calculate the time-dependent coefficients (3.31) and (3.32):

$$A_q(\lambda_k^+;t)=\left[\frac{d^q}{d\lambda^q}\frac{1}{e_1^{(1)}(0,\lambda;t)}\right]_{\lambda=\lambda_k^+},\quad q=0,1,...,m_k-1,\quad k=1,...,\gamma_1 \tag{3.81}$$

$$B_p(\lambda_j^-;t)=\left[\frac{d^p}{d\lambda^p}\frac{1}{e_2^{(2)}(0,\lambda;t)}\right]_{\lambda=\lambda_j^-},\quad p=0,1,...,n_j-1,\quad j=1,...,\gamma_2, \tag{3.82}$$

where the functions $e_1^{(1)}(0,\lambda;t)$ and $e_2^{(2)}(0,\lambda;t)$ are calculated by (3.77) for $n=1$ and (3.78) for $n=2$, respectively.

Substituting (3.81) and (3.82) into (3.34) and (3.35), respectively, we determine the time-dependent standardized polynomials (3.36):

$$\begin{aligned} g_k^+(-x;t) &= p_k(x;t)e^{i\lambda_k^+ x}, \quad k=1,...,\gamma_1; \\ f_j^-(x;t) &= \tilde{q}_j(x;t)e^{-i\lambda_j^- x}, \quad k=1,...,\gamma_2. \end{aligned} \tag{3.83}$$

The functions (3.83) characterize the scattering problem (3.42)*, (3.43) on the discrete spectrum. The scattering function (3.50) possessing the properties enumerated in Lemma 3.12 characterizes this problem on the continuous spectrum. Hence, the time-dependence of the scattering data of the scattering problem (3.42)*, (3.43) on the complete spectrum is determined by the functions:

$$\begin{aligned} f(x;t) &= f_S(x;t) - \sum_{j=1}^{\gamma_2} f_j^-(x;t), \\ g(-x;t) &= g_S(-x;t) - \sum_{k=1}^{\gamma_1} g_k^+(-x;t), \end{aligned} \qquad 0 \le x < \infty, \quad -\infty < t < \infty \tag{3.84}$$

where

$$\begin{aligned} f_S(x;t) &= \frac{1}{2\pi}\int_{-\infty-i\eta}^{\infty-i\eta} [S(\lambda)e^{-8i\lambda^3 t} - 1]e^{-i\lambda x}d\lambda, \\ g_S(x;t) &= \frac{1}{2\pi}\int_{-\infty+i\eta}^{\infty+i\eta} [S^{-1}(\lambda)e^{8i\lambda^3 t} - 1]e^{-i\lambda x}d\lambda. \end{aligned} \qquad 0 < \eta < \varepsilon_0 \tag{3.85}$$

Thus, the difficulty mentioned in the Remark 3.1 is overcome. The following lemma is proved.

Lemma 3.13. *(1) The unknown BVs $e^{(1)}(0,\lambda;t)$ and $e^{(2)}(0,\lambda;t)$ of the Jost solutions of the problem (3.42)*, (3.49) obey the evolutionary equations (3.75) and (3.76), respectively. The solutions of these equations, i.e., the BVs are found by the explicit formulas (3.77) and (3.78), which are expressible in terms of quantities calculated from the known initial and boundary conditions (3.61) and (3.62);*

(2) The time-dependence of the scattering data of the problem (3.42), (3.43) on the discrete spectrum and complete spectrum is determined by the functions (3.81)–(3.83) and (3.84)–(3.85), respectively.*

3.2.5 The solution of the IBVPs for mKdV equations

Let the time-dependent potential matrix C of the scattering problem (3.42)*, (3.43) associated with the mKdV equation (3.59) be given by

$$C = C_1(x,t) = \begin{pmatrix} 0 & -c_1(x,t) \\ c_2(x,t) & 0 \end{pmatrix} = \begin{pmatrix} 0 & q(x,t) \\ q(x,t) & 0 \end{pmatrix},$$
$$0 \leq x < \infty, \quad -\infty < t < \infty, \tag{3.86}$$

while the potential matrix C of the scattering problem (3.42)*–(3.43) associated with the mKdV equation (3.60) is

$$C = C_2(x,t) = \begin{pmatrix} 0 & -c_1(x,t) \\ -c_2(x,t) & 0 \end{pmatrix} = \begin{pmatrix} 0 & q(x,t) \\ -q(x,t) & 0 \end{pmatrix},$$
$$0 \leq x < \infty, \quad -\infty < t < \infty, \tag{3.87}$$

where the potential $q(x,t)$ belongs to the class P and satisfies estimate (3.44).

We consider the scattering problem for the system (3.42) with the potential matrix $C_1(x,0)$ and the scattering problem for the system (3.42) with the potential matrix $C_2(x,0)$. The scattering data sets of these scattering problems are calculated according to Definition 3.2. We denote these sets by s_1 and s_2, respectively. Every set $s_n = s_n(0)$, $n = 1,2$, consists of the quantities of the type (a), (b) and (c) enumerated in Theorem 3.1. These quantities satisfy conditions which are similar to the conditions (1)–(5) enumerated in this theorem. Then making use of Lemmas 3.12 and 3.13, we construct the time-dependent sets $s_n(t)$ from the calculated sets s_n and known initial and boundary conditions (3.61) and (3.62):

$$s_n(t) = \{S_n(\lambda)e^{-8i\lambda^3 t};\ \lambda^+_{kn},\ \mathrm{Im}\lambda^+_{kn} \geq \varepsilon_0,\ \lambda^-_{jn},\ \mathrm{Im}\lambda^-_{jn} \leq -\varepsilon_0;$$
$$p_{kn}(x;t),\ \tilde{q}_{jn}(x;t),\ k = 1,...,\gamma_{1n}\ j = 1,...,\gamma_{2n}\}\ \varepsilon_0 > 0,\ \ n = 1,2. \tag{3.88}$$

Since the potential matrices (3.86) and (3.87) are particular cases of the potential matrix of the problem (3.4)–(3.5), then the Theorem 3.1 for the description of scattering data of the problem (3.42)*–(3.43) is valid.

Thus, Theorem 3.1 and Lemmas 3.12 and 3.13 reduce the problem of solving IBVPs (3.59), (3.61), (3.62) and (3.60), (3.61), (3.62) to that of solving the systems of fundamental equations (3.38) and (3.39) for every $x \geq 0$, $\xi > 0$:

$$\begin{cases} H^{[n]}_{11}(x, x+\xi;t) + \displaystyle\int_0^\infty H^{[n]}_{12}(x, x+\tau;t) f_n(2x+\xi+\tau;t)d\tau = 0, \\ g_n(-2x-\xi;t) + H^{[n]}_{12}(x,x+\xi;t) + \displaystyle\int_0^\infty H^{[n]}_{11}(x,x+\tau;t) g_n(-2x-\xi-\tau;t)d\tau = 0, \end{cases} \tag{3.89}$$

$$\begin{cases} f_n(2x+\xi;t)+H_{21}^{[n]}(x,x+\xi;t)+\int\limits_0^\infty H_{22}^{[n]}(x,x+\tau;t)f_n(2x+\xi+\tau;t)d\tau=0, \\ H_{22}^{[n]}(x,x+\xi;t)+\int\limits_0^\infty H_{21}^{[n]}(x,x+\tau;t)g_n(-2x-\xi-\tau;t)d\tau=0, \end{cases} \tag{3.90}$$

where both x and t enter the systems (3.89) and (3.90) merely as parameters, $f_n(x;t)$ and $g_n(-x;t)$ are known functions constructed by the formulas (3.84) and (3.85) from the data set $s_n(t), n=1,2$ given by (3.88); $H_{jk}^{[n]}(x,x+\xi;t)$, $j,k=1,2$, are unknown functions of ξ for every $(x,t)\in[0,\infty)\times(-\infty,\infty)$.

The systems (3.89) and (3.90) for $n=1$ are designed for the problem of reconstructing potential matrix (3.86). These systems for $n=2$ are designed for the problem of reconstructing potential matrix (3.87).

The potential $q(x,t)$ in (3.86) and (3.87) belongs to the class P, therefore from integral equations of the form (3.13)–(3.16) with the potential $q(x,t)=-c_1(x,t)=c_2(x,t)$ belonging to the class P, it follows that the solutions $H_{ij}(x,s)$, $0\le x\le s$, $i,j=1,2$ of these integral equations belong also to this class. Further, using systems (3.89) and (3.90), we can verify that the functions $f_n(2x+\xi;t)$ and $g_n(-2x-\xi;t)$ in these systems possess the properties, which are analogous with the properties of the functions $H_{jk}^{[n]}(x,x+\xi;t)$, $j,k=1,2$, i.e., the functions $f_n(2x+\xi;t)$ and $g_n(-2x-\xi;t)$, $n=1,2$ in systems (3.89) and (3.90) belong also to the class P. Owing to this fact and the condition (5) of the Theorem 3.1, the constructed functions $f_n(x;t)$ and $g_n(-x;t)$ satisfy the estimate for any $t\in(-\infty,\infty)$:

$$|f_n(x;t)|\le\tilde{C}\exp\left\{\frac{-\varepsilon x}{4}\right\},\quad |g_n(-x;t)|\le\tilde{C}\exp\left\{\frac{-\varepsilon x}{4}\right\}, \tag{3.91}$$

where $\tilde{C}$ is a constant, $\varepsilon>0$, $n=1,2$, and $f_n(x;t)$ and $g_n(-x;t)$ belong to the class P, i.e., they are continuous functions possessing the first partial continuous derivative with respect to t and the third partial continuous derivative with respect to x.

By using the estimate (3.91) and the properties of functions $f_n(x;t)$, $g_n(-x;t)$, arguing as in the proof of Theorem 1.1 and Lemmas 3.12 and 3.13, we can prove the following:

(I). The systems (3.89) and (3.90) are the Fredholm systems of integral equations in $L_1[0,\infty)$ and have for every $(x,t)\in[0,\infty)\times(-\infty,\infty)$ a unique solution in $L_1[0,\infty)$: $(H_{11}^{[n]}(x,x+\xi;t),H_{12}^{[n]}(x,x+\xi;t))$ and $(H_{21}^{[n]}(x,x+\xi;t),H_{22}^{[n]}(x,x+\xi;t)), n=1,2$, respectively. The obtained solution satisfies an estimate of the type (3.10);

(II). The vector functions

$$(e^{(1)}_{1[n]}(x,\lambda;t), e^{(1)}_{2[n]}(x,\lambda;t)) = (e^{i\lambda x} + e^{i\lambda x}\int_0^\infty H^{[n]}_{11}(x,x+\tau;t)e^{i\lambda\tau}d\tau,$$
$$e^{i\lambda x}\int_0^\infty H^{[n]}_{21}(x,x+\tau;t)e^{i\lambda\tau}d\tau),\quad n=1,2, \tag{3.92}$$

$$(e^{(2)}_{1[n]}(x,\lambda;t), e^{(2)}_{2[n]}(x,\lambda;t)) = (e^{-i\lambda x}\int_0^\infty H^{[n]}_{12}(x,x+\tau;t)e^{-i\lambda\tau}d\tau,\ e^{-i\lambda x}+$$
$$e^{-i\lambda x}\int_0^\infty H^{[n]}_{22}(x,x+\tau;t)e^{-i\lambda\tau}d\tau),\quad n=1,2, \tag{3.93}$$

are the Jost solutions of a system of the form (3.42)* with boundary condition (3.49) at infinity.

The functions (3.92) and (3.93) for $n=1$ satisfy the system (3.42)* with the reconstructed potential matrix (3.86) for the mKdV equation (3.59):

$$C_1(x,t) = \begin{pmatrix} 0 & -c_1(x,t) \\ c_2(x,t) & 0 \end{pmatrix},$$
$$-c_1(x,t) = c_2(x,t) = 2iH^{[1]}_{12}(x,x;t) = 2iH^{[1]}_{21}(x,x;t). \tag{3.94}$$

While the functions (3.92) and (3.93) for $n=2$ satisfy the system (3.42)* with the reconstructed potential matrix (3.87) for the mKdV equation (3.60):

$$C_2(x,t) = \begin{pmatrix} 0 & -c_1(x,t) \\ -c_2(x,t) & 0 \end{pmatrix},$$
$$-c_1(x,t) = -c_2(x,t) = 2iH^{[2]}_{12}(x,x;t) = -2iH^{[2]}_{21}(x,x;t). \tag{3.95}$$

In addition the found potential function $c_j(x,t), j=1,2$ satisfies an estimate of the type (3.44) and has the first partial continuous derivative with respect to t and the third partial continuous derivative with to x;

(III). The sets $s_1(t)$ and $s_2(t)$ given by (3.88) constructed from known initial and boundary conditions (3.61) and (3.62) are the scattering data of the scattering problems for a system of the form (3.42)* with the reconstructed potentials (3.94) and with (3.95), respectively. In fact, according to (3.50) and (3.57), the BVs of the Jost solutions (3.92) and (3.93) evaluated at $x=0$ are related to the scattering function by the identities:

$$e^{(1)}_{1[n]}(0,\lambda;t) - e^{(1)}_{2[n]}(0,\lambda;t) - \left[e^{(2)}_{2[n]}(0,\lambda;t) - e^{(2)}_{1[n]}(0,\lambda;t)\right] S_n(\lambda)e^{-8i\lambda^3 t} = 0, \tag{3.96}$$

$|\mathrm{Im}\lambda| < \varepsilon_0, \ \ \varepsilon_0 > 0, \ \ -\infty < t < \infty, \ \ n = 1, 2.$

Owing to (3.55) and (3.56), the BVs of the Jost solutions at $x = 0$ and the non-real singular numbers satisfy the conditions (1.32):

$$\begin{aligned}
&\left\{\frac{d^q}{d\lambda^q}\left[e^{(1)}_{1[n]}(0,\lambda;t) - e^{(1)}_{2[n]}(0,\lambda;t)\right]\right\}_{\lambda=\lambda^+_{kn}} = \\
&\left\{\frac{d^q}{d\lambda^q}\left[e^{(1)}_{1[n]}(0,\lambda;0) - e^{(1)}_{2[n]}(0,\lambda;0)\right] e^{-4i\lambda^3 t}\right\}_{\lambda=\lambda^+_{kn}} = 0, \\
&\mathrm{Im}\lambda^+_{kn} \geq \varepsilon_0, \ \ q = 0, 1, ..., m_k - 1, \ \ k = 1, ..., \gamma_{1n}, \ \ n = 1, 2; \\
&\left\{\frac{d^p}{d\lambda^p}\left[e^{(2)}_{2[n]}(0,\lambda;t) - e^{(2)}_{1[n]}(0,\lambda;t)\right]\right\}_{\lambda=\lambda^-_{jn}} = \\
&\left\{\frac{d^p}{d\lambda^p}\left[e^{(2)}_{2[n]}(0,\lambda;0) - e^{(2)}_{1[n]}(0,\lambda;0)\right] e^{4i\lambda^3 t}\right\}_{\lambda=\lambda^-_{jn}} = 0 \\
&\mathrm{Im}\lambda^-_{jn} \leq -\varepsilon_0, \ \ p = 0, 1, ..., n_j - 1, \ \ j = 1, ..., \gamma_{2n}, \ \ n = 1, 2,
\end{aligned} \tag{3.97}$$

where m_k is the multiplicity of the zero λ^+_{kn} of the numerator, n_j is the multiplicity of the zero λ^-_{jn} of the denominator of the scattering function $S_n(\lambda), n = 1, 2$.
The identity (3.96) characterizes the problem (3.42)*, (3.43) on the continuous spectrum, while the identities (3.97) characterize this problem on the discrete spectrum;

(IV). A correspondence between the scattering data set $s_1(t)$ ($s_2(t)$) given by (3.88) and the reconstructed potential matrix (3.94) ((3.95)) is one-to-one, and the matrices $Q(x,t)$ and $\tilde{V}(x,t)$ in systems (3.42) and (3.45) are reconstructed uniquely from the potential matrix (3.94) ((3.95)). In fact, due to the estimate (3.91) and properties of functions $f_n(x,t)$ and $g_n(-x,t)$, the solutions of systems (3.89) and (3.90) are found uniquely from the given data set (3.88). By Lemmas 1.11 and 1.12 the obtained solutions satisfy an estimate of the type (1.63) and systems of integral equations of the form (3.13), (3.15) and (3.14), (3.16) with the reconstructed potential matrix (3.94) ((3.95)). Owing to the unique solvability of these systems of integral equations, the solutions of these systems coincide with the solutions of systems (3.89) and (3.90) and are kernels of the transformation operator. The found solutions $H^{[n]}_{jk}(x, x+\xi; t), j, k = 1, 2$, possess the properties which are similar to properties of functions $f_n(x,t)$ and $g_n(-x,t)$, *i.e*, they are continuous functions satisfying an estimate of the type (3.10) and possessing the first partial continuous derivative with respect to t and the third partial continuous derivative with respect to x. Hence, the eigenvalue equations (3.42) and (3.45) with the reconstructed potential matrix (3.94) ((3.95)) are compatible, i.e., $u_{xt} = u_{tx}$. Then by Lemma 3.11 the recovered potential (3.94) ((3.95)) is the solution of the IBVP (3.59), (3.61), (3.62) ((3.60), (3.61), (3.62)).

Thus, the IBVPs for the mKdV equations are solved.

Theorem 3.2. *The IBVPs* (3.59), (3.61), (3.62) *and* (3.60), (3.61), (3.62) *are solved by the ISM in the class of isospectral potentials which are continuous functions belonging to the class P and satisfying estimate* (3.44). *Namely,*

(1) The quantities of the data sets $s_1(t)$ and $s_2(t)$ of the type (3.88) *constructed from the known initial and boundary conditions* (3.61) *and* (3.62) *are the scattering data of the reconstructed problems* (3.42)*, (3.43) *with the reconstructed potential matrix* (3.94) *and* (3.95), *respectively. The time-dependence of these sets is determined by the Lemmas 3.12 and 3.13;*

(2) Every scattering data set $s_n(t), n = 1, 2$ of the type (3.88) *corresponds to a unique solution of the considered IBVP. This solution is found by formula* (3.94) *with $n = 1$ and* (3.95) *with $n = 2$ in terms of solutions of systems of fundamental equations* (3.89) *and* (3.90) *for $n = 1, 2$. The obtained solutions are the kernels of the transformation operator and are constructed from the data set, which is calculated from the known initial and boundary conditions* (3.61) *and* (3.62).

Remark 3.2. The potentials $q_j(x,t), j = 1, 2$ in the mKdV equations (3.59) and (3.60) depend on an additional parameter t and satisfy constraint (3.44). The initial point is $t = 0$, therefore, it is more natural to confine the t-values belonging to $(-\infty, \infty)$ to values $t \in [0, \infty)$. Due to this fact, the IBVPs (3.59)–(3.61)–(3.62) and (3.60)–(3.61)–(3.62) can be solved for the region $(x, t) \in [0, \infty) \times [0, \infty)$.

Figure 3.1 illustrating our method for solving the IBVPs for mKdV equations (3.59) and (3.60) on a half-line can be outlined as follows:

3.2.6 Relation between solutions of the mKdV and KdV equations

We prove that there exists a relation between the solution of the IBVP (3.59), (3.61)–(3.62) and the solution of the KdV equation on the half-line:

$$r_t - 6rr_x + r_{xxx} = 0, \quad 0 \le x < \infty, \quad -\infty < t < \infty. \tag{3.98}$$

By using the simple substitution:

$$\psi_1(x,\lambda;t) = \frac{1}{2}[u_1(x,\lambda;t) + u_2(x,\lambda;t)], \; \psi_2(x,\lambda;t) = \frac{1}{2i}[u_2(x,\lambda;t) - u_1(x,\lambda;t)],$$

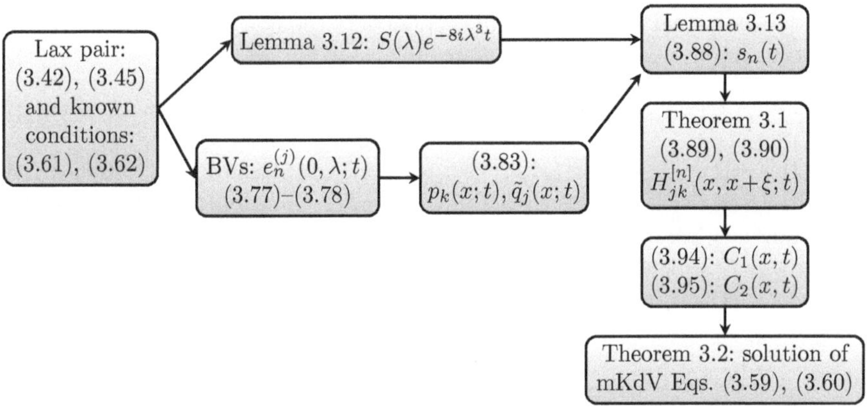

FIGURE 3.1: The steps in solving the IBVPs for the mKdV equations (3.59) and (3.60).

we can reduce the system (3.42) to the more usual form of the so-called Dirac system of equations with mass $m = 0$:

$$T\tilde{\psi}_x(x,\lambda;t) + \Omega(x,t)\tilde{\psi}(x,\lambda;t) = \lambda\tilde{\psi}(x,\lambda;t), \;\; 0 \le x < \infty, \;\; -\infty < t < \infty, \tag{3.99}$$

where

$$T = \begin{pmatrix} 0 & -1 \\ 1 & 0 \end{pmatrix}, \;\; \tilde{\psi}(x,\lambda;t) = \begin{pmatrix} \psi_1(x,\lambda;t) \\ \psi_2(x,\lambda;t) \end{pmatrix}, \;\; \Omega(x,t) = \begin{pmatrix} \alpha(x,t) & \beta(x,t) \\ \beta(x,t) & -\alpha(x,t) \end{pmatrix},$$

and $\alpha(x,t) = \dfrac{1}{2}[c_1(x,t) + c_2(x,t)]$, $\beta(x,t) = \dfrac{i}{2}[c_1(x,t) - c_2(x,t)]$.

From (3.86) we have $-c_1(x,t) = c_2(x,t) = c(x,t)$, therefore

$$\alpha(x,t) \equiv 0, \;\; \beta(x,t) = -ic_2(x,t) = -ic(x,t),$$

and the system (3.99) is reduced to the system of the form:

$$\psi_{1x}(x,\lambda;t) + \beta(x,t)\psi_1(x,\lambda;t) = \lambda\psi_2(x,\lambda;t), \tag{3.100}$$

$$0 \le x < \infty, \;\; -\infty < t < \infty$$

$$-\psi_{2x}(x,\lambda;t) + \beta(x,t)\psi_2(x,\lambda;t) = \lambda\psi_1(x,\lambda;t), \tag{3.101}$$

where the potential function $\beta(x,t)$ satisfies estimate (3.44).

Substituting (3.101) into (3.100), we obtain that the function $\psi(x,\lambda;t) = \psi_2(x,\lambda;t)$ satisfies a Schrödinger equation on the half-line:

$$-\psi_{xx}(x,\lambda;t) + r(x,t)\psi(x,\lambda;t) = \lambda^2\psi(x,\lambda;t), \;\; 0 \le x < \infty, \; -\infty < t < \infty, \tag{3.102}$$

where the potential function $r(x,t)$ is defined by the non-linear Miura's transformation [1]:

$$r(x,t) = \beta^2(x,t) + \beta_x(x,t) = -c^2(x,t) - ic_x(x,t). \tag{3.103}$$

Since the solution $u = (u_1(x,\lambda;t),\ u_2(x,\lambda;t))$ of system (3.42) satisfies the boundary condition (3.43), then the solution of equation (3.102) obeys the zero boundary condition at the origin $x = 0$:

$$\psi(0,\lambda;t) = \frac{1}{2i}[u_2(0,\lambda;t) - u_1(0,\lambda;t)] = 0. \tag{3.104}$$

Let $c(x,t)$ be a solution of the IBVP (3.59), (3.61)–(3.62):

$$Pc = c_t + c_{xxx} + 6c^2c_x = 0 \quad 0 \le x < \infty, \quad -\infty < t < \infty, \tag{3.105}$$

then the function $r(x,t)$ defined by the non-linear Miura's transformation (3.103) is a solution of the KdV equation (3.98). Indeed, substituting (3.103) into the left-hand side of the equation (3.98), we obtain

$$Rr = r_t - 6rr_x + r_{xxx} = -(2c + i\frac{\partial}{\partial x})Pc, \tag{3.106}$$

so that if $c(x,t)$ satisfies the mKdV equation (3.105), i.e., $Pc = 0$, then the function $r(x,t)$ defined by (3.103) satisfies the KdV equation (3.98):

$$Rr = R(-c^2 - ic_x) = 0.$$

If the potential $c(x,t)$ is a pure imaginary-valued function, i.e., $c(x,t) = i\mu(x,t)$, wherein $\mu(x,t)$ is a real-valued function, then the function $r(x,t)$ is also real:

$$r(x,t) = -(i\mu(x,t))^2 - i(i\mu(x,t))_x = \mu^2(x,t) + \mu_x(x,t), \tag{3.107}$$

and then $\mu(x,t)$ satisfies the mKdV equation (3.60):

$$P(i\mu) = i\left(\mu_t + \mu_{xxx} - 6\mu^2\mu_x\right) = 0. \tag{3.108}$$

The function $r(x,t)$ defined by (3.107) is a solution of the KdV equation (3.108). Indeed, instead of the function $r(x,t)$ in (3.98), we substitute the function $\mu^2(x,t) + \mu_x(x,t)$ and get

$$Rr = R(\mu^2 + \mu_x) = \left(2\mu + \frac{\partial}{\partial x}\right)\left(\mu_t + \mu_{xxx} - 6\mu^2\mu_x\right) = 0.$$

Hence, if the real-valued function $\mu(x,t)$ satisfies the mKdV equation (3.60), then the function $r(x,t)$ defined by (3.107) satisfies the KdV equation (3.98).

We arrive now at the following theorem.

Theorem 3.3. *Under the reduction $c_1(x,t) = -c_2(x,t) = -c(x,t)$ the system* (3.42) *with the boundary condition* (3.43) *is reduced to the Schrödinger equation* (3.102) *with the zero boundary condition* (3.104). *The Miura's transformation* (3.103) *transforms the solution $c(x,t)$ of the mKdV equation* (3.59) *into the solution of the KdV equation* (3.98).

If $c_1(x,t) = -c_2(x,t) = -i\mu(x,t)$, where $\mu(x,t)$ is a real-valued solution of equation (3.60), *then the transformation* (3.107) *transforms the solution $\mu(x,t)$ of the mKdV equation* (3.60) *into the real solution $\mu^2(x,t) + \mu_x(x,t)$ of the KdV equation* (3.98).

The inverse scattering problem for the Shrödinger equation (3.102) with the real potential and with zero boundary condition (3.104) is associated with the IBVP for the KdV equation (3.98). This IBVP will be studied in Chapter 6.

3.3 Non-scattering potentials and exact solutions

3.3.1 Exact solutions of systems of fundamental equations

As has been shown in Section 2.2 of Chapter 2, the scattering problem with the non-scattering potentials is completely determined by the characteristics of the discrete spectrum. The non-scattering data of the scattering problem (3.4)–(3.5) with the non-scattering potentials in system (3.4) are completely determined by the discrete spectrum and can be formulated in the form:

$$s = \Bigg\{ S(\lambda;t) \equiv 1, \ \ |\mathrm{Im}\lambda| < \varepsilon_0; \ \ \lambda_k^+, \ \mathrm{Im}\lambda_k^+ \geq \varepsilon_0, \ \ p_k(x), \ \ k = 1, ..., \gamma_1;$$

$$\lambda_j^-, \ \mathrm{Im}\lambda_j^- \leq -\varepsilon_0, \ \ \varepsilon_0 > 0, \ \ \tilde{q}_j(x), \ \ j = 1, ..., \gamma_2 \Bigg\}. \tag{3.109}$$

Under condition (3.109) the systems of fundamental equations (3.38) and (3.39) can be solved exactly. The solutions of these systems are to be sought in the form:

$$H_{n1}(x,u) = \sum_{j=1}^{\gamma_2} H_{n1,j}(x) e^{-i\lambda_j^- u}, \ \ H_{n2}(x,u) = \sum_{k=1}^{\gamma_1} H_{n2,k}(x) e^{i\lambda_k^+ u} \tag{3.110}$$

For simplicity, we suppose that the given non-scattering data (3.109) consist of only two discrete spectra: $\lambda^+, \mathrm{Im}\lambda^+ \geq \varepsilon_0$ and $\lambda^-, \mathrm{Im}\lambda^- \leq -\varepsilon_0, \varepsilon_0 > 0$, which are simple poles. Using the formulas (3.34) and (3.35), we calculate the functions $g^+(-x)$ and $f^-(x)$ for this case [114]:

$$f^-(x) = i\frac{e^{-i\lambda^- x}}{D_\lambda(\lambda^-)e_2^{(2)}(0,\lambda^-)}, \quad g^+(-x) = i\frac{e^{i\lambda^+ x}}{N_\lambda(\lambda^+)e_1^{(1)}(0,\lambda^+)}, \quad 0 \le x < \infty, \tag{3.111}$$

where

$$\begin{aligned}
&N(\lambda^+;0) = N(\lambda^+) = e_1^{(1)}(0,\lambda^+) - e_2^{(1)}(0,\lambda^+) = 0,\\
&D(\lambda^-;0) = D(\lambda^-) = e_2^{(2)}(0,\lambda^-) - e_1^{(2)}(0,\lambda^-) = 0,\\
&N_\lambda(\lambda^+;0) = N_\lambda(\lambda^+) \neq 0, \quad D_\lambda(\lambda^-;0) = D_\lambda(\lambda^-) \neq 0,\\
&e_1^{(1)}(0,\lambda^+) \neq 0, \quad e_2^{(2)}(0,\lambda^-) \neq 0.
\end{aligned}$$

With due regard for (3.110), we write the systems (3.38) and (3.39) for this case:

$$\begin{cases}
H_{11}(x) + \dfrac{e^{i(\lambda^+ - \lambda^-)x}}{(\lambda^+ - \lambda^-)e_2^{(2)}(0,\lambda^-)D_\lambda(\lambda^-)} H_{12}(x) = 0,\\
\dfrac{e^{i(\lambda^+ - \lambda^-)x}}{(\lambda^+ - \lambda^-)e_1^{(1)}(0,\lambda^+)N_\lambda(\lambda^+)} H_{11}(x) + H_{12}(x) = i\dfrac{e^{i\lambda^+ x}}{e_1^{(1)}(0,\lambda^+)N_\lambda(\lambda^+)},
\end{cases} \quad 0 \le x < \infty \tag{3.112}$$

$$\begin{cases}
\dfrac{e^{i(\lambda^+ - \lambda^-)x}}{(\lambda^+ - \lambda^-)e_1^{(1)}(0,\lambda^+)N_\lambda(\lambda^+)} H_{21}(x) + H_{22}(x) = 0,\\
H_{21}(x) + \dfrac{e^{i(\lambda^+ - \lambda^-)x}}{(\lambda^+ - \lambda^-)e_2^{(2)}(0,\lambda^-)D_\lambda(\lambda^-)} H_{22}(x) = i\dfrac{e^{-i\lambda^- x}}{e_2^{(2)}(0,\lambda^-)D_\lambda(\lambda^-)}.
\end{cases} \tag{3.113}$$

We calculate $H_{12}(x)$ and $H_{21}(x)$ from the systems (3.112) and (3.113), and then with the help of the formulas (3.9) and (3.110) we find exact non-scattering potentials $c_1(x)$ and $c_2(x)$, $0 \le x < \infty$, i.e., the exact solutions of the system of fundamental equations (3.38) and (3.39) for this case:

$$\begin{aligned}
c_1(x) =& -2iH_{12}(x,x) = -2iH_{12}(x)e^{i\lambda^+ x} =\\
&2\frac{(\lambda^+ - \lambda^-)^2 e_2^{(2)}(0,\lambda^-)D_\lambda(\lambda^-)e^{2i\lambda^+ x}}{(\lambda^+ - \lambda^-)^2 e_1^{(1)}(0,\lambda^+)N_\lambda(\lambda^+)e_2^{(2)}(0,\lambda^-)D_\lambda(\lambda^-) - e^{2i(\lambda^+ - \lambda^-)x}},
\end{aligned} \tag{3.114}$$

$$\begin{aligned}
c_2(x) =& 2iH_{21}(x,x) = 2iH_{21}(x)e^{-i\lambda^- x} =\\
&-2\frac{(\lambda^+ - \lambda^-)^2 e_1^{(1)}(0,\lambda^+)N_\lambda(\lambda^+)e^{-2i\lambda^- x}}{(\lambda^+ - \lambda^-)^2 e_1^{(1)}(0,\lambda^+)N_\lambda(\lambda^+)e_2^{(2)}(0,\lambda^-)D_\lambda(\lambda^-) - e^{2i(\lambda^+ - \lambda^-)x}}.
\end{aligned} \tag{3.115}$$

3.3.2 The time-dependence of standardized multipliers and an exact solution of system (3.41)

The time-evolution of non-scattering potentials (3.114), (3.115) is obtained with the help of the substitution:

$$\begin{aligned} &e_1^{(1)}(0,\lambda^+) \to e_1^{(1)}(0,\lambda^+;t), \;\; e_2^{(2)}(0,\lambda^-) \to e_2^{(2)}(0,\lambda^-;t), \\ &N_\lambda(\lambda^+) \to N_\lambda(\lambda^+;t), \;\; D_\lambda(\lambda^-) \to D_\lambda(\lambda^-;t). \end{aligned} \tag{3.116}$$

In order to find the time-dependence of non-scattering potentials (3.114) and (3.115), we need to derive the time-dependence of the following standardized multipliers:

$$i\left(e_1^{(1)}(0,\lambda^+;t)N_\lambda(\lambda^+;t)\right)^{-1}, \;\; i\left(e_2^{(2)}(0,\lambda^-;t)D_\lambda(\lambda^-;t)\right)^{-1}. \tag{3.117}$$

Suppose that for any time t the function $N(\lambda;t)$ has only one simple zero λ^+, $\text{Im}\lambda^+ \geq \varepsilon_0$ and the function $D(\lambda;t)$ has only one simple zero λ^-, $\text{Im}\lambda^- \leq -\varepsilon_0$, $\varepsilon_0 > 0$, then by Lemma 3.4, these simple zeroes are simple non-real eigenvalues of the problem (3.4)–(3.5), and

$$\begin{aligned} &N(\lambda^+;t) = e_1^{(1)}(0,\lambda^+;t) - e_2^{(1)}(0,\lambda^+;t) = 0, \\ &D(\lambda^-;t) = e_2^{(2)}(0,\lambda^-;t) - e_1^{(2)}(0,\lambda^-;t) = 0, \\ &N_\lambda = N_\lambda(\lambda^+;t) \neq 0, D_\lambda = D_\lambda(\lambda^-;t) \neq 0, \\ &e_1^{(1)} = e_1^{(1)}(0,\lambda^+;t) \neq 0, e_2^{(2)} = e_2^{(2)}(0,\lambda^-;t) \neq 0. \end{aligned} \tag{3.118}$$

After substitution (3.116) into the formulas (3.114) and (3.115), using (3.118), we obtain the time-dependent potentials:

$$q_1(x,t) = -c_1(x,t) = \frac{2}{(\lambda^+ - \lambda^-)^{-2}\beta - \alpha^{-1}}, \tag{3.119}$$

$$q_2(x,t) = c_2(x,t) = \frac{2}{(\lambda^+ - \lambda^-)^{-2}\alpha - \beta^{-1}} = \alpha^{-1}\beta q_1(x,t), \tag{3.120}$$

where

$$\alpha = (e_1^{(1)}N_\lambda)^{-1}e^{2i\lambda^+x} \neq 0, \;\; \beta = (e_2^{(2)}D_\lambda)^{-1}e^{-2i\lambda^-x} \neq 0. \tag{3.121}$$

Owing to (3.118) and (3.121), we have for all $t \in (-\infty;\infty)$:

$$\begin{aligned} &(\lambda^+ - \lambda^-)^{-2}\alpha \neq \beta^{-1} \;\text{ or }\; (\lambda^+ - \lambda^-)^{-2}\beta \neq \alpha^{-1}, \textit{i.e.}, \\ &(\lambda^+ - \lambda^-)^{-2}\alpha\beta \neq 1. \end{aligned} \tag{3.122}$$

Substituting (3.119) and (3.120) into the system (3.41), we obtain

$$q_{1t} + q_{1xxx} + 6q_1q_2q_{1x} = \frac{1}{2}q_1^2\left\{(\lambda^+ - \lambda^-)^{-2}\beta\left[\frac{\partial}{\partial t}\ln\left(e_2^{(2)}D_\lambda\right) - i(2\lambda^-)^3\right]\right.$$
$$\left. + \alpha^{-1}\left[\frac{\partial}{\partial t}\ln\left(e_1^{(1)}N_\lambda\right) + i(2\lambda^+)^3\right]\right\} = 0, \tag{3.123}$$

$$q_{2t} + q_{2xxx} + 6q_1q_2q_{2x} = \frac{1}{2}q_2^2\left\{(\lambda^+ - \lambda^-)^{-2}\alpha\left[\frac{\partial}{\partial t}\ln\left(e_1^{(1)}N_\lambda\right) + i(2\lambda^+)^3\right]\right.$$
$$\left. + \beta^{-1}\left[\frac{\partial}{\partial t}\ln\left(e_2^{(2)}D_\lambda\right) - i(2\lambda^-)^3\right]\right\} = 0. \tag{3.124}$$

From (3.123) and (3.124) it follows that

$$(\lambda^+ - \lambda^-)^{-2}\alpha\beta\left\{\frac{\partial}{\partial t}\ln\left(e_2^{(2)}D_\lambda\right) - i(2\lambda^-)^3\right\} = -i(2\lambda^+)^3 - \frac{\partial}{\partial t}\ln\left(e_1^{(1)}N_\lambda\right), \tag{3.125}$$

$$(\lambda^+ - \lambda^-)^{-2}\alpha\beta\left\{\frac{\partial}{\partial t}\ln\left(e_1^{(1)}N_\lambda\right) + i(2\lambda^+)^3\right\} = i(2\lambda^-)^3 - \frac{\partial}{\partial t}\ln\left(e_2^{(2)}D_\lambda\right). \tag{3.126}$$

Substituting the left-hand side of (3.125) (and of (3.126), respectively) into the left-hand side of (3.126) (and of (3.125)), we obtain:

$$(\lambda^+ - \lambda^-)^{-4}\alpha^2\beta^2\left\{\frac{\partial}{\partial t}\ln\left(e_2^{(2)}D_\lambda\right) - i(2\lambda^-)^3\right\} = \frac{\partial}{\partial t}\ln\left(e_2^{(2)}D_\lambda\right) - i(2\lambda^-)^3, \tag{3.127}$$

$$(\lambda^+ - \lambda^-)^{-4}\alpha^2\beta^2\left\{\frac{\partial}{\partial t}\ln\left(e_1^{(1)}N_\lambda\right) + i(2\lambda^+)^3\right\} = \frac{\partial}{\partial t}\ln\left(e_1^{(1)}N_\lambda\right) + i(2\lambda^+)^3. \tag{3.128}$$

Due to (3.122) the equalities (3.127) and (3.128) are fulfilled if and only if

$$\frac{\partial}{\partial t}\ln\left(e_1^{(1)}N_\lambda\right) = -i(2\lambda^+)^3, \quad \frac{\partial}{\partial t}\ln\left(e_2^{(2)}D_\lambda\right) = i(2\lambda^-)^3. \tag{3.129}$$

The time-dependence of standardized multipliers (3.117) are deduced from (3.129):

$$i\left(e_1^{(1)}(0,\lambda^+;t)N_\lambda(\lambda^+;t)\right)^{-1} = i\left(e_{10}^{(1)}N_{\lambda 0}\right)^{-1} e^{8i(\lambda^+)^3 t}, \tag{3.130}$$

$$i\left(e_2^{(2)}(0,\lambda^-;t)D_\lambda(\lambda^-;t)\right)^{-1} = i\left(e_{20}^{(2)}D_{\lambda 0}\right)^{-1} e^{-8i(\lambda^-)^3 t}, \tag{3.131}$$

where

$$e_{10}^{(1)} = e_1^{(1)}(0,\lambda^+;0) \neq 0, \quad e_{20}^{(2)} = e_2^{(2)}(0,\lambda^-;0) \neq 0,$$
$$N_{\lambda 0} = N_\lambda(\lambda^+;0) \neq 0, \quad D_{\lambda 0} = D_\lambda(\lambda^-;0) \neq 0.$$

If the vector solution $(\varphi_1(x,\lambda;t),\ \varphi_2(x,\lambda;t))$ of the problem (3.42), (3.47) satisfies the condition:

$$\frac{d}{dt}\left(\lim_{x\to\infty}\varphi_1(x,\lambda;t)e^{-i\lambda x}\right) = \frac{d}{dt}\left(\lim_{x\to\infty}\varphi_2(x,\lambda;t)e^{i\lambda x}\right)$$
$$\text{for } |\mathrm{Im}\lambda| < \varepsilon_0 \text{ and any } t \in (-\infty,\infty), \tag{3.132}$$

then using (3.48) and (3.49), from (3.132) we have the equality:

$$\frac{d}{dt}D(\lambda;t) = \frac{d}{dt}N(\lambda;t) \text{ for } |\mathrm{Im}\lambda| < \varepsilon_0 \text{ and any } t \in (-\infty,\infty). \tag{3.133}$$

Since the potentials are non-scattering, for which $S(\lambda;0) = S(\lambda) = 1$, i.e., $N(\lambda) = D(\lambda)$ for $|\mathrm{Im}\lambda| < \varepsilon_0$, then from (3.133) it follows that $D(\lambda;t) = N(\lambda;t)$ for $|Im\lambda| < \varepsilon_0$ and any $t \in (-\infty,\infty)$, i.e.,

$$S(\lambda;t) \equiv 1 \text{ for } |\mathrm{Im}\lambda| < \varepsilon_0,\ \varepsilon_0 > 0 \text{ and any } t \in (-\infty,\infty). \tag{3.134}$$

This means that if the condition (3.132) is satisfied, then the non-scattering property (3.134) of the scattering function $S(\lambda;t)$ is conserved in time when these potentials change according to system of cubic NLEEs (3.41).

Hence, the exact non-scattering potentials (3.114), (3.115) are a solution of the system of NLEEs (3.41), if and only if the condition (3.132) is satisfied and the time-dependence of the standardized multipliers obeys the evolution equations (3.130) and (3.131). The time-dependent non-scattering potentials (3.114) and (3.115) with standardized multipliers defined by (3.130) and (3.131) are:

$$q_1(x,t) = -c_1(x,t) = 2\frac{(\lambda^+-\lambda^-)^2 e_{20}^{(2)} D_{\lambda_0} e^{2i\lambda^+ x + 8i(\lambda^-)^3 t}}{e^{2i(\lambda^+-\lambda^-)x} - (\lambda^+-\lambda^-)^2 e_{10}^{(1)} e_{20}^{(2)} N_{\lambda_0} D_{\lambda_0} e^{8i[(\lambda^-)^3-(\lambda^+)^3]t}}, \tag{3.135}$$

$$q_2(x,t) = c_2(x,t) = 2\frac{(\lambda^+-\lambda^-)^2 e_{10}^{(1)} N_{\lambda_0} e^{-2i\lambda^- x - 8i(\lambda^+)^3 t}}{e^{2i(\lambda^+-\lambda^-)x} - (\lambda^+-\lambda^-)^2 e_{10}^{(1)} e_{20}^{(2)} N_{\lambda_0} D_{\lambda_0} e^{8i[(\lambda^-)^3-(\lambda^+)^3]t}}. \tag{3.136}$$

The non-scattering potentials (3.135) and (3.136) are calculated from the given non-scattering data consisting of two simple poles λ^+, $\mathrm{Im}\lambda^+ > \varepsilon_0$, λ^-, $\mathrm{Im}\lambda^- < -\varepsilon_9$ and two standardized multipliers (3.62), (3.63). These given scattering data are calculated from the known initial condition:

$$q_j(x,t)|_{t=0} = q_j(x), \quad j = 1, 2, \tag{3.137}$$

where $q_j(x)$ is a continuous function possessing the third continuous derivative and satisfying the estimate (3.44).

Thus, we arrive at the following theorem.

Theorem 3.4. *In the class of non-scattering potentials belonging to the class* P *and satisfying estimate* (3.44)*, the Cauchy initial-value problem for the system of cubic NLEEs* (3.41) *with the known initial condition* (3.137) *is solved exactly by the ISM. If the non-scattering data consist of only two simple poles:* $\lambda^+, \mathrm{Im}\lambda^+ \geq \varepsilon_0$ *and* $\lambda^-, \mathrm{Im}\lambda^- \leq -\varepsilon_0, \varepsilon_0 > 0$ *and two standardized multipliers* (3.117)*, then the non-scattering potentials are constructed by formulas* (3.114) *and* (3.115) *in terms of solutions of systems* (3.112) *and* (3.113)*. These constructed potentials are a solution of the system* (3.41) *if and only if the condition* (3.132) *is satisfied and the time-dependence of standardized multipliers* (3.117) *obeys evolution equations* (3.130) *and* (3.131)*. Then the time-dependent potentials* (3.135) *and* (3.136) *are an exact solution of the Cauchy problem* (3.41), (3.137).

3.3.3 Exact solutions of equations mKdV and KdV

In the class of non-scattering potentials we can find exact soliton-solutions of the Cauchy initial-value problem for mKdV equations (3.59) and (3.60). For simplicity we suppose that the scattering data of the scattering problem associated with the mKdV equation (3.59) or (3.60) consist of only one pair of simple poles: $\lambda^+, \lambda^- = -\lambda^+, \mathrm{Im}\lambda^+ > \varepsilon_0, \varepsilon_0 > 0$. Using this supposition and (3.130), (3.131), we calculate the quantities (3.121) for this case:

$$\alpha = \left(e_{10}^{(1)} N_{\lambda 0}\right)^{-1} e^{8i(\lambda^+)^3 t + 2i\lambda^+ x}, \quad \beta = \left(e_{20}^{(2)} D_{\lambda 0}\right)^{-1} e^{8i(\lambda^+)^3 t + 2i\lambda^+ x}, \tag{3.138}$$
$$e_{10}^{(1)} = e_1^{(1)}(0, \lambda^+; 0) \neq 0, \quad e_{20}^{(2)} = e_2^{(2)}(0, -\lambda^+; 0) \neq 0,$$
$$N_{\lambda 0} = N_\lambda(\lambda^+; 0) \neq 0, \quad D_{\lambda 0} = D_\lambda(-\lambda^+; 0) \neq 0.$$

Making use of (3.59), (3.60) and (3.121), (3.122), from (3.120) we obtain that if $q_1 = q_2$ then $\alpha = \beta$, and if $q_1 = -q_2$ then $\alpha = -\beta$.

Hence, in this case the non-scattering data for the mKdV equation (3.59) or (3.60) are

$$S(\lambda; t) \equiv 1 \text{ for } |\mathrm{Im}\lambda| < \epsilon_0, \quad \epsilon_0 > 0,$$
$$\lambda^+, \ -\lambda^+, \ \mathrm{Im}\lambda^+ \geq \varepsilon_0, \ \mathrm{i}\left(\mathrm{e}_{10}^{(1)} \mathrm{N}_{\lambda 0}\right)^{-1} \mathrm{e}^{8\mathrm{i}(\lambda^+)^3 \mathrm{t}}. \tag{3.139}$$

Then with due regard for $\alpha = \beta$ and (3.138), from (3.119) and (3.135) we obtain the exact solution $q(x,t)$ of the mKdV equation (3.59) for the considered case:

$$q(x,t)=q_1(x,t)=\frac{2}{(2\lambda^+)^{-2}(e_{10}^{(1)}N_{\lambda 0})^{-1}e^{8i(\lambda^+)^3t+2i\lambda^+x}-e_{10}^{(1)}N_{\lambda 0}e^{-8i(\lambda^+)^3t-2i\lambda^+x}}, \tag{3.140}$$

From (3.120) on putting $\alpha = -\beta$, we find the exact solution $q_2(x,t)$ of the mKdV equation (3.60):

$$q_2(x,t) = -q_1(x,t) = -q(x,t). \tag{3.141}$$

We now construct the solutions of the KdV from solutions (3.140) and (3.141) of the mKdV equations. Let the solution $q_1(x,t)$ of the mKdV equation (3.59) be constructed from the given non-scattering data (3.139) by the formula (3.140). This solution is related to the potential $c(x,t)$ from (3.86) by the equality: $q_1(x,t) = -c(x,t)$. If $c(x,t)$ is a solution of the mKdV equation (3.59), then by Theorem 3.3 the function $r(x,t)$ defined by the Miura's transformation (3.103):

$$r(x,t) = -\,c^2(x,t) - ic_x(x,t) = -q_1^2(x,t) + iq_{1x}(x,t), \tag{3.142}$$

is the exact solution of the KdV equation (3.98).

Let $q_2(x,t) = c(x,t) = i\mu(x,t)$, where $\mu(x,t)$ is a real-valued soliton-solution of the mKdV equation (3.60), then by Theorem 3.3 the function defined by the non-linear transformation (3.107):

$$r(x,t) = \mu^2(x,t) + \mu_x(x,t), \tag{3.143}$$

is the exact real soliton-solution of the KdV equation (3.98).

Thus, the exact solutions (3.142) and (3.143) of the KdV equation (3.98) are constructed from solutions (3.140) and (3.141) of mKdV equations (3.59) and (3.60), respectively. They are expressible through the explicit formula (3.140) for the solution of the mKdV equation (3.59) in terms of the non-scattering data (3.139).

3.4 The Cauchy problem for cubic non-linear equation (3.3)

We now turn attention to the reduction: $q_1(x,t) = -\overline{q_2(x,t)}$. Under this reduction the system (3.41) reduces to the cubic non-linear equation:

$$q_t + q_{xxx} - 6|q|^2 q_x = 0, \quad 0 \le x < \infty, \quad -\infty < t < \infty. \tag{3.144}$$

Due to (3.42), the reduction $q_1(x,t) = -\overline{q_2(x,t)} = q(x,t)$ is equivalent to the condition:

$$q(x,t) = -c_1(x,t) = -\overline{c_2(x,t)}, \quad 0 \leq x < \infty, \quad -\infty < t < \infty. \tag{3.145}$$

Due to (3.145), the potential matrix $C(x,t)$ in system (3.4) is self-adjoint:

$$C(x,t) = \begin{pmatrix} 0 & c_1(x,t) \\ c_2(x,t) & 0 \end{pmatrix} = \begin{pmatrix} 0 & -q(x,t) \\ -\overline{q(x,t)} & 0 \end{pmatrix} = \overline{C(x,t)}.$$

The scattering problem (3.4)–(3.5) with this potential matrix is associated with the Cauchy problem for equation (3.144) with the given initial condition:

$$q(x,0) = -c_1(x) = -\overline{c_2(x)}, \quad 0 \leq x < \infty, \tag{3.146}$$

The Cauchy problem (3.144), (3.146) is associated with the self-adjoint scattering problem (3.4)–(3.5) with the potentials (3.145) belonging to the class P and satisfying the estimate of the form:

$$|q(x,t)| \leq \frac{\tilde{C}}{1 + x^{1+\varepsilon}} \quad \text{for any } \ t \in (-\infty, \infty), \tag{3.147}$$

here $\tilde{C}$ is a constant, $\varepsilon > 0$.

Under reduction (3.145) and estimate (3.147), the scattering problem (3.4)– (3.5) has no discrete spectrum. Note that the potentials of the non-self-adjoint problem (3.4)–(3.5) are required to decrease exponentially as $x \to \infty$, while the potentials of the self-adjoint problem are power decreasing.

The integral equations (3.13)–(3.16) with the potentials belonging to the class P and satisfying conditions (3.145), (3.147) have unique solutions. These solutions are kernels of the transformation operator of the associated self-adjoint problem. According to the relations (1.94), the kernel functions satisfy the relation equalities:

$$H_{11}(x,s;t) = \overline{H_{22}(x,s;t)}; \quad H_{12}(x,s;t) = \overline{H_{21}(x,s;t)}. \tag{3.148}$$

Furthermore, these kernel functions belong to the class P and satisfy the estimate of the form [119]:

$$|H_{jk}(x,s;t)| \leq \frac{\tilde{C}}{1 + (x+s)^{1+\varepsilon}} \quad \text{for} \quad 0 \leq x < s, \ -\infty < t < \infty \quad j,k = 1,2, \tag{3.149}$$

$\tilde{C}$ is a constant, $\varepsilon > 0$.

Owing to condition (3.145) and estimate (3.147), using Lemma 1.13, we easily prove that the correspondence between the quantities $A(\lambda;t)$ and $B(\lambda;t)$ established by asymptotic expressions (3.7) is one-to-one. Therefore, the scattering function $S(\lambda;t)$ for the scattering problem (3.4)–(3.5) with the potential self-adjoint matrix is uniquely determined by the relation:

$$S(\lambda;t)A(\lambda;t) = B(\lambda;t), \quad \text{where } t \text{ is a parameter.}$$

By an argument analogous to that used for the proof of Theorem 1.2 in Section 1.2 of Chapter 1, we can prove the following theorem regarding the complete description of the scattering function for the self-adjoint problem (3.4)–(3.5).

Theorem 3.5. *For a given function $S(\lambda;t)$ (t is a parameter) to be the scattering function for system* (3.4) *on the half-line with boundary condition* (3.5) *and with potentials belonging to the class P and satisfying the self-adjoint condition* (3.145) *and estimate* (3.147)*, it is necessary and sufficient that for any $x \geq 0$ there exists a function $S(x,\lambda;t)$ such that $S(\lambda;t) = S(0,\lambda;t)$ and*

(1) $S(x,\lambda;t)$ admits the factorization:

$$S(x,\lambda;t) = \frac{1 + \int\limits_0^\infty H_+(x,\xi;t) e^{i\lambda\xi} d\xi}{1 + \int\limits_0^\infty \overline{H_+(x,\xi;t)} e^{-i\lambda\xi} d\xi},$$

where $H_+(x,\xi;t) = H_{11}(x,x+\xi;t) - H_{21}(x,x+\xi;t)$, the function $1 + \int\limits_0^\infty H_+(x,\xi;t)e^{i\lambda\xi}d\xi$ is analytic in λ and does not have zero in the upper half-plane $\mathrm{Im}\lambda \geq 0$, and therefore, there exist the absolutely integrable functions $K_+(x,\xi;t)$ and $K_-(x,\xi;t)$ such that

$$\left(1 + \int\limits_0^\infty H_+(x,\xi;t)e^{i\lambda\xi}d\xi\right)^{-1} = 1 + \int\limits_0^\infty K_+(x,\xi,t)e^{i\lambda\xi}d\xi,$$

$$\left(1 + \int\limits_0^\infty \overline{H_+(x,\xi;t)}e^{-i\lambda\xi}d\xi\right)^{-1} = 1 + \int\limits_0^\infty K_-(x,\xi,t)e^{-i\lambda\xi}d\xi,$$

$$\overline{S(x,\lambda;t)} = S^{-1}(x,\lambda;t), \quad S(0,\lambda;t) = S(\lambda;t) \ \text{for} \ |Im\lambda| < \varepsilon, \ \ \varepsilon > 0;$$

(2) For any $x \geq 0$ the functions $S(x,\lambda;t) - 1$ and $S^{-1}(x,\lambda;t) - 1$ are the Fourier transformations of the absolutely integrable functions $f(x,\xi;t)$ and $g(x,\xi;t)$ with respect to ξ:

$$S(x,\lambda;t) = 1 + \int\limits_{-\infty}^{\infty} f(x,\xi;t)e^{i\lambda\xi}d\xi,$$

$$f(x,\xi;t) = \frac{1}{2\pi}\int\limits_{-\infty}^{\infty} [S(x,\lambda;t) - 1]\, e^{-i\lambda\xi}d\xi,$$

$$S^{-1}(x,\lambda;t) = 1 + \int_{-\infty}^{\infty} g(x,\xi;t)e^{i\lambda\xi}d\xi,$$

$$g(x,\xi;t) = \frac{1}{2\pi}\int_{-\infty}^{\infty} \left[S^{-1}(x,\lambda;t) - 1\right] e^{-i\lambda\xi}d\xi$$

where

$$f(x,\xi;t) = H_+(x,\xi;t) + K_-(x,-\xi;t)+$$
$$\int_{-\infty}^{0} K_-(x,-\tau;t)H_+(x,\xi-\tau;t)d\tau, \quad f(0,\xi;t) = f(\xi;t),$$

$$g(x,\xi;t) = H_-(x,-\xi;t) + K_+(x,\xi;t)+$$
$$\int_{0}^{\infty} K_+(x,\tau;t)H_-(x,\tau-\xi;t)d\tau, \quad g(0,\xi;t) = g(\xi;t);$$

(3) For any $x \geq 0$ and $\xi > 0$ the functions $f(x,\xi;t)$ and $g(x,\xi;t)$ are closely related to the functions $f(\xi;t)$ and $g(\xi;t)$ by the equalities:

$$f(x,\xi;t) = f(\xi+2x;t), \quad g(x,-\xi;t) = g(-\xi-2x;t),$$

where $f(x;t) = \overline{g(-x;t)}$;

(4) The following estimates hold for any $t \in (-\infty,\infty)$ and $x \geq 0$:

$$|f(x;t)| \leq \frac{\tilde{C}}{1+x^{1+\varepsilon}}, \quad |g(-x;t)| \leq \frac{\tilde{C}}{1+x^{1+\varepsilon}},$$

$\tilde{C}$ is a constant, $\varepsilon > 0$.

By Lemma 3.12 and Theorem 3.5, the scattering function $S(x,\lambda;t)$ possesses the properties enumerated in this lemma and obeys the evolutionary equation for any $x \geq 0$:

$$S(x,\lambda;t) = S(x,\lambda)e^{-8i\lambda^3 t}, \quad S^{-1}(x,\lambda;t) = S^{-1}(x,\lambda)e^{8i\lambda^3 t} = \overline{S(x,\lambda)}e^{8i\lambda^3 t}$$
$$\text{for } |\mathrm{Im}\lambda| < \varepsilon, \quad -\infty < t < \infty. \tag{3.150}$$

Thus, Theorem 3.5 and Lemma 3.12 reduce the problem of solving the Cauchy problem (3.144), (3.146) to that of solving the system of fundamental equations (3.39). Indeed, the functions $f(x,\xi;t)$ and $g(x,-\xi;t)$ appeared in the system (3.39) are constructed from the given scattering function $S(x, \lambda)$ by the formulas:

$$
\begin{aligned}
f(x, \xi; t) &= \frac{1}{2\pi} \int_{-\infty}^{\infty} [S(x, \lambda; t) - 1] \, e^{-i\lambda\xi} d\lambda \\
&= \frac{1}{2\pi} \int_{-\infty}^{\infty} \left[S(x, \lambda) e^{-8i\lambda^3 t} - 1\right] e^{-i\lambda\xi} d\lambda,
\end{aligned}
\tag{3.151}
$$

$$
\begin{aligned}
g(x, -\xi; t) &= \frac{1}{2\pi} \int_{-\infty}^{\infty} \left[S^{-1}(x, \lambda; t) - 1\right] e^{i\lambda\xi} d\lambda \\
&= \frac{1}{2\pi} \int_{-\infty}^{\infty} \left[S^{-1}(x, \lambda) e^{8i\lambda^3 t} - 1\right] e^{i\lambda\xi} d\lambda = \overline{f(x, \xi; t)},
\end{aligned}
$$

where $S(x, \lambda; t)$ is the scattering function of the self-adjoint problem (3.4)–(3.5). It possesses the properties (1)–(4) enumerated in Theorem 3.5.

Using system (3.39), we can verify that the functions $f(x, \xi; t)$ and $g(x, -\xi; t)$ in this system possess properties, which are analogous with the properties of the kernels of the transformation operator; i.e., these functions are continuous functions having the first partial continuous derivative with respect to t and the third partial continuous derivative with respect to x.

Owing to the condition (4) of Theorem 3.5 and by an argument analogous to that used for the proof of Lemmas 1.18, 1.20 and 1.21 in Chapter 1, we can prove that the system of fundamental equations (3.39) has for every $(x, t) \in [0, \infty) \times (-\infty, \infty)$ a unique solution $(H_{21}(x, s; t), H_{22}(x, s; t))\,, 0 \le x \le s$. Since the known functions $f(2x + \xi; t)$ and $g(-2x - \xi; t)$ in system (3.39) possess the properties (3) and (4) enumerated in Theorem 3.5, then the found solution $(H_{21}(x, s; t), H_{22}(x, s; t))$, $0 \le x \le s$, of system (3.39) has properties, which are analogous with the properties of $f(2x + \xi; t)$ and $g(-2x - \xi; t)$, i.e., the found solution belongs to the class P and satisfies relations (3.148), the estimate of the type (3.149) and integral equations of the form (3.14) and (3.16), respectively for the kernels of the transformation operator. These kernels are related to the reconstructed potential matrix in these integral equations by the formula:

$$
C(x, t) = \begin{pmatrix} 0 & -c_1(x, t) \\ -c_2(x, t) & 0 \end{pmatrix} = \begin{pmatrix} 0 & 2iH_{12}(x, x; t) \\ 2i\, H_{21}(x, x; t) & 0 \end{pmatrix},
\tag{3.152}
$$

where the found potential $q(x, t) = -c_1(x, t) = -\overline{c_2(x, t)}$ belongs to the class P and satisfies estimate (3.147).

Since the solutions of equations (3.14) and (3.16) are unique, then the potential self-adjoint matrix (3.152) is uniquely recovered from the given $S(\lambda)$ satisfying the conditions (1)–(4) enumerated in the Theorem 3.5. Hence, the eigenvalue equations (3.42) and (3.45) with the reconstructed potential matrix (3.152) are compatible, i.e., $u_{xt} = u_{tx}$. Then by Lemma 3.11 the potential $q(x,t) = -c_1(x,t)$ defined by the formula (3.152) satisfies the cubic NLEEs (3.144) i.e., the function

$$\begin{aligned} q(x,t) &= -c_1(x,t) = -\overline{c_2(x,t)} \\ &= 2iH_{12}(x,x;t) = 2\,i\,\overline{H_{21}(x,x;t)}, \quad 0 \le x < \infty, \quad -\infty < t < \infty. \end{aligned} \tag{3.153}$$

is the solution of the Cauchy problem (3.144), (3.146). Thus, the Cauchy problem (3.144), (3.146) is solved by the ISM.

Theorem 3.6. *Under reduction* (3.145) *and estimate* (3.147), *the self-adjoint problem* (3.4)–(3.5) *has no discrete spectrum. The scattering function* $S(\lambda)$ *of this problem is calculated from the known initial condition* (3.146) *and possesses the properties (1)–(4) enumerated in Theorem 3.5. By Lemma 3.12 the time-dependence of the scattering function* $S(x,\lambda;t)$ *for real* λ *obeys the evolutionary equation* (3.150). *A correspondence between* $S(x,\lambda;t)$ *and the reconstructed potential matrix* (3.152) *is one-to-one. By Lemma 3.11 the found potential* (3.153) *satisfies the cubic NLEE* (3.144). *Thus, the Cauchy problem for the cubic NLEE* (3.144) *with the known initial condition* (3.146) *has the unique solution* (3.153) *in the class of isospectral potentials satisfying estimate* (3.147) *and possessing the third partial continuous derivative with respect to* x *and the first partial continuous derivative with respect to* t.

Chapter 4

The Dirichlet IBVPs for sine and sinh-Gordon Equations

In Chapter 4 we consider the initial-boundary value problems (IBVPs) for the sine-Gordon (sG) and sinh-Gordon (shG) equations:

$$v_{xt} = \sin v, \quad 0 \le x < \infty, \quad 0 \le t \le T,$$
$$v_{xt} = \mathrm{sh} v, \quad 0 \le x < \infty, \quad 0 \le t \le T,$$

with the known initial and boundary conditions:

$$v(x, 0) = v(x),$$
$$v(0, t) = h(t),$$

where $v(x, t)$ and $v_x(x, t)$ are real-valued continuous functions, and $v_x(x, t)$ rapidly decreases.

The Dirichlet IBVP for the sG equation is associated with the non-self-adjoint scattering problem, whereas the Dirichlet IBVP for the shG equation is associated with the self-adjoint scattering problem.

The Volterra integral equations for the unknown boundary data are derived. The solutions of these derived equations are found by the method of successive approximations in terms of known initial and boundary conditions of the considered IBVP. Then the solutions of the Dirichlet IBVPs are found and expressed through the solution of the system of fundamental equations in the inverse problem. The known functions in this system are constructed from the scattering data. These scattering data depend on the found boundary data, which are calculated from the known initial and boundary conditions. Thereby, every scattering data set of the associated scattering problem corresponds a unique solution of the considered IBVP.
The chapter is based on the published papers [97, 119, 138, 143].

We consider the initial-boundary value problems (IBVPs) for the sine-Gordon (sG) and sinh-Gordon (shG) equations:

$$v_{xt} = \sin v, \quad 0 \le x < \infty, \quad 0 \le t \le T, \tag{4.1}$$
$$v_{xt} = \mathrm{sh} v, \quad 0 \le x < \infty, \quad 0 \le t \le T, \tag{4.2}$$

with the known initial and boundary conditions:

$$v(x, 0) = v(x), \tag{4.3}$$
$$v(0, t) = h(t), \tag{4.4}$$

DOI: 10.1201/9781003370543-4

where $v(x,t)$ and $v_x(x,t)$ are real-valued continuous functions, and $v_x(x,t)$ rapidly decreases.

The Dirichlet IBVP (4.1), (4.3)–(4.4) is associated with the inverse scattering problem (ISP) for the system of first-order ordinary differential equations (ODEs):

$$u_x = U_1 u, \; U_1 = i\lambda J_1 + \frac{i}{2}\begin{pmatrix} 0 & v_x \\ v_x & 0 \end{pmatrix}, \; 0 \le x < \infty, \; 0 \le t < T, \tag{4.5}$$
$$J_1 = diag(1,-1), \quad u = (u_1(x,\lambda;t), u_2(x,\lambda;t)),$$

with the boundary condition at the origin $x = 0$:

$$u_1(0,\lambda;t) = u_2(0,\lambda;t). \tag{4.6}$$

The Dirichlet IBVP (4.2)–(4.3)–(4.4) is associated with the ISP for the system of first-order ODEs:

$$u_x = U_2 u, \; U_2 = i\lambda J_1 + \frac{i}{2}\begin{pmatrix} 0 & v_x \\ -v_x & 0 \end{pmatrix}, \; 0 \le x < \infty, \; 0 \le t \le T, \tag{4.7}$$

with the boundary condition (4.6) at $x = 0$.

The sG equation in light-cone coordinates (dimensionless variables) is one of the most widely studied non-linear wave equations, because of its intrinsic mathematical properties and its wide applicability in physics. The initial-value problem for the sG equation on an infinite line has been solved by the inverse scattering method (ISM) [3]. The Dirichlet IBVP for the sG equation is solved by the inverse scattering transform for the Schrödinger spectral problem on the half-line solved as a Hilbert boundary value problem [70, 71]. The time-dependence of the spectral data is derived, where the evolution of the reflection coefficient eventually reads as the non-linear Ricatti differential equation [70].

We consider the sG equation (4.1) and shG equation (4.2) in light-cone coordinates. The subscript stands for partial derivative for the field $v(x,t)$.

$O_1(t)$ and $O_2(t)$ denote the operators defined in $L_2[0,\infty)$ by the eigenvalue problems (4.5)–(4.6) and (4.7)–(4.6), respectively. The potentials in the systems (4.5) and (4.7) are called isospectral if the spectrum of the operators $O_1(t)$ and $O_2(t)$ is invariant with time, i.e., $\lambda_t = 0$.

Denote by P the class of real-valued continuous functions rapidly decreasing and possessing the continuously mixed derivative of the second-order.

We consider the IBVPs for the sG equation (4.1) and for shG equation (4.2) in the class of isospectral potentials which are real-valued continuous functions belonging to the class P. We apply the results of the associated ISPs to solving the considered IBVPs. The difficulty in solving these IBVPs is that the time-dependence of the scattering data is determined by the unknown boundary values (BVs) at $x = 0$ of the Jost solutions of the system of first order ODEs, and moreover, the evolution equations for the unknown BVs contain the unknown boundary data at $x = 0$. We overcome this difficulty by deriving the Volterra integral equations for the unknown boundary data, whose solutions are found by the method of successive approximations in terms of the known initial and boundary conditions (4.3) and (4.4). Consequently, the unknown BVs are calculated in terms of the boundary data, and those are expressed through the known conditions (4.3) and (4.4). The solution of the Dirichlet IBVPs is expressed through the solution of the system of fundamental equations in the inverse problem. The solution of this system is constructed from the scattering data, which are constructed from the calculated BVs. Thus, every scattering data set corresponds to a unique solution of the considered IBVP.

The sG equation (the shG) considered in this chapter can be written as the compatibility condition of two linear eigenvalue equations, i.e., of the x part and the t part of the Lax pair. The x part is the eigenvalue problem generated by the system of first-order ODEs (4.5) ((4.7)) with boundary condition (4.6) and the t part is defined by the equation governing the time-dependence of the eigenfunctions of the eigenvalue problem. The problem of solving the Dirichlet IBVP for the sG equation (the shG equation) is reduced to deriving the time-dependence of the scattering data, and then to reconstructing the potential matrix which is expressed through the solution of the system of fundamental equations in the inverse problem.

In Section 3.1 of Chapter 3, we study the scattering problem for the system of first-order ODEs:

$$-iJ_1u_x + C(x)u = \lambda u, \quad u = (u_1(x,\lambda),\, u_2(x,\lambda)), \quad 0 \le x < \infty, \tag{4.8}$$

with the boundary condition at the origin $x = 0$:

$$u_1(0,\lambda) = u_2(0,\lambda), \tag{4.9}$$

where λ is a complex-valued parameter, $J_1 = \begin{pmatrix} 1 & 0 \\ 0 & -1 \end{pmatrix}$, $C(x) = \begin{pmatrix} 0 & c_1(x) \\ c_2(x) & 0 \end{pmatrix}$, the potentials $c_1(x)$ and $c_2(x)$ are complex-valued continuous and rapidly decreasing functions.

In this chapter, we consider two special cases of the potential matrix $C(x)$ in system (4.8). Namely, the Dirichlet IBVP for the sG equation is associated with the scattering problem (4.8)–(4.9) with the potential non-self-adjoint matrix [97]:

$$C(x) = \begin{pmatrix} 0 & c_1(x) \\ c_2(x) & 0 \end{pmatrix} = \begin{pmatrix} 0 & -\frac{1}{2}v_x(x) \\ \frac{1}{2}v_x(x,t) & 0 \end{pmatrix}, \tag{4.10}$$

where $v_x(x)$ is a continuous real-valued function satisfying the estimate:

$$|v_x(x)| \le \tilde{C}e^{-\varepsilon x}, \quad \tilde{C} \text{ is a constant}, \ \varepsilon > 0. \tag{4.11}$$

The Dirichlet IBVP for the shG equation is associated with the scattering problem (4.8)–(4.9) with the potential self-adjoint matrix [119]:

$$C(x) = \begin{pmatrix} 0 & c_1(x) \\ c_2(x) & 0 \end{pmatrix} = \begin{pmatrix} 0 & -\frac{1}{2}v_x(x) \\ -\frac{1}{2}v_x(x) & 0 \end{pmatrix}, \tag{4.12}$$

where $v_x(x)$ is a real-valued continuous function satisfying the inequality:

$$|v_x(x)| \le \frac{\tilde{C}}{1+x^{1+\varepsilon}}, \quad \tilde{C} \text{ is a constant, } \varepsilon > 0. \tag{4.13}$$

The potential matrices (4.10) and (4.12) are special cases of the potential matrix $C(x)$ in system (4.8). The results of the non-self-adjoint and of self-adjoint problems (4.8)–(4.9) are presented in Sections 3.1 and 3.4, respectively of Chapter 3. In this chapter we apply the obtained results of these ISPs to solving the Dirichlet IBVPs for the sG and shG equations.

4.1 The IBVP for the sG equation

We recall the results of the ISP (4.8)–(4.9) with the non-self-adjoint matrix (4.10) satisfying estimate (4.11) from Section 3.1 of Chapter 3, and apply the results of this ISP to solving the IBVP for the sG equation [143]:

$$v_{xt} = \sin v, \quad 0 \le x < \infty, \quad 0 \le t \le T \tag{4.14}$$

with the known initial and boundary conditions:

$$v(x,0) = v(x), \tag{4.15}$$

$$v(0,t) = h(t), \tag{4.16}$$

where $h(t), v(x,t)$ and $v_x(x,t)$ are continuous real-valued functions, $h(t)$ has the continuous derivative, and $v_x(x,t)$ possesses the continuous partial derivative with respect to t and satisfies the estimate:

$$|v_x(x,t)| \le \tilde{C}e^{-\varepsilon x} \quad \text{for any} \quad (x,t) \in [0,\infty) \times [0,T], \tag{4.17}$$

$\tilde{C}$ is a constant, $\varepsilon > 0$.

4.1.1 The Jost solutions

Let the potentials $c_1(x,t)$ and $c_2(x,t)$ in system (4.8) depend on an additional parameter t (time). We write the time-dependent system (4.8) for the vector-function $u = (u_1(x,\lambda;t), u_2(x,\lambda;t))$ in the operator form:

$$u_x = U_1 u, \quad 0 \le x < \infty, \quad 0 \le t \le T \tag{4.18}$$

with the boundary condition at the origin $x = 0$:

$$u_1(0, \lambda; t) = u_2(0, \lambda; t), \tag{4.19}$$

where U_1 is the operator defined by system (4.8) with potential matrix (4.10):

$$U_1 = i\lambda J_1 + iQ, \;\; J_1 = \begin{pmatrix} 1 & 0 \\ 0 & -1 \end{pmatrix},$$

$$Q = -J_1 C(x,t) = \begin{pmatrix} 0 & -c_1(x,t) \\ c_2(x,t) & 0 \end{pmatrix}.$$

Putting $-c_1(x,t) = c_2(x,t) = \frac{1}{2}v_x(x,t)$, we have

$$Q = \begin{pmatrix} 0 & \frac{1}{2}v_x(x,t) \\ \frac{1}{2}v_x(x,t) & 0 \end{pmatrix}, \tag{4.20}$$

and $v_x(x,t)$ is a real-valued continuous function possessing the continuous partial derivative with respect to t and satisfying the estimate (4.17).

The time-dependence of eigenfunctions of the problem (4.18)–(4.19) obeys the system of evolutionary equations:

$$u_t = V_1 u, \;\; V_1 = -\frac{i}{4\lambda}\begin{pmatrix} \cos v(x,t) + \alpha(t) & -i\sin v(x,t) \\ i\sin v(x,t) & -\cos v(x,t) + \alpha(t) \end{pmatrix}, \tag{4.21}$$

where $\alpha(t)$ is any arbitrary bounded function of t.

The Jost solutions $e^{(1)}(x,\lambda;t)$ and $e^{(2)}(x,\lambda;t)$ of system (4.18) satisfy the system of integral equations (3.17) and (3.18), respectively. Under the reduction (4.20), from systems (3.17) and (3.18) there obviously arises a relationship between the Jost solutions $e^{(1)}(x,\lambda;t)$ and $e^{(2)}(x,\lambda;t)$:

$$e_1^{(1)}(x,\lambda;t) = e_2^{(2)}(x,-\lambda;t), \;\; e_2^{(1)}(x,\lambda;t) = e_1^{(2)}(x,-\lambda;t)$$
$$\text{for } (x,t) \in [0,\infty) \times [0,T). \tag{4.22}$$

By Lemma 3.1 the Jost solutions $e^{(1)}(x,\lambda;t)$ and $e^{(2)}(x,\lambda;t)$ of system (4.18) have the representations of the form (3.21) and (3.22), respectively:

$$e^{(1)}(x,\lambda;t) = \left(e_1^{(1)}(x,\lambda;t),\; e_2^{(1)}(x,\lambda;t)\right) =$$
$$\left(e^{i\lambda x} + e^{i\lambda x}\int_0^\infty H_{11}(x,x+\tau;t)e^{i\lambda\tau}d\tau,\; e^{i\lambda x}\int_0^\infty H_{21}(x,x+\tau;t)e^{i\lambda\tau}d\tau\right), \tag{4.23}$$

$$e^{(2)}(x,\lambda;t) = \left(e_1^{(2)}(x,\lambda;t),\; e_2^{(2)}(x,\lambda;t)\right) =$$
$$\left(e^{-i\lambda x}\int_0^\infty H_{12}(x,x+\tau;t)e^{-i\lambda\tau}d\tau,\; e^{-i\lambda x} + e^{-i\lambda x}\int_0^\infty H_{22}(x,x+\tau;t)e^{-i\lambda\tau}d\tau\right), \tag{4.24}$$

where the kernels $H_{jk}(x,s), 0 \leq x \leq s, j,k = 1,2$ satisfy the integral equations (3.13)–(3.16).

Under reduction (4.20) and estimate (4.17), from systems of integral equations (3.13)–(3.16) it follows that the kernels $H_{11}(x,s;t), H_{22}(x,s;t)$ are real-valued functions and $H_{12}(x,x;t), H_{21}(x,x;t)$ are pure imaginary-valued functions, moreover

$$H_{11}(x,s;t) = H_{22}(x,s;t), \;\; H_{12}(x,s;t) = H_{21}(x,s;t), \; 0 \leq x \leq s, \tag{4.25}$$

$$|H_{jk}(x,s;t)| \leq \tilde{C}e^{-\varepsilon\frac{x+s}{2}} \text{ for any } t \in [0,T], \; 0 \leq x \leq s, \; j,k = 1,2, \tag{4.26}$$

$\tilde{C}$ is a constant, $\varepsilon > 0$.

Since the Jost solutions $e^{(1)}(x,\lambda;t)$ and $e^{(2)}(x,\lambda;t)$ of system (4.18) satisfy the asymptotic conditions (3.19) and (3.20), then the asymptotic behavior of $e^{(1)}(x,\lambda;t)$ and $e^{(2)}(x,\lambda;t)$ as $x \to \infty$ does not depend on time:

$$\lim_{x\to\infty} e_{jt}^{(1)}(x,\lambda;t) = 0, \;\; \lim_{x\to\infty} e_{jt}^{(2)}(x,\lambda;t) = 0, \; j = 1,2. \tag{4.27}$$

The time-dependence of the Jost solutions $e^{(1)}(x,\lambda;t)$ and $e^{(2)}(x,\lambda;t)$ of system (4.18) is defined by the system of evolutionary equations (4.21) with any arbitrary bounded function $\alpha(t)$. Therefore, we can choose the system (4.21) with $\alpha(t) \equiv -1$ as the system of evolutionary equations for $e^{(1)}(x,\lambda;t)$, and the system (4.21) with $\alpha(t) \equiv 1$ as the system of evolutionary equations for $e^{(2)}(x,\lambda;t)$, i.e.,

$$e_t^{(1)} = V_1 e^{(1)}, \;\; V_1 = -\frac{i}{4\lambda}\begin{pmatrix} \cos v(x,t) - 1 & -i\sin v(x,t) \\ i\sin v(x,t) & -\cos v(x,t) - 1 \end{pmatrix}, \tag{4.28}$$

$$e_t^{(2)} = V_1 e^{(2)}, \;\; V_1 = -\frac{i}{4\lambda}\begin{pmatrix} \cos v(x,t) + 1 & -i\sin v(x,t) \\ i\sin v(x,t) & 1 - \cos v(x,t) \end{pmatrix}. \tag{4.29}$$

The natural boundary condition for the real-valued solution $v(x,t)$ of the sG equation (4.24) is [70]:

$$v(x,t) \to 0 \pmod{2\pi} \text{ as } x \to \infty. \tag{4.30}$$

Due to (3.19), (3.20) and (4.30) the functions on the right-hand side of (4.28) and (4.29) tend to zero uniformly with respect to t as $x \to \infty$, i.e., the derived equations (4.28) and (4.29) satisfy the required asymptotic conditions (4.27) as $x \to \infty$. Consequently, the systems (4.28) and (4.29) are the operator equations governing the time-evolution of the Jost solutions $e^{(1)}(x,\lambda;t)$ and $e^{(2)}(x,\lambda;t)$, respectively of the system (4.18).

4.1.2 The Lax compatibility condition

Let the potentials in system (4.18) be isospectral, i.e., the spectrum of the operator O_1 is invariant with time, and therefore $\lambda_t = 0$.

Lemma 4.1. *Let the isospectral potential $v_x(x,t)$ in system (4.18) be a real-valued continuous function belonging to the class P, then the systems (4.18) and (4.21) are compatible, i.e., $u_{xt} = u_{tx}$, and the function $v(x,t)$ varies with time as the solution of the sG equation (4.14). Conversely, if the function $v(x,t)$ satisfies the sG equation (4.14), then $\lambda_t = 0$, and the systems (4.18) and (4.21) are compatible.*

Proof. Let $\lambda_t = 0$, then differentiating (4.18) with respect to t and (4.21) with respect to x gives

$$u_{xt} = (U_{1t} + U_1 V_1)\, u, \tag{4.31}$$

$$u_{tx} = (V_{1x} + V_1 U_1)\, u. \tag{4.32}$$

By the supposition that the mixed partial derivative $v_{xt}(x,t)$ is a real-valued continuous function, therefore the functions $u_{xt}(x,t)$ and $u_{tx}(x,t)$ are also continuous. Hence, the systems (4.18) and (4.21) are compatible, i.e., $u_{xt} = u_{tx}$. Then from (4.31) and (4.32) we obtain the Lax compatibility condition for matrix equations (4.18) and (4.21):

$$U_{1t} - V_{1x} + U_1 V_1 - V_1 U_1 = 0. \tag{4.33}$$

The matrix equation (4.33) is equivalent to the sG equation (4.14).

The converse assertion is proved in the same way as in the proof of the Lemma 2.1. □

Remark 4.1. Due to (3.23) and (3.34)–(3.36) the time-dependence of the scattering function and standardized polynomials is determined by the unknown BVs $e^{(1)}(0,\lambda;t)$ and $e^{(2)}(0,\lambda;t)$ of the Jost solutions $e^{(1)}(x,\lambda;t)$ and $e^{(2)}(x,\lambda;t)$ of systems (3.17) and (3.18). It is the main difficulty in solving the IBVP (4.14)–(4.15)–(4.16).

4.1.3 Evaluation of unknown BVs

We introduce the following notations:

$$\begin{aligned}
&e_n^{(j)}(0,\lambda;0) = e_{no}^{(j)}, \quad n,j = 1,2,\\
&e_1^{(1)}(0,\lambda;t) - e_2^{(1)}(0,\lambda;t) = N(\lambda;t),\\
&e_2^{(2)}(0,\lambda;t) - e_1^{(2)}(0,\lambda;t) = D(\lambda;t),\\
&e_1^{(j)}(0,\lambda;t) + e_2^{(j)}(0,\lambda;t) = M_j(\lambda;t) \ \ j = 1,2.
\end{aligned} \tag{4.34}$$

We call $N(\lambda;t)$, $D(\lambda;t)$ $M_1(\lambda;t)$ and $M_2(\lambda;t)$ the boundary data at $x = 0$ of the boundary problem (4.18)–(4.19).

Using notations (4.34), from (4.28) and (4.29) we derive the evolutionary equations for the unknown BVs $e_n^{(1)}(0,\lambda;t)$ and $e_n^{(2)}(0,\lambda;t)$:

$$e_{1t}^{(1)}(0,\lambda;t) = \frac{i}{4\lambda}\left[1-e^{-iv(0,t)}\right]e_1^{(1)}(0,\lambda;t) + \frac{1}{4\lambda}\sin v(0,t)N(\lambda;t), \tag{4.35}$$

$$\text{Im}\lambda \geq -\varepsilon_0,\ 0 \leq t \leq T,\ \varepsilon_0 > 0$$

$$e_{2t}^{(1)}(0,\lambda;t) = \frac{i}{4\lambda}\left[1+e^{-iv(0,t)}\right]e_2^{(1)}(0,\lambda;t) + \frac{1}{4\lambda}\sin v(0,t)N(\lambda;t), \tag{4.36}$$

$$e_{1t}^{(2)}(0,\lambda;t) = -\frac{i}{4\lambda}\left[1+e^{-iv(0,t)}\right]e_1^{(2)}(0,\lambda;t) - \frac{1}{4\lambda}\sin v(0,t)D(\lambda;t), \tag{4.37}$$

$$\text{Im}\lambda \leq \varepsilon_0,\ 0 \leq t \leq T,\ \varepsilon_0 > 0$$

$$e_{2t}^{(2)}(0,\lambda;t) = -\frac{i}{4\lambda}\left[1-e^{-iv(0,t)}\right]e_2^{(2)}(0,\lambda;t) - \frac{1}{4\lambda}\sin v(0,t)D(\lambda;t). \tag{4.38}$$

The equations (4.35)–(4.38) contain the unknown boundary data $N(\lambda;t)$ and $D(\lambda;t)$. To calculate the unknown BVs $e_n^{(j)}(0,\lambda;t)$, $n,j=1,2$, we first need to find these boundary data, so that these desired boundary data are expressible in terms of the known initial and boundary conditions (4.15) and (4.16). To this purpose, from (4.35)–(4.38) we derive the systems of equations for the unknown boundary data $N(\lambda;t)$, $M_1(\lambda;t)$ and $D(\lambda;t)$, $M_2(\lambda;t)$:

$$\begin{cases} N_t(\lambda;t) = \dfrac{i}{4\lambda}N(\lambda;t) - \dfrac{i}{4\lambda}e^{-iv(0,t)}M_1(\lambda;t), \\ \qquad\qquad\qquad\qquad \text{Im}\lambda \geq -\varepsilon_0,\ 0 \leq t \leq T, \varepsilon_0 > 0 \\ M_{1t}(\lambda;t) = \dfrac{i}{4\lambda}M_1(\lambda;t) - \dfrac{i}{4\lambda}e^{iv(0,t)}N(\lambda;t), \end{cases} \tag{4.39}$$

$$\begin{cases} D_t(\lambda;t) = -\dfrac{i}{4\lambda}D(\lambda;t) + \dfrac{i}{4\lambda}e^{-iv(0,t)}M_2(\lambda;t), \\ \qquad\qquad\qquad\qquad \text{Im}\lambda \leq \varepsilon_0,\ 0 \leq t \leq T, \varepsilon_0 > 0 \\ M_{2t}(\lambda;t) = -\dfrac{i}{4\lambda}M_2(\lambda;t) + \dfrac{i}{4\lambda}e^{iv(0,t)}D(\lambda;t). \end{cases} \tag{4.40}$$

We find the solutions of systems (4.39) and (4.40) by the method of variation constants with due regard for the time invariance of the spectrum of the operator $O_1(t)$:

$$N(\lambda;t) = \left\{e_{10}^{(1)} - e_{20}^{(1)} - \frac{i}{4\lambda}\int_0^t M_1(\lambda;\tau)e^{-\frac{i}{4\lambda}\tau - iv(0,\tau)}d\tau\right\}e^{\frac{i}{4\lambda}t}, \tag{4.41}$$

$$\text{Im}\lambda \geq -\varepsilon_0,\ \ 0 \leq t \leq T, \varepsilon_0 > 0$$

$$M_1(\lambda;t) = \left\{e_{10}^{(1)} + e_{20}^{(1)} - \frac{i}{4\lambda}\int_0^t N(\lambda;\tau)e^{iv(0,\tau) - \frac{i}{4\lambda}\tau}d\tau\right\}e^{\frac{i}{4\lambda}t}, \tag{4.42}$$

$$D(\lambda;t)=\left\{e_{20}^{(2)}-e_{10}^{(2)}+\frac{i}{4\lambda}\int_0^t M_2(\lambda;\tau)e^{\frac{i}{4\lambda}\tau-iv(0,\tau)}d\tau\right\}e^{-\frac{i}{4\lambda}t}, \tag{4.43}$$

$$\mathrm{Im}\lambda\geq-\varepsilon_0,\quad 0\leq t\leq T,\varepsilon_0>0$$

$$M_2(\lambda;t)=\left\{e_{20}^{(2)}+e_{10}^{(2)}+\frac{i}{4\lambda}\int_0^t D(\lambda;\tau)e^{iv(0,\tau)+\frac{i}{4\lambda}\tau}d\tau\right\}e^{-\frac{i}{4\lambda}t}. \tag{4.44}$$

Putting

$$\tilde{N}(\lambda;t)=N(\lambda;t)e^{-\frac{i}{4\lambda}t},\quad \tilde{D}(\lambda;t)=D(\lambda;t)e^{\frac{i}{4\lambda}t}, \tag{4.45}$$

and substituting the expression (4.42) for $M_1(\lambda;t)$ into the right-hand side of equality (4.41) and the expression (4.44) for $M_2(\lambda;t)$ into the right-hand side of (4.43), we derive the Volterra integral equations of the second kind for the unknown boundary data (4.45) [112, 116]:

$$\tilde{N}(\lambda;t)=h_1(\lambda;t)+\left(\frac{i}{4\lambda}\right)^2\int_0^t e^{-iv(0,\tau)}d\tau\int_0^\tau e^{iv(0,\tau_1)}\tilde{N}(\lambda;\tau_1)d\tau_1, \tag{4.46}$$

$$\tilde{D}(\lambda;t)=h_2(\lambda;t)+\left(\frac{i}{4\lambda}\right)^2\int_0^t e^{-iv(0,\tau)}d\tau\int_0^\tau e^{iv(0,\tau_1)}\tilde{D}(\lambda;\tau_1)d\tau_1, \tag{4.47}$$

where $v(0,t)$, $h_1(\lambda;t)$ and $h_2(\lambda;t)$ are the real-valued continuous functions calculated from the known conditions (4.15) and (4.16). These functions are defined by

$$h_1(\lambda;t)=e_{10}^{(1)}-e_{20}^{(1)}-\frac{i}{4\lambda}\left(e_{10}^{(1)}+e_{20}^{(1)}\right)\int_0^t e^{-iv(0,\tau)}d\tau,$$

$$h_1(\lambda;0)=\tilde{N}(\lambda;0)=N(\lambda),\quad \mathrm{Im}\lambda\geq-\varepsilon_0, \tag{4.48}$$

$$h_2(\lambda;t)=e_{20}^{(2)}-e_{10}^{(2)}+\frac{i}{4\lambda}\left(e_{20}^{(2)}+e_{10}^{(2)}\right)\int_0^t e^{-iv(0,\tau)}d\tau,$$

$$h_2(\lambda;0)=\tilde{D}(\lambda;0)=D(\lambda),\quad \mathrm{Im}\lambda\leq\varepsilon_0. \tag{4.49}$$

We write equations (4.46) and (4.47) in the operator form:

$$\tilde{N}=h_1+K_1\tilde{N}, \tag{4.50}$$

$$\tilde{D}=h_2+K_2\tilde{D}, \tag{4.51}$$

where

$$(K_1h_1)(\lambda;t) = \left(\frac{i}{4\lambda}\right)^2 \int_0^t d\tau \int_0^\tau \mathcal{K}(\tau,\tau_1)h_1(\lambda;\tau_1)d\tau_1, \ \text{Im}\lambda \geq -\varepsilon_0, \quad (4.52)$$

$$(K_2h_2)(\lambda;t) = \left(\frac{i}{4\lambda}\right)^2 \int_0^t d\tau \int_0^\tau \mathcal{K}(\tau,\tau_1)h_2(\lambda;\tau_1)d\tau_1, \ \text{Im}\lambda \leq \varepsilon_0, \quad (4.53)$$

the kernel $\mathcal{K}(\tau,\tau_1) = e^{-iv(0,\tau)}e^{iv(0,\tau_1)}$, $0 \leq \tau_1 \leq \tau \leq t \leq T$, is determined for any positive number T and $|\mathcal{K}(\tau,\tau_1)| = 1$.

The Volterra integral operator equations (4.50) and (4.51) map the space $C[0,T]$ into $C[0,T]$. We solve these Volterra integral operator equations by the method of successive approximations. By (4.50) and (4.52), the successive approximations $\tilde{N}^{(p)}$ are determined by the formulas: $\tilde{N}^{(0)} = h_1$,

$$\tilde{N}^{(p)}(\lambda;t) = h_1(\lambda;t) + \left(\frac{i}{4\lambda}\right)^2 \int_0^t d\tau \int_0^\tau \mathcal{K}(\tau,\tau_1)\tilde{N}^{(p-1)}(\lambda;\tau_1)d\tau_1 =$$

$$h_1(\lambda;t) + K_1\tilde{N}^{(p-1)}(\lambda;t), \ p = 1,2,... \quad (4.54)$$

We prove by induction that

$$\tilde{N}^{(p)} = \sum_{k=0}^{p} K_1^k h_1, \ \ p = 0,1,... \quad (4.55)$$

where the term K_1^k denotes a power of the operator K_1.

In fact, the formula (4.55) is true for $p = 0 : \tilde{N}^{(0)} = h_1$. Supposing that this formula is true for $k = p$, then in the recurrence relation (4.54) replacing p by $p+1$, we obtain the formula (4.55) for $p+1$:

$$\tilde{N}^{(p+1)} = h_1 + K_1 \sum_{k=0}^{p} K_1^k h_1 = h_1 + \sum_{k=0}^{p} K_1^{k+1} h_1 = \sum_{k=0}^{p+1} K_1^k h_1.$$

In this way, the formula (4.55) is seen to be true for all p.

We prove by induction that the iterations $(K_1^p h_1)(\lambda;t)$ of the function h_1 satisfy the estimate:

$$|(K_1^p h_1)(\lambda;t)| \leq \left|e_{10}^{(1)} - e_{20}^{(1)}\right| \left(\frac{t}{4|\lambda|}\right)^{2p} \frac{1}{(2p)!} +$$

$$\left|e_{10}^{(1)} + e_{20}^{(1)}\right| \left(\frac{t}{4|\lambda|}\right)^{2p+1} \frac{1}{(2p+1)!}, \ p = 0,1,... \quad (4.56)$$

In fact, since $v(0,t)$ is a real-valued continuous function, then $\left|e^{-iv(0,\tau)}\right| = 1$, and from (4.48) it follows that the estimate (4.56) for $p = 0$ is true. Supposing that the estimate (4.56) is true for $k = p - 1$, then due to $|\mathcal{K}(\tau, \tau_1)| = 1$ we obtain

$$|(K_1^p h_1)(\lambda; t)| = \left|K_1 \left(K_1^{p-1} h_1\right)(\lambda; t)\right| =$$

$$\left|\left(\frac{i}{4\lambda}\right)^2 \int_0^t d\tau \int_0^\tau \mathcal{K}(\tau, \tau_1) \left(K^{p-1} h_1\right)(\lambda; \tau_1) d\tau_1\right| \leq$$

$$\frac{1}{(4|\lambda|)^2} \int_0^t d\tau \int_0^\tau d\tau_1 \left\{ \left|e_{10}^{(1)} - e_{20}^{(1)}\right| \frac{\tau_1^{2(p-1)}}{(4|\lambda|)^{2(p-1)}(2(p-1))!} + \right.$$

$$\left. \left|e_{10}^{(1)} + e_{20}^{(1)}\right| \frac{\tau_1^{2(p-1)+1}}{(4|\lambda|)^{2(p-1)+1}(2(p-1)+1)!} \right\} =$$

$$\left|e_{10}^{(1)} - e_{20}^{(1)}\right| \left(\frac{t}{4|\lambda|}\right)^{2p} \frac{1}{(2p)!} + \left|e_{10}^{(1)} + e_{20}^{(1)}\right| \left(\frac{t}{4|\lambda|}\right)^{2p+1} \frac{1}{(2p+1)!}.$$

Hence, the induction is proved, and the functions $(K_1^p h_1)(\lambda; t) \in C[0, T]$. From the proved estimate (4.56) it follows that the Neumann series $\sum_{k=0}^{\infty} \left(K_1^k h_1\right)(\lambda; t)$ is majorized by the uniformly convergent numerical series:

$$\left|\sum_{k=0}^{\infty} \left(K_1^k h_1\right)(\lambda; t)\right| \leq \left|e_{10}^{(1)} - e_{20}^{(1)}\right| \sum_{k=0}^{\infty} \frac{T^{2k}}{(4|\lambda|)^{2k}(2k)!} +$$

$$\left|e_{10}^{(1)} + e_{20}^{(1)}\right| \sum_{k=0}^{\infty} \frac{T^{2k+1}}{(4|\lambda|)^{2k+1}(2k+1)!} =$$

$$\left|e_{10}^{(1)} - e_{20}^{(1)}\right| \mathrm{ch}\left\{\frac{T}{4|\lambda|}\right\} + \left|e_{10}^{(1)} + e_{20}^{(1)}\right| \mathrm{sh}\left\{\frac{T}{4|\lambda|}\right\}. \tag{4.57}$$

where T is taken so that $0 < T < 4|\lambda|$, $\mathrm{Im}\lambda > 0$.

Hence, the constructed Neumann series converges uniformly with respect to $t \in [0, T]$ for $0 < T < 4|\lambda|$, $\mathrm{Im}\lambda > 0$, and its sum is a continuous function $\tilde{N}(\lambda; t)$. The successive approximations (4.55) converge uniformly with respect to $t \in [0, T]$ and tend to the continuous function $\tilde{N}(\lambda; t)$:

$$\lim_{p \to \infty} \tilde{N}^{(p)}(\lambda; t) = \sum_{k=0}^{\infty} \left(K_1^k h_1\right)(\lambda; t) = \tilde{N}(\lambda; t).$$

In the recurrence relation (4.54) passing to the limit as $p \to \infty$ and using the uniform convergence of the sequence $\left\{\tilde{N}^{(p)}(\lambda; t)\right\}$ to $\tilde{N}(\lambda; t)$, we obtain that the function $\tilde{N}(\lambda;t)$ satisfies the Volterra integral equation (4.46):

$$\tilde{N}(\lambda;t) = \lim_{p\to\infty} \tilde{N}^{(p)}(\lambda;t) = h_1(\lambda;t)+$$

$$\left(\frac{i}{4\lambda}\right)^2 \int_0^t d\tau \int_0^\tau \mathcal{K}(\tau,\tau_1) \lim_{p\to\infty} \tilde{N}^{(p-1)}(\lambda;\tau_1)d\tau_1 =$$

$$h_1(\lambda;t) + \left(\frac{i}{4\lambda}\right)^2 \int_0^t e^{-iv(0,\tau)}d\tau \int_0^\tau e^{iv(0,\tau_1)}\tilde{N}(\lambda;\tau_1)d\tau_1,$$

for $t \in [0,T]\,, 0 < T < 4\,|\lambda|\,, \mathrm{Im}\lambda > 0$.

Hence, the solution $\tilde{N}(\lambda;t)$ of equation (4.46) is the sum of the constructed Neumann series. Owing to estimate (4.57), the function $\tilde{N}(x;t)$ satisfies the estimate:

$$\max_{0\le t\le T}\left|\tilde{N}(\lambda;t)\right| \le \left|e_{10}^{(1)}-e_{20}^{(1)}\right| \mathrm{ch}\left\{\frac{T}{4|\lambda|}\right\} + \left|e_{10}^{(1)}+e_{20}^{(1)}\right| \mathrm{sh}\left\{\frac{T}{4|\lambda|}\right\} \text{ for } t\in[0,T]\,, \tag{4.58}$$

where T is taken so that $0 < T < 4|\lambda|$, $\mathrm{Im}\lambda > 0$.

To prove the uniqueness of the solution of equation (4.50) in the space $C[0,T]$, it suffices to show that the homogeneous equation corresponding to equation (4.50) has only a zero solution in $C[0,T]$. In fact, if the function $\tilde{N}_0 \in C[0,T]$ is the solution of this homogeneous equation, i.e., $\tilde{N}_0 = K_1\tilde{N}_0$, then $\tilde{N}_0$ satisfies the equations: $\tilde{N}_0 = K_1(K_1\tilde{N}_0) = K_1^2\tilde{N}_0 = ... = K_1^p\tilde{N}_0$. The iterations $K_1^p\tilde{N}_0(\lambda;t)$ satisfy the homogeneous equation corresponding with the equation (4.54):

$$\left|\tilde{N}_0(\lambda;t)\right| = \left|\left(K_1^p\tilde{N}_0\right)(\lambda;t)\right| = \left(\frac{i}{4\lambda}\right)^2 \int_0^t d\tau \int_0^\tau \mathcal{K}(\tau,\tau_1)\tilde{N}^{(p-1)}(\lambda;\tau_1)d\tau_1.$$

By induction we can verify that the iterations $K_1^{(p)}\tilde{N}_0(\lambda;t)$ satisfy the estimate:

$$\left|\tilde{N}_0(\lambda;t)\right| = \left|\left(K_1^p\tilde{N}_0\right)(\lambda;t)\right| \le \max_{0\le\tau\le T}|\tilde{N}_0(\lambda;\tau)|\left(\frac{t}{4|\lambda|}\right)^{2p}\frac{1}{(2p)!},\quad p=0,1,...,$$

where $t\in[0,T]$ and the number T is taken so that $0 < T < 4|\lambda|$, $\mathrm{Im}\lambda > 0$.

The expression on the right-hand side of the last inequality tends to zero as $p\to\infty$, hence $\tilde{N}_0(\lambda;t)\equiv 0$. Thus, the uniqueness of solution of equation (4.50) is proved.

By using (4.49), (4.51) and (4.53), we prove analogously that the Volterra integral equation (4.47) has in $C[0,T]$ a unique solution $\tilde{D}(\lambda;t)$ satisfying the estimate:

$$\max_{0\le t\le T}\left|\tilde{D}(\lambda;t)\right| \le \left|e_{10}^{(2)} - e_{20}^{(2)}\right| \mathrm{ch}\left\{\frac{T}{4|\lambda|}\right\} + \left|e_{20}^{(2)} + e_{20}^{(2)}\right| \mathrm{sh}\left\{\frac{T}{4|\lambda|}\right\}, \tag{4.59}$$

where T is taken so that $0 < T < 4|\lambda|$, $\mathrm{Im}\lambda < 0$.

Due to estimate (4.26), it follows from (4.23) and (4.24) that the initial BVs of the Jost solutions satisfy the asymptotic conditions as $|\lambda| \to \infty$:

$$e_{10}^{(1)} = e_1^{(1)}(0,\lambda;0) = 1 + o(1), \;\; e_{20}^{(1)} = e_2^{(1)}(0,\lambda;0) = o(1), \tag{4.60}$$

$$e_{10}^{(2)} = e_1^{(2)}(0,\lambda;0) = o(1), \;\; e_{20}^{(2)} = e_2^{(1)}(0,\lambda;0) = 1 + o(1). \tag{4.61}$$

Using (4.60) and (4.61), from (4.46)–(4.49), we obtain the asymptotic conditions of the solutions $\tilde{N}(\lambda;t)$ and $\tilde{D}(\lambda;t)$ of Volterra integral equations (4.46) and (4.47) as $|\lambda| \to \infty$:

$$\lim_{|\lambda|\to\infty} \tilde{N}(\lambda;t) = \lim_{|\lambda|\to\infty} h_1(\lambda;t) = \lim_{|\lambda|\to\infty} N(\lambda) = 1, \;\; \text{Im}\lambda \geq -\varepsilon_0,$$

$$\lim_{|\lambda|\to\infty} \tilde{D}(\lambda;t) = \lim_{|\lambda|\to\infty} h_2(\lambda;t) = \lim_{|\lambda|\to\infty} D(\lambda) = 1, \;\; \text{Im}\lambda \leq \varepsilon_0.$$

Knowing the solutions $\tilde{N}(\lambda;t)$ and $\tilde{D}(\lambda;t)$ of equations (4.46) and (4.47), we calculate the unknown $N(\lambda;t)$ for any λ, $\text{Im}\lambda > 0$ and $D(\lambda;t)$ for any λ, $\text{Im}\lambda < 0$ by the formula (4.45). Knowing $N(\lambda;t)$ and $D(\lambda;t)$, we can solve equations (4.35)–(4.38). In fact, the solutions of these equations, i.e., the unknown BVs $e^{(1)}(0,\lambda;t)$ and $e^{(2)}(0,\lambda;t)$ of the Jost solutions evaluated at $x = 0$, are found by the method of variation constants with due regard for the time-invariance of the spectrum of the operator $O_1(t)$:

$$\begin{aligned} &e_1^{(1)}(0,\lambda;t) = \\ &\left\{ e_{10}^{(1)} + \frac{1}{4\lambda}\int_0^t \sin v(0,\tau) N(\lambda;\tau) \exp\left\{ \frac{i}{4\lambda}\left[\int_0^\tau e^{-iv(0,\xi)} d\xi - \tau \right] \right\} d\tau \right\} \times \\ &\exp\left\{ \frac{i}{4\lambda}\left[t - \int_0^t e^{-iv(0,\xi)} d\xi \right] \right\} \quad \text{for } t \in [0,T], \; 0 < T < 4|\lambda|, \; \text{Im}\lambda > 0, \end{aligned} \tag{4.62}$$

$$\begin{aligned} &e_2^{(1)}(0,\lambda;t) = \\ &\left\{ e_{20}^{(1)} + \frac{1}{4\lambda}\int_0^t \sin v(0,\tau) N(\lambda;\tau) \exp\left\{ -\frac{i}{4\lambda}\left[\int_0^\tau e^{-iv(0,\xi)} d\xi + \tau \right] \right\} d\tau \right\} \times \\ &\exp\left\{ \frac{i}{4\lambda}\left[t + \int_0^t e^{-iv(0,\xi)} d\xi \right] \right\} \quad \text{for } t \in [0,T], \; 0 < T < 4|\lambda|, \; \text{Im}\lambda > 0, \end{aligned} \tag{4.63}$$

$$e_1^{(2)}(0,\lambda;t) =$$
$$\left\{e_{10}^{(2)} - \frac{1}{4\lambda}\int_0^t \sin v(0,\tau)D(\lambda;\tau)\exp\left\{\frac{i}{4\lambda}\left[\int_0^\tau e^{-iv(0,\xi)}d\xi+\tau\right]\right\}d\tau\right\}\times$$
$$\exp\left\{-\frac{i}{4\lambda}\left[t+\int_0^t e^{-iv(0,\xi)}d\xi\right]\right\} \text{ for } t\in[0,T],\ 0<T<4|\lambda|,\ \text{Im}\lambda<0, \tag{4.64}$$

$$e_2^{(2)}(0,\lambda;t) =$$
$$\left\{e_{20}^{(2)} - \frac{1}{4\lambda}\int_0^t \sin v(0,\tau)D(\lambda;\tau)\exp\left\{\frac{i}{4\lambda}\left[\tau-\int_0^\tau e^{-iv(0,\xi)}d\xi\right]\right\}d\tau\right\}\times$$
$$\exp\left\{\frac{i}{4\lambda}\left[\int_0^t e^{-iv(0,\xi)}d\xi-t\right]\right\} \text{ for } t\in[0,T],\ 0<T<4|\lambda|,\ \text{Im}\lambda<0. \tag{4.65}$$

Here, due to the reduction (4.20) and equations (3.17) and (3.18), the following relations are satisfied:

$$e_{10}^{(1)} = e_1^{(1)}(0,\lambda;0) = e_2^{(2)}(0,-\lambda;0),\quad e_{20}^{(1)} = e_2^{(1)}(0,\lambda;0) = e_1^{(2)}(0,-\lambda;0).$$

Using these relations, we can verify that the calculated BVs (4.62)–(4.65) possess the properties for any $t\in[0,T], 0<T<4|\lambda|$, $\text{Im}\lambda>0$:

$$\begin{aligned}&e_1^{(1)}(0,\lambda;t) = e_2^{(2)}(0,-\lambda;t),\quad e_2^{(1)}(0,\lambda;t) = e_1^{(2)}(0,-\lambda;t),\\ &N(\lambda;t) = D(-\lambda;t),\quad M_1(\lambda;t) = M_2(-\lambda;t).\end{aligned} \tag{4.66}$$

Due to (4.60) and (4.61), the calculated BVs (4.62)–(4.65) for any $t\in[0,T]$ satisfy the asymptotic conditions as $|\lambda|\to\infty$:

$$\begin{aligned}&e_1^{(1)}(0,\lambda;t) = 1+o(1),\quad e_2^{(1)}(0,\lambda;t) = o(1),\\ &e_1^{(2)}(0,\lambda;t) = o(1),\quad e_2^{(2)}(0,\lambda;t) = 1+o(1).\end{aligned}$$

Consequently, the calculated BVs (4.62)–(4.65) satisfy the required asymptotic conditions (4.60), (4.61) and relation equalities (4.66).

4.1.4 The time-dependence of the scattering data

We prove that if λ_k^+, $k=1,...,\gamma_1$, are the zeros of the numerator $N(\lambda)$, then the same λ_k^+, $k=1,...,\gamma_1$ also are the zeros of $N(\lambda;t)$, and vice versa.

Indeed, from system (4.39) and using (4.66) we obtain the relation between $N(\lambda;t)$ and $N(\lambda)$ for any $t \in [0,T]$, $0 < T < 4|\lambda|$, $\mathrm{Im}\lambda > 0$:

$$N(\lambda;t) = N(\lambda)\exp\left\{\frac{i}{4\lambda}t - \frac{i}{4\lambda}\int_0^t e^{-iv(0,\tau)}\frac{M_1(\lambda;\tau)}{N(\lambda;\tau)}d\tau\right\} = D(-\lambda;t),\ \mathrm{Im}\lambda \geq \varepsilon_0. \tag{4.67}$$

Let λ_k^+, $k = 1,...,\gamma_1$ be the zeros of $N(\lambda)$, then using the continuity of functions $N(\lambda)$ and $N(\lambda;t)$ of λ, from (4.67) we get the relation:

$$\lim_{\lambda\to\lambda_k^+}\exp\left\{\frac{i}{4\lambda}\int_0^t e^{-iv(0,\tau)}\frac{M_1(\lambda;\tau)}{N(\lambda;\tau)}d\tau\right\}N(\lambda_k^+;t) = \exp\left\{\frac{i}{4\lambda_k^+}t\right\}N(\lambda_k^+) = 0,$$
$$k = 1,...,\gamma_1 \tag{4.68}$$

From (4.67) and (4.68) it is clear that the zeros λ_k^+, $k = 1,...,\gamma_1$ of $N(\lambda)$ coincide with zeros of $N(\lambda;t)$ for any $t \in [0,T]$, $0 < T < 4|\lambda|$, $\mathrm{Im}\lambda > 0$, and moreover, the zeros of $N(\lambda;t)$ are the same zeros of $D(-\lambda;t)$. This means that if λ_k^+, $k = 1,...,\gamma_1$ are the zeros of $N(\lambda;t)$, then $-\lambda_k^+$, $k = 1,...,\gamma_1$ are the zeros of $D(\lambda;t)$, and the multiplicity m_k of the zero λ_k^+ of $N(\lambda;t)$ is the same multiplicity of the corresponding zero $-\lambda_k^+$ of $D(\lambda;t)$. As a consequence, the zeros of $N(\lambda;t)$ and $D(\lambda;t)$ come in pairs. Therefore, the singular numbers of the operator $O_1(t)$ are the pairs of numbers: λ_k^+, $-\lambda_k^+$, $\mathrm{Im}\lambda_k^+ \geq \varepsilon_0 > 0$, $k = 1,...,\gamma_1$.

Having the formulas (4.62), (4.65) and (4.66), we calculate the time-dependent coefficients of the form (3.81) and (3.82):

$$A_q(\lambda_k^+;t) = \frac{d^q}{d\lambda^q}\left[\frac{1}{e_1^{(1)}(0,\lambda;t)}\right]_{\lambda=\lambda_k^+},\ q = 1,...,m_{k-1},\ k = 1,...,\gamma_1, \tag{4.69}$$

$$B_q(-\lambda_k^+;t) = \frac{d^q}{d\lambda^q}\left[\frac{1}{e_2^{(2)}(0,\lambda;t)}\right]_{\lambda=-\lambda_k^+} =$$
$$\frac{d^q}{d\lambda^q}\left[\frac{1}{e_1^{(1)}(0,-\lambda;t)}\right]_{\lambda=\lambda_k^+} = A_q(-\lambda_k^+;t),\ q = 1,...,m_{k-1},\ k = 1,...,\gamma_1. \tag{4.70}$$

Substituting (4.69) and (4.70) into the formulas (3.34) and (3.35), respectively, using (4.66), we calculate the time-dependent standardized polynomials of the form (3.36):

$$g_k^+(-x;t) = p_k(x;t)e^{i\lambda_k^+x},\ \ f_k^-(x;t) = \tilde{p}_k(x;t)e^{-i(-\lambda_k^+)x} = \tilde{p}_k(x;t)e^{i\lambda_k^+x},$$
$$k = 1,...,\gamma_1,$$

i.e.,

$$g_k^+(-x;t) = f_k^-(x;t), \quad k = 1, ..., \gamma_1, \tag{4.71}$$

where $p_k(x;t)$ and $\tilde{p}_k(x;t)$ are the polynomials in x of degree $m_k - 1$. These polynomials depend on t as parameter.

The functions (4.71) characterize the operator $O_1(t)$ on the discrete spectrum.

The scattering function $S(\lambda;t)$ for the operator $O_1(t)$ is formulated from the calculated boundary data $N(\lambda;t)$ and $D(\lambda;t)$:

$$S(\lambda;t) = \frac{N(\lambda;t)}{D(\lambda;t)} = \frac{N(\lambda;t)}{N(-\lambda;t)}, \quad S(\lambda;0) = S(\lambda), \; |\mathrm{Im}\lambda| < \varepsilon_0 \tag{4.72}$$

for $t \in [0,T], \quad 0 < T < 4|\lambda|$.

Substituting (4.67) into (4.72), we obtain the relation between $S(\lambda;t)$ and $S(\lambda)$:

$$S(\lambda;t) = S(\lambda) \exp\left\{ \frac{i}{2\lambda} t - \frac{i}{4\lambda} \int_0^t e^{-iv(0,\tau)} \left[\frac{M_1(\lambda;\tau)}{N(\lambda;\tau)} + \frac{M_1(-\lambda;\tau)}{N(-\lambda;\tau)} \right] d\tau \right\}, \tag{4.73}$$

for $t \in [0,T], \quad 0 < T < 4|\lambda|, \; |\mathrm{Im}\lambda| < \varepsilon_0$.

Owing to estimate (4.26) for the kernels $H_{jk}(x,s), \quad 0 \le x \le s, \quad j,k = 1,2,$ the Jost solutions (4.23) and (4.24) are holomorphic in λ for $\mathrm{Im}\lambda > -\varepsilon_0$ and for $\mathrm{Im}\lambda < \varepsilon_0$, respectively. Hence, from (4.22), (4.34) it follows that the boundary data $N(\lambda;t)$, $M_1(\lambda;t)$ and $N(-\lambda;t)$, $M_1(-\lambda;t)$ are analytic in λ in the upper half-plane $\mathrm{Im}\lambda > -\varepsilon_0$ and in the lower half-plane $\mathrm{Im}\lambda < \varepsilon_0$, respectively. From the above discussion we deduce that the scattering function $S(\lambda;t)$ possesses the same properties of $S(\lambda)$ enumerated in Lemma 3.3; i.e., $S(\lambda;t)$ is the function of the type (S), and due to (4.66)

$$S(-\lambda;t) = S^{-1}(\lambda;t) \;\text{ for }\; t \in [0,T], \;\; 0 < T < 4|\lambda|, \;\; |\mathrm{Im}\lambda| < \varepsilon_0. \tag{4.74}$$

The scattering function $S(\lambda;t)$ gives the complete characterization of the operator $O_1(t)$ on the continuous spectrum.

Thus, the time-dependent scattering data set $s(t)$ for the problem (4.18)–(4.19) is constructed from the discrete and continuous spectra defined by (4.71) and (4.73), respectively:

$$s(t) = \Big\{ S(\lambda;t) \;\text{ for }\; |\mathrm{Im}\lambda| < \varepsilon_0, \; (\varepsilon_0 > 0); \; \lambda_k^+, \; -\lambda_k^+, \; \mathrm{Im}\lambda_k^+ \ge \varepsilon_0; \\ p_k(x,t), \tilde{p}_k(x,t), k = 1, ..., \gamma_1 \Big\}, \tag{4.75}$$

where for every $t \in [0,T]$, $0 < T < 4|\lambda|$, $|\mathrm{Im}\lambda| < \varepsilon_0$, the quantities of the set $s(t)$ satisfy the conditions which are analogous with the conditions (1)–(5) enumerated in Theorem 3.1.

The time-dependent functions of the form (3.84) are constructed from (4.71) and (4.73) for the systems of fundamental equations in the inverse scattering problem:

$$f(x;t) = f_S(x;t) - \sum_{k=1}^{\gamma_1} f_k^-(x;t), \quad g(-x;t) = g_S(-x;t) - \sum_{k=1}^{\gamma_1} g_k^+(-x;t), \tag{4.76}$$

where

$$f_S(x;t) = \frac{1}{2\pi} \int\limits_{-\infty-i\eta}^{\infty-i\eta} [S(\lambda;t) - 1]\, e^{-i\lambda x} d\lambda,$$

$$g_S(x;t) = \frac{1}{2\pi} \int\limits_{-\infty+i\eta}^{\infty+i\eta} \left[S^{-1}(\lambda;t) - 1\right] e^{-i\lambda x} d\lambda,$$

where by (4.74), the functions $f_S(x;t)$ and $g_S(x;t)$ satisfy the relation equalities:

$$g_S(-x;t) = f_S(x;t).$$

The functions defined by (4.76) characterize the operator $O_1(t)$ on the complete spectrum. Thus, the theorem about the time-dependence of the scattering data of the problem (4.18)–(4.19) is proved.

Theorem 4.1. *We consider the boundary problem* (4.18)–(4.19) *with the potential matrix* (4.20) *in the class of isospectral potentials belonging to the class* P *and satisfying the estimate* (4.17). *The following assertions for considered problem are proved:*

(1) The Volterra integral equations (4.46) *and* (4.47) *for the boundary data* $\tilde{N}(\lambda;t)$, $\mathrm{Im}\lambda > 0$, *and* $\tilde{D}(\lambda;t)$, $\mathrm{Im}\lambda < 0$, $t \in [0,T], 0 < T < 4|\lambda|$ *have unique solutions in the space* $C[0,T]$. *The solutions of these equations are constructed by the Neumann uniformly convergent series. The sums of these Neumann series are expressible in terms of the known initial and boundary conditions* (4.15) *and* (4.16), *and satisfy estimates* (4.58) *and* (4.59), *respectively;*

(2) The boundary data $N(\lambda;t)$, $\mathrm{Im}\lambda > 0$ *and* $D(\lambda;t)$, $\mathrm{Im}\lambda < 0$, *are related to the found functions* $\tilde{N}(\lambda;t)$ *and* $\tilde{D}(\lambda;t)$ *by the formula* (4.45) *for* $t \in [0,T]$, $0 < T < 4|\lambda|$. *Consequently, the unknown BVs at* $x = 0$ *of the Jost solutions are found by the formulas* (4.62)–(4.65) *which are expressible in terms of known conditions* (4.15) *and* (4.16);

(3) The relationships between the boundary data $N(\lambda;t)$ and $N(\lambda)$, and between the scattering functions $S(\lambda;t)$ and $S(\lambda)$ are established by (4.67) and (4.72), (4.73), respectively. Thus, the boundary data $N(t,\lambda)$ and $D(t,\lambda)$ possess the same properties of $N(\lambda)$ and $D(\lambda)$, respectively, and $S(\lambda;t)$ is a scattering function of the type (S) possessing the property (4.74) and the properties of $S(\lambda)$ enumerated in Lemma 3.3;

(4) The time-dependence of the scattering data (4.75) of the associated problem (4.18)–(4.19) is derived from (4.69)–(4.71) for the discrete spectrum and from (4.73) for the continuous spectrum of the operator $O_1(t)$.

4.1.5 The IBVP (4.14)–(4.16)

The scattering problem (4.18)–(4.19) is associated with the IBVP (4.14)–(4.15)–(4.16). Let the elements of the potential matrix Q in system (4.18) be isospectral potentials belonging to the class P and satisfying the reduction (4.20) and estimate (4.17). Then according to Definition 3.2 in Chapter 3, we calculate a data set $s = s(0)$ defined by the set (4.75) for $t = 0$ from the known real-valued continuous function $v_x(x, 0)$ satisfying estimate (4.17). Since the potential matrix (4.20) is the particular case of the potential matrix $C(x)$ in system (3.4), then Theorem 3.1 about the description of the scattering data of the problem (4.18)–(4.19) is applicable, whence it follows that the conditions (1)–(5) enumerated in Theorem 3.1 are the necessary and sufficient conditions for the quantities of the calculated set s to be the scattering data for the scattering problem (4.18)–(4.19) for every fixed $t \in [0, T]$. Further, based on Theorem 4.1 we construct a time-dependent data set $s(t)$ of the type (4.75) from the constructed data set s and the known initial and boundary conditions (4.15), (4.16). Thus, Theorems 3.1 and 4.1 reduce the problem of solving the IBVP (4.14)–(4.15)–(4.16) to that of solving the systems of time-dependent equations of the form (3.38) and (3.39) for every $x \geq 0, \xi > 0$:

$$\begin{cases} H_{11}(x, x+\xi; t) + \int\limits_0^\infty H_{12}(x, x+\tau; t) f(2x+\xi+\tau; t) d\tau = 0, \\ g(-2x-\xi; t) + H_{12}(x, x+\xi; t) + \int\limits_0^\infty H_{11}(x, x+\tau; t) g(-2x-\xi-\tau; t) d\tau = 0, \end{cases} \tag{4.77}$$

$$\begin{cases} f(2x+\xi; t) + H_{21}(x, x+\xi; t) + \int\limits_0^\infty H_{22}(x, x+\tau; t) f(2x+\xi+\tau; t) d\tau = 0, \\ H_{22}(x, x+\xi; t) + \int\limits_0^\infty H_{21}(x, x+\tau; t) g(-2x-\xi-\tau; t) d\tau = 0, \end{cases} \tag{4.78}$$

where x and t enter systems (4.77) and (4.78) merely as parameter, $f(x;t)$ and $g(-x;t)$ are known functions constructed by formulas (4.71) and (4.76) from the given scattering data set (4.75), $H_{jk}(x, x+\xi; t), j, k = 1, 2$, are unknown functions of ξ for every $(x, t) \in [0, \infty) \times [0, T]$.

The elements of the potential matrix (4.20) belong to the class P, therefore from integral equations (3.13)–(3.16) it follows that the solutions of these equations, i.e., the kernels of the transformation operator are continuous functions satisfying estimate (4.26) and possessing the first continuous partial derivative with respect to t. Further, using systems (4.77) and (4.78), we can verify that the known functions $f(2x+\xi;t)$ and $g(-2x-\xi;t)$ in these systems possess properties, which are analogous with properties of the kernels; i.e., these known functions are also continuous and have the first continuous partial derivative with respect to t. Furthermore, due to the condition (5) of Theorem 3.1, the constructed functions (4.76) satisfy the estimate for any $t \in [0, T]$:

$$|f(x;t)| \leq \tilde{C}e^{-\frac{1}{4}\varepsilon x}, \quad |g(-x;t)| \leq \tilde{C}e^{-\frac{1}{4}\varepsilon x}, \tag{4.79}$$

where $\tilde{C}$ is a constant, $\varepsilon > 0$.

Using properties of constructed functions $f(2x+\xi;t)$, $g(-2x-\xi;t)$, and by arguing as in the proof of Theorems 1.1 and 4.1, we can prove the following:

(I). The systems of fundamental equations (4.77) and (4.78) have for every $(x, t) \in [0, \infty) \times [0, T]$ a unique solution $(H_{11}(x, x+\xi; t)), (H_{12}(x, x+\xi; t)) \in L_1[0, \infty), j = 1, 2$. These solutions satisfy estimate of the type (4.26) and integral equations (3.13)–(3.16). Owing to the uniqueness of solution of these equations, the solutions of systems (4.77) and (4.78) are the kernels of the transformation operator of the scattering problem generated by the system (4.18) with the constructed potential matrix:

$$\begin{aligned} &2i\begin{pmatrix} 0 & -H_{12}(x,x;t) \\ H_{21}(x,x;t) & 0 \end{pmatrix} \\ &= \begin{pmatrix} 0 & c_1(x,t) \\ c_2(x,t) & 0 \end{pmatrix} = \begin{pmatrix} 0 & -\frac{1}{2}v_x(x,t) \\ \frac{1}{2}v_x(x,t) & 0 \end{pmatrix}. \end{aligned} \tag{4.80}$$

Hence, the kernels, i.e., the solutions of systems (4.77) and (4.78) satisfy relations (4.25) and estimates (4.26). Then from (3.14) and (3.15) it follows that $H_{12}(x, x; t)$ and $H_{21}(x, x; t)$ are pure imaginary-valued continuous functions possessing the continuous partial derivative with respect to t. Therefore, the reconstructed potential $v_x(x, t)$ is a real-valued continuous function satisfying estimate (4.17), and the recovered potential matrix (4.80) satisfies reduction (4.20);

(II). The vector functions $e^{(1)}(x, \lambda; t)$ and $e^{(2)}(x, \lambda; t) = e^{(1)}(x, -\lambda; t)$ constructed by formulas (4.23) and (4.24) from the found solutions of systems (4.77) and (4.78) are the Jost solutions of a system of the form (4.18) with the constructed potential matrix (4.80);

(III). The quantities of the data set (4.75) constructed from the known initial and boundary conditions (4.15) and (4.16) coincide with the respective quantities of the scattering data of the problem for a system of the form (4.18) with the recovered potential matrix (4.80) and with the boundary condition (4.19). Hence, the BVs at $x = 0$ of the constructed Jost solutions $e^{(1)}(x, \lambda; t)$ and $e^{(1)}(x, -\lambda; t)$ satisfy the following relations:

$$e_1^{(1)}(0, \lambda; t) - e_2^{(1)}(0, \lambda; t) - \left[e_1^{(1)}(0, -\lambda; t) - e_2^{(1)}(0, -\lambda; t)\right] S(\lambda; t) = 0$$
$$\text{for } t \in [0, T],\ 0 < T < 4|\lambda|,\ \text{Im}\lambda| < \varepsilon_0,\ \varepsilon_0 > 0, \tag{4.81}$$

$$\left[\frac{d^q}{d\lambda^q} N(\lambda; t)\right]_{\lambda=\lambda_k^+} =$$
$$\frac{d^q}{d\lambda^q}\left[N(\lambda)\exp\left\{\frac{i}{4\lambda}t - \frac{i}{4\lambda}\int_0^t e^{-iv(0,\tau)}\frac{M_1(\lambda;\tau)}{N(\lambda;\tau)}d\tau\right\}\right]_{\lambda=\lambda_k^+} = 0,$$
$$\text{Im}\lambda_k^+ \geq \varepsilon_0,\ q = 1, ..., m_k - 1,\ k = 1, ..., \gamma_1, \tag{4.82}$$

$$\left[\frac{d^q}{d\lambda^q} D(\lambda; t)\right]_{\lambda=-\lambda_k^+} = \left[\frac{d^q}{d\lambda^q} N(-\lambda; t)\right]_{\lambda=\lambda_k^+} =$$
$$\left[\frac{d^q}{d\lambda^q} N(-\lambda)\exp\left\{\frac{i}{4\lambda}\int_0^t e^{-iv(0,\tau)}\frac{M_1(-\lambda;\tau)}{N(-\lambda;\tau)}d\tau - \frac{i}{4\lambda}t\right\}\right]_{\lambda=\lambda_k^+} = 0;$$

where m_k is the multiplicity of the zero λ_k^+ of the numerator and the zero $-\lambda_k^+$ of the denominator of the scattering function $S(\lambda; t)$;

(IV). A correspondence between the scattering data set (4.75) and the reconstructed potential matrix (4.80) is one-to-one. Therefore, the potential $v_x(x, t)$ in system (4.18) is recovered uniquely from the known initial and boundary conditions (4.15) and (4.16) by formula (4.80), that is expressed through the solution of system (4.77). Hence, an anti-derivative of $\frac{1}{2}v_x(x, t) = 2iH_{12}(x, x; t)$ is obtained by the formula:

$$v(x, t) = 4i\int_0^x H_{12}(\xi, \xi; t)d\xi + h(t), \tag{4.83}$$

where $h(t)$ has the continuous derivative. $4iH_{12}(x, x; t)$ is the real-valued continuous function satisfying estimate of the type (4.26) and possessing the first continuous partial derivative with respect to t.

Hence, the found function (4.83) is real-valued continuous and has the continuously mixed derivative of the second-order.

By properties of the found function $v(x,t)$ and by Lemma 4.1, the function (4.83) is the solution of the considered IBVP. The theorem about the solution of the IBVP (4.14)–(4.15)–(4.16) is deduced from the enumerated results (I)–(IV).

Theorem 4.2. *The Dirichlet IBVP* (4.14)–(4.15)–(4.16) *can be solved by the ISM in the class of isospectral potentials, belonging to the class P and satisfying estimate* (4.17) *and reduction* (4.20). *Namely,*

(1) The vector functions $\mathrm{e}^{(1)}(x,\lambda;t)$ *and* $\mathrm{e}^{(1)}(x,-\lambda;t)$ *constructed by the formulas* (4.23) *and* (4.24) *from the solutions of systems* (4.77) *and* (4.78) *are the Jost solutions of system* (4.18) *with the reconstructed potential matrix* (4.80) *satisfying estimate* (4.17)*;*

(2) The quantities of the data set of the form (4.75) *constructed from the known initial and boundary conditions* (4.15) *and* (4.16) *coincide with the scattering data of the associated scattering problem for the system of the form* (4.18) *with the reconstructed potential matrix* (4.80) *and with boundary condition* (4.19). *Hence, the BVs* $\mathrm{e}^{(1)}(0,\lambda;t)$ *and* $\mathrm{e}^{(1)}(0,-\lambda;t)$ *of the Jost solutions of this problem satisfy the identities* (4.81) *and* (4.82)*;*

(3) Owing to Lemma 4.1 the function $v(x,t)$ *determined by formula* (4.83) *is the solution of the Dirichlet IBVP* (4.14)–(4.15)–(4.16). *This solution is the real-valued continuous function and is expressed through the solution of the system of fundamental equations* (4.77) *and* (4.78) *by formula* (4.80). *Thus, every scattering data set of the form* (4.75) *corresponds a unique solution of the considered IBVP.*

Thus, the sketch illustrating our method for solving the IBVP (4.14)-(4.15)-(4.16) is shown in Figure 4.1.

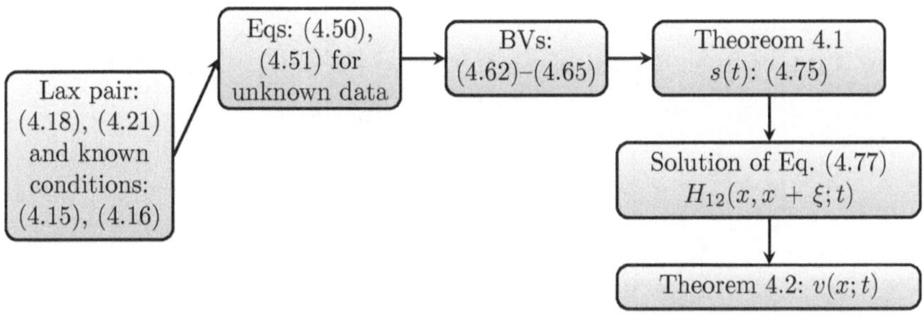

FIGURE 4.1: The steps in solving the IBVPs (4.14)–(4.15)–(4.16).

4.2 The IBVP for the shG equation

In this section we consider the IBVP for the shG equation [143]:

$$v_{xt} = \text{sh} v, \quad 0 \le x < \infty, \quad 0 \le t < T, \tag{4.84}$$

with the known initial and boundary conditions:

$$v(x,0) = v(x), \tag{4.85}$$
$$v(0,t) = h(t), \tag{4.86}$$

where $h(t), v(x,t)$ and $v_x(x,t)$ are real-valued continuous functions, and $v_x(x,t)$ possesses the continuous partial derivative with respect to t and satisfies the estimate:

$$|v_x(x,t)| \le \frac{\tilde{C}}{1+x^{1+\varepsilon}} \text{ for any } (x,t) \in [0,\infty) \times [0,T], \tag{4.87}$$

$\tilde{C}$ is a constant, $\varepsilon > 0$.

4.2.1 The self-adjoint problem associated with the shG equation

The Dirichlet IBVP (4.84)–(4.85)–(4.86) is associated with the scattering problem for the system of the form (4.8):

$$u_x = U_2 u, \quad u = (u_1(x,\lambda;t),\ u_2(x,\lambda;t)),\ 0 \le x < \infty,\ 0 \le t \le T \tag{4.88}$$

with the boundary condition at $x = 0$:

$$u_1(0,\lambda;t) = u_2(0,\lambda;t), \tag{4.89}$$

where λ is a complex-valued parameter, $J_1 = \begin{pmatrix} 1 & 0 \\ 0 & -1 \end{pmatrix}$, $U_2 = i\lambda J_1 + iQ$,

$$Q = -J_1 C(x,t) = \begin{pmatrix} 0 & -c_1(x,t) \\ c_2(x,t) & 0 \end{pmatrix}.$$

Putting $c_1(x,t) = c_2(x,t) = -\frac{1}{2}v_x(x,t)$, we have

$$Q = \begin{pmatrix} 0 & \frac{1}{2}v_x(x,t) \\ -\frac{1}{2}v_x(x,t) & 0 \end{pmatrix}, \tag{4.90}$$

and the potential $\frac{1}{2}v_x(x,t)$ is a real-valued continuous function possessing the continuous partial derivative with respect to t and satisfying the estimate (4.87).

Under the reduction (4.90) the matrix $C(x,t)$ in (4.88) is self-adjoint:

$$C(x,t) = \begin{pmatrix} 0 & c_1(x,t) \\ c_2(x,t) & 0 \end{pmatrix} = \begin{pmatrix} 0 & -\frac{1}{2}v_x(x,t) \\ -\frac{1}{2}v_x(x,t) & 0 \end{pmatrix}.$$

The IBVP (4.84)–(4.85)–(4.86) is associated with the self-adjoint problem (4.88)–(4.89). We recall the results of this self-adjoint problem from Section 3.4 [119]. This problem has no discrete spectrum, and the potential element $\frac{1}{2}v_x(x)$ of the self-adjoint matrix (4.90) is a power decreasing function as $x \to \infty$, while the potential element of the non-self-adjoint matrix (4.20) is required to exponentially decrease as $x \to \infty$.

Due to reduction (4.90) and estimate (4.87), from (3.13)–(3.16) it follows that the kernels of the transformation operator of the self-adjoint problem (4.88)–(4.89) satisfy the relations:

$$H_{11}(x,s;t) = H_{22}(x,s;t); \quad H_{12}(x,s;t) = \overline{H_{21}(x,s;t)}, \quad 0 \le x \le s, \tag{4.91}$$

and the estimate:

$$|H_{jk}(x,s;t)| \le \frac{\tilde{C}}{1+(x+s)^{1+\varepsilon}} \text{ for any } t \in [0,T], \ 0 \le x < s, \ j,k = 1,2, \tag{4.92}$$

where $H_{11}(x,s;t)$ is a real-valued function, $H_{12}(x,x;t)$ is a pure imaginary-valued function, and $\tilde{C}$ is a constant, $\varepsilon > 0$.

The Jost solutions $e^{(1)}(x,\lambda;t)$ and $e^{(2)}(x,\lambda;t)$ of system (4.88) satisfy the asymptotic conditions (3.19) and (3.20) as $x \to \infty$. Therefore, they do not depend on time at infinity:

$$\lim_{x\to\infty} e^{(1)}_{jt}(x,\lambda;t) = 0, \quad \lim_{x\to\infty} e^{(2)}_{jt}(x,\lambda;t) = 0, \ j = 1,2 \tag{4.93}$$

Under the reduction (4.90), from (3.17) and (3.18) we obtain the relations between the Jost solutions $e^{(1)}(x,\lambda;t)$ and $e^{(2)}(x,\lambda;t)$ of system (4.88):

$$\overline{e^{(1)}_1(x,\lambda;t)} = e^{(2)}_2(x,\lambda;t), \ \overline{e^{(1)}_2(x,\lambda;t)} = e^{(2)}_1(x,\lambda;t). \tag{4.94}$$

Under the reduction (4.90) and estimate (4.92), the Lemma 1.13 is valid, therefore the correspondence between the numbers $A(\lambda;t)$ and $B(\lambda;t)$ established by the asymptotic (3.7) is one-to-one. Thus, the scattering function $S(\lambda;t)$ for the self-adjoint problem (4.88)–(4.89) is uniquely defined by the relation:

$$S(\lambda;t)A(\lambda;t) = B(\lambda;t), \text{ where } t \text{ is a parameter.}$$

The scattering function $S(a,\lambda;t)$ for the self-adjoint problem on the half-line $a \le x < \infty, \ \ a \ge 0$, is determined analogously: $S(a,\lambda;t)A_a(\lambda;t) = B_a(\lambda;t)$. Hence, the assertions enumerated in Theorem 3.5 [119] in Chapter 3 for the function $S(\lambda;t)$ are valid, i.e., for any $x \ge 0$ there exists the function $S(x,\lambda;t)$ such that $S(\lambda;t) = S(0,\lambda;t)$ and this function satisfies conditions which are analogous to the conditions (1)–(4) enumerated in the Theorem 3.5.

4.2.2 The Lax compatibility condition

The time-dependence of the eigenfunctions of the problem (4.88)–(4.89) is defined by the system of evolutionary equations:

$$u_t = V_2 u, \quad V_2 = -\frac{i}{4\lambda}\begin{pmatrix} \mathrm{ch}v(x,t)+\alpha(t) & -i\mathrm{sh}v(x,t) \\ -i\mathrm{sh}v(x,t) & -\mathrm{ch}v(x,t)+\alpha(t) \end{pmatrix}, \tag{4.95}$$

where $\alpha(t)$ is any arbitrary bounded function.

The time-dependence of the Jost solutions $e^{(1)}(x,\lambda;t)$ and $e^{(2)}(x,\lambda;t)$ of system (4.88) is defined by system (4.95) with $\alpha \equiv -1$ for $e^{(1)}(x,\lambda;t)$ and with $\alpha \equiv 1$ for $e^{(2)}(x,\lambda;t)$:

$$e_t^{(1)} = V_2 e^{(1)}, \quad V_2 = -\frac{i}{4\lambda}\begin{pmatrix} \mathrm{ch}v(x,t)-1 & -i\mathrm{sh}v(x,t) \\ -i\mathrm{sh}v(x,t) & -\mathrm{ch}v(x,t)-1 \end{pmatrix}, \tag{4.96}$$

$$e_t^{(2)} = V_2 e^{(2)}, \quad V_2 = -\frac{i}{4\lambda}\begin{pmatrix} \mathrm{ch}v(x,t)+1 & -i\mathrm{sh}v(x,t) \\ -i\mathrm{sh}v(x,t) & 1-\mathrm{ch}v(x,t) \end{pmatrix}. \tag{4.97}$$

The natural boundary condition for the real-valued solution $v(x,t)$ of the shG equation (4.84) [70] is

$$v(x,t) \to 0 \ (\mathrm{mod}\ 2\pi) \text{ as } x \to \infty. \tag{4.98}$$

Owing to (3.19), (3.20) and (4.98) the functions on the right-hand sides of (4.96) and (4.97) tend to zero uniformly with respect to t as $x \to \infty$. Hence, the derived equations (4.96) and (4.97) satisfy the required asymptotic conditions (4.93) as $x \to \infty$. Consequently, the system of evolutionary equations (4.96) and (4.97) are the operator equations governing the time-evolution of the Jost solutions of the system (4.88).

By an argument analogous to that used for the proof of Lemma 4.1, we can prove the following lemma.

Lemma 4.2. *Let the isospectral potential $v_x(x,t)$ in system* (4.88) *be a real-valued continuous function belonging to the class P, then the systems* (4.88) *and* (4.95) *are compatible, i.e., $u_{xt} = u_{tx}$, and the function $v(x,t)$ varies with time as the solution of the shG equation* (4.84). *Conversely, if the function $v(x,t)$ satisfies the shG equation* (4.84), *then $\lambda_t = 0$, and the systems* (4.88) *and* (4.95) *are compatible.*

The shG equation (4.84) is equivalent to the Lax compatibility condition:

$$U_{2t} - V_{2x} + U_2 V_2 - V_2 U_2 = 0, \tag{4.99}$$

where the matrices U_2 and V_2 are defined by (4.90) and (4.95), respectively.

Thus, the shG equation (4.84) on the half-line can be written as the compatibility condition (4.99) of two linear eigenvalue equations (4.88) as x part, and (4.95) as t part, i.e., the shG equation (4.84) admits by these appropriate linear equations.

Note that due to the definition (3.50) the time-dependence of the scattering function $S(\lambda;t)$ is determined by the unknown BVs $e^{(1)}(0,\lambda;t)$ and $e^{(2)}(0,\lambda;t)$ of the Jost solutions $e^{(1)}(x,\lambda;t)$ and $e^{(2)}(x,\lambda;t)$ of systems (3.17), (3.18). The presence of these unknown BVs is the main difficulty in solving the IBVP (4.84)–(4.85)–(4.86). We deal with this difficultly in the next two subsections.

4.2.3 Evaluation of unknown BVs

Owing to reduction (4.90) and relations (4.94), it suffices to derive the system of evolutionary equations for $e^{(1)}(0,\lambda;t)$ or $e^{(2)}(0,\lambda;t)$. So we write up the system of evolutionary equations (4.96) for the unknown BVs $\left(e_1^{(1)}(0,\lambda;t),\ e_2^{(1)}(0,\lambda;t)\right)$:

$$\begin{cases} e_{1t}^{(1)}(0,\lambda;t) = -\dfrac{i}{4\lambda}\left\{[chv_0-1]\,e_1^{(1)}(0,\lambda;t) - i\mathrm{sh}v_0 e_2^{(1)}(0,\lambda;t)\right\}, \\ e_{2t}^{(1)}(0,\lambda;t) = -\dfrac{i}{4\lambda}\left\{-i\mathrm{sh}v_0 e_1^{(1)}(0,\lambda;t) - (1+chv_0)\,e_2^{(1)}(0,\lambda;t)\right\}, \end{cases}$$

where $v_0 = v(0,t) = h(t),\ h(t)$ is a known real-valued continuous function.

Using the known formula: $e^{-v_0} = chv_0 - shv_0$, we write the above system of evolutionary equations in the form:

$$e_{1t}^{(1)}(0,\lambda;t) = -\frac{i}{4\lambda}\left\{\left[e^{-v_0}-1\right]e_1^{(1)}(0,\lambda;t)+\mathrm{sh}v_0\left[e_1^{(1)}(0,\lambda;t)-ie_2^{(1)}(0,\lambda;t)\right]\right\}, \tag{4.100}$$

$$e_{2t}^{(1)}(0,\lambda;t) = \frac{i}{4\lambda}\left\{\left[1+e^{-v_0}\right]e_2^{(1)}(0,\lambda;t)+i\mathrm{sh}v_0\left[e_1^{(1)}(0,\lambda;t)-ie_2^{(1)}(0,\lambda;t)\right]\right\}, \tag{4.101}$$

whereas using the formula: $e^{v_0} = \mathrm{ch}v_0 + \mathrm{sh}v_0$, we write up this system of evolutionary equations as follows:

$$e_{1t}^{(1)}(0,\lambda;t) = -\frac{i}{4\lambda}\left\{\left[e^{v_0}-1\right]e_1^{(1)}(0,\lambda;t)-\mathrm{sh}v_0\left[e_1^{(1)}(0,\lambda;t)+ie_2^{(1)}(0,\lambda;t)\right]\right\}, \tag{4.102}$$

$$e_{2t}^{(1)}(0,\lambda;t) = \frac{i}{4\lambda}\left\{\left[1+e^{v_0}\right]e_2^{(1)}(0,\lambda;t)+i\mathrm{sh}v_0\left[e_1^{(1)}(0,\lambda;t)+ie_2^{(1)}(0,\lambda;t)\right]\right\}. \tag{4.103}$$

We put

$$B(\lambda;t) = e_1^{(1)}(0,\lambda;t) - ie_2^{(1)}(0,\lambda;t),\ \ B(\lambda;0) = B(\lambda) = e_{10}^{(1)} - ie_{20}^{(1)},$$
$$\mathrm{Im}\lambda \geq 0,\ \ e_{j0}^{(1)} = e_j^{(1)}(0,\lambda;0),\ j=1,2,$$
$$M(\lambda;t) = e_1^{(1)}(0,\lambda;t) + ie_2^{(1)}(0,\lambda;t),\ \ M(\lambda,0) = M(\lambda) = e_{10}^{(1)} + ie_{20}^{(1)}.$$

We call $B(\lambda;t)$ and $M(\lambda;t)$ the boundary data at $x=0$ of the problem (4.88)–(4.89). To find the unknown BVs $e_n^{(j)}(0,\lambda;t)$, $j,n=1,2$, we first calculate the unknown boundary data $B(\lambda;t)$. To this purpose, from (4.100), (4.101) and (4.102), (4.103) we derive the system for the unknown boundary data $B(\lambda;t)$ and $M(\lambda;t)$:

$$B_t(\lambda;t)=\frac{i}{4\lambda}B(\lambda;t)-\frac{i}{4\lambda}e^{-v_0}M(\lambda;t),$$
$$\mathrm{Im}\lambda\geq 0 \tag{4.104}$$
$$M_t(\lambda;t)=\frac{i}{4\lambda}M(\lambda;t)-\frac{i}{4\lambda}e^{v_0}B(\lambda;t),$$

where λ enters system (4.104) merely as parameter. Using the time-invariance of the spectrum of the problem (4.88)–(4.89) and the method of variation constants, we solve the first equation, and then the second equation of the system (4.104). As a result, $B(\lambda;t)\,(M(\lambda;t))$ is expressed through $M(\lambda;t)\,(B(\lambda;t))$:

$$B(\lambda;t)=\left\{B(\lambda)-\frac{i}{4\lambda}\int_0^t M(\lambda;\tau)e^{-\frac{i}{4\lambda}\tau-v(0,\tau)}d\tau\right\}e^{\frac{i}{4\lambda}t}, \tag{4.105}$$

$$M(\lambda;t)=\left\{M(\lambda)-\frac{i}{4\lambda}\int_0^t B(\lambda;\tau)e^{-\frac{i}{4\lambda}\tau+v(0,\tau)}d\tau\right\}e^{\frac{i}{4\lambda}t}, \tag{4.106}$$

where $B(\lambda)=B(\lambda;0)$, $M(\lambda)=M(\lambda;0)$.
Putting

$$\tilde{B}(\lambda;t)=B(\lambda;t)e^{-\frac{i}{4\lambda}t}, \tag{4.107}$$

and substituting the right-hand side of (4.106) into (4.105), we derive the linear Volterra integral equation of the second kind for the unknown boundary data $\tilde{B}(\lambda;t)$:

$$\tilde{B}(\lambda;t)=r(\lambda;t)+\left(\frac{i}{4\lambda}\right)^2\int_0^t e^{-v(0,\tau)}d\tau\int_0^\tau e^{v(0,\tau_1)}\tilde{B}(\lambda;\tau_1)d\tau_1, \tag{4.108}$$

where

$$r(\lambda;t)=e_{10}^{(1)}-ie_{20}^{(1)}-\frac{i}{4\lambda}\left[e_{10}^{(1)}+ie_{20}^{(1)}\right]\int_0^t e^{-v(0,\tau)}d\tau,\quad r(\lambda;0)=B(\lambda). \tag{4.109}$$

In the same way, as in the case of equation (4.46), we solve the Volterra equation (4.108) by the method of successive approximations [112, 116]. We first write equation (4.108) in the operator form:

$$\tilde{B}=r+R\tilde{B}, \tag{4.110}$$

where R is the integral operator defined by

$$(Rr)(\lambda;t) = \left(\frac{i}{4\lambda}\right)^2 \int_0^t d\tau \int_0^\tau \mathcal{R}(\tau,\tau_1) r(\lambda,\tau_1) d\tau_1, \tag{4.111}$$

the kernel $\mathcal{R}(\tau,\tau_1) = e^{-v(0,\tau)} e^{v(0,\tau_1)}, 0 \leq \tau_1 \leq \tau \leq t \leq T$, is a continuous function defined in $C[0,T]$ for any positive number T and satisfies the estimate:

$$|\mathcal{R}(\tau,\tau_1)| \leq \max_{0\leq\tau\leq T} e^{2|v(0,\tau)|}. \tag{4.112}$$

The integral operator equation (4.110) transforms the space $C[0,T]$ into $C[0,T]$. We solve this equation by the method of successive approximations. The successive approximations $\tilde{B}^{(p)}$ are determined by the formulas: $\tilde{B}^{(0)} = r$,

$$\begin{aligned}\tilde{B}^{(p)}(\lambda;t) &= r(\lambda;t) + \left(\frac{i}{4\lambda}\right)^2 \int_0^t d\tau \int_0^\tau \mathcal{R}(\tau,\tau_1) \tilde{B}^{(p-1)}(\lambda;\tau_1) d\tau_1 \\ &= r(\lambda;t) + R\tilde{B}^{(p-1)}(\lambda;t), \ p = 1, 2, ...\end{aligned} \tag{4.113}$$

We prove by induction that

$$\tilde{B}^{(p)}(\lambda;t) = \sum_{k=0}^{p} R^k r(\lambda;t), \ p = 0, 1, ... \tag{4.114}$$

where the term R^k denotes a power of the operator R.

Indeed, by an argument analogous to that used for the proof of formula (4.55), we prove that the successive approximations $\tilde{B}^{(p)}$ given by (4.114) is true for all p. Further, in the same way as the case of estimate (4.56), taking (4.111) and (4.112) into account, we prove by induction that the iterations $(R^p r)(\lambda;t)$ of the function r satisfy the estimate:

$$\begin{aligned}|(R^p r)(\lambda;t)| &= \left| \left(\frac{i}{4\lambda}\right)^2 \int_0^t d\tau \int_0^\tau \mathcal{R}(\tau,\tau_1) \left(R^{p-1} r\right)(\lambda;\tau_1) d\tau_1 \right| \\ &\leq \left| e_{10}^{(1)} - ie_{20}^{(1)} \right| \left(\max_{0\leq\tau\leq T} e^{|v(0,\tau)|} \right)^{2p} \left(\frac{t}{4|\lambda|}\right)^{2p} \frac{1}{(2p)!} + \\ &\left| e_{10}^{(1)} + ie_{20}^{(1)} \right| \left(\max_{0\leq\tau\leq T} e^{|v(0,\tau)|} \right)^{2p+1} \left(\frac{t}{4|\lambda|}\right)^{2p+1} \frac{1}{(2p+1)!}, \\ &p = 1, 2, ...\end{aligned}$$

Thereby, the induction is proved, and the functions $R^p r \in C[0,T]$. The constructed Neumann series $\sum_{k=0}^{\infty} (R^k r)(\lambda;t)$ is majorized by the uniformly convergent numerical series:

$$\left|\sum_{k=0}^{\infty} (R^k r)(\lambda;t)\right| \le \left|e_{10}^{(1)} - ie_{20}^{(1)}\right| \sum_{k=0}^{\infty} \left(\frac{T}{4|\lambda|} \times \max_{0\le\tau\le T} e^{|v(0,\tau)|}\right)^{2k} \frac{1}{(2k)!} +$$
$$\left|e_{10}^{(1)} + ie_{20}^{(1)}\right| \sum_{k=0}^{\infty} \left(\frac{T}{4|\lambda|} \times \max_{0\le\tau\le T} e^{|v(0,\tau)|}\right)^{2k+1} \frac{1}{(2k+1)!} =$$
$$\left|e_{10}^{(1)} - ie_{20}^{(1)}\right| \mathrm{ch}\left\{\frac{T}{4|\lambda|} \times \max_{0\le\tau\le T} e^{|v(0,\tau)|}\right\} +$$
$$\left|e_{10}^{(1)} + ie_{20}^{(1)}\right| \mathrm{sh}\left\{\frac{T}{4|\lambda|} \times \max_{0\le\tau\le T} e^{|v(0,\tau)|}\right\}, \qquad (4.115)$$

where T is taken so that $0 < T < 4|\lambda|$, $\mathrm{Im}\lambda > 0$.

The successive approximations (4.114) converge uniformly with respect to $t \in [0,T]$, $0 < T < 4|\lambda|$, $\mathrm{Im}\lambda > 0$, and tend to a continuous function $\tilde{B}(\lambda;t)$:

$$\lim_{p\to\infty} \tilde{B}^{(p)}(\lambda;t) = \sum_{k=0}^{\infty} (R^k r)(\lambda;t) = \tilde{B}(\lambda;t).$$

Thus, the constructed Neumann series converges uniformly with respect to $t \in [0,T]$, $0 < T < 4|\lambda|$, $\mathrm{Im}\lambda > 0$, and its sum is a continuous function $\tilde{B}(\lambda;t)$. Then in the recurrence relation (4.113) passing to the limit as $p \to \infty$ and using the uniform convergence of the sequence $\left\{\tilde{B}^{(p)}(\lambda;t)\right\}$ to $\tilde{B}(\lambda;t)$, we obtain that the function $\tilde{B}(\lambda;t)$ satisfies the Volterra integral equation (4.108):

$$\tilde{B}(\lambda;t) = \lim_{p\to\infty} \tilde{B}^{(p)}(\lambda;t)$$
$$= r(\lambda;t) + \left(\frac{i}{4\lambda}\right)^2 \int_0^t e^{-v(0,\tau)}d\tau \int_0^\tau e^{v(0,\tau_1)} \lim_{p\to\infty} \tilde{B}^{(p-1)}(\lambda;\tau_1)d\tau$$
$$= r(\lambda;t) + \left(\frac{i}{4\lambda}\right)^2 \int_0^t e^{-v(0,\tau)}d\tau \int_0^\tau e^{v(0,\tau_1)} \tilde{B}(\lambda;\tau_1)d\tau_1,$$

for $t \in [0,T], 0 < T < 4|\lambda|$, $\mathrm{Im}\lambda > 0$.

The solution $\tilde{B}(\lambda;t)$ of equation (4.108) is the sum of the constructed Neumann series. Due to the estimate (4.115), this solution $\tilde{B}(x;t)$ satisfies the estimate:

$$\max_{0\le t\le T}\left|\tilde{B}(\lambda;t)\right| \le \left|e_{10}^{(1)} - ie_{20}^{(1)}\right| \operatorname{ch}\left\{\frac{T}{4|\lambda|} \times \max_{0\le\tau\le T} e^{|v(0,\tau)|}\right\} +$$

$$\left|e_{10}^{(1)} + ie_{20}^{(1)}\right| \operatorname{sh}\left\{\frac{T}{4|\lambda|} \times \max_{0\le\tau\le T} e^{|v(0,\tau)|}\right\} \quad \text{for } t\in[0,T], \tag{4.116}$$

where T is taken so that $0 < T < 4|\lambda|$.

By an argument analogous to that used for the proof of uniqueness of solution of operator equation (4.50), we show that the solution $\tilde{B}(\lambda;t)$ of operator equation (4.110) is unique. Using (4.60) and (4.61), from (4.108) and (4.109) we obtain the asymptotic condition of the solution $\tilde{B}(\lambda;t)$ of equation (4.108) as $|\lambda|\to\infty$:

$$\lim_{|\lambda|\to\infty}\tilde{B}(\lambda;t) = \lim_{|\lambda|\to\infty} r(\lambda;t) = \lim_{|\lambda|\to\infty} B(\lambda) = 1.$$

Knowing the solution $\tilde{B}(\lambda;t)$ of equation (4.108), we calculate the unknown boundary data $B(\lambda;t)$ by the formula (4.107). Knowing $B(\lambda;t)$ for $t\in[0,T]$, $0 < T < 4|\lambda|$, $\mathrm{Im}\lambda > 0$. we can solve equations (4.100) and (4.101) by the method of variation constants with due regard for the time-invariance of the spectrum of the operator O_2. The solutions of these equations, i.e., the BVs of the Jost solutions at $x=0$ of system (4.88) are represented in the form:

$$e_1^{(1)}(0,\lambda;t) =$$

$$\left\{e_{10}^{(1)} - \frac{i}{4\lambda}\int_0^t \operatorname{sh}v(0,\tau)B(\lambda;\tau)\exp\left\{-\frac{i}{4\lambda}\left[\tau - \int_0^\tau e^{-v(0,\xi)}d\xi\right]\right\}d\tau\right\}\times$$

$$\exp\left\{\frac{i}{4\lambda}\left[t - \int_0^t e^{-v(0,\xi)}d\xi\right]\right\}, \quad t\in[0,T], 0<T<4|\lambda|, \quad \mathrm{Im}\lambda>0, \tag{4.117}$$

$$e_2^{(1)}(0,\lambda;t) =$$

$$\left\{e_{20}^{(1)} - \frac{1}{4\lambda}\int_0^t \operatorname{sh}v(0,\tau)B(\lambda;\tau)\exp\left\{-\frac{i}{4\lambda}\left[\tau + \int_0^\tau e^{-v(0,\xi)}d\xi\right]\right\}d\tau\right\}\times$$

$$\exp\left\{\frac{i}{4\lambda}\left[t + \int_0^t e^{-v(0,\xi)}d\xi\right]\right\}, \quad t\in[0,T], 0<T<4|\lambda| \quad \mathrm{Im}\lambda>0, \tag{4.118}$$

where the initial BVs $e_{10}^{(1)}$ and $e_{20}^{(1)}$ of the Jost solution satisfy the asymptotic condition (4.60), whence it follows that the calculated BVs (4.117) and (4.118) satisfy the required asymptotic condition as $|\lambda|\to\infty$:

$$e_1^{(1)}(0,\lambda;t) = 1 + o(1), \quad e_2^{(1)}(0,\lambda;t) = o(1) \text{ as } |\lambda|\to\infty.$$

With the help of the relation (4.94) we find the BVs at $x = 0$ of the Jost solution $\left(e_1^{(2)}(x,\lambda;t),\ e_2^{(2)}(x,\lambda;t)\right)$ from the calculated BVs (4.117) and (4.118):

$$e_2^{(2)}(0,\lambda;t) = \overline{e_1^{(1)}(0,\lambda;t)},\quad e_1^{(2)}(0,\lambda;t) = \overline{e_2^{(1)}(0,\lambda;t)}. \tag{4.119}$$

4.2.4 The time-dependence of the scattering function

Substituting the calculated BVs (4.117), (4.118) and (4.119) into the definition (3.50) of the scattering function, we find the factorization of the time-dependent scattering function $S(\lambda;t)$ for the self-adjoint problem (4.88)–(4.89):

$$S(\lambda;t) = \frac{e_1^{(1)}(0,\lambda;t) - e_2^{(1)}(0,\lambda;t)}{\overline{e_1^{(1)}(0,\lambda;t) - e_2^{(1)}(0,\lambda;t)}} = \frac{N(\lambda;t)}{\overline{N(\lambda;t)}} \tag{4.120}$$

for $t \in [0,T], 0 < T < 4|\lambda|, |\text{Im}\lambda| < \epsilon_0$, where $S(\lambda;0) = S(\lambda) = \frac{N(\lambda)}{\overline{N(\lambda)}}, N(\lambda) = N(\lambda;0), S(\lambda;t) = 1 + o(1)$ as $|\lambda| \to \infty$.

Since $S(\lambda;t)$ is the scattering function of the self-adjoint problem (4.88)–(4.89), then by Theorem 3.5 for any $x \geq 0$ and $t \in [0,T], 0 < T < 4|\lambda|$, there exists a function $S(x,\lambda;t)$ such that $S(\lambda;t) = S(0,\lambda;t)$, and this function satisfies the conditions (1)–(4) enumerated in this theorem. By the condition 1 in Theorem 3.5, the function $S(x,\lambda;t)$ admits the factorization:

$$S(x,\lambda;t) = \frac{N(x,\lambda;t)}{\overline{N(x,\lambda;t)}}, \tag{4.121}$$

for any $x \geq 0$ and $t \in [0,T], 0 < T < 4|\lambda|, |\text{Im}\lambda| < \epsilon_0$, where the numerator $N(x,\lambda;t)$ of factorization (4.121) is analytic in λ, $\text{Im}\lambda \geq 0$ and does not have zero in the upper half-plane $\text{Im}\lambda > 0$.

Thus, the scattering function (4.120) characterizes the self-adjoint problem (4.88)–(4.89) on the continuous spectrum. Moreover, owing to Theorem 3.5 and by using the scattering function (4.121), we can reconstruct uniquely the potential self-adjoint matrix (4.90), i.e., the self-adjoint problem (4.88)–(4.89) is reconstructed uniquely from the known scattering function (4.120).

4.2.5 The IBVP for the shG equation

The Theorem 3.5 about the description of the scattering function of problem (4.88)–(4.89) with the potential self-adjoint matrix (4.90) is valid. Hence, the Theorem 3.5 and the time-dependent scattering function (4.120) reduce the problem of solving the IBVP (4.84)–(4.85)–(4.86) to that of solving the system of fundamental equations (4.77). In fact, the functions $f(x,\xi;t)$ and $g(x,-\xi;t)$ appearing in system (4.77) are constructed from the calculated scattering function (4.121) by the formulas:

$$\begin{aligned} f(x,\xi;t) &= \frac{1}{2\pi}\int_{-\infty}^{\infty}\left[S(x,\lambda;t)-1\right]e^{-i\lambda\xi}d\lambda \\ &= \frac{1}{2\pi}\int_{-\infty}^{\infty}\left[\frac{N(x,\lambda;t)}{\overline{N(x,\lambda;t)}}-1\right]e^{-i\lambda\xi}d\lambda, \\ g(x,-\xi;t) &= \frac{1}{2\pi}\int_{-\infty}^{\infty}\left[S^{-1}(x,\lambda;t)-1\right]e^{i\lambda\xi}d\lambda \\ &= \frac{1}{2\pi}\int_{-\infty}^{\infty}\left[\frac{\overline{N(x,\lambda;t)}}{N(x,\lambda;t)}-1\right]e^{i\lambda\xi}d\lambda, \end{aligned} \tag{4.122}$$

where the functions $f(x,\xi;t)$ and $g(x,-\xi;t)$ satisfy the conditions (2), (3) and (4) enumerated in the Theorem 3.5. Namely,

$$\begin{aligned} &f(x,\xi;t) = f(\xi+2x;t), \quad g(x,-\xi;t) = g(-\xi-2x;t), \\ &f(\xi+2x;t) = \overline{g(-\xi-2x;t)}, \text{ for any } x \geq 0, \\ &|f(x;t)| \leq \frac{\tilde{C}}{1+x^{1+\varepsilon}}, |g(-x;t)| \leq \frac{\tilde{C}}{1+x^{1+\varepsilon}} \text{ for any } (x,t)\in[0,\infty)\times[0,T], \end{aligned} \tag{4.123}$$

where $\tilde{C}$ is a constant, $\varepsilon > 0$.

The elements of the potential matrix (4.90) belong to the class P, therefore from integral equations (3.13) and (3.15) it follows that the solutions of these equations, i.e., the kernels of the transformation operator also belong to this class. Further, by using system (4.77), we can verify that the known functions $f(2x+\xi;t)$ and $g(-2x-\xi;t)$ in this system possess the properties analogous with properties of the kernel functions $H_{11}(x,x+\xi;t)$ and $H_{12}(x,x+\xi;t)$, i.e., the functions $f(2x+\xi;t)$ and $g(-2x-\xi;t)$ are continuous and have the first continuous partial derivative with respect to t.

Using the properties of the constructed functions $f(2x+\xi;t)$, $g(-2x-\xi;t)$ and estimate (4.123), by an argument analogous to that used for the proof of Lemmas 1.18, 1.20 and 1.21 we can prove the following:

(I). The system (4.77), in which the functions $f(\xi+2x;t)$, $g(-\xi-2x;t)$ are constructed by formulas (4.122) from the known function $S(\lambda;t)$, has for every $(x,t)\in[0,\infty)\times[0,T]$ a unique solution $(H_{11}(x,x+\xi;t),\ H_{12}(x,x+\xi;t))\in L_1[0,\infty)$. This solution satisfies the integral equations (3.13) and (3.15) and the estimate of the type (4.92). Owing to the uniqueness of solution of these equations, the solution of system (4.77) is the kernels of the transformation operator of the scattering problem generated by system (4.88) with boundary condition (4.89) and the reconstructed self-adjoint potential matrix:

$$-2i\begin{pmatrix} 0 & H_{12}(x,x;t) \\ H_{12}(x,x;t) & 0 \end{pmatrix}$$
$$= \begin{pmatrix} 0 & c_1(x,t) \\ c_2(x,t) & 0 \end{pmatrix} = -\frac{1}{2}\begin{pmatrix} 0 & v_x(x,t) \\ v_x(x,t) & 0 \end{pmatrix} \tag{4.124}$$

Hence, the kernels, i.e., the solution of system (4.77) satisfies relations (4.91) and estimate (4.92). Then, from (3.13) and (3.15) it follows that the found function $H_{11}(x,x;t)$ is real-valued continuous and the function $H_{12}(x,x;t)$ is pure imaginary-valued continuous. These functions possess the continuous partial derivative with respect to t. Hence, the reconstructed potential $-\frac{1}{2}v_x(x,t) = -2iH_{12}(x,x;t)$ is a real-valued function satisfying estimate (4.87) and the potential matrix (4.124) satisfies reduction (4.90).

(II). The vector functions $e^{(1)}(x,\lambda;t)$ and $\overline{e^{(1)}(x,\lambda;t)}$ constructed by (4.23) and (4.24) from the solutions $(H_{11}(x,s;t),\ H_{12}(x,s;t))$ and $\left(H_{11}(x,s;t),\ \overline{H_{12}(x,s;t)}\right)$, $0 \le x < s,\ t \in [0,T]$ of system (4.77) are the Jost solutions of a system of the form (4.88) with the reconstructed self-adjoint potential matrix (4.124);

(III). The functions (4.120) and (4.121) are the scattering functions for the self-adjoint problems for system of the form (4.88) with the reconstructed self-adjoint potential matrix (4.124) on the half-line $[0,\infty)$ and $[x,\infty)$, respectively for any $x \ge 0$ and $t \in [0,T]$;

(IV). The correspondence between the scattering function (4.120) and the reconstructed potential matrix (4.124) is one-to-one. Therefore, the potential $-\frac{1}{2}v_x(x,t)$ in system (4.88) is recovered uniquely from the calculated scattering function (4.120) by formula (4.124) in terms of the solution of system (4.77). Hence, the anti-derivative of the reconstructed continuous function $v_x(x,t)$ is found uniquely from the known initial and boundary conditions (4.85) and (4.86):

$$v(x,t) = 4i\int_0^x H_{12}(x_1,x_1;t)dx_1 + h(t), \tag{4.125}$$

where the real-valued function $h(t)$ given by (4.86) has the continuous derivative, $4iH_{12}(x_1,x_1;t)$ is the real-valued continuous function possessing the first continuous partial derivative with respect to t. Hence, the found function $v(x,t)$ is real-valued continuous and has continuously mixed derivatives of second-order.

Owing to properties of the function (4.125), and by Lemma 4.2 the constructed function (4.125) is the real solution of the IVBP (4.84)–(4.85)–(4.86).

The results obtained in this section can be formulated in the main theorem.

Theorem 4.3. *The IBVP* (4.84)–(4.85)–(4.86) *can be solved by ISM in the class of isospectral potentials, which are real-valued continuous functions possessing continuously mixed derivatives of the second order, and satisfying estimate* (4.87) *and reduction* (4.90). *Namely,*

(1) The scattering function $S(\lambda; t)$ of the self-adjoint scattering problem (4.88)–(4.89) *is defined by the factorization* (4.120), *in which the unknown BVs* (4.117)–(4.119) *are calculated from the solution of linear Volterra integral equation* (4.108) *in terms of the known initial and boundary conditions* (4.85) *and* (4.86). *This scattering function $S(\lambda; t)$ possesses the properties (1)–(4) enumerated in Theorem 3.5;*

(2) The correspondence between $S(x, \lambda; t)$ for any $x \geq 0$ defined by the factorization (4.121) *and the reconstructed potential matrix* (4.124) *is one-to-one. The potential $v_x(x, t)$ is recovered uniquely by formula* (4.124) *in terms of the solution of the system of fundamental equations* (4.77). *It satisfies estimate* (4.87). *The solution of system* (4.77) *is found uniquely from the scattering function $S(\lambda; t)$, which is constructed from the BVs. The BVs are calculated from the known conditions* (4.85) *and* (4.86). *By Lemma 4.2, the found function* (4.125) *is the real solution of the Dirichlet IBVP* (4.84)–(4.85)–(4.86).

Remark 4.2. (1) The Cauchy problem (3.144), (3.146) considered in Chapter 3 and the IBVP (4.84)–(4.85)–(4.86) are associated with the scattering problem (3.4)–(3.5) with the potential self-adjoint matrix. In this case the known functions $f(2x + \xi; t)$ and $g(-2x - \xi; t)$ in systems (4.77) and (3.39) satisfy the condition (3) of Theorem 3.5, therefore the system (4.77) coincides with system (3.39).

(2) The solution (3.152) of the Cauchy problem (3.144), (3.146) and the solution (4.125) of the IBVP (4.84)–(4.85)–(4.86) are expressed through the solution of the system of fundamental equations (4.77). But the known functions (3.151) in system (4.77) are defined by the scattering function (3.150), which is constructed from only the known initial condition (3.146). Whereas the known functions (4.122) in system (4.77) are defined by the scattering function (4.121), which is constructed from not only the known initial condition (4.85) but also the known boundary one (4.86).

4.3 Exact soliton-solutions of the sG and shG equations

As shown in Section 3.3, the scattering problem with the non-scattering potential is completely determined by the characteristics of the discrete spectrum. The non-scattering data of this problem are the set of the type (3.109). In this case the system of fundamental equations of the form (4.77) constructed from the set (3.109) can be solved exactly, and this system readily yields to solution in explicit form, because it is a system of equations with degenerated kernels. Some explicit forms of exact solutions of the sG and shG on the half-line are found in [138].

For simplicity we suppose that the non-scattering data of the non-scattering problem associated with the sG or shG equation consist of only one pair of simple poles: λ^+, $-\lambda^+$, $\mathrm{Im}\lambda^+ > \varepsilon_0$, $\varepsilon_0 > 0$. We find explicit solutions of the sG and shG equations for this case. Now we construct the solution of the sG equation from the non-scattering data of the form (3.139):

$$\begin{aligned} &S(\lambda;t) \equiv 1 \ \text{ for } \ |\mathrm{Im}\lambda| < \varepsilon_0, \varepsilon_0 > 0, \\ &\lambda^+, \ -\lambda^+, \ \mathrm{Im}\lambda^+ \geq \varepsilon_0, \ i\left(e_1^{(1)} N_\lambda\right)^{-1}, \end{aligned} \tag{4.126}$$

where $e_1^{(1)} = e_1^{(1)}(0,\lambda^+;t) \neq 0, N_\lambda = N_\lambda(\lambda^+;t) \neq 0$.

Under the reduction (4.20) for the sG equation (4.14), it follows from (3.119) and (3.120) that $-c_1(x,t) = c_2(x,t) = \frac{1}{2}v_x(x,t)$ and $\alpha = \beta$. Hence, in this case the non-scattering potential for the sG equation (4.14) is obtained from (3.119) subject to the condition: $\alpha = \beta = (e_1^{(1)} N_\lambda)^{-1} e^{2i\lambda^+ x}$:

$$v_x(x,t) = \frac{4}{(2\lambda^+)^{-2}(e_1^{(1)} N_\lambda)^{-1} e^{2i\lambda^+ x} - e_1^{(1)} N_\lambda e^{-2i\lambda^+ x}}. \tag{4.127}$$

Note that the potential $v_x(x,t)$ is constructed from the non-scattering data set (4.126). This data set is similar to the data set (3.139) for constructing the solution of the mKdV equation (3.59). But the time-dependences of the standardized multipliers in the sets (3.139) and (4.126) are different from each other.

In fact, we now derive the time-dependence of the standardized multiplier $i\left(e_1^{(1)} N_\lambda\right)^{-1}$ in (4.126). For this purpose, we rewrite the function (4.127) in the form:

$$v_x(x,t) = \frac{4}{(2\lambda^+)^{-1}} \cdot \frac{(2\lambda^+)^{-1}(e_1^{(1)} N_\lambda)^{-1} e^{2i\lambda^+ x}}{(2\lambda^+)^{-2}(e_1^{(1)} N_\lambda)^{-2} e^{4i\lambda^+ x} - 1}.$$

Taking the integral of the function defined by the last equality, we get

$$v(x,t) = -2i\ln\frac{(2\lambda^+)^{-1}\left(e_1^{(1)}N_\lambda\right)^{-1}e^{2i\lambda^+x}-1}{(2\lambda^+)^{-1}\left(e_1^{(1)}N_\lambda\right)^{-1}e^{2i\lambda^+x}+1}\ (\text{mod}2\pi i)\quad 0\le x<\infty. \tag{4.128}$$

Using (4.128), from the sG equation (4.14) we have

$$v_{xt}(x,t) = \sin v(x,t) = \frac{1}{2i}\left\{e^{iv(x,t)} - e^{-iv(x,t)}\right\} =$$
$$-\frac{4e^{2i\lambda^+x}}{2i\lambda^+\left(e_1^{(1)}N_\lambda\right)}\times\frac{(2\lambda^+)^{-2}\left(e_1^{(1)}N_\lambda\right)^{-2}e^{4i\lambda^+x}+1}{\left[(2\lambda^+)^{-2}\left(e_1^{(1)}N_\lambda\right)^{-2}e^{4i\lambda^+x}-1\right]^2}\quad 0\le x<\infty. \tag{4.129}$$

On the other hand, differentiating of the equality (4.127) with respect to t, yields:

$$v_{xt}(x,t) = \frac{4e^{2i\lambda^+x}}{\left(e_1^{(1)}N_\lambda\right)^2}\times\frac{(2\lambda^+)^{-2}\left(e_1^{(1)}N_\lambda\right)^{-2}e^{4i\lambda^+x}+1}{\left[(2\lambda^+)^{-2}\left(e_1^{(1)}N_\lambda\right)^{-2}e^{4i\lambda^+x}-1\right]^2}\times\frac{d}{dt}\left(e_1^{(1)}N_\lambda\right),$$
$$0\le x<\infty. \tag{4.130}$$

Equating the expressions (4.129) and (4.130), we find:

$$\frac{d}{dt}\left(\ln e_1^{(1)}N_\lambda\right) = -\frac{1}{2i\lambda^+}.$$

Hence, the standardized multiplier $i\left(e_1^{(1)}N_\lambda\right)^{-1}$ of the non-scattering data (4.126) for the sG equation (4.14) obeys the evolution equation:

$$i\left(e_1^{(1)}N_\lambda\right)^{-1} = i\left(e_{10}^{(1)}N_{\lambda_0}\right)^{-1}\exp\left\{\frac{t}{2i\lambda^+}\right\},\ \ \text{Im}\lambda^+\ge\varepsilon_0,\ \varepsilon>0. \tag{4.131}$$

By (4.131) the exact solution (4.128) of the sG equation (4.14) for the considered case is represented in the explicit form:

$$v(x,t) = -2i\ln\frac{(2\lambda^+)^{-1}\left(e_{10}N_{\lambda_0}\right)^{-1}\exp\left\{\frac{t}{2i\lambda^+}+2i\lambda^+x\right\}-1}{(2\lambda^+)^{-1}\left(e_{10}N_{\lambda_0}\right)^{-1}\exp\left\{\frac{t}{2i\lambda^+}+2i\lambda^+x\right\}+1}\ (\text{mod}2\pi i),$$
$$0\le x<\infty. \tag{4.132}$$

Let the element $v_x(x,t)$ of the potential matrix (4.90) be a real-valued function satisfying estimate (4.87), then the scattering problem (4.88)–(4.89) is self-adjoint. We can construct the exact solution of the shG equation (4.84) from the same non-scattering data (4.126). Under the reduction (4.90), it follows from (3.120) that $c_1(x,t) = c_2(x,t) = \frac{1}{2}v_x(x,t)$ and $\alpha = -\beta$. Hence, in this case the non-scattering potential for the shG equation (4.84) is obtained from (3.120) and (3.121):

$$v_x(x,t) = \frac{4}{(2\lambda^+)^{-2}(e_1^{(1)}N_\lambda)^{-1}e^{2i\lambda^+x} + e_1^{(1)}N_\lambda e^{-2i\lambda^+x}}. \tag{4.133}$$

The evolutionary equation for the standardized multiplier $i\left(e_1^{(1)}N_\lambda\right)^{-1}$ of the non-scattering data (4.126) for the shG equation (4.84) is derived by an equation analogous to the previous one. Indeed, we rewrite the function (4.133) in the form:

$$v_x(x,t) = \frac{4}{(2\lambda^+)^{-1}} \cdot \frac{(2\lambda^+)^{-1}(e_1^{(1)}N_\lambda)^{-1}e^{2i\lambda^+x}}{(2\lambda^+)^{-2}(e_1^{(1)}N_\lambda)^{-2}e^{4i\lambda^+x} + 1}.$$

Taking the integral of the function defined by the last equality, we have

$$v(x,t) = -4i\text{arctg}\left\{(2\lambda^+)^{-1}\left(e_1^{(1)}N_\lambda\right)^{-1}e^{2i\lambda^+x}\right\},$$

whence it follows that

$$-\text{tg}\left(\frac{v(x,t)}{4i}\right) = (2\lambda^+)^{-1}(e_1^{(1)}N_\lambda)^{-1}e^{2i\lambda^+x}$$

$$= i\frac{\exp\left\{\frac{1}{4}v(x,t)\right\} - \exp\left\{-\frac{1}{4}v(x,t)\right\}}{\exp\left\{\frac{1}{4}v(x,t)\right\} + \exp\left\{-\frac{1}{4}v(x,t)\right\}}.$$

Using the last equality, we obtain the representation of the function $\exp\left\{\frac{1}{4}v(x,t)\right\}$ in terms of the given non-scattering data:

$$e^{\frac{1}{2}v(x,t)} = \frac{1 - i(2\lambda^+)^{-1}(e_1^{(1)}N_\lambda)^{-1}e^{2i\lambda^+x}}{1 + i(2\lambda^+)^{-1}(e_1^{(1)}N_\lambda)^{-1}e^{2i\lambda^+x}}.$$

Then the desired solution $v(x,t)$ takes the form:

$$v(x,t) = 2\ln\frac{1 - i(2\lambda^+)^{-1}(e_1^{(1)}N_\lambda)^{-1}e^{2i\lambda^+x}}{1 + i(2\lambda^+)^{-1}(e_1^{(1)}N_\lambda)^{-1}e^{2i\lambda^+x}} \ (\text{mod}2\pi i), \quad 0 \le x < \infty. \tag{4.134}$$

From the shG equation (4.84) with due regard for (4.134), we have

$$v_{xt} = \text{sh}v(x,t) = \frac{1}{2}\left\{e^{iv(x,t)} - e^{-iv(x,t)}\right\} =$$

$$\frac{4e^{2i\lambda^+ x}}{2i\lambda^+ \left(e_1^{(1)} N_\lambda\right)} \times \frac{1 - (2\lambda^+)^{-2}\left(e_1^{(1)} N_\lambda\right)^{-2} e^{4i\lambda^+ x}}{\left[1 + (2\lambda^+)^{-2}\left(e_1^{(1)} N_\lambda\right)^{-2} e^{4i\lambda^+ x}\right]^2}, \quad 0 \le x < \infty. \quad (4.135)$$

Next, differentiating of the equality (4.133) with respect to t, yields

$$v_{xt}(x,t) = \frac{4e^{2i\lambda^+ x}}{\left(e_1^{(1)} N_\lambda\right)^2} \times \frac{1 - (2\lambda^+)^{-2}\left(e_1^{(1)} N_\lambda\right)^{-2} e^{4i\lambda^+ x}}{\left[1 + (2\lambda^+)^{-2}\left(e_1^{(1)} N_\lambda\right)^{-2} e^{4i\lambda^+ x}\right]^2} \times \frac{d}{dt}\left(e_1^{(1)} N_\lambda\right),$$

$$0 \le x < \infty. \quad (4.136)$$

Equating the expressions (4.135) and (4.136), we find:

$$\frac{d}{dt}\left(\ln e_1^{(1)} N_\lambda\right) = \frac{1}{2i\lambda^+}.$$

Hence, the standardized multiplier $i(e_1^{(1)} N_\lambda)^{-1}$ of the non-scattering data (4.126) for the shG equation (4.84) obeys the evolution equation:

$$i\left(e_1^{(1)} N_\lambda\right)^{-1} = i\left(e_{10}^{(1)} N_{\lambda_0}\right)^{-1} \exp\left\{-\frac{t}{2i\lambda^+}\right\}, \quad \text{Im}\lambda^+ > \varepsilon_0,\ \varepsilon > 0. \quad (4.137)$$

Thus, the exact solution (4.134) of the shG equation (4.84) is represented in the explicit form:

$$v(x,t) = 2\ln \frac{1 - i(2\lambda^+)^{-1}\left(e_{10}^{(1)} N_{\lambda_0}\right)^{-1} \exp\left\{-\frac{t}{2i\lambda^+} + 2i\lambda^+ x\right\}}{1 + i(2\lambda^+)^{-1}\left(e_{10}^{(1)} N_{\lambda_0}\right)^{-1} \exp\left\{-\frac{t}{2i\lambda^+} + 2i\lambda^+ x\right\}} \ \text{mod}(2\pi i), \quad (4.138)$$

where the standardized multiplier $i\left(e_{10}^{(1)} N_{\lambda_0}\right)^{-1}$ is determined by the BVs: $e_{10}^{(1)}$ and N_{λ_0} of the Jost solution $e^{(1)}(x,\lambda;t)$ at $x = 0$ and at $\lambda = \lambda^+$, $\text{Im}\lambda^+ > \varepsilon_0$, $\varepsilon_0 > 0$. In addition, these BVs are calculated from the initial condition (4.85).

The results obtained in this section can be formulated as follows.

Lemma 4.3. *The Cauchy initial-value problem for the sG equation* (4.14) *(the shG equation* (4.84)*) on a half-line can be solved exactly by the ISM in the class of non-scattering potentials satisfying the estimate* (4.17), ((4.87)). *If the* *non-scattering data set of the scattering problem associated with the Cauchy problem is given by the non-scattering data set* (4.126)*, then the Cauchy initial-value problem for the sG equation* (4.14) *(the shG equation* (4.84)*) with the known initial condition* (4.15)*),* ((4.85)) *has the explicit soliton-solution* (4.132)*,* ((4.138))*, if and only if the time-dependence of the standardized multiplier of the non-scattering data set* (4.126) *obeys the evolution equation* (4.131)*,* ((4.137))*. The found soliton-solutions* (4.132) ((4.138)) *are expressible in terms of the given non-scattering time-dependent data* (4.126).

Chapter 5

Inverse Scattering for Integration of The Continual System of Non-Linear Interaction Waves

In this chapter, we consider the Cauchy initial-value problem for the continual system of non-linear interaction waves. This continual system describes a non-linear interaction of waves. We prove that the continual system can be integrated by the inverse scattering method (ISM). The method is based on applying the results of the associated inverse scattering problem for the transport equation on the whole line. This transport equation represents the continual generalization from the system of n first-order ordinary differential equations.

The chapter is based on the published papers [124, 126, 145].

Consider the continual system of non-linear interaction waves:

$$\frac{\partial q(x,y,\tau;t)}{\partial t}+w(y,\tau)\frac{\partial q(x,y,\tau;t)}{\partial x}$$
$$+i\int_{-1}^{1}\beta(y,\tau,\tau_1)q(x,y,\tau_1;t)q(x,\tau_1,\tau;t)d\tau_1=0, \tag{5.1}$$

where $x\in(-\infty,\infty); y,\tau\in(-1,1), w(y,\tau)=\frac{\alpha(\tau)-\alpha(y)}{y-\tau}, \beta(y,\tau,\tau_1)=\alpha(\tau_1)-\alpha(\tau)+\frac{(\tau-\tau_1)}{y-\tau}(\alpha(y)-\alpha(\tau)), \alpha(y)$ is any continuous function, $\alpha(y)\neq 0$ for any $y\in(-1,1)$. The continual system (5.1) can be represented by the Lax condition of compatibility of the operator transport equation:

$$i\frac{\partial\varphi}{\partial x}+U\varphi=\lambda J\varphi,$$
$$U\varphi(x,y;t)=\int_{-1}^{1}u(x,y,\tau;t)\varphi(x,\tau;t)d\tau,\ J\varphi(x,y;\tau)=y\varphi(x,y;\tau), \tag{5.2}$$

DOI: 10.1201/9781003370543-5

and the operator governing the time-evolution of eigenfunctions of the transport equation (5.2):

$$i\frac{\partial \varphi}{\partial t} + V\varphi = \lambda T\varphi, \text{ where } \lambda \text{ is a complex-valued spectral parameter,}$$

$$V\varphi(x,y;t) = \int_{-1}^{1} v(x,y,\tau;t)\varphi(x,\tau;t)d\tau,\ T\varphi(x,y;\tau) = \alpha(y)\varphi(x,y;\tau). \quad (5.3)$$

The potentials u and v in equations (5.2) and (5.3) are related to the coefficient q in the continual system (5.1) by the commutators $[J,Q]$ and $[T,Q]$, respectively:

$$U\varphi = [J,Q]\varphi = \int_{-1}^{1} q(x,y,\tau;t)(y-\tau)\varphi(x,\tau;t)d\tau,$$

$$V\varphi = [T,Q]\varphi = \int_{-1}^{1} q(x,y,\tau;t)(\alpha(y)-\alpha(\tau))\varphi(x,\tau;t)d\tau,$$

whence it follows that

$$u(x,y,\tau;t) = (y-\tau)q(x,y,\tau;t),$$
$$v(x,y,\tau;t) = (\alpha(y)-\alpha(\tau))\, q(x,y,\tau;t).$$

The Cauchy problem for the continual system (5.1) of non-linear interaction waves is associated with the inverse scattering problem (ISP) for the transport equation (5.2) with the potential $c(x,y,\tau) = u(x,y,\tau;0)$ satisfying the estimate:

$$|c(x,y,\tau)| \leq \tilde{C}\frac{|y-\tau|}{(1+|x|)^{2(1+\varepsilon)}}, \quad (5.4)$$

where $\tilde{C}$ is a constant, $\varepsilon > 0$ and $c(x,y,\tau)$ is a complex-valued measurable function of its variables.

The estimate (5.4) for the potential $c(x,y,\tau)$ in the transport equation (5.2) ensures the existence and uniqueness of the solution of the direct and ISP for the transport equation. A rigorous mathematical theory of the boundary-value problems for the transport equations can be found in [115].

We prove in Section 5.2 that the transition operator for the scattering problem for the transport equation (5.2) admits the right and left Volterra factorization and connects with an intermediate operator. The intermediate operator is constructed from the known transition operator and admits not only the right and left Volterra factorizations but also the analytic factorization. Then the functions appearing in the fundamental equation in the inverse problem are constructed from the intermediate operator. By this fact the potential in the transport equation is uniquely reconstructed through the solutions of the fundamental equations.

Further in Section 5.3 of this chapter the generalized Lax equation for the transport equation is introduced. This enables us to derive the time-evolution of the transition operator. Thus, the solution of the considered Cauchy problem for the continual system (5.1) is found uniquely from the known initial condition in terms of solutions of the fundamental equations in the inverse problem.

5.1 The direct and ISP for a system of n first-order ODEs

The transport equation (5.2) is the continual generalization from the system of n first-order ordinary differential equation (ODEs). In fact, if in the transport equation the parameter y takes on a finite number of values $y_j, j = 1, ..., n$, and if the measure $d\tau$ is concentrated at the points $y_j, j = 1, ..., n$, then the transport equation represents the system of n first-order ODEs:

$$-i\frac{d}{dx}u(x,\lambda) + C(x)u(x,\lambda) = \lambda J_n u(x,\lambda) \quad -\infty < x < \infty, \tag{5.5}$$

where λ is a complex valued spectral parameter, $u(x,\lambda) = (u_1(x,\lambda), \ldots, u_n(x,\lambda))$, $C(x)$ is a potential matrix $C(x) = [c_{jk}(x)]$ with the zero elements along the principal diagonal, i.e., $c_{jj}(x) = 0$, and the elements $c_{jk}(x)$, $j \neq k$ are the complex-valued measurable functions satisfying the estimate:

$$|c_{jk}(x)| \leq \frac{\tilde{C}}{(1+|x|)^{2(1+\varepsilon)}} < \infty, \ \tilde{C} \text{ is a positive number}, \ \varepsilon > 0, \ j,k = 1, ..., n, \tag{5.6}$$

$J_n = J_n^* = \mathrm{diag}(a_1, ..., a_n)$, a_j, $j = 1, ..., n$, are real regulated numbers: $a_1 > ... > a_n$, J_n^* denotes the complex conjugate of J_n.

Hence, the transport equation (5.2) is the continual generalization from the system of n first-order ODEs (5.5).

The ISP for the system of the form (5.5) for the whole line has been investigated in [81], [117, 118, 121], [109, 110] and for the half-line in [97], [119].

With the purpose of preparation for solving the ISP for the transport equation (5.2), we present in this section the solution of the direct and ISP for system (5.5) that is quoted from the dissertation [126]. This solution is represented in terms that differ slightly from terms used in [81], [109, 110]. In Remark 5.1 at the end of Section 5.2, we show that the results of the direct and ISP for the equation (5.2) are generalized from corresponding results of the direct and ISP, respectively for system (5.5). There exists a deep analogy between the obtained results of the considered problems.

5.1.1 The transition matrix $S(\lambda)$

Determine the matrices $u^+(x,\lambda) = [u^+_{jk}(x,\lambda)]$ and $u^-(x,\lambda) = [u^-_{jk}(x,\lambda)], j,k = 1,...,n$, from the systems of Volterra integral equations:

$$u^+_{jk}(x,\lambda) = \delta_{jk}e^{i\lambda a_k x} + i\int_x^{+\infty}\sum_m c_{jm}(s)u^+_{mk}(s,\lambda)e^{i\lambda a_j(x-s)}ds, \tag{5.7}$$

$$u^-_{jk}(x,\lambda) = \delta_{jk}e^{i\lambda a_k x} - i\int_{-\infty}^{x}\sum_m c_{jm}(s)u^-_{mk}(s,\lambda)e^{i\lambda a_j(x-s)}ds, \tag{5.8}$$

where δ_{jk} is the Kronecker symbol, $\delta_{jk} = \begin{cases} 0 & \text{for } j \neq k, \\ 1 & \text{for } j = k \end{cases}, j,k = 1,...,n,$ and λ is a real-valued parameter.

By straight differentiating equations (5.7) and (5.8), we are easily persuaded that the solution of Volterra integral equations (5.7) and (5.8) satisfies the matrix equation (5.5). Thus, two sets of matrix solutions $u^+(x,\lambda)$ and $u^-(x,\lambda)$ of system (5.5) are determined by the asymptotic conditions:

$$u^+_{jk}(x,\lambda) \to \delta_{jk}e^{i\lambda a_k x} \quad \text{as } x \to +\infty, \tag{5.9}$$

$$u^-_{jk}(x,\lambda) \to \delta_{jk}e^{i\lambda a_k x} \quad \text{as } x \to -\infty. \tag{5.10}$$

Each of these sets is fundamental, therefore there exists a matrix $S(\lambda)$ such that

$$u^+(x,\lambda) = u^-(x,\lambda)S(\lambda). \tag{5.11}$$

The matrix function $S(\lambda)$ is called the transition matrix for system (5.5). The sets of solutions $u^+(x,\lambda)$ and $u^-(x,\lambda)$ are called the Jost solutions of system (5.5).

5.1.2 Representations of solutions of system (5.5)

The analyticity of the Jost functions determined by the asymptotic types (5.9) and (5.10) plays a basic role in solving the inverse problem. However, it is easily seen that in general the Jost solutions $u^+(x,\lambda)$ and $u^-(x,\lambda)$ defined by

(5.7) and (5.8), respectively do not admit any analytic continuation from the real axis of the parameter λ. Using systems (5.7) and (5.8), we determine the matrices $\psi^+(x,\lambda) = [\psi^+_{jk}(x,\lambda)]$ and $\psi^-(x,\lambda) = [\psi^-_{jk}(x,\lambda)]$ from the systems of integral equations:

$$\begin{aligned}
\psi^+_{jk}(x,\lambda) &= -i\int_{-\infty}^{x}\sum_m c_{jm}(s)\psi^+_{mk}(s,\lambda)e^{i\lambda(a_j-a_k)(x-s)}ds,\ j<k,\\
\psi^+_{kk}(x,\lambda) &= 1+i\int_{x}^{\infty}\sum_m c_{km}(s)\psi^+_{mk}(s,\lambda)ds,\\
\psi^+_{jk}(x,\lambda) &= i\int_{x}^{\infty}\sum_m c_{jm}(s)\psi^+_{mk}(s,\lambda)e^{i\lambda(a_j-a_k)(x-s)}ds,\ j>k,
\end{aligned}\tag{5.12}$$

$$\begin{aligned}
\psi^-_{jk}(x,\lambda) &= i\int_{x}^{\infty}\sum_m c_{jm}(s)\psi^-_{mk}(s,\lambda)e^{i\lambda(a_j-a_k)(x-s)}ds,\ j<k,\\
\psi^-_{kk}(x,\lambda) &= 1-i\int_{-\infty}^{x}\sum_m c_{km}(s)\psi^-_{mk}(s,\lambda)ds,\\
\psi^-_{jk}(x,\lambda) &= -i\int_{-\infty}^{x}\sum_m c_{jm}(s)\psi^-_{mk}(s,\lambda)e^{i\lambda(a_j-a_k)(x-s)}ds,\ j>k.
\end{aligned}\tag{5.13}$$

Since the real numbers a_j are regulated ($a_j > a_k$ for $j < k$) and the potentials $c_{jk}(x)$ satisfy estimate (5.6), then the integrals (5.12) converge for $\mathrm{Im}\lambda \geq 0$, and the integrals (5.13) converge for $\mathrm{Im}\lambda \leq 0$, i.e., the functions $\psi^+_{jk}(x,\lambda)$ admit an analytic continuation into the upper half-plane $\mathrm{Im}\lambda \geq 0$ and $\psi^-_{jk}(x,\lambda)_-$ an analytic continuation into the lower half-plane $\mathrm{Im}\lambda \leq 0$. By straightforward differentiation of (5.12) and (5.12), we easily verify that the solutions of systems (5.12) and (5.13) satisfy the differential matrix equation:

$$-i\frac{d}{dx}\psi^{\pm}(x,\lambda) + C(x)\psi^{\pm}(x,\lambda) = \lambda(J_n\psi^{\pm}(x,\lambda) - \psi^{\pm}(x,\lambda)J_n). \tag{5.14}$$

Setting

$$\varphi^+_{jk}(x,\lambda) = \psi^+_{jk}(x,\lambda)e^{i\lambda a_k x}, \quad \varphi^-_{jk}(x,\lambda) = \psi^-_{jk}(x,\lambda)e^{i\lambda a_k x}, \tag{5.15}$$

we obtain that the functions $\varphi^+_{jk}(x,\lambda)$ and $\varphi^-_{jk}(x,\lambda)$ satisfy the systems of integral equations:

$$\varphi^+_{jk}(x,\lambda) = -i\int_{-\infty}^{x}\sum_m c_{jm}(s)\varphi^+_{mk}(s,\lambda)e^{i\lambda a_j(x-s)}ds,\ j<k,$$

$$\varphi^+_{kk}(x,\lambda) = e^{i\lambda a_k x} + i\int_{x}^{\infty}\sum_m c_{km}(s)\varphi^+_{mk}(s,\lambda)e^{i\lambda a_k(x-s)}ds, \tag{5.16}$$

$$\varphi^+_{jk}(x,\lambda) = i\int_{x}^{\infty}\sum_m c_{jm}(s)\varphi^+_{mk}(s,\lambda)e^{i\lambda a_j(x-s)}ds,\ j>k,$$

$$\varphi^-_{jk}(x,\lambda) = i\int_{x}^{\infty}\sum_m c_{jm}(s)\varphi^-_{mk}(s,\lambda)e^{i\lambda a_j(x-s)}ds,\ j<k,$$

$$\varphi^-_{kk}(x,\lambda) = e^{i\lambda a_k x} - i\int_{-\infty}^{x}\sum_m c_{km}(s)\varphi^-_{mk}(s,\lambda)e^{i\lambda a_k(x-s)}ds, \tag{5.17}$$

$$\varphi^-_{jk}(x,\lambda) = -i\int_{-\infty}^{x}\sum_m c_{jm}(s)\varphi^-_{mk}(s,\lambda)e^{i\lambda a_j(x-s)}ds,\ j>k.$$

By straightforward differentiation of systems (5.16) and (5.17), we are easily persuaded that the solution of systems (5.16) and (5.17) satisfies the differential matrix equation (5.5).

Lemma 5.1. *Every bounded solution $\varphi^\pm(x,\lambda)$ of system (5.5) with the matrix potential $C(x) = [c_{jk}(x)]$ satisfying estimate (5.6) is represented in the form:*

$$\varphi^+(x,\lambda) = \left(I + \int_0^\infty H_+(x,s)e^{i\lambda s}ds\right)e^{i\lambda J_n x}, \tag{5.18}$$

$$\varphi^-(x,\lambda) = \left(I + \int_{-\infty}^0 H_-(x,s)e^{i\lambda s}ds\right)e^{i\lambda J_n x}, \tag{5.19}$$

where λ is a real valued parameter, $e^{i\lambda J_n x} = diag(e^{i\lambda a_1 x}, ..., e^{i\lambda a_n x})$, $a_1 > ... > a_n$.

The matrix kernels $H_+(x,s) = [H^+_{jk}(x,s)]$ and $H_-(x,s) = [H^-_{jk}(x,s)]$ are uniquely determined by the system (5.5) and are related to the potentials by the formulas:

$$c_{jk}(x) = -i(a_j - a_k)H^+_{jk}(x,0) = i(a_j - a_k)H^-_{jk}(x,0),\ j<k, \tag{5.20}$$

$$c_{jk}(x) = -i(a_k - a_j)H^+_{jk}(x,0) = i(a_k - a_j)H^-_{jk}(x,0),\ j>k, \tag{5.21}$$

and satisfy the estimates:

$$|H_{jk}^{\pm}(x,s)| \leq \frac{\tilde{C}}{(1+|(a_j-a_k)x-s|)^{1+\varepsilon}},\ j<k, \tag{5.22}$$

$$|H_{kk}^{\pm}(x,s)| \leq \frac{\tilde{C}}{(1+|s|)^{1+\varepsilon}}, \tag{5.23}$$

$$|H_{jk}^{\pm}(x,s)| \leq \frac{\tilde{C}}{(1+|(a_k-a_j)x+s|)^{1+\varepsilon}},\ j>k. \tag{5.24}$$

Proof. First, we find the solution of the system (5.12) in the form:

$$\psi^+(x,\lambda) = I + \int_0^{\infty} H_+(x,s)e^{i\lambda s}ds. \tag{5.25}$$

Substituting formula (5.25) into system (5.12) with due regard for the uniqueness of the Fourier transform, we obtain for $s \geq 0$:

$$\begin{aligned} H_{jk}^{+}(x,s) &= \frac{i}{a_j-a_k}c_{jk}\left(x-\frac{s}{a_j-a_k}\right) - \\ &\quad i\int_{x-\frac{s}{a_j-a_k}}^{x}\sum_m c_{jm}(\xi)H_{mk}^{+}(\xi,s-(a_j-a_k)(x-\xi))d\xi,\ j<k, \\ H_{kk}^{+}(x,s) &= i\int_x^{\infty}\sum_m c_{km}(\xi)H_{mk}^{+}(\xi,s)d\xi, \\ H_{jk}^{+}(x,s) &= \frac{i}{a_k-a_j}c_{jk}\left(x+\frac{s}{a_k-a_j}\right) + \\ &\quad i\int_x^{x+\frac{s}{a_k-a_j}}\sum_m c_{jm}(\xi)H_{mk}^{+}(\xi,s-(a_k-a_j)(\xi-x))d\xi,\ j>k. \end{aligned} \tag{5.26}$$

We proceed to find the solution of system (5.13) in the form:

$$\psi^-(x,\lambda) = I + \int_{-\infty}^{0} H_-(x,s)e^{i\lambda s}ds. \tag{5.27}$$

Substituting (5.27) into (5.13) and using the uniqueness of the Fourier transform yields for $s \le 0$:

$$H^-_{jk}(x,s) = -\frac{i}{a_j - a_k} c_{jk}\left(x - \frac{s}{a_j - a_k}\right) + i \int_x^{x-\frac{s}{a_j-a_k}} \sum_m c_{jm}(\xi) H^-_{mk}\left(\xi, s - (a_j - a_k)(x-\xi)\right) d\xi,\ j<k$$

$$H^-_{kk}(x,s) = -i \int_{-\infty}^{x} \sum_m c_{km}(\xi) H^-_{mk}(\xi,s) d\xi,\ s \le 0, \tag{5.28}$$

$$H^-_{jk}(x,s) = -\frac{i}{a_k - a_j} c_{jk}\left(x + \frac{s}{a_k - a_j}\right) - i \int_{x+\frac{s}{a_k-a_j}}^{x} \sum_m c_{jm}(\xi) H^-_{mk}\left(\xi, s - (a_k - a_j)(\xi - x)\right) d\xi,\ j>k.$$

The equalities (5.26) and (5.28) are considered as the systems of integral equations for the matrix functions $H^+(x,s) = [H^+_{jk}(x,s)]$ and $H^-(x,s) = [H^-_{jk}(x,s)]$, respectively.

Owing to estimate (5.6) for the potentials, we prove that the method of successive approximations applied to the systems (5.26) and (5.28) converges, i.e., the solutions of these systems exist uniquely in the space $C(E^2; E^n)$, where E^2 and E^n denote the two–dimensional Euclidean and n-dimensional Euclidean spaces, respectively. In fact, we write the system (5.26) in the operator form:

$$H^+ = h + AH^+, \quad H^+ = [H^+_{jk}], \quad h = [h_{jk}], \tag{5.29}$$

where for $s \ge 0$:

$$h_{jk}(x,s) = \frac{i}{a_j - a_k} c_{jk}\left(x - \frac{s}{a_j - a_k}\right) \text{ for } j<k, \tag{5.30}$$

$$h_{kk}(x,s) = 0,$$

$$h_{jk}(x,s) = \frac{i}{a_k - a_j} c_{jk}\left(x + \frac{s}{a_k - a_j}\right) \text{ for } j>k. \tag{5.31}$$

$$(Ah)_{jk}(x,s) = -i \int_{x-\frac{s}{a_j-a_k}}^{x} \sum_m c_{jm}(\xi) h_{mk}\left(\xi, s - (a_j - a_k)(x-\xi)\right) d\xi,\ j<k, \tag{5.32}$$

$$(Ah)_{kk}(x,s) = i \int_x^{\infty} \sum_m c_{km}(\xi) h_{mk}(\xi,s) d\xi, \tag{5.33}$$

$$(Ah)_{jk}(x,s) = i \int\limits_{x}^{x+\frac{s}{a_k-a_j}} \sum_m c_{jm}(\xi) h_{mk}\left(\xi, s-(a_k-a_j)(\xi-x)\right) d\xi, \; j>k. \tag{5.34}$$

We need to prove that the series:

$$H = (h)_{jk} + (A^1h)_{jk} + ... + (A^ph)_{jk} + ... \tag{5.35}$$

converges. First, we estimate the free member of series (5.35) with due regard for estimate (5.6). For $j<k$ and $s \geq 0$:

$$|(h)_{jk}(x,s)| \leq \frac{\tilde{C}}{(1+|(a_j-a_k)x-s|)^{2(1+\varepsilon)}}. \tag{5.36}$$

For $j>k$ and $s \geq 0$:

$$|(h)_{jk}(x,s)| \leq \frac{\tilde{C}}{(1+|(a_k-a_j)x+s|)^{2(1+\varepsilon)}}. \tag{5.37}$$

Suppose that the following estimates hold for all x and $s \geq 0$:

$$|(A^ph)_{jk}(x,s)| \leq \frac{\tilde{C}}{(1+|(a_j-a_k)x-s|)^{1+\varepsilon}} \int\limits_{T}^{+\infty} \frac{d\xi_1}{(1+|\xi_1|)^{1+\varepsilon}} \times$$
$$\int\limits_{\xi_1}^{+\infty} \frac{d\xi_2}{(1+|\xi_2|)^{1+\varepsilon}} \cdots \int\limits_{\xi_{p-1}}^{+\infty} \frac{d\xi_p}{(1+|\xi_p|)^{1+\varepsilon}} \text{ for } j<k, \tag{5.38}$$

$$|(A^ph)_{kk}(x,s)| \leq \frac{\tilde{C}}{(1+s)^{1+\varepsilon}} \int\limits_{T}^{+\infty} \frac{d\xi_1}{(1+|\xi_1|)^{1+\varepsilon}} \times$$
$$\int\limits_{\xi_1}^{+\infty} \frac{d\xi_2}{(1+|\xi_2|)^{1+\varepsilon}} \cdots \int\limits_{\xi_{p-1}}^{+\infty} \frac{d\xi_p}{(1+|\xi_p|)^{1+\varepsilon}}, \tag{5.39}$$

$$|(A^ph)_{jk}(x,s)| \leq \frac{\tilde{C}}{(1+|(a_k-a_j)x+s|)^{1+\varepsilon}} \int\limits_{T}^{+\infty} \frac{d\xi_1}{(1+|\xi_1|)^{1+\varepsilon}} \times$$
$$\int\limits_{\xi_1}^{+\infty} \frac{d\xi_2}{(1+|\xi_2|)^{1+\varepsilon}} \cdots \int\limits_{\xi_{p-1}}^{+\infty} \frac{d\xi_p}{(1+|\xi_p|)^{1+\varepsilon}} \text{ for } j>k, \tag{5.40}$$

where T is any number such that $T \leq x - \frac{s}{|a_j-a_k|}$, $j \neq k$.

We prove by induction that the estimates (5.38)–(5.40) must be satisfied for $p+1$. In fact, from the definition (5.32) we have for $j<k$:

$$\left|(A^{p+1}h)_{jk}(x,s)\right| = \left| -i \int\limits_{x-\frac{s}{a_j-a_k}}^{x} d\xi \left\{ \sum_{m<k} c_{jm}(\xi)(A^p h)_{mk}(\xi, s-(a_j-a_k)(x-\xi)) + c_{jk}(\xi)(A^p h)_{kk}(\xi, s-(a_j-a_k)(x-\xi)) + \sum_{m>k} c_{jm}(\xi)(A^p h)_{mk}(\xi, s-(a_j-a_k)(x-\xi)) \right\} \right|.$$

If $x-\frac{s}{a_j-a_k}\geq 0$, then by the induction supposition (5.38)–(5.40) with due regard for estimate (5.6), from the last equality we obtain the estimate for all x and $s\geq 0$:

$$\left|(A^{p+1}h)_{jk}(x,s)\right| \leq \frac{\tilde{C}}{(1+|(a_j-a_k)x-s|)^{1+\varepsilon}} \int\limits_{T}^{+\infty} \frac{d\xi_1}{(1+|\xi_1|)^{1+\varepsilon}} \times \int\limits_{\xi_1}^{+\infty} \frac{d\xi_2}{(1+|\xi_2|)^{1+\varepsilon}} \cdots \int\limits_{\xi_p}^{+\infty} \frac{d\xi_{p+1}}{(1+|\xi_{p+1}|)^{1+\varepsilon}},$$

where T is any number such that $T\leq x-\frac{s}{a_j-a_k}$ for $j<k$.

If $x-\frac{s}{a_j-a_k}<0$, then by splitting the interval of integration and using the supposition (5.38)–(5.40) with due regard for (5.6), we get estimates for all x and $s\geq 0$:

$$\left|(A^{p+1}h)_{jk}(x,s)\right| \leq \sum_{m<j} \left(\int\limits_{x-\frac{s}{a_j-a_k}}^{\frac{(a_j-a_k)x-s}{a_m-a_j}+1} d\xi_1 + \int\limits_{\frac{(a_j-a_k)x-s}{a_m-a_j}+1}^{x} d\xi_1 \right) \frac{\tilde{C}}{(1+|\xi_1|)^{2(1+\varepsilon)}} \times \frac{\tilde{C}}{(1+|(a_m-a_j)\xi_1+(a_j-a_k)x-s|)^{1+\varepsilon}} \times \int\limits_{\xi_1}^{+\infty} \frac{d\xi_2}{(1+|\xi_2|)^{1+\varepsilon}} \cdots \int\limits_{\xi_p}^{+\infty} \frac{d\xi_{p+1}}{(1+|\xi_{p+1}|)^{1+\varepsilon}} +$$

$$\sum_{j<m<k} \left(\int\limits_{x-\frac{s}{a_j-a_k}}^{\frac{(a_j-a_k)x-s}{a_j-a_m}+1} d\xi_1 + \int\limits_{\frac{(a_j-a_k)x-s}{a_j-a_m}+1}^{x} d\xi_1 \right) \frac{\tilde{C}}{(1+|\xi_1|)^{2(1+\varepsilon)}} \times \frac{\tilde{C}}{(1+|(a_m-a_j)\xi_1+(a_j-a_k)x-s|)^{1+\varepsilon}} \times$$

$$\int_{\xi_1}^{+\infty} \frac{d\xi_2}{(1+|\xi_2|)^{1+\varepsilon}} \cdots \int_{\xi_p}^{+\infty} \frac{d\xi_{p+1}}{(1+|\xi_{p+1}|)^{1+\varepsilon}} +$$

$$\left(\int_{x-\frac{s}{a_j-a_k}}^{\frac{(a_j-a_k)x-s}{a_j-a_k+1}} d\xi_1 + \int_{\frac{(a_j-a_k)x-s}{a_j-a_k+1}}^{x} d\xi_1 \right) \times$$

$$\frac{\tilde{C}}{(1+|\xi_1|)^{2(1+\varepsilon)}} \frac{\tilde{C}}{(1+|(a_j-a_k)\xi_1+s-(a_j-a_k)x|)^{1+\varepsilon}} \times$$

$$\int_{\xi_1}^{+\infty} \frac{d\xi_2}{(1+|\xi_2|)^{1+\varepsilon}} \cdots \int_{\xi_p}^{+\infty} \frac{d\xi_{p+1}}{(1+|\xi_{p+1}|)^{1+\varepsilon}} +$$

$$\sum_{m>k} \left(\int_{x-\frac{s}{a_j-a_k}}^{\frac{(a_j-a_k)x-s}{a_j-a_m+1}} d\xi_1 + \int_{\frac{(a_j-a_k)x-s}{a_j-a_m+1}}^{x} d\xi_1 \right) \times$$

$$\frac{\tilde{C}}{(1+|\xi_1|)^{2(1+\varepsilon)}} \frac{\tilde{C}}{(1+|(a_j-a_m)\xi_1+s-(a_j-a_k)x|)^{1+\varepsilon}} \times$$

$$\int_{\xi_1}^{+\infty} \frac{d\xi_2}{(1+|\xi_2|)^{1+\varepsilon}} \cdots \int_{\xi_p}^{+\infty} \frac{d\xi_{p+1}}{(1+|\xi_{p+1}|)^{1+\varepsilon}} \le \frac{\tilde{C}}{(1+|(a_j-a_k)x-s|)^{1+\varepsilon}} \times$$

$$\int_{T}^{+\infty} \frac{d\xi_1}{(1+|\xi_1|)^{1+\varepsilon}} \int_{\xi_1}^{+\infty} \frac{d\xi_2}{(1+|\xi_2|)^{1+\varepsilon}} \cdots \int_{\xi_p}^{+\infty} \frac{d\xi_{p+1}}{(1+|\xi_{p+1}|)^{1+\varepsilon}},$$

where T is any number such that $T \le x - \frac{s}{a_j-a_k}$, $j<k$.

Similarly, using (5.6) and (5.38)–(5.40), from the definitions (5.33) and (5.34) we can prove by induction that for all x and $s \ge 0$:

$$|(A^{p+1}h)_{kk}(x,s)| \le \frac{\tilde{C}}{(1+s)^{1+\varepsilon}} \int_{T}^{+\infty} \frac{d\xi_1}{(1+|\xi_1|)^{1+\varepsilon}} \times$$

$$\int_{\xi_1}^{+\infty} \frac{d\xi_2}{(1+|\xi_2|)^{1+\varepsilon}} \cdots \int_{\xi_p}^{+\infty} \frac{d\xi_{p+1}}{(1+|\xi_{p+1}|)^{1+\varepsilon}},$$

$$|(A^{p+1}h)_{jk}(x,s)| \leq \frac{\tilde{C}}{(1+|(a_k-a_j)x+s|)^{1+\varepsilon}} \int\limits_T^{+\infty} \frac{d\xi_1}{(1+|\xi_1|)^{1+\varepsilon}} \times$$

$$\int\limits_{\xi_1}^{+\infty} \frac{d\xi_2}{(1+|\xi_2|)^{1+\varepsilon}} \cdots \int\limits_{\xi_p}^{+\infty} \frac{d\xi_{p+1}}{(1+|\xi_{p+1}|)^{1+\varepsilon}} \text{ for } j>k,$$

where T is any number such that $T \leq x - \frac{s}{a_k-a_j}$, $k<j$.

Thus, the induction is proved, i.e., the estimates (5.38)–(5.40) for iterations $(A^p h)_{jk}(x,s)$ are seen to be satisfied for all p.

Making use of estimates (5.36)–(5.40), we construct the following majorant series of series (5.35):

$$\frac{\tilde{C}}{(1+|(a_j-a_k)x-s|)^{1+\varepsilon}} \Bigg\{ 1 + \int\limits_T^{+\infty} \frac{d\xi_1}{(1+|\xi_1|)^{1+\varepsilon}} + \cdots + \int\limits_T^{+\infty} \frac{d\xi_1}{(1+|\xi_1|)^{1+\varepsilon}} \times$$

$$\int\limits_{\xi_1}^{+\infty} \frac{d\xi_2}{(1+|\xi_2|)^{1+\varepsilon}} \cdots \int\limits_{\xi_p}^{+\infty} \frac{d\xi_{p+1}}{(1+|\xi_{p+1}|)^{1+\varepsilon}} + \cdots \Bigg\}, \text{ for } j<k, \tag{5.41}$$

$$\frac{C}{(1+s)^{1+\varepsilon}} \Bigg\{ 1 + \int\limits_T^{+\infty} \frac{d\xi_1}{(1+|\xi_1|)^{1+\varepsilon}} + \cdots + \int\limits_T^{+\infty} \frac{d\xi_1}{(1+|\xi_1|)^{1+\varepsilon}} \times$$

$$\int\limits_{\xi_1}^{+\infty} \frac{d\xi_2}{(1+|\xi_2|)^{1+\varepsilon}} \cdots \int\limits_{\xi_p}^{+\infty} \frac{d\xi_{p+1}}{(1+|\xi_{p+1}|)^{1+\varepsilon}} + \cdots \Bigg\}, \tag{5.42}$$

$$\frac{\tilde{C}}{(1+|(a_k-a_j)x+s|)^{1+\varepsilon}} \Bigg\{ 1 + \int\limits_T^{+\infty} \frac{d\xi_1}{(1+|\xi_1|)^{1+\varepsilon}} + \cdots + \int\limits_T^{+\infty} \frac{d\xi_1}{(1+|\xi_1|)^{1+\varepsilon}} \times$$

$$\int\limits_{\xi_1}^{+\infty} \frac{d\xi_2}{(1+|\xi_2|)^{1+\varepsilon}} \cdots \int\limits_{\xi_p}^{+\infty} \frac{d\xi_{p+1}}{(1+|\xi_{p+1}|)^{1+\varepsilon}} + \cdots \Bigg\}, \text{ for } j>k. \tag{5.43}$$

The series (5.35) is majorized by the constructed uniformly convergent series (5.41), (5.42) and (5.43) for the case $j<k, j=k$ and $j>k$, respectively. In fact, denote the expression in the bracket (5.41) or (5.42), (5.43) by the function $\Phi(T)$. Then, it is easy to verify that $\Phi(T) = \exp\left\{ - \int\limits_T^{+\infty} \frac{d\xi}{(1+|\xi|)^{1+\varepsilon}} \right\}$, i.e., the series (5.41), (5.42) and (5.43) are summarized, and their corresponding sums are equal to:

$$\frac{\tilde{C}}{\left(1+\left|(a_j-a_k)x-s\right|\right)^{1+\varepsilon}}\exp\left\{-\int_T^{+\infty}\frac{d\xi}{(1+|\xi|)^{1+\varepsilon}}\right\} \text{ for } j<k,$$

$$\frac{\tilde{C}}{(1+s)^{1+\varepsilon}}\exp\left\{-\int_T^{+\infty}\frac{d\xi}{(1+|\xi|)^{1+\varepsilon}}\right\},$$

and

$$\frac{\tilde{C}}{\left(1+\left|(a_k-a_j)x+s\right|\right)^{1+\varepsilon}}\exp\left\{-\int_T^{+\infty}\frac{d\xi}{(1+|\xi|)^{1+\varepsilon}}\right\} \text{ for } j>k,$$
$$-\infty<x<\infty,\ s\geq 0.$$

Hence, the series (5.35) which is majorized by the majorants (5.41), (5.42) and (5.43) converges. Thus, the method of successive approximations applied to the operator equation (5.29) converges, whence it follows that there exists in the space $C(E^2;E^n)$ a unique matrix solution $H^+(x,s)$ for all $x\in(-\infty,\infty)$ and $s\geq 0$ of system (5.26), and the estimates (5.22)–(5.24) for the found matrix solution $H^+(x,s)$ are satisfied.

Similarly, we prove that the system (5.28) for $H^-(x,s)$ for all $x\in(-\infty,\infty)$ and $s\leq 0$ has in $C(E^2;E^n)$ a unique solution satisfying estimates (5.22)–(5.24).

Using (5.15), from (5.25) and (5.27) we obtain the representations of solutions (5.18) and (5.19) of system (5.5). These representations can be written in the operator form:

$$\varphi^+(x,\lambda)=\left(I+\tilde{H}_+(x,\lambda)\right)e^{i\lambda J_n x},\quad \varphi^-(x,\lambda)=\left(I+\tilde{H}_-(x,\lambda)\right)e^{i\lambda J_n x}, \tag{5.44}$$

where $\tilde{H}_+(x,\lambda)=\int_0^\infty H_+(x,s)e^{i\lambda s}ds$, $\tilde{H}_-(x,\lambda)=\int_{-\infty}^0 H_-(x,s)e^{i\lambda s}ds$. The kernels of the operators $\tilde{H}_+(x,\lambda)$ and $\tilde{H}_-(x,\lambda)$ are the matrix functions: $H_+(x,s)=\left[H_{jk}^+(x,s)\right], s\geq 0$ and $H_-(x,s)=\left[H_{jk}^-(x,s)\right], s\leq 0$, respectively. $H_+(x,\lambda)$ and $H_-(x,\lambda)$ are the analog of the kernels of the transformation operators that are usually utilized in scattering problems [7].

Further, in systems (5.26) and (5.28) by setting $s=0$, we obtain the formulas (5.20) and (5.21) for the relations between the potentials and the kernels of the transformation operators. This completes the proof of Lemma 5.1. □

5.1.3 The intermediate matrix $\tilde{S}(\lambda)$

The determinant functions $\det[I + \tilde{H}_+(x,\lambda)]$, $\mathrm{Im}\lambda \geq 0$ and $\det[I + \tilde{H}_-(x,\lambda)]$, $\mathrm{Im}\lambda \leq 0$, in general, may have zeros in their half-plane. Owing to analyticity of these determinant functions and the estimates for the kernel functions $H_+(x,s)$, $s \geq 0$ and $H_-(x,s)$, $s \leq 0$, the numbers of zeros of the determinant functions of the matrices $I + \tilde{H}_+(x,\lambda)$, $\mathrm{Im}\lambda \geq 0$ and $I + \tilde{H}_-(x,\lambda)$ $\mathrm{Im}\lambda \leq 0$, are finite.

Lemma 5.2. *If the determinant functions* $\det[I + \tilde{H}_+(x,\lambda)]$, $\mathrm{Im}\lambda > 0$ *and* $\det[I + \tilde{H}_-(x,\lambda)]$, $\mathrm{Im}\lambda < 0$ *have no zeros in their half-plane, then there exist matrix functions* $K_+(x,s)$ $s \geq 0$ *and* $K_-(x,s)$, $s \leq 0$, *such that they are absolutely integrable with respect to* s *and for any* x:

$$\left(I + \int_0^\infty H_+(x,s)e^{i\lambda s}ds\right)^{-1} = I + \int_0^\infty K_+(x,s)e^{i\lambda s}ds, \ \mathrm{Im}\lambda \geq 0, \tag{5.45}$$

$$\left(I + \int_{-\infty}^0 H_-(x,s)e^{i\lambda s}ds\right)^{-1} = I + \int_{-\infty}^0 K_-(x,s)e^{i\lambda s}ds, \ \mathrm{Im}\lambda \leq 0. \tag{5.46}$$

Proof. Let $\varphi^+(x,\lambda)$ be a bounded matrix solution of system (5.16), then $\varphi^+(x,\lambda)$ satisfies system (5.5). Using system (5.5) for $\varphi^+(x,\lambda)$ and the Jacobi's identity of differentiation of determinants, we find for a real parameter λ:

$$-i\frac{d}{dx}\ln\det[\varphi^+(x,\lambda)] = S_p(\lambda J - C(x)) = S_p(\lambda J_n) = \lambda\sum_{k=1}^n a_k,$$

whence it follows that for any x and real λ:

$$\det[\varphi^+(x,\lambda)] = \det[\varphi^+(0,\lambda)]\exp\left\{i\lambda\left(\sum_{k=1}^n a_k\right)x\right\}. \tag{5.47}$$

We prove that $\det[\varphi^+(0,\lambda)] \neq 0$ for any real parameter λ. In fact, suppose that for some real λ_0 : $\det|\varphi^+(0,\lambda_0)| = 0$. Then from (5.47) it follows that for any x :

$$\det[\varphi^+(x,\lambda_0)] \equiv 0, \ \ \mathrm{Im}\lambda_0 = 0. \tag{5.48}$$

Since $\varphi^+(x,\lambda)$ satisfies system (5.16) and $\psi^+(x,\lambda)$ satisfies system (5.12), then

$$\lim_{x\to+\infty}\det[\varphi^+_{jk}(x,\lambda_0)e^{-i\lambda_0 a_k x}] = \det I = 1,$$

which contradicts (5.48). Thus, $\det[\varphi^+(0,\lambda)] \neq 0$ for any real parameter λ. Owing to this fact, it follows from (5.44) and (5.47) that for any x and real λ:

$$\det[I + \tilde{H}_+(x,\lambda)] = det[I + \int_0^\infty H_+(x,s)e^{i\lambda s}ds] \neq 0. \tag{5.49}$$

Due to (5.49) and the supposition that the matrix function $I + \tilde{H}_+(x,\lambda)$ has no zeros in half-plane $\text{Im}\lambda > 0$, we have

$$\det\left[I + \int_0^\infty H_+(x,s)e^{i\lambda s}ds\right] \neq 0 \text{ for any } x \text{ and } \text{Im}\lambda \geq 0.$$

Owing to this fact and by a theorem of Wiener [66], there exists a matrix function $K_+(x,s)$, $s \geq 0$ such that this function is absolutely integrable with respect to s and satisfies the relation equality (5.45). The assertion (5.46) is proved similarly. □

Hence, all sets of solutions $\varphi^+(x,\lambda)$ and $\varphi^-(x,\lambda)$ of system (5.5) is fundamental, therefore there exists a matrix function $\tilde{S}(\lambda)$ such that

$$\varphi^+(x,\lambda) = \varphi^-(x,\lambda)\tilde{S}(\lambda). \tag{5.50}$$

$\tilde{S}(\lambda)$ is called the intermediate matrix for system (5.5).

From (5.50) with due regard for (5.44)–(5.46), we obtain the analytic factorization of the intermediate matrix $\tilde{S}(\lambda)$:

$$\tilde{S}(\lambda) = e^{-i\lambda J_n x}(I + \tilde{H}_-(x,\lambda))^{-1}(I + \tilde{H}_+(x,\lambda))e^{i\lambda J_n x}, \tag{5.51}$$

where the matrix functions $I + \tilde{H}_+(x,\lambda)$ and $I + \tilde{H}_-(x,\lambda)$ admit an analytic continuation from the real axis into the upper half-plane $\text{Im}\lambda \geq 0$ and into the lower half-plane $\text{Im}\lambda \leq 0$, respectively.

5.1.4 The bilateral factorization of the transition matrix $S(\lambda)$

Let two sets of the Jost solutions $u^+(x,\lambda)$ and $u^-(x,\lambda)$ of system (5.5) determined by the asymptotic conditions (5.9) and (5.10) be given. Then, taking (5.47) into account, we have for any x and real λ:

$$\det[u^+(x,\lambda)] = \det[u^+(0,\lambda)]\exp\left\{i\lambda\left(\sum_{k=1}^n a_k\right)x\right\}, \tag{5.52}$$

$$\det[u^-(x,\lambda)] = \det[u^-(0,\lambda)]\exp\left\{i\lambda\left(\sum_{k=1}^n a_k\right)x\right\}, \tag{5.53}$$

where due to the systems (5.7) and (5.8):

$$\det[u^+(0,\lambda)] = \det[u^+(x,\lambda)] \exp\left\{-i\lambda\left(\sum_{k=1}^{n} a_k\right)x\right\} =$$
$$\det[u^+(x,\lambda)e^{-i\lambda J_n x}] = 1 + o(1) \text{ as } x \to +\infty,$$
$$\det[u^-(0,\lambda)] = \det[u^-(x,\lambda)e^{-i\lambda J_n x}] = 1 + o(1) \text{ as } x \to -\infty.$$

Therefore, the formulas (5.52) and (5.53) take the form for any x and real λ:

$$\det[u^+(x,\lambda)] = \det[u^-(x,\lambda)] = \exp\{i\lambda S_p(J_n)x\} \neq 0. \tag{5.54}$$

Using (5.54), from the definition (5.11), we have:

$$\det[S(\lambda)] = 1 \text{ for real } \lambda. \tag{5.55}$$

Making use of (5.54) and (5.55), we can solve the problem of triangular factorization of the transition matrix $S(\lambda)$ for system (5.5). In fact, we consider the matrix equations:

$$\left[\varphi^+_{jk}(x,\lambda)\right] = \left[u^+_{jp}(x,\lambda)\right]\left[s^+_{pk}(\lambda)\right], \tag{5.56}$$
$$j,k = 1,...,n.$$
$$\left[\varphi^+_{jk}(x,\lambda)\right] = \left[u^-_{jp}(x,\lambda)\right]\left[s^-_{pk}(\lambda)\right]. \tag{5.57}$$

In (5.56) and (5.57) the matrix functions $\left[u^+_{jk}(x,\lambda)\right]$, $\left[u^-_{jk}(x,\lambda)\right]$ and $\left[\varphi^+_{jk}(x,\lambda)\right]$ are known, $\left[\varphi^+_{jk}(x,\lambda)\right]$ satisfies system of equations (5.16), and $[s^+_{jk}(\lambda)]$, $[s^-_{jk}(\lambda)]$ are unknown matrix functions. By using the asymptotic conditions of the matrix functions $\left[u^+_{jk}(x,\lambda)\right]$, $\left[\varphi^+_{jk}(x,\lambda)\right]$ and $\left[u^-_{jk}(x,\lambda)\right]$ as $x \to +\infty$ and as $x \to -\infty$, from (5.56) and (5.57) we obtain

$$s^+_{jk} = 0 \text{ for } j > k, \quad s^-_{jk} = 0 \text{ for } j < k, \tag{5.58}$$
$$s^+_{jj} = 1,\ j = 1,...,n, \tag{5.59}$$
$$\lim_{x\to-\infty} \varphi^+_{jj}(x,\lambda)e^{-i\lambda a_j x} = \lim_{x\to-\infty} \psi^+_{jj}(x,\lambda) = s^-_{jj}(\lambda),\ j = 1,...,n, \tag{5.60}$$

where the matrix function $\left[\psi^+_{jk}(x,\lambda)\right]$ satisfies system (5.12) and admits an analytic continuation into the upper half-plane $\text{Im}\lambda \geq 0$.

We need to consider the property of $s^-_{jj}(\lambda)$ defined by (5.60). It follows from (5.44) and (5.45) that for any x and real λ:

$$\det[\varphi^+(x,\lambda)e^{-i\lambda J_n x}] = \det[I + \tilde{H}_+(x,\lambda)] \neq 0,$$

where the matrix kernel $H_+(x,s)$ satisfies system (5.26).

Since the functions $\varphi^+(x,\lambda)$ and $\psi^+(x,\lambda)$ satisfy systems (5.16) and (5.12), respectively, and the potentials $c_{jk}(x)$ obey estimate (5.6), then for any x and real λ:

$$\lim_{x\to-\infty}\det\left[\varphi^+(x,\lambda)e^{-i\lambda J_n x}\right]=\prod_{j=1}^{n}\lim_{x\to-\infty}\varphi^+_{jj}(x,\lambda)e^{-i\lambda a_j x}\neq 0.$$

Due to this fact, the functions $s^-_{jj}(\lambda)$ defined by (5.60) for real λ, $j=1,...,n$ are different from zero and admit an analytic continuation into the upper half-plane $\mathrm{Im}\lambda\geq 0$.

Further, we consider the matrix equations:

$$\left[\varphi^-_{jk}(x,\lambda)\right]=\left[u^-_{jp}(x,\lambda)\right]\left[R^+_{pk}(\lambda)\right], \tag{5.61}$$

$$\left[\varphi^-_{jk}(x,\lambda)\right]=\left[u^+_{jp}(x,\lambda)\right]\left[R^-_{pk}(\lambda)\right]. \tag{5.62}$$

In (5.61) and (5.62) the matrix functions $\left[u^-_{jk}(x,\lambda)\right]$, $\left[u^+_{jk}(x,\lambda)\right]$ and $\left[\varphi^-_{jk}(x,\lambda)\right]$ are known, $\left[\varphi^-_{jk}(x,\lambda)\right]$ satisfies system of equations (5.17), and $\left[R^+_{jk}(\lambda)\right]$, $\left[R^-_{jk}(\lambda)\right]$ are unknown matrix functions. Using the asymptotic conditions of the matrix functions $[u^-_{jk}(x,\lambda)]$, $\left[\varphi^-_{jk}(x,\lambda)\right]$ and $\left[u^+_{jk}(x,\lambda)\right]$ as $x\to-\infty$ and as $x\to+\infty$, from (5.61) and (5.62) we have

$$R^+_{jk}=0 \text{ for } j>k,\quad R^-_{jk}=0 \text{ for } j<k, \tag{5.63}$$

$$R^+_{jj}=1,\ j=1,...,n, \tag{5.64}$$

$$\lim_{x\to+\infty}\varphi^-_{jj}(x,\lambda)e^{-i\lambda a_j x}=\lim_{x\to+\infty}\psi^-_{jj}(x,\lambda)=R^-_{jj}(\lambda),\ j=1,...,n, \tag{5.65}$$

where the matrix function $\left[\psi^-_{jk}(x,\lambda)\right]$ satisfies system (5.13) and admits an analytic continuation into the lower half-plane $\mathrm{Im}\lambda\leq 0$.

Taking (5.44) and (5.46) into account, we have for any x and real λ:

$$\det[\varphi^-(x,\lambda)e^{-i\lambda J_n x}]=\det[I+\tilde{H}_-(x,\lambda)]\neq 0,$$

where the matrix kernel $H_-(x,s)$ satisfies the system (5.28), the functions $\varphi^-(x,\lambda)$ and $\psi^-(x,\lambda)$ satisfy the system (5.17) and (5.13), respectively, and the potentials $c_{jk}(x)$ obey the estimate (5.6), therefore for any x and real λ:

$$\lim_{x\to+\infty}\det\left[\varphi^-(x,\lambda)e^{-i\lambda J_n x}\right]=\prod_{j=1}^{n}\lim_{x\to+\infty}\varphi^-_{jj}(x,\lambda)e^{-i\lambda a_j x}\neq 0\,.$$

Due to this fact, the functions $R^-_{jj}(\lambda)$ defined by (5.65) for real λ, $j=1,...,n$ are different from zero and admit an analytic continuation into the lower half-plane $\mathrm{Im}\lambda\leq 0$.

Owing to (5.54), for any x and real λ there exist the inverse matrices $[u^+(x,\lambda)]^{-1}$ and $[u^-(x,\lambda)]^{-1}$, i.e., the matrix equations (5.56), (5.57) and (5.61), (5.62) are uniquely solvable.

Denote the solutions of equations (5.56), (5.57) and (5.61), (5.62) by $S^+(\lambda)$, $S^-(\lambda)$ and $R^+(\lambda)$, $R^-(\lambda)$, respectively. According to (5.58)–(5.60) and (5.63)–(5.65), the obtained solutions are represented in the triangular form:

$$S^+(\lambda) = \begin{pmatrix} 1, & s_{12}^+(\lambda), & \cdots, & s_{1n}^+(\lambda) \\ 0, & 1, & \cdots, & s_{2n}^+(\lambda) \\ \cdot, & \cdot, & \cdots, & \cdot \\ 0, & 0, & \cdots, & 1 \end{pmatrix},$$

$$S^-(\lambda) = \begin{pmatrix} s_{11}^-(\lambda), & 0, & \cdots, & 0 \\ s_{21}^-(\lambda), & s_{22}^-(\lambda), & \cdots, & 0 \\ \cdot, & \cdot, & \cdots, & \cdot \\ s_{n1}^-(\lambda), & s_{n2}^-(\lambda), & \cdots, & s_{nn}^-(\lambda) \end{pmatrix}, \tag{5.66}$$

$$R^+(\lambda) = \begin{pmatrix} 1, & R_{12}^+(\lambda), & \cdots, & R_{1n}^+(\lambda) \\ 0, & 1, & \cdots, & R_{2n}^+(\lambda) \\ \cdot, & \cdot, & \cdots, & \cdot \\ \cdot, & \cdot, & \cdots, & 1 \end{pmatrix},$$

$$R^-(\lambda) = \begin{pmatrix} R_{11}^-(\lambda), & \cdot, & \cdots, & 0 \\ R_{21}^-(\lambda), & R_{22}^-(\lambda), & \cdots, & 0 \\ \cdot, & \cdot, & \cdots, & \cdot \\ R_{n1}^-(\lambda), & R_{n2}^-(\lambda), & \cdots, & R_{nn}^-(\lambda) \end{pmatrix}. \tag{5.67}$$

Comparing (5.56) with (5.57) and (5.61) with (5.62), then using the definition (5.11), we obtain the bilateral factorization of the transition matrix $S(\lambda)$, i.e., its right and left factorizations:

$$S(\lambda) = R^+(\lambda)(R^-(\lambda))^{-1} = S^-(\lambda)(S^+(\lambda))^{-1}. \tag{5.68}$$

Due to (5.55) and (5.68):

$$\det|S(\lambda)S^+(\lambda)| = \det|S^+(\lambda)| = \det|S^-(\lambda)|,$$
$$\det|S(\lambda)R^-(\lambda)| = \det|R^-(\lambda)| = \det|R^+(\lambda)|.$$

Making use of the above equalities and (5.66)–(5.68) yields

$$\det[S^+(\lambda)] = \det[R^+(\lambda)] = 1, \tag{5.69}$$
$$\det[S^-(\lambda)] = \prod_{j=1}^{n} s_{jj}^-(\lambda) = 1, \quad \det[R^-(\lambda)] = \prod_{j=1}^{n} R_{jj}^-(\lambda) = 1.$$

The obtained properties of $S(\lambda)$ can be formulated as follows.

Theorem 5.1. *Consider the system of n first-order ODEs* (5.5) *with the potential matrix $C(x) = [c_{jk}(x)]$. Let the elements $c_{jk}(x)$, $j \neq k$ be complex* *measurable functions satisfying estimate* (5.6), *and the elements along the principal diagonal be zero:* $c_{jj}(x) = 0$.

Then, the transition matrix $S(\lambda)$ *for system* (5.5) *admits the right and left factorizations* (5.68) *by the triangular matrix multipliers* $S^+(\lambda)$, $S^-(\lambda)$ *and* $R^+(\lambda)$, $R^-(\lambda)$ *determined by formulas* (5.66) *and* (5.67). *The diagonal elements* $s_{jj}^-(\lambda)$ *and* $R_{jj}^-(\lambda)$, $j = 1, ..., n$ *of the triangular matrices* $S^-(\lambda)$ *and* $R^-(\lambda)$ *are different from zero for real* λ *and admit the analytic continuation into the upper half-plane* $Im\lambda > 0$ *and into the lower half-plane* $Im\lambda < 0$, *respectively.*

We now find triangular matrix multipliers from the known transition matrix $S(\lambda)$.

Lemma 5.3. *The triangular matrix multipliers* $S^+(\lambda)$, $S^-(\lambda)$ *and* $R^+(\lambda)$, $R^-(\lambda)$ *of the bilateral factorization* (5.68) *are found uniquely from the known elements* $s_{jk}(\lambda)$ *of the transition matrix* $S(\lambda)$ *for system* (5.5).

Proof. Consider the following matrix equations derived from (5.68):

$$S(\lambda)S^+(\lambda) = S^-(\lambda), \tag{5.70}$$

$$S(\lambda)R^-(\lambda) = R^+(\lambda), \tag{5.71}$$

where the elements $s_{jk}(\lambda)$ of the transition matrix $S(\lambda) = [s_{jk}(\lambda)]$ for system (5.5) are known, the triangular matrices $S^+(\lambda)$, $S^-(\lambda)$ and $R^+(\lambda)$, $R^-(\lambda)$ of the forms (5.66) and (5.67), respectively, are unknown.

From the matrix equations (5.70) and (5.71) we obtain the formulas for calculating the diagonal elements of the matrices $S^-(\lambda)$ and $R^-(\lambda)$:

$$s_{jj}^-(\lambda) = \frac{\Delta_j^+(\lambda)}{\Delta_{j-1}^+(\lambda)}, \; R_{jj}^-(\lambda) = \frac{\Delta_{n-j}^-(\lambda)}{\Delta_{n-j+1}^-(\lambda)}, \; j = 1, ..., n, \tag{5.72}$$

where $\Delta_j^+(\lambda)$ and $\Delta_j^-(\lambda)$ are the j-th upper and lower principal minors of the known transition matrix $S(\lambda)$, wherein: $\Delta_0^+ = \Delta_0^- = 1$, $\Delta_1^+(\lambda) = s_{11}(\lambda)$, $\Delta_1^-(\lambda) = s_{nn}(\lambda)$, $\Delta_n^+(\lambda) = \Delta_n^-(\lambda) = \det S(\lambda)$.

Owing to the properties of the diagonal elements $s_{jj}^-(\lambda)$ and $R_{jj}^-(\lambda)$, $j = 1, ..., n$ of triangular matrices $S^-(\lambda)$ and $R^-(\lambda)$, we obtain from (5.72) that the j-th upper and lower principal minors $\Delta_j^+(\lambda)$ and $\Delta_j^-(\lambda)$ $j = 1, ..., n$, are different from zero for real λ, and admit an analytic continuation into the upper half-plane $\mathrm{Im}\lambda \geq 0$ and into the lower half-plane $\mathrm{Im}\lambda \leq 0$, respectively. Due to this fact, from the considered matrix equations (5.70) and (5.71) we uniquely express the elements of the unknown triangular matrix multipliers $S^+(\lambda)$, $S^-(\lambda)$ and $R^+(\lambda)$, $R^-(\lambda)$ through the known elements of the transition matrix $S(\lambda)$, i.e., the elements of the unknown triangular matrices are found uniquely in terms of the known elements of $S(\lambda)$. Thus, Lemma 5.3 is proved. □

5.1.5 The analytic and bilateral factorizations of $\tilde{S}(\lambda)$

According to the definition (5.50) of the intermediate matrix $\tilde{S}(\lambda)$ and using the relation equalities (5.56), (5.57) and (5.61), (5.62), we have:

$$\begin{aligned}\varphi^+(x,\lambda) = u^+(x,\lambda)S^+(\lambda) = \varphi^-(x,\lambda)\tilde{S}(\lambda) = u^+(x,\lambda)R^-(\lambda)\tilde{S}(\lambda),\\ \varphi^+(x,\lambda) = u^-(x,\lambda)S^-(\lambda) = \varphi^-(x,\lambda)\tilde{S}(\lambda) = u^-(x,\lambda)R^+(\lambda)\tilde{S}(\lambda).\end{aligned} \tag{5.73}$$

From (5.73) with due regard for (5.54) we obtain that the intermediate matrix $\tilde{S}(\lambda)$ admits the bilateral factorization, i.e., the right and left factorizations:

$$\tilde{S}(\lambda) = [R^+(\lambda)]^{-1}S^-(\lambda) = [R^-(\lambda)]^{-1}S^+(\lambda), \tag{5.74}$$

where $S^+(\lambda)$, $S^-(\lambda)$ and $R^+(\lambda)$, $R^-(\lambda)$ are the triangular matrices determined by formulas (5.66) and (5.67), respectively.

Thus, the lemma for the analytic and bilateral factorization of $\tilde{S}(\lambda)$ is proved.

Lemma 5.4. *Let the elements $c_{jk}(x)$ of the potential matrix $C(x)$ in system* (5.5) *satisfy estimate* (5.6) *and $c_{jj}(x) = 0$. Then the intermediate matrix $\tilde{S}(\lambda)$ for system* (5.5) *admits the analytic factorization* (5.51) *and the right and left factorizations* (5.74) *by the matrix triangular multipliers of the forms* (5.66) *and* (5.67).

Based on the formulas (5.44) and (5.51), we introduce the following concept.

Definition 5.1. *(1) A complex number λ_0, $\mathrm{Im}\lambda_0 \neq 0$, for which the determinant of the matrix $I + \tilde{H}_+(x,\lambda_0)$ or $I + \tilde{H}_-(x,\lambda_0)$ is equal to zero, is called a singular number of system* (5.5).

(2) A complex number λ_0, $\mathrm{Im}\lambda_0 \neq 0$, is called a generalized eigenvalue for system (5.5)*, if this system has a nontrivial generalized solution $\Phi(x,\lambda_0)$ such that $\Phi(x,\lambda_0)e^{-i\lambda_0 bx} \to 0$ for $x \to \pm\infty$ with some choice of a number $b \in (a_n, a_1)$.*

Lemma 5.5. *If a complex number λ_0, $\mathrm{Im}\lambda_0 \neq 0$, is a singular number of system* (5.5)*, then it is a generalized eigenvalue for this system.*

Proof. Let λ_0, $\mathrm{Im}\lambda_0 > 0$ be a singular number of system (5.5), then

$$\det[I + \tilde{H}_+(x,\lambda_0)] = 0, \tag{5.75}$$

whence it follows that

$$\det[(I + \tilde{H}_+(x,\lambda_0))e^{i\lambda_0 Jx}] = 0. \tag{5.76}$$

It follows from (5.76) that there exist numbers b_j, $j = 1, ..., n$, which are not simultaneously zero such that

$$\sum_{j=1}^{n} b_j \left(\delta_{ij} + \tilde{H}_{+ij}(x,\lambda_0)\right) e^{i\lambda_0 a_j x} = 0,\ i = 1, ..., n, \tag{5.77}$$

where $\tilde{H}_{+ij}(x,\lambda_0)$ are the operators with the kernels $H_{+ij}(x,s),\ s\geq 0,\ i,j=1,...,n,\ \delta_{ij}$ is the Kronecker symbol, $\delta_{ij}=\begin{cases}0 \text{ for } i\neq j,\\ 1 \text{ for } i=j.\end{cases}\quad i,j=1,...,n.$

From (5.77) we have

$$\sum_{j=1}^{m} b_j(\delta_{ij}+\tilde{H}_{+ij}(x,\lambda_0))e^{i\lambda_0 a_j x}=-\sum_{j=m+1}^{n} b_j(\delta_{ij}+\tilde{H}_{+ij}(x,\lambda_0))e^{i\lambda_0 a_j x},\quad i=1,...,n, \tag{5.78}$$

where m is any natural number among the natural numbers $1,...,n$, such that $1<m<n$.

Denote the left and the right parts of equality (5.78) by $P_1(x,\lambda_0)$ and $-P_2(x,\lambda_0)$, respectively. Putting

$$P(x,\lambda_0)=P_1(x,\lambda_0)=-P_2(x,\lambda_0), \tag{5.79}$$

we easily see that $P(x,\lambda_0)$ is a solution of system (5.5). If we take a number $b\in(a_{m+1},a_m)$, so that $a_{m+1}<b<a_m$, then

$$e^{i\lambda_0(a_j-b)x}\to 0 \text{ as } x\to+\infty, \text{ for } j=1,\cdots,m,$$
$$e^{i\lambda_0(a_j-b)x}\to 0 \text{ as } x\to-\infty, \text{ for } j=m+1,\cdots,n,$$

whence it follows from (5.78) that the solutions $P_1(x,\lambda_0)$ and $P_2(x,\lambda_0)$ of system (5.5) satisfy the asymptotic conditions:

$$P_1(x,\lambda_0)e^{-i\lambda_0 bx}\to 0 \text{ as } x\to+\infty,$$
$$P_2(x,\lambda_0)e^{-i\lambda_0 bx}\to 0 \text{ as } x\to-\infty,$$

i.e., there exists some number b such that

$$P(x,\lambda_0)e^{-i\lambda_0 bx}\to 0 \text{ as } x\to\pm\infty.$$

It remains to be proved that for a certain of choice of a number $b\in(a_n,a_1)$ the solution $P(x,\lambda_0)$ of system (5.5) is nontrivial. Suppose the contrary that for any number $b\in(a_n,a_1):\ \ P(x,\lambda_0)\equiv 0$. Then from (5.78) and (5.79) we have:

$$
\begin{aligned}
&b_{m+1}\tilde{H}_{+n,m+1}(x,\lambda_0)e^{i\lambda_0 a_{m+1}x}+\cdots\\
&+b_{n-1}\tilde{H}_{+n,n-1}(x,\lambda_0)e^{i\lambda_0 a_{n-1}x}+b_n(1+\tilde{H}_{+n,n}(x,\lambda_0))e^{i\lambda_0 a_n x}=0,\\
&b_{m+1}\tilde{H}_{+n-1,m+1}(x,\lambda_0)e^{i\lambda_0 a_{m+1}x}+\cdots\\
&+b_{n-1}\left(1+\tilde{H}_{+n-1,n-1}(x,\lambda_0)\right)e^{i\lambda_0 a_{n-1}x}+b_n\tilde{H}_{+n-1,n}(x,\lambda_0)e^{i\lambda_0 a_n x}=0,\\
&\cdots\ \cdots\ \cdots\quad \cdots\ \cdots\ \cdots\quad \cdots\ \cdots\ \cdots\quad \cdots\ \cdots\ \cdots \qquad (5.80)\\
&b_{m+1}\left(1+\tilde{H}_{+m+1,m+1}(x,\lambda_0)\right)e^{i\lambda_0 a_{m+1}x}+\cdots\\
&+b_{n-1}\tilde{H}_{+m+1,n-1}(x,\lambda_0)e^{i\lambda_0 a_{n-1}x}+b_n\tilde{H}_{+m+1,n}(x,\lambda_0)e^{i\lambda_0 a_n x}=0.\\
&\cdots\ \cdots\ \cdots\quad \cdots\ \cdots\ \cdots\quad \cdots\ \cdots\ \cdots\quad \cdots\ \cdots\ \cdots
\end{aligned}
$$

$$
\begin{aligned}
&b_1\tilde{H}_{+m,1}(x,\lambda_0)e^{i\lambda_0 a_1 x}+\cdots\\
&+b_{m-1}\tilde{H}_{+m,m-1}(x,\lambda_0)e^{i\lambda_0 a_{m-1}x}+b_m\left(1+\tilde{H}_{+m,m}(x,\lambda_0)\right)e^{i\lambda_0 a_m x}=0,\\
&b_1\tilde{H}_{+m-1,1}(x,\lambda_0)e^{i\lambda_0 a_1 x}+\cdots\\
&+b_{m-1}\left(1+\tilde{H}_{+m-1,m-1}(x,\lambda_0)\right)e^{i\lambda_0 a_{m-1}x}+b_m\tilde{H}_{+m-1,m}(x,\lambda_0)e^{i\lambda_0 a_m x}=0,\\
&\cdots\ \cdots\ \cdots\quad \cdots\ \cdots\ \cdots\quad \cdots\ \cdots\ \cdots\quad \cdots\ \cdots\ \cdots \qquad (5.81)\\
&b_1\left(1+\tilde{H}_{+1,1}(x,\lambda_0)\right)e^{i\lambda_0 a_1 x}+\cdots\\
&+b_{m-1}\tilde{H}_{+1,m-1}(x,\lambda_0)e^{i\lambda_0 a_{m-1}x}+b_m\tilde{H}_{+1,m}(x,\lambda_0)e^{i\lambda_0 a_m x}=0.
\end{aligned}
$$

Multiplying the first equality of (5.80) by $e^{-i\lambda_0 a_n x}$, then passing to the limit as $x \to +\infty$ with due regard for estimates (5.22)–(5.24) we obtain

$$\lim_{x\to\infty} b_n\left(1+\tilde{H}_{+n,n}(x,\lambda_0)\right)=b_n=0.$$

Multiplying the second equality of (5.80) by $e^{-i\lambda_0 a_{n-1}x}$, then passing to the limit to $x \to +\infty$ with due regard for (5.22)–(5.24) and for $b_n = 0$, we show that $b_{n-1} = 0$. Analogously, multiplying the last equality of (5.80) by $e^{-i\lambda_0 a_{m+1}x}$ with due regard for $b_n = b_{n-1} = \cdots = b_{m+2} = 0$, we obtain that $b_{m+1} = 0$. Using this fact, from (5.81) we prove analogously that $b_m = b_{m-1} = \cdots = b_1 = 0$. Thus, in (5.80) and, (5.81) for $\lambda = \lambda_0$, $\mathrm{Im}\lambda_0 > 0$ all numbers b_i, $i = 1,...,n$ are simultaneously equal to zero, which contradicts the condition (5.75). Hence, for a certain of choice of a number $b \in (a_n, a_1)$: $P(x,\lambda_0) \not\equiv 0$, i.e., $P(x,\lambda_0)$ is the nontrivial solution of systems (5.5), hence λ_0, $\mathrm{Im}\lambda_0 > 0$ is a generalized eigenvalue for system (5.5).

Let λ_0, $\mathrm{Im}\lambda_0 < 0$ be a given singular number of system (5.5), then

$$\det[I+\tilde{H}_-(x,\lambda_0)]=0. \tag{5.82}$$

Using the condition (5.82) and by an argument analogous to the previous one, we prove that the given singular number λ_0, $\mathrm{Im}\lambda_0 < 0$ of system (5.5) is a generalized eigenvalue for this system. Thus, Lemma 5.5 is proved. □

5.1.6 The inverse scattering problem

We introduce the intermediate operators $F(x,\lambda)$ and $G(x,\lambda)$ by the equalities:

$$F(x,\lambda) = e^{i\lambda J_n x}\tilde{S}(\lambda)e^{-i\lambda J_n x} - I = e^{i\lambda J_n x}F(\lambda)e^{-i\lambda J_n x}, \tag{5.83}$$

$$G(x,\lambda) = e^{i\lambda J_n x}\tilde{S}^{-1}(\lambda)e^{-i\lambda J_n x} - I = e^{i\lambda J_n x}G(\lambda)e^{-i\lambda J_n x}, \tag{5.84}$$

where $e^{i\lambda J_n x} = \operatorname{diag}(e^{i\lambda a_1 x}, ..., e^{i\lambda a_n x})$, $a_1 > ... > a_n$.

The intermediate matrix operators given by (5.83) and (5.84) for $x = 0$ and real λ are:

$$F(\lambda) = F(0,\lambda) = \tilde{S}(\lambda) - I,$$
$$G(\lambda) = G(0,\lambda) = \tilde{S}^{-1}(\lambda) - I.$$

Owing to (5.45) and (5.46) the intermediate matrices $\tilde{S}(\lambda)$ and $\tilde{S}^{-1}(\lambda)$ admit the factorization (5.51) for real parameter λ:

$$e^{i\lambda J_n x}\tilde{S}(\lambda)e^{-i\lambda J_n x} = \left(I + \tilde{K}_-(x,\lambda)\right)\left(I + \tilde{H}_+(x,\lambda)\right), \tag{5.85}$$

$$e^{i\lambda J_n x}\tilde{S}^{-1}(\lambda)e^{-i\lambda J_n x} = \left(I + \tilde{K}_+(x,\lambda)\right)\left(I + \tilde{H}_-(x,\lambda)\right), \tag{5.86}$$

wherein

$$I + \tilde{K}_-(x,\lambda) = I + \int_{-\infty}^{0} K_-(x,s)e^{i\lambda s}ds = \left(I + \int_{-\infty}^{0} H_-(x,s)e^{i\lambda s}ds\right)^{-1},$$

$$I + \tilde{K}_+(x,\lambda) = I + \int_{0}^{\infty} K_+(x,s)e^{i\lambda s}ds = \left(I + \int_{0}^{\infty} H_+(x,s)e^{i\lambda s}ds\right)^{-1}.$$

Lemma 5.6. *The intermediate matrix operators $F(\lambda)$ and $G(\lambda)$ are the Fourier transforms of the matrix absolutely integrable functions: $f(\xi) = [f_{jk}(\xi)]$ and $g(\xi) = [g_{jk}(\xi)]$, $j,k = 1,...,n$.*

Proof. From the factorizations of the matrices $\tilde{S}(\lambda)$ and $\tilde{S}^{-1}(\lambda)$ defined by (5.85) and (5.86) for $x = 0$ and real λ, using the definitions of the operators $F(\lambda)$ and $G(\lambda)$, we obtain

$$\tilde{S}(\lambda) = I + F(\lambda) = I + \int_{-\infty}^{\infty} f(\xi)e^{i\lambda\xi}d\xi, \tag{5.87}$$

$$\tilde{S}^{-1}(\lambda) = I + G(\lambda) = I + \int_{-\infty}^{\infty} g(\xi)e^{i\lambda\xi}d\xi, \tag{5.88}$$

where the matrix functions $f(\xi)$ and $g(\xi)$ are determined by:

$$f(\xi) = K_-(0,\xi) + H_+(0,\xi) + \int_{-\infty}^{\xi} K_-(0,t)H_+(0,\xi-t)dt, \tag{5.89}$$

$$g(\xi) = K_+(0,\xi) + H_-(0,\xi) + \int_{\xi}^{\infty} K_+(0,t)H_-(0,\xi-t)dt. \tag{5.90}$$

Owing to Lemma 5.2 and estimates (5.22)–(5.24) for the matrix kernel functions $H_+(x,\xi)$ and $H_-(x,\xi)$, the matrix functions $f(\xi)$ and $g(\xi)$ are absolutely integrable. Thus, Lemma 5.6 is proved. □

Using (5.83) and (5.84), we obtain from factorization (5.51):

$$\tilde{H}_-(x,\lambda) + F(x,\lambda) + \tilde{H}_-(x,\lambda)F(x,\lambda) = \tilde{H}_+(x,\lambda),$$
$$\tilde{H}_+(x,\lambda) + G(x,\lambda) + \tilde{H}_+(x,\lambda)G(x,\lambda) = \tilde{H}_-(x,\lambda),$$

whence it follows that

$$H_-(x,\xi) + f(x,\xi) + \int_{\xi}^{\infty} H_-(x,\xi-\eta)f(x,\eta)d\eta = H_+(x,\xi), \tag{5.91}$$

$$H_+(x,\xi) + g(x,\xi) + \int_{-\infty}^{\xi} H_+(x,\xi-\eta)g(x,\eta)d\eta = H_-(x,\xi). \tag{5.92}$$

By virtue of the formulas (5.83), (5.84), (5.87) and (5.88), the matrix functions $f(x,\xi)$ and $g(x,\xi)$ are represented in the form:

$$f(x,\xi) = \begin{pmatrix} f_{11}(\xi), & f_{12}(\xi-(a_1-a_2)x), & \cdots, & f_{1n}(\xi-(a_1-a_n)x) \\ f_{21}(\xi-(a_2-a_1)x), & f_{22}(\xi), & \cdots, & f_{2n}(\xi-(a_2-a_n)x) \\ ----, & ----, & \cdots, & ---- \\ f_{n1}(\xi-(a_n-a_1)x), & f_{n2}(\xi-(a_n-a_2)x), & \cdots, & f_{nn}(\xi) \end{pmatrix},$$

$$g(x,\xi) = \begin{pmatrix} g_{11}(\xi), & g_{12}(\xi-(a_1-a_2)x), & \cdots, & g_{1n}(\xi-(a_1-a_n)) \\ g_{21}(\xi-a_2-a_1)x), & g_{22}(\xi), & \cdots, & g_{2n}(\xi-(a_2-a_n)) \\ ----, & ----, & \cdots, & ---- \\ g_{n1}(\xi-(a_n-a_1)x), & g_{n2}(\xi-(a_n-a_2)x), & \cdots, & g_{nn}(\xi) \end{pmatrix}.$$

Since $H_-(x,\xi) = 0$ for $\xi > 0$ and $H_+(x,\xi) = 0$ for $\xi < 0$, then from (5.91) and (5.92) we derive the fundamental matrix equations connecting the matrix kernels $H_+(x,\xi)$ and $H_-(x,\xi)$ of the transformation operators $\tilde{H}_+(x,\lambda)$ and $\tilde{H}_-(x,\lambda)$ with the matrix kernels $f(x,\xi)$ and $g(x,\xi)$ of the intermediate operators $F(x,\lambda)$ and $G(x,\lambda)$:

$$H_-(x,\xi) + f(x,\xi) + \int_{\xi}^{\infty} H_-(x,\xi-\eta) f(x,\eta) d\eta = 0, \quad \xi < 0, \tag{5.93}$$

$$H_+(x,\xi) + g(x,\xi) + \int_{-\infty}^{\xi} H_+(x,\xi-\eta) g(x,\eta) d\eta = 0, \quad \xi > 0. \tag{5.94}$$

Lemma 5.7. *Suppose that the system* (5.5) *has no generalized eigenvalues. Then for each fixed x, the fundamental matrix equations* (5.93) *and* (5.94) *have a unique solution in the space $L_1(-\infty,\infty)$.*

Proof. By using Lemma 5.6 and the Fredholm's Theorems [101, 116], we can verify that the equations (5.93) and (5.94) are Fredholm integral equations of the second kind in the space $L_1(-\infty,\infty)$. Therefore, the unique solvability of these equations is deduced from the uniqueness of the solution. Then, it is sufficient to show that the homogeneous equations:

$$h_-(s) + \int_{s}^{\infty} h_-(s-\eta) f(\eta) d\eta = 0, \quad s < 0, \tag{5.95}$$

$$h_+(s) + \int_{-\infty}^{s} h_+(s-\eta) g(\eta) d\eta = 0, \quad s > 0, \tag{5.96}$$

have in $L_1(-\infty,\infty)$ only the trivial solution $h_\pm(s) \equiv 0$.

We write equations (5.95) and (5.96) in the form:

$$h_-(s) + \int_{-\infty}^{0} h_-(z) f(s-z) dz = 0, \quad s < 0, \tag{5.97}$$

$$h_+(s) + \int_{0}^{\infty} h_+(z) g(s-z) dz = 0, \quad s > 0. \tag{5.98}$$

In equation (5.98), instead of $g(s-z)$ we substitute the expression (5.90) for $\xi = s - z$, then after some transformations, we obtain

$$h_+(s) + \int_0^s h_+(z)K_+(0, s-z)dz + \int_s^\infty h_+(p)H_-(0, s-p)dp +$$

$$\int_0^\infty h_+(z)dz \int_{\max(s,z)}^\infty K_+(0, p-z)H_-(0, s-p)dp =$$

$$Q_+(s) + \int_s^\infty Q_+(p)H_-(0, s-p)dp = 0, \quad s > 0, \tag{5.99}$$

where

$$Q_+(s) = h_+(s) + \int_0^s h_+(z)K_+(0, s-z)dz \ \text{ for } \ s > 0. \tag{5.100}$$

If we set $Q_+(s) = 0$ for $s < 0$, and

$$U_-(s) = \begin{cases} \int_s^\infty Q_+(p)H_-(0, s-p)dp \ \text{ for } \ s < 0, \\ 0 \ \text{ for } \ s > 0, \end{cases}$$

then from (5.99) we obtain the equation for any real s:

$$Q_+(s) + \int_s^\infty Q_+(p)H_-(0, s-p)dp = U_-(s) \ \text{ for } \ -\infty < s < \infty. \tag{5.101}$$

Applying the Fourier transform to equality (5.101), we get

$$\int_0^\infty Q_+(s)e^{i\lambda s}ds \left(1 + \int_{-\infty}^0 H_-(0,t)e^{i\lambda t}dt\right) = \int_{-\infty}^0 U_-(s)e^{i\lambda s}ds. \tag{5.102}$$

If the system (5.5) has no generalized eigenvalues, then according to Lemma 5.5 it has no singular numbers. Hence, by Lemma 5.2 there exists the bounded operator $\left(I + \tilde{H}_-(0,\lambda)\right)^{-1}$, $\operatorname{Im}\lambda \le 0$. Using this fact and equality (5.102), we set

$$N(\lambda) = \int_0^\infty Q_+(s)e^{i\lambda s}ds = \int_{-\infty}^0 U_-(s)e^{i\lambda s}ds \left(I + \int_{-\infty}^0 H_-(0,t)e^{i\lambda t}dt\right)^{-1}.$$

Then the operator function $N(\lambda)$ admits an analytic continuation into the entire plane and is bounded uniformly with respect to λ, therefore, by the Liouville's Theorem, $N(\lambda)$ is a constant, but

$$\lim_{\mathrm{Im}\lambda\to\infty}\int_0^\infty Q_+(s)e^{i\lambda s}ds = 0,$$

consequently, $N(\lambda) = \int_0^\infty Q_+(s)e^{i\lambda s}ds \equiv 0$. Owing to the uniqueness of the Fourier transform, we obtain that $Q_+(s) \equiv 0, \ \ s > 0$. Due to this fact, from (5.100) we have a homogeneous Volterra equation for $h_+(s)$, which, as known, has in the space $L_1(0, \infty)$ only the zero solution $h_+(s) \equiv 0, \ \ s > 0$.

Similarly one shows that the equation (5.97) has in the space $L_1(-\infty, 0)$ only the zero solution $h_-(s) \equiv 0, \ \ s < 0$. This proves the lemma. □

We turn to solving the ISP, which consists of the reconstruction of the potential matrix $C(x)$ in system (5.5) from the known transition matrix $S(\lambda)$. Suppose that the system (5.5) has no generalized eigenvalues, and the transition matrix $S(\lambda)$ for this system is known. Then, according to Lemma 5.3, one can find uniquely the triangular matrix functions $R^+(\lambda)$ and $S^-(\lambda)$, which are expressible through the elements of the known matrix $S(\lambda)$. Further, according to the right factorization (5.74), we construct the intermediate matrix $\tilde{S}(\lambda)$ from the found triangular multipliers $R^+(\lambda)$ and $S^-(\lambda)$. The following intermediate matrices are constructed:

$$F(x,\lambda) = e^{i\lambda J_n x}\tilde{S}(\lambda)e^{-i\lambda J_n x} - I, \ \ G(x,\lambda) = e^{i\lambda J_n x}\tilde{S}^{-1}(\lambda)e^{-i\lambda J_n x} - I,$$

where $e^{i\lambda J_n x} = \mathrm{diag}\left(e^{i\lambda a_1 x}, ..., e^{i\lambda a_n x}\right)$, a_j, $j = 1, ..., n$, are given real regulated numbers: $a_1 > \cdots > a_n$.

Consequently, the functions $f(x,\xi)$ and $g(x,\xi)$ appearing in fundamental equations (5.93) and (5.94) are constructed from the intermediate matrix $\tilde{S}(\lambda)$. According to Lemma 5.7, these equations have a unique solution $H_-(x,\xi), \xi < 0$ and $H_+(x,\xi), \xi > 0$. The obtained matrix functions $H_-(x,\xi)$ and $H_+(x,\xi)$ uniquely determine a system of equations of the form (5.5) with the reconstructed potential matrix:

$$c_{jk}(x) = -i(a_j - a_k)H^+_{jk}(x,0) = i(a_j - a_k)H^-_{jk}(x,0), \ \ j < k, \tag{5.103}$$

$$c_{jk}(x) = -i(a_k - a_j)H^+_{jk}(x,0) = i(a_k - a_j)H^-_{jk}(x,0), \ \ j > k. \tag{5.104}$$

We arrive at the theorem about the solution of the ISP for the system (5.5).

Theorem 5.2. *Suppose that we are given the transition matrix* $S(\lambda)$ *for system* (5.5), *which has no generalized eigenvalues. Then the matrix functions* $f(x,\xi)$ *and* $g(x,\xi)$ *in the fundamental matrix equations* (5.93) *and* (5.94) *are constructed uniquely from the intermediate matrix* $\tilde{S}(\lambda)$ *which is calculated* *from the known transition matrix $S(\lambda)$ by Lemmas 5.3, 5.4 and 5.5. By Lemma 5.7 for fixed x, the fundamental equations* (5.93) *and* (5.94) *have unique solutions $H_-(x,\xi)$, $\xi<0$ and $H_+(x,\xi)$, $\xi>0$, respectively. The potential matrix $C(x)$ in system* (5.5) *is uniquely reconstructed and is expressible through the found solutions of equations* (5.93) *and* (5.94) *by means of the formulas* (5.103) *and* (5.104).

5.2 The direct and ISP for the transport equation

In this section we present the direct and ISP for the one-dimensional transport equation on the whole line:

$$i\frac{\partial}{\partial x}\varphi(x,y)+\int_{-1}^{1}c(x,y,\tau)\varphi(x,\tau)d\tau=\lambda y\varphi(x,y), \tag{5.105}$$

where $x\in(-\infty,\infty)$, $y,\tau\in(-1,1)$, while λ is a complex parameter and the potential $c(x,y,\tau)$ is a complex-valued measurable function of its variables satisfying the estimate:

$$|c(x,y,\tau)|\le\tilde{C}\frac{|y-\tau|}{(1+|x|)^{2(1+\varepsilon)}}, \tag{5.106}$$

where $\tilde{C}$ is a constant, $\varepsilon>0$.

5.2.1 The transition operator $S(\lambda)$

Lemma 5.8. *For every given bounded solution $\varphi(x,y)$ of the one-dimensional transport equation* (5.105) *with the potential satisfying estimate* (5.106)*, there exist bounded functions for a real parameter λ:*

$$A(y,\lambda)=\lim_{x\to-\infty}\varphi(x,y)e^{i\lambda yx},\quad B(y,\lambda)=\lim_{x\to\infty}\varphi(x,y)e^{i\lambda yx}. \tag{5.107}$$

The solution $\varphi(x,y)$ is determined uniquely by each of these functions.

Proof. Let $\varphi(x,y)$ be a bounded solution of equation (5.105), then owing to the uniqueness of the solution of the Cauchy problem for equation (5.105):

$$\varphi(x,y)=\varphi(0,y)e^{-i\lambda yx}-i\int_{x}^{0}e^{-i\lambda y(x-\xi)}d\xi\int_{-1}^{1}c(\xi,y,\tau)\varphi(\xi,\tau)d\tau,$$

$$\varphi(x,y)=\varphi(0,y)e^{-i\lambda yx}+i\int_{0}^{x}e^{-i\lambda y(x-\xi)}d\xi\int_{-1}^{1}c(\xi,y,\tau)\varphi(\xi,\tau)d\tau,$$

whence it follows that the functions (5.107) are defined by the equalities:

$$A(y,\lambda) = \varphi(0,y) - i \int_{-\infty}^{0} e^{i\lambda y\xi} d\xi \int_{-1}^{1} c(\xi,y,\tau)\varphi(\xi,\tau)d\tau,$$

$$B(y,\lambda) = \varphi(0,y) + i \int_{0}^{+\infty} e^{i\lambda y\xi} d\xi \int_{-1}^{1} c(\xi,y,\tau)\varphi(\xi,\tau)d\tau.$$

The second assertion is deduced from the fact that the finding of a solution of equation (5.105) with one of the asymptotic conditions (5.107) is equivalent to the solving the Volterra integral equations:

$$\varphi(x,y) = A(y,\lambda)e^{-i\lambda yx} + i \int_{-\infty}^{x} e^{-i\lambda y(x-\xi)} d\xi \int_{-1}^{1} c(\xi,y,\tau)\varphi(\xi,\tau)d\tau, \quad (5.108)$$

$$\varphi(x,y) = B(y,\lambda)e^{-i\lambda yx} - i \int_{x}^{\infty} e^{-i\lambda y(x-\xi)} d\xi \int_{-1}^{1} c(\xi,y,\tau)\varphi(\xi,\tau)d\tau. \quad (5.109)$$

□

We determine the transition operator for the transport equation (5.105).

Definition 5.2. *To each function $A(y,\lambda)$ giving an incident wave, there corresponds, according to Lemma 5.8, a unique solution $\varphi(x,y)$ of equation* (5.105) *with the boundary condition:*

$$\lim_{x\to-\infty} \varphi(x,y)e^{i\lambda yx} = A(y,\lambda).$$

By this lemma, for the found solution $\varphi(x,y)$ there exists a bounded function $B(y,\lambda)$ defining a scattered wave so that

$$B(y,\lambda) = \lim_{x\to\infty} \varphi(x,y)e^{i\lambda yx}.$$

The one-to-one correspondence between the incident wave $A(y,\lambda)$ and the scattered wave $B(y,\lambda)$ defines the $S(\lambda)$-operator, which is called the transition operator for the transport equation (5.105)*:*

$$S(\lambda)A(y,\lambda) = B(y,\lambda). \quad (5.110)$$

This operator will be considered in the space $L_2(-1,1)$.

5.2.2 Volterra integral representations of solutions

The role of the special Volterra integral representations of solutions of the transport equation (5.105) is analogous to the role of transformation operators which are utilized in scattering problems for the ODEs. Both of them are significant and important in solving the ISPs. These representations are introduced by considering the following Volterra's integral equations of the type:

$$\begin{aligned}\psi_1(x,y,y_0) &= i\theta(y_0-y)\int_{-\infty}^{x} e^{-i\lambda(y-y_0)(x-\xi)}c(\xi,y,y_0)d\xi \\ &+ i\theta(y_0-y)\int_{-\infty}^{x} e^{-i\lambda(y-y_0)(x-\xi)}d\xi\int_{-1}^{1} c(\xi,y,\tau)\psi_1(\xi,\tau,y_0)d\tau \\ &- i\theta(y-y_0)\int_{x}^{\infty} e^{-i\lambda(y-y_0)(x-\xi)}c(\xi,y,y_0)d\xi \\ &- i\theta(y-y_0)\int_{x}^{\infty} e^{-i\lambda(y-y_0)(x-\xi)}d\xi\int_{-1}^{1} c(\xi,y,\tau)\psi_1(\xi,\tau,y_0)d\tau,\end{aligned} \tag{5.111}$$

$$\begin{aligned}\psi_2(x,y,y_0) &= i\theta(y-y_0)\int_{-\infty}^{x} e^{-i\lambda(y-y_0)(x-\xi)}c(\xi,y,y_0)d\xi \\ &+ i\theta(y-y_0)\int_{-\infty}^{x} e^{-i\lambda(y-y_0)(x-\xi)}d\xi\int_{-1}^{1} c(\xi,y,\tau)\psi_2(\xi,\tau,y_0)d\tau \\ &- i\theta(y_0-y)\int_{x}^{\infty} e^{-i\lambda(y-y_0)(x-\xi)}c(\xi,y,y_0)d\xi \\ &- i\theta(y_0-y)\int_{x}^{\infty} e^{-i\lambda(y-y_0)(x-\xi)}d\xi\int_{-1}^{1} c(\xi,y,\tau)\psi_2(\xi,\tau,y_0)d\tau,\end{aligned} \tag{5.112}$$

where y_0 is any fixed point from the interval $(-1,1)$, while $\theta(y)$ is the Heaviside function, i.e., $\theta(y) = \begin{cases} 1 & \text{for } y>0, \\ 0 & \text{for } y\le 0. \end{cases}$ and the potential function $c(x,y,\tau)$ satisfies the constraint (5.106).

We set:

$$\chi_1(x,y,y_0) = \psi_1(x,y,y_0)e^{-i\lambda y_0 x} + \delta(y-y_0)e^{-i\lambda y x}, \tag{5.113}$$

$$\chi_2(x,y,y_0) = \psi_2(x,y,y_0)e^{-i\lambda y_0 x} + \delta(y-y_0)e^{-i\lambda y x}, \tag{5.114}$$

where $\delta(x)$ is the Dirac function.

Instead of $\psi_1(x, y, y_0)$ and $\psi_2(x, y, y_0)$ in (5.113) and (5.114), we substitute (5.111) and (5.112), respectively. As a result of this, we obtain the following integral equations for the functions $\chi_1(x, y, y_0)$ and $\chi_2(x, y, y_0)$:

$$\chi_1(x, y, y_0) = \delta(y - y_0)e^{-i\lambda yx} + \\ i\theta(y_0 - y) \int_{-\infty}^{x} e^{-i\lambda y(x-\xi)} d\xi \int_{-1}^{1} c(\xi, y, \tau)\chi_1(\xi, \tau, y_0)d\tau \\ - i\theta(y - y_0) \int_{x}^{\infty} e^{-i\lambda y(x-\xi)} d\xi \int_{-1}^{1} c(\xi, y, \tau)\chi_1(\xi, \tau, y_0)d\tau,$$

$$\chi_2(x, y, y_0) = \delta(y - y_0)e^{-i\lambda yx} + \\ i\theta(y - y_0) \int_{-\infty}^{x} e^{-i\lambda y(x-\xi)} d\xi \int_{-1}^{1} c(\xi, y, \tau)\chi_2(\xi, \tau, y_0)d\tau \\ - i\theta(y_0 - y) \int_{x}^{\infty} e^{-i\lambda y(x-\xi)} d\xi \int_{-1}^{1} c(\xi, y, \tau)\chi_2(\xi, \tau, y_0)d\tau.$$

It is easy to verify that the functions $\chi_1(x, y, y_0)$ and $\chi_2(x, y, y_0)$ defined by the above integral equations satisfy the transport equation (5.105) in the generalized sense.

Let the operator T_x be defined by the equality:

$$T_x f(y, \lambda) = e^{-i\lambda yx} f(y, \lambda), \tag{5.115}$$

where $f(y, \lambda)$ is any bounded function.

Lemma 5.9. *Every bounded solution $\varphi(x, y)$ of the transport equation* (5.105) *with a potential satisfying the estimate* (5.106) *can be represented in the form:*

$$\varphi(x, y) = (I + V_+(x, \lambda))T_x f_+(y, \lambda), \tag{5.116}$$

$$\varphi(x, y) = (I + V_-(x, \lambda))T_x f_-(y, \lambda), \tag{5.117}$$

where λ is a real parameter, $f_+(y, \lambda)$, $f_-(y, \lambda)$ are arbitrary bounded functions, $V_+(x, \lambda)$, $V_-(x, \lambda)$ are Hilbert–Schmidt operators. The kernels $v_+(x, \lambda; y, y_0)$ and $v_-(x, \lambda; y, y_0)$ of these operators have the representations:

$$v_+(x, \lambda; y, y_0) = \int_0^{\infty} H_+(x, s; y, y_0)e^{i\lambda|y-y_0|s} ds,$$

and

$$v_-(x, \lambda; y, y_0) = \int_{-\infty}^{0} H_-(x, s; y, y_0)e^{i\lambda|y-y_0|s} ds.$$

The functions $H_+(x,s;y,y_0)$ and $H_-(x,s;y,y_0)$ are uniquely determined by the transport equation (5.105) *and are related to the potential by the formulas:*

$$H_+(x,0;y,y_0) = ic(x,y,y_0),\ y_0 > y;$$
$$H_+(x,0,y,y_0) = -ic(x,y,y_0),\ y_0 < y, \tag{5.118}$$
$$H_-(x,0;y,y_0) = ic(x,y,y_0),\ y_0 < y;$$
$$H_-(x,0,y,y_0) = -ic(x,y,y_0),\ y_0 > y, \tag{5.119}$$

and satisfy the estimates:

$$|H_\pm(x,s;y,y_0)| \le \tilde{C}\frac{y_0 - y}{(1 + (y_0 - y)|x - s|)^{1+\varepsilon}},\ y < y_0,$$
$$|H_\pm(x,s;y,y_0)| \le \tilde{C}\frac{y - y_0}{(1 + (y - y_0)|x + s|)^{1+\varepsilon}},\ y > y_0. \tag{5.120}$$

Proof. First, we find the solution of equation (5.111) in the form:

$$\psi_1(x,y,y_0) = \int\limits_0^\infty H_+(x,s;y,y_0)e^{i\lambda|y-y_0|s}ds. \tag{5.121}$$

Substituting $\psi_1(x,y,y_0)$ from (5.121) into (5.111) with due regard for the uniqueness of the Fourier transform, we obtain

$$H_+(x,s;y,y_0) = ic(x-s,y,y_0)+$$
$$i\int\limits_{x-s}^{x} d\xi \int\limits_{-1}^{1} c(\xi,y,\tau)H_+\left(\xi, \frac{y_0 - y}{|y_0 - \tau|}(s+\xi-x);\tau,y_0\right)\frac{y_0 - y}{|y_0-\tau|}d\tau,$$
$$\text{for } y_0 > y \text{ and } s \ge 0. \tag{5.122}$$

Analogously, we get:

$$H_+(x,s;y,y_0) = -ic(x+s,y,y_0)-$$
$$i\int\limits_{x}^{x+s} d\xi \int\limits_{-1}^{1} c(\xi,y,\tau)H_+\left(\xi, \frac{y - y_0}{|\tau - y_0|}(s+x-\xi);\tau,y_0\right)\frac{y - y_0}{|\tau - y_0|}d\tau,$$
$$\text{for } y_0 < y \text{ and } s \ge 0. \tag{5.123}$$

We continue to find the solution of equation (5.112) in the form:

$$\psi_2(x,y,y_0) = \int\limits_{-\infty}^0 H_-(x,s;y,y_0)e^{i\lambda|y-y_0|s}ds. \tag{5.124}$$

Substituting $\psi_2(x, y, y_0)$ from (5.124) into (5.112) with due regard for the uniqueness of the Fourier transform, we obtain:

$$H_-(x, s; y, y_0) = -ic(x - s, y, y_0)-$$
$$i\int_x^{x-s} d\xi \int_{-1}^{1} c(\xi, y, \tau)H_-\left(\xi, \frac{y_0 - y}{|y_0 - \tau|}(s + \xi - x); \tau, y_0\right) \frac{y_0 - y}{|y_0 - \tau|} d\tau,$$
$$\text{for } y_0 > y \text{ and } s \leq 0. \qquad (5.125)$$

Analogously, we have:

$$H_-(x, s; y, y_0) = ic(x + s, y, y_0)+$$
$$i\int_{x+s}^{x} d\xi \int_{-1}^{1} c(\xi, y, \tau)H_-\left(\xi, \frac{y - y_0}{|\tau - y_0|}(s + x - \xi); \tau, y_0\right) \frac{y - y_0}{|\tau - y_0|} d\tau,$$
$$\text{for } y_0 < y \text{ and } s \leq 0. \qquad (5.126)$$

To investigate the solvability of systems (5.122)–(5.123) and (5.125)–(5.126), it is convenient to consider the functions:

$$H_{+1}(x, s; y, y_0) = \frac{1}{y_0 - y} H_+\left(x, \frac{s}{y_0 - y}; y, y_0\right), \ y < y_0,$$
$$\text{for } s > 0 \qquad (5.127)$$
$$H_{+2}(x, s; y, y_0) = \frac{1}{y - y_0} H_+\left(x, \frac{s}{y - y_0}; y, y_0\right), \ y > y_0,$$

$$H_{-1}(x, s; y, y_0) = \frac{1}{y_0 - y} H_-\left(x, \frac{s}{y_0 - y}; y, y_0\right), \ y < y_0,$$
$$\text{for } s < 0 \qquad (5.128)$$
$$H_{-2}(x, s; y, y_0) = \frac{1}{y - y_0} H_-\left(x, \frac{s}{y - y_0}; y, y_0\right), \ y > y_0.$$

Using estimate (5.106), we prove that the method of successive approximations applied to systems (5.122)–(5.123) and (5.125)–(5.126) converges [112, 116]. Indeed, using(5.127), we write system (5.122)–(5.123) in the operator form:

$$H_+ = h + AH_+, \qquad (5.129)$$

where $H_+ = (H_{+1}, H_{+2})$, $AH_+ = (A_1H_+, A_2H_+)$, $h = (h_1, h_2)$,

$$H_{+1} = H_{+1}(x, s; y, y_0), \ H_{+2} = H_{+2}(x, s; y, y_0),$$
$$h_1 = \frac{i}{y_0 - y} c(x - \frac{s}{y_0 - y}; y, y_0), \ y < y_0, \ s \geq 0,$$
$$h_2 = -\frac{i}{y - y_0} c(x + \frac{s}{y - y_0}; y, y_0), \ y > y_0, \ s \geq 0,$$

$$A_1H_+ = i\int\limits_{x-\frac{s}{y_0-y}}^{x} d\xi \int\limits_{y_0}^{1} c(\xi,y,\tau)H_{+2}(\xi, s+(y_0-y)(\xi-x);\tau,y_0)d\tau +$$

$$i\int\limits_{x-\frac{s}{y_0-y}}^{x} d\xi \int\limits_{-1}^{y_0} c(\xi,y,\tau)H_{+1}(\xi, s+(y_0-y)(\xi-x);\tau,y_0)d\tau,\ y<y_0,\ s\geq 0, \tag{5.130}$$

$$A_2H_+ = -i\int\limits_{x}^{x+\frac{s}{y-y_0}} d\xi \int\limits_{y_0}^{1} c(\xi,y,\tau)H_{+2}(\xi, s+(y-y_0)(x-\xi);\tau,y_0)d\tau -$$

$$i\int\limits_{x}^{x+\frac{s}{y-y_0}} d\xi \int\limits_{-1}^{y_0} c(\xi,y,\tau)H_{+1}(\xi, s+(y-y_0)(x-\xi);\tau,y_0)d\tau,\ y>y_0,\ s\geq 0. \tag{5.131}$$

We now prove that the series:

$$H_{+1} = h_1 + (A^1h)_1 + \cdots + (A^nh)_1 + \cdots \tag{5.132}$$

$$H_{+2} = h_2 + (A^1h)_2 + \cdots + (A^nh)_2 + \cdots \tag{5.133}$$

converge. First, using estimate (5.106) for the potential, we estimate the free terms of the above series:

$$|h_1| = \left|\frac{i}{y_0-y}c\left(x-\frac{s}{y_0-y};y,y_0\right)\right| \leq \frac{C_1}{(1+|(y_0-y)x-s|)^{2(1+\varepsilon)}},\ \text{for } y<y_0, \tag{5.134}$$

$$|h_2| = \left|-\frac{i}{y-y_0}c\left(x+\frac{s}{y-y_0};y,y_0\right)\right| \leq \frac{C_1}{(1+|(y-y_0)x+s|)^{2(1+\varepsilon)}}\ \text{for } y>y_0, \tag{5.135}$$

where C_1 is a constant.

Due to estimate (5.106) and definitions (5.129)–(5.131), the term $(A^1h)_1$ of series (5.132) satisfies the inequality:

$$|(A^1h)_1(x,s;y,y_0)| =$$

$$\left| -i^2 \int\limits_{x-\frac{s}{y_0-y}}^{x} d\xi \int\limits_{y_0}^{1} \frac{1}{\tau-y_0} c(\xi,y,\tau) c\left(\xi + \frac{s+(y_0-y)(\xi-x)}{\tau-y_0}; \tau, y_0\right) d\tau + \right.$$

$$\left. i^2 \int\limits_{x-\frac{s}{y_0-y}}^{x} d\xi \int\limits_{-1}^{y_0} c(\xi,y,\tau) c\left(\xi - \frac{s+(y_0-y)(\xi-x)}{y_0-\tau}; \tau, y_0\right) \frac{1}{y_0-\tau} d\tau \right|$$

$$\leq \int\limits_{y_0}^{1} d\tau \int\limits_{x-\frac{s}{y_0-y}}^{x} \frac{\tilde{C}}{(1+|\xi|)^{2(1+\varepsilon)}} \cdot \frac{C_1}{\left(1+\frac{1}{2}|(\tau-y)\xi - [(y_0-y)x-s]|\right)^{2(1+\varepsilon)}} d\xi$$

$$+ \int\limits_{-1}^{y_0} d\tau \int\limits_{x-\frac{s}{y_0-y}}^{x} \frac{\tilde{C}}{(1+|\xi|)^{2(1+\varepsilon)}} \cdot \frac{C_1}{\left(1+\frac{1}{2}|(y-\tau)\xi + (y_0-y)x-s|\right)^{2(1+\varepsilon)}} d\xi,$$

where $\tilde{C}$ and C_1 are constants.

Using the last inequality, we estimate the term $(A^1h)_1(x,s;y,y_0)$. If $x - \frac{s}{y-y_0} \geq 0$, then from the last inequality it follows that

$$|(A^1h)_1(x,s;y,y_0)| \leq \frac{C_2}{(1+(y_0-y)x-s)^{1+\varepsilon}} \int\limits_{x-\frac{s}{y_0-y}}^{x} \frac{d\xi}{(1+|\xi|)^{1+\varepsilon}} \text{ for } y < y_0,$$

where C_2 is a constant.

If $x - \frac{s}{y_0 - y} \leq 0$, then by splitting the interval of integration, we get the estimate:

$$|(A^1 h)_1(x,s;y,y_0)| \leq \int\limits_{y_0}^{1} d\tau \left\{ \int\limits_{x-\frac{s}{y_0-y}}^{\frac{(y_0-y)x-s}{\tau-y+1}} d\xi + \int\limits_{\frac{(y_0-y)x-s}{\tau-y+1}}^{x} d\xi \right\} \frac{\tilde{C}}{(1+|\xi|)^{2(1+\varepsilon)}} \times$$

$$\frac{C_1}{\left(1+\frac{1}{2}|(\tau-y)\xi - [(y_0-y)x-s]|\right)^{2(1+\varepsilon)}} +$$

$$\int\limits_{y}^{y_0} d\tau \left\{ \int\limits_{x-\frac{s}{y_0-y}}^{\frac{(y_0-y)x-s}{y-\tau+2}} d\xi + \int\limits_{\frac{(y_0-y)x-s}{y-\tau+2}}^{x} d\xi \right\} \frac{\tilde{C}}{(1+|\xi|)^{2(1+\varepsilon)}} \times$$

$$\frac{C_1}{\left(1+\frac{1}{2}|(y-\tau)\xi + [(y_0-y)x-s]|\right)^{2(1+\varepsilon)}} +$$

$$\int\limits_{-1}^{y} d\tau \left\{ \int\limits_{x-\frac{s}{y_0-y}}^{\frac{(y_0-y)x-s}{y-\tau+1}} d\xi + \int\limits_{\frac{(y_0-y)x-s}{y-\tau+1}}^{x} d\xi \right\} \frac{\tilde{C}}{(1+|\xi|)^{2(1+\varepsilon)}} \times$$

$$\frac{C_1}{\left(1+\frac{1}{2}|(y-\tau)\xi + [(y_0-y)x-s]|\right)^{2(1+\varepsilon)}}$$

$$\leq C_1 \int\limits_{y_0}^{1} \frac{d\tau}{\left(1+\frac{|(y_0-y)x-s|}{\tau-y+1}\right)^{1+\varepsilon}} \left\{ \int\limits_{x-\frac{s}{y_0-y}}^{\frac{(y_0-y)x-s}{\tau-y+1}} \frac{d\xi}{(1+|\xi|)^{1+\varepsilon}} + \right.$$

$$\left. \int\limits_{\frac{(y_0-y)x-s}{\tau-y+1}}^{x} \frac{d\xi}{(1+|\xi|)^{2(1+\varepsilon)}} \right\} +$$

$$C_1 \int\limits_{y}^{y_0} \frac{d\tau}{\left(1+\frac{|(y_0-y)x-s|}{2+y-\tau}\right)^{1+\varepsilon}} \left\{ \int\limits_{x-\frac{s}{y_0-y}}^{\frac{(y_0-y)x-s}{y-\tau+2}} \frac{d\xi}{(1+|\xi|)^{2(1+\varepsilon)}} + \right.$$

$$\left. \int\limits_{\frac{(y_0-y)x-s}{y-\tau+2}}^{x} \frac{d\xi}{(1+|\xi|)^{1+\varepsilon}} \right\} +$$

$$C_1 \int\limits_{-1}^{y} \frac{d\tau}{\left(1+\frac{|(y_0-y)x-s|}{y-\tau+1}\right)^{1+\varepsilon}} \left\{ \int\limits_{x-\frac{s}{y_0-y}}^{\frac{(y_0-y)x-s}{y-\tau+1}} \frac{d\xi}{(1+|\xi|)^{2(1+\varepsilon)}} + \right.$$

$$\left. \int\limits_{\frac{(y_0-y)x-s}{y-\tau+1}}^{x} \frac{d\xi}{(1+|\xi|)^{1+\varepsilon}} \right\}$$

$$\leq \frac{C_2}{(1+|(y_0-y)x-s|)^{1+\varepsilon}} \int\limits_{x-\frac{s}{y_0-y}}^{x} \frac{d\xi}{(1+|\xi|)^{1+\varepsilon}} \text{ for } y < y_0.$$

where $\tilde{C}, C_1$ and C_2 are constants.

Thus, for any $x \in (-\infty, \infty)$ and $s \geq 0$ the term $\left(A^1 h\right)_1$ of series (5.132) satisfies the estimate:

$$|(A^1 h)_1(x, s; y, y_0)| \leq \frac{C_2}{(1+|(y_0-y)x-s|)^{1+\varepsilon}} \int\limits_{T}^{+\infty} \frac{d\xi}{(1+|\xi|)^{1+\varepsilon}} \text{ for } y < y_0, \tag{5.136}$$

where T is any number such that $T \leq x - \frac{s}{y_0-y}$, $y < y_0$.

By an argument analogous to the previous one, we show that the term $(A^1 h)_2$ of series (5.133) satisfies the estimate for any $x \in (-\infty, \infty)$ and $s \geq 0$:

$$|(A^1 h)_2(x, s; y, y_0)| \leq \frac{\tilde{C}}{(1+|(y-y_0)x+s|)^{1+\varepsilon}} \int\limits_{T}^{+\infty} \frac{d\xi}{(1+|\xi|)^{1+\varepsilon}} \text{ for } y > y_0, \tag{5.137}$$

where T is any number such that $T \leq x + \frac{s}{y-y_0}$, $y_0 < y$.

Suppose that for any $x \in (-\infty, +\infty)$, $s \geq 0$ the following estimates are satisfied for $q \leq k$:

$$|(A^q h)_1(x, s; y, y_0)| \leq \frac{\tilde{C}}{(1+|(y_0-y)x-s|)^{1+\varepsilon}} \int\limits_{T}^{+\infty} \frac{d\xi_1}{(1+|\xi_1|)^{1+\varepsilon}} \times \int\limits_{\xi_1}^{+\infty} \frac{d\xi_2}{(1+|\xi_2|)^{1+\varepsilon}} \cdots \int\limits_{\xi_{k-1}}^{+\infty} \frac{d\xi_q}{(1+|\xi_q|)^{1+\varepsilon}}, \tag{5.138}$$

for any number T such that. $T \leq x - \frac{s}{y_0-y}$ $y_0 > y$, and

$$|(A^q h)_2(x, s; y, y_0)| \leq \frac{\tilde{C}}{(1+|(y-y_0)x+s|)^{1+\varepsilon}} \int\limits_{T}^{+\infty} \frac{d\xi_1}{(1+|\xi_1|)^{1+\varepsilon}} \times \int\limits_{\xi_1}^{+\infty} \frac{d\xi_2}{(1+|\xi_2|)^{1+\varepsilon}} \cdots \int\limits_{\xi_{k-1}}^{+\infty} \frac{d\xi_q}{(1+|\xi_q|)^{1+\varepsilon}}, \tag{5.139}$$

for any number T such that $T \leq x + \frac{s}{y-y_0}$ and $y > y_0$.

By an argument analogous to the proof of the estimate (5.136), one can prove that the estimates (5.138) and (5.139) are satisfied for $q = k + 1$.

Thus, the induction is proved, and the estimates (5.138) and (5.139) for $(A^q h)_1(x, s; y, y_0)$ and $(A^q h)_2(x, s; y, y_0)$, respectively, are seen to be satisfied for all q. Using the proved estimates (5.134)–(5.139), we construct the following majorant series:

$$\frac{\tilde{C}}{(1+|(y_0-y)x-s|)^{1+\varepsilon}}\Big\{1+\int_T^{+\infty}\frac{d\xi_1}{(1+|\xi_1|)^{1+\varepsilon}}+\int_T^{+\infty}\frac{d\xi_1}{(1+|\xi_1|)^{1+\varepsilon}}\times$$
$$\int_{\xi_1}^{+\infty}\frac{d\xi_2}{(1+|\xi_2|)^{1+\varepsilon}}+\cdots+\int_T^{+\infty}\frac{d\xi_1}{(1+\xi_1)^{1+\varepsilon}}\times$$
$$\int_{\xi_1}^{+\infty}\frac{d\xi_2}{(1+|\xi_2|)^{1+\varepsilon}}\cdots\int_{\xi_k}^{+\infty}\frac{d\xi_{k+1}}{(1+|\xi_{k+1}|)^{1+\varepsilon}}+\cdots\Big\},\quad y<y_0 \tag{5.140}$$

$$\frac{\tilde{C}}{(1+|(y_0-y)x+s|)^{1+\varepsilon}}\Big\{1+\int_T^{+\infty}\frac{d\xi_1}{(1+|\xi_1|)^{1+\varepsilon}}+\int_T^{+\infty}\frac{d\xi_1}{(1+|\xi_1|)^{1+\varepsilon}}\times$$
$$\int_{\xi_1}^{+\infty}\frac{d\xi_2}{(1+|\xi_2|)^{1+\varepsilon}}+\cdots+\int_T^{+\infty}\frac{d\xi_1}{(1+\xi_1)^{1+\varepsilon}}\times$$
$$\int_{\xi_1}^{+\infty}\frac{d\xi_2}{(1+|\xi_2|)^{1+\varepsilon}}\cdots\int_{\xi_k}^{+\infty}\frac{d\xi_{k+1}}{(1+|\xi_{k+1}|)^{1+\varepsilon}}+\cdots\Big\},\quad y>y_0. \tag{5.141}$$

The expressions in the both brackets in (5.140) and (5.141) are the same, so we denote these expressions by the function $\Phi(T)$. It is easy to verify that

$$\Phi(T)=\exp\Big\{\int_T^{+\infty}\frac{d\xi}{(1+|\xi|)^{1+\varepsilon}}\Big\}.$$

Hence, the series (5.140) and (5.141) are summable, and their corresponding sums are equal to:

$$\frac{\tilde{C}}{(1+|(y_0-y)x-s|)^{1+\varepsilon}}\exp\Big\{\int_T^{+\infty}\frac{d\xi}{(1+|\xi|)^{1+\varepsilon}}\Big\}$$

for $y<y_0$, $-\infty<x<\infty$, $s\geq 0$, and $T\leq x-\dfrac{s}{y_0-y}$,

$$\frac{\tilde{C}}{(1+|(y-y_0)x+s|)^{1+\varepsilon}}\exp\Big\{\int_T^{+\infty}\frac{d\xi_1}{(1+|\xi_1|)^{1+\varepsilon}}\Big\}$$

for $y_0<y$, $-\infty<x<\infty$, $s\geq 0$, and $T\leq x+\dfrac{s}{y_0-y}$.

The convergence of the series (5.132) and (5.133) is deduced from the convergence of the majorant series (5.140) and (5.141), respectively. Hence, the method of successive approximations that is applied to the operator equation (5.129), is convergent, and the system (5.122)–(5.123) has in $C(E^3)$ a unique solution satisfying estimate (5.120).

Similarly, we prove that the method of successive approximations that is applied to the system (5.125)–(5.126) is convergent, and this system has in $C(E^3)$ a unique solution satisfying the estimate (5.120).

Further, in systems (5.122)–(5.123) and (5.125)–(5.126) putting $s = 0$, we obtain the formulas (5.118) and (5.119).

Since the solutions $\psi_1(x, y, y_0)$ and $\psi_2(x, y, y_0)$ of equations (5.111) and (5.112) are found in the form (5.121) and (5.124), respectively, then the functions $\chi_1(x, y, y_0)$ and $\chi_2(x, y, y_0)$ defined by (5.113) and (5.114), respectively, are solutions of system (5.105).

Consider the functions:

$$\varphi(x, y) = \int_{-1}^{1} f_+(y_0, \lambda)\chi_1(x, y, y_0)dy_0; \quad \varphi(x, y) = \int_{-1}^{1} f_-(y_0, \lambda)\chi_2(x, y, y_0)dy_0, \tag{5.142}$$

where $\chi_1(x, y, y_0)$ and $\chi_2(x, y, y_0)$ are determined by (5.113) and (5.114), respectively, and $f_+(y_0, \lambda)$ and $f_-(y_0, \lambda)$ are any bounded functions.

Since $\chi_1(x, y, y_0)$ and $\chi_2(x, y, y_0)$ are the solutions of the transport equation (5.105), then the functions $\varphi(x, y)$ defined by (5.142) are also the solutions of this equation.

Making use of (5.113), (5.114) and (5.121), (5.124), we write the solutions (5.142) of the transport equation (5.105) in the form:

$$\varphi(x, y) = f_+(y, \lambda)e^{-i\lambda yx} + \int_{-1}^{1} f_+(y_0, \lambda)e^{-i\lambda y_0 x}dy_0 \int_{0}^{\infty} H_+(x, s; y, y_0)e^{i\lambda|y-y_0|s}ds, \tag{5.143}$$

$$\varphi(x, y) = f_-(y, \lambda)e^{-i\lambda yx} + \int_{-1}^{1} f_-(y_0, \lambda)e^{-i\lambda y_0 x}dy_0 \int_{-\infty}^{0} H_-(x, s; y, y_0)e^{i\lambda|y-y_0|s}ds. \tag{5.144}$$

With due regard for (5.115), from (5.143) and (5.144) we obtain the integral Volterra representations of solutions (5.116) and (5.117) of the transport equation (5.105).

Due to estimates (5.120), for fixed x and y the kernels:

$$v_-(x,\lambda;y,y_0) = \int_{-\infty}^{0} H_-(x,s;y,y_0)e^{i\lambda|y-y_0|s}ds$$

$$v_+(x,\lambda;y,y_0) = \int_{0}^{\infty} H_+(x,s;y,y_0)e^{i\lambda|y-y_0|s}ds,$$

are bounded functions of y_0, therefore they are the kernels of the Hilbert–Schmidt operators $V_-(x,\lambda)$ and $V_+(x,\lambda)$, respectively. This completes the proof of Lemma 5.9. □

The functions $f_+(y,\lambda)$ and $f_-(y,\lambda)$ will be called, respectively, the "$+$" and "$-$" preimages of the solutions $\varphi(x,y)$ of the transport equation (5.105).

5.2.3 Bilateral Volterra factorization of the S-operator

We now establish the connection between the preimages and asymptotics (5.107) of the solutions of the transport equation (5.105).

Lemma 5.10. *The incident wave $A(y,\lambda)$ is related to the preimages $f_+(y,\lambda)$ and $f_-(y,\lambda)$ of the solutions of the transport equation* (5.105) *by the formulas:*

$$A(y,\lambda)e^{-i\lambda yx} = (I + K_1^+(x,\lambda))T_x f_+(y,\lambda), \tag{5.145}$$

$$A(y,\lambda)e^{-i\lambda yx} = (I + K_2^-(x,\lambda))T_x f_-(y,\lambda), \tag{5.146}$$

while the scattered wave $B(y,\lambda)$ is related to $f_+(y,\lambda)$ and $f_-(y,\lambda)$ by the formulas:

$$B(y,\lambda)e^{-i\lambda yx} = (I + K_1^-(x,\lambda))T_x f_+(y,\lambda), \tag{5.147}$$

$$B(y,\lambda)e^{-i\lambda yx} = (I + K_2^+(x,\lambda))T_x f_-(y,\lambda), \tag{5.148}$$

where λ is a real parameter, $K_j^-(x,\lambda)$ and $K_j^+(x,\lambda)$, $j = 1,2$, are Volterra integral Hilbert–Schmidt operators with variable lower and upper limits, respectively.

Proof. Instead of the function $\varphi(x,y)$ in equation (5.108), substituting the Volterra representation of $\varphi(x,y)$ from (5.143) with due regard for equations (5.122) and (5.123), we obtain:

$$A(y,\lambda)e^{-i\lambda yx} = f_+(y,\lambda)e^{-i\lambda yx} + \int_{-1}^{y} k_1^+(x,\lambda;y,y_0)f_+(y_0,\lambda)e^{-i\lambda y_0 x}dy_0, \tag{5.149}$$

where

$$k_1^+(x,\lambda;y,y_0) = -i\int_{-\infty}^{\infty} e^{i\lambda(y-y_0)s}ds\Bigg\{c(x+s,y,y_0)$$

$$+\int_{-\infty}^{x+s}d\xi\int_{-1}^{1}c(\xi,y,\tau)H_+\left(\xi,\frac{y-y_0}{|y_0-\tau|}(s+x-\xi);\tau,y_0\right)\frac{y-y_0}{|y_0-\tau|}d\tau\Bigg\},\ y_0<y,$$

Further instead of the function $\varphi(x,y)$ in equation (5.108), substituting the Volterra representation of $\varphi(x,y)$ from (5.144), using equations (5.125) and (5.126), we get:

$$A(y,\lambda)e^{-i\lambda yx} = f_-(y,\lambda)e^{-i\lambda yx} + \int_y^1 k_2^-(x,\lambda;y,y_0)f_-(y_0,\lambda)e^{-i\lambda y_0 x}dy_0, \tag{5.150}$$

where

$$k_2^-(x,\lambda;y,y_0) = -i\int_{-\infty}^{\infty} e^{i\lambda(y_0-y)s}ds\Bigg\{c(x-s,y,y_0)$$

$$+\int_{-\infty}^{x-s}d\xi\int_{-1}^{1}c(\xi,y,\tau)H_-\left(\xi,\frac{y_0-y}{|y_0-\tau|}(s+\xi-x);\tau,y_0\right)\frac{y_0-y}{|y_0-\tau|}d\tau\Bigg\},\ y_0>y.$$

Similarly, instead of $\varphi(x,y)$ in (5.109) substituting the Volterra representation of $\varphi(x,y)$ from (5.143) with due regard for equations (5.122) and (5.123) gives:

$$B(y,\lambda)e^{-i\lambda yx} = f_+(y,\lambda)e^{-i\lambda yx} + \int_y^1 k_1^-(x,\lambda;y,y_0)f_+(y_0,\lambda)e^{-i\lambda y_0 x}dy_0, \tag{5.151}$$

where

$$k_1^-(x,\lambda;y,y_0) = i\int_{-\infty}^{\infty} e^{i\lambda(y_0-y)s}ds\Bigg\{c(x-s,y,y_0)+$$

$$+\int_{x-s}^{\infty}d\xi\int_{-1}^{1}c(\xi,y,\tau)H_+\left(\xi,\frac{y_0-y}{|\tau-y_0|}(s+\xi-x);\tau,y_0\right)\frac{y_0-y}{|\tau-y_0|}d\tau\Bigg\},\ y_0>y.$$

Further, instead of the function $\varphi(x,y)$ in equation (5.109), substituting the Volterra representation of $\varphi(x,y)$ from (5.144), using equations (5.125) and (5.126), yields:

$$B(y,\lambda)e^{-i\lambda yx} = f_-(y,\lambda)e^{-i\lambda yx} + \int_{-1}^{y} k_2^+(x,\lambda;y,y_0)f_-(y_0,\lambda)e^{-i\lambda y_0 x}dy_0, \tag{5.152}$$

where

$$k_2^+(x,\lambda;y,y_0) = i\int_{-\infty}^{\infty} e^{i\lambda(y-y_0)s}ds\Bigg\{c(x+s,y,y_0)+$$

$$+\int_{x+s}^{\infty} d\xi \int_{-1}^{1} c(\xi,y,\tau)H_-\left(\xi, \frac{y-y_0}{|\tau-y_0|}(s+x-\xi);\tau,y_0\right)\frac{y-y_0}{|\tau-y_0|}d\tau\Bigg\},\ y_0<y.$$

Making use of estimates (5.106) and (5.120), we easily prove that the functions $k_j^-(x,\lambda;y,y_0)$ and $k_j^+(x,\lambda;y,y_0)$, $j=1,2$ are bounded, therefore they are kernels of the Volterra integral Hilbert–Schmidt operators $K_j^-(x,\lambda)$ and $K_j^+(x,\lambda)$, $j=1,2$ with variable lower and upper limits, respectively. This fact proves the formulas (5.145)-(5.148). □

Theorem 5.3. *The transition operator $S(\lambda)$ for the transport equation* (5.105) *with a potential satisfying the estimate* (5.106) *admits the bilateral factorization, i.e., the right and left Volterra factorizations:*

$$S(\lambda) = T_{-x}(I+K_2^+(x,\lambda))(I+K_2^-(x,\lambda))^{-1}T_x, \tag{5.153}$$
$$S(\lambda) = T_{-x}(I+K_1^-(x,\lambda))(I+K_1^+(x,\lambda))^{-1}T_x, \tag{5.154}$$

where the operators $K_j^{\pm}(x,\lambda)$, $j=1,2$ are the Volterra integral Hilbert–Schmidt operators defined by Lemma 5.10 and the operator T_x is defined by formula (5.115).

Proof. We consider equations (5.149)–(5.152). The unique solvability of these equations in the space $L_2(-1,1)$ is deduced from the fact that these equations are Volterra integral equations with respect to the variable y and from the boundedness of the functions $A(y,\lambda)$, $B(y,\lambda)$, $k_j^{\pm}(x,\lambda;y,y_0)$, $j=1,2$. Taking this fact into account, from (5.145)–(5.148), we have

$$f_+(y,\lambda)e^{-i\lambda yx} = (I+K_1^+(x,\lambda))^{-1}T_xA(y,\lambda), \tag{5.155}$$
$$f_-(y,\lambda)e^{-i\lambda yx} = (I+K_2^-(x,\lambda))^{-1}T_xA(y,\lambda), \tag{5.156}$$
$$f_+(y,\lambda)e^{-i\lambda yx} = (I+K_1^-(x,\lambda))^{-1}T_xB(y,\lambda), \tag{5.157}$$
$$f_-(y,\lambda)e^{-i\lambda yx} = (I+K_2^+(x,\lambda))^{-1}T_xB(y,\lambda). \tag{5.158}$$

Comparing (5.156) with (5.158) and (5.155) with (5.157) with due regard for the definition of the transition operator $S(\lambda)$, we obtain the right and left Volterra factorizations (5.153) and (5.154) of the transition operator $S(\lambda)$. The theorem is proved. □

We introduce the operator $\tilde{\mathcal{F}}(\lambda)$ defined in the space $L_2(-1,1)$ by the equalities:

$$\tilde{\mathcal{F}}(\lambda) = S(\lambda) - I,\ \tilde{\mathcal{F}}(\lambda)a(y,\lambda) = \int_{-1}^{1} \mathcal{F}(y,\tau;\lambda)a(\tau,\lambda)d\lambda, \tag{5.159}$$

where $a(y,\lambda)$ is any bounded function.

Lemma 5.11. *The Volterra integral Hilbert-Schmidt operators $K_j^+(x,\lambda)$ and $K_j^-(x,\lambda)$, $j=1,2$, are found uniquely from the left and right of factorization of the known transition operator $S(\lambda)$ for the transport equation* (5.105).

Proof. Let the transition operator $S(\lambda)$ for the transport equation (5.105) be known, then according to Theorem 5.3 this operator admits the right and left factorizations (5.153) and (5.154), wherein $K_j^+(x,\lambda)$ and $K_j^-(x,\lambda)$, $j=1,2$ are unknown operators. From (5.153) and (5.154) with due regard for (5.159), we have

$$\mathcal{F}(y,y_0;\lambda) + k_2^-(0,y,y_0;\lambda) + \int_{-1}^{y_0} \mathcal{F}(y,\tau;\lambda)k_2^-(0,\tau,y_0;\lambda)d\tau = k_2^+(0,y,y_0;\lambda), \tag{5.160}$$

$$\mathcal{F}(y,y_0;\lambda) + k_1^+(0,y,y_0;\lambda) + \int_{y_0}^{1} \mathcal{F}(y,\tau;\lambda)k_1^+(0,\tau,y_0;\lambda)d\tau = k_1^-(0,y,y_0;\lambda). \tag{5.161}$$

Since $k_2^+(0,y,y_0;\lambda)=0$ for $y<y_0$ and $k_1^-(0,y,y_0;\lambda)=0$ for $y>y_0$, then from (5.160) and (5.161) we obtain the equations connecting the known transition operator $S(\lambda)$ with the unknown operators $K_2^-(0,\lambda)$ and $K_1^+(0,\lambda)$:

$$\mathcal{F}(y,y_0;\lambda) + k_2^-(0,y,y_0;\lambda) + \int_{-1}^{y_0} \mathcal{F}(y,\tau;\lambda)k_2^-(0,\tau,y_0;\lambda)d\tau = 0,\ y<y_0, \tag{5.162}$$

$$\mathcal{F}(y,y_0;\lambda) + k_1^+(0,y,y_0;\lambda) + \int_{y_0}^{1} \mathcal{F}(y,\tau;\lambda)k_1^+(0,\tau,y_0;\lambda)d\tau = 0,\ y>y_0. \tag{5.163}$$

We write the equation (5.162) in the operator form:

$$(I + \tilde{\mathcal{F}}(\lambda)Q_{y_0})k_{y_0}(y) = -\mathcal{F}_{y_0}(y;\lambda),\ y < y_0, \tag{5.164}$$

where Q_{y_0} is the designing operator defined by:

$$Q_{y_0}a(y,\lambda) = \begin{cases} 0 & \text{for } y > y_0, \\ a(y,\lambda) & \text{for } y < y_0. \end{cases}$$

In equation (5.164) y is a variable, λ and y_0 are parameters, $k_{y_0}(y)$ is an unknown function and $\mathcal{F}_{y_0}(y,\lambda) = \mathcal{F}(y, y_0;\lambda)$ is the known function. We prove that the equation (5.164) is uniquely solvable in the space $L_2(-1, y_0)$. Indeed, according to the right Volterra factorization (5.153) of $S(\lambda)$, we can put: $I + R_2^-(0,\lambda) = (I + K_2^-(0,\lambda))^{-1}$. Then using (5.159), from (5.153) we have:

$$\tilde{\mathcal{F}}(\lambda) = K_2^+(0,\lambda) + R_2^-(0,\lambda) + K_2^+(0,\lambda)R_2^-(0,\lambda). \tag{5.165}$$

It follows from (5.165) that for any $y_0 \in (-1,1)$:

$$I + \tilde{\mathcal{F}}(\lambda)Q_{y_0} = I + K_2^+(0,\lambda)Q_{y_0} + R_2^-(0,\lambda)Q_{y_0} + K_2^+(0,\lambda)R_2^-(0,\lambda)Q_{y_0}. \tag{5.166}$$

Since $R_2^-(0,\lambda)Q_{y_0} = Q_{y_0}R_2^-(0,\lambda)Q_{y_0}$, then the equality (5.166) is rewritten in the form:

$$I + \tilde{\mathcal{F}}(\lambda)Q_{y_0} = (I + K_2^+(0,\lambda)Q_{y_0})(I + R_2^-(0,\lambda)Q_{y_0}).$$

From the existence of operators $(I + K_2^+(0,\lambda))^{-1}$ and $(I + R_2^-(0,\lambda))^{-1}$ it follows that the operators $(I + K_2^+(0,\lambda)Q_{y_0})^{-1}$ and $(I + R_2^-(0,\lambda)Q_{y_0})^{-1}$ also exist, therefore

$$(I + \tilde{\mathcal{F}}(\lambda)Q_{y_0})^{-1} = (I + R_2^-(0,\lambda)Q_{y_0})^{-1}(I + K_2^+(0,\lambda)Q_{y_0})^{-1}.$$

The existence of the operator $\left(I + \tilde{\mathcal{F}}(\lambda)Q_{y_0}\right)^{-1}$ is equivalent to the unique solvability of equation (5.164) in the space $L_2(-1, y_0)$. Hence, the right Volterra factorization (5.153) of $S(\lambda)$ leads to the unique solvability of (5.162) in $L_2(-1, y_0)$.

By an argument analogous to the previous one, using the left Volterra factorization (5.154) of $S(\lambda)$, we obtain the existence of the following operator:

$$(I + \tilde{\mathcal{F}}(\lambda)P_{y_0})^{-1} = (I + R_1^+(0,\lambda)P_{y_0})^{-1}(I + K_1^-(0,\lambda)P_{y_0})^{-1},$$

where P_{y_0} is the designing operator defined by

$$P_{y_0}a(y,\lambda) = \begin{cases} a(y,\lambda) & \text{for } y > y_0, \\ 0 & \text{for } y < y_0, \end{cases}$$

$I + R_1^+(0,\lambda) = (I + K_1^+(0,\lambda)^{-1}$ and $a(y,\lambda)$ is any bounded function.

The existence of the operator $(I+\tilde{\mathcal{F}}(\lambda)P_{y_0})^{-1}$ is equivalent to the unique solvability of equation (5.163). Hence, the left Volterra factorization (5.154) of $S(\lambda)$ leads to the unique solvability of equation (5.163) in the space $L_2(y_0,1)$.

Thus, if the transition operator $S(\lambda)$ for (5.105) is known, then the functions $k_2^-(0,y,y_0;\lambda)$, $y<y_0$ and $k_1^+(0,y,y_0;\lambda)$, $y>y_0$ are found uniquely from equations (5.162) and (5.163), respectively. Hence, the Volterra Hilbert-Schmidt operators $K_2^-(0,\lambda)$ and $K_1^+(0,\lambda)$ are found uniquely from the given operator $\tilde{\mathcal{F}}(\lambda)$. Knowing $k_2^-(0,y,y_0;\lambda)$ and $k_1^+(0,y,y_0;\lambda)$, from (5.160) and (5.161) we determine functions $k_2^+(0,y,y;\lambda)$ and $k_1^-(0,y,y_0;\lambda)$ by means of the equalities:

$$k_2^+(0,y,y_0;\lambda)=\mathcal{F}(y,y_0;\lambda)+\int_{-1}^{y_0}\mathcal{F}(y,\tau;\lambda)k_2^-(0,\tau,y_0;\lambda)d\tau,\quad y>y_0,$$

$$k_1^-(0,y,y_0;\lambda)=\mathcal{F}(y,y_0;\lambda)+\int_{y_0}^{1}\mathcal{F}(y,\tau;\lambda)k_1^+(0,\tau,y_0;\lambda)d\tau,\quad y<y_0.$$

Hence, the operators $K_j^+(x,\lambda)$ and $K_j^-(x,\lambda)$, $j=1,2$ are found uniquely from the known transition operator $S(\lambda)$. □

5.2.4 Analytic and bilateral Volterra factorizations of the intermediate operator $\tilde{S}(\lambda)$

Lemma 5.12. *For any $x\in(-1,1)$ and real λ there exist in the space $L_2(-1,1)$ the bounded operators $(I+V_+(x,\lambda))^{-1}$, $(I+V_-(x,\lambda))^{-1}$, where $V_\pm(x,\lambda)$ are the Hilbert-Schmidt operators defined by Lemma 5.9.*

Proof. We consider the operator equation (5.116). In this equation x and λ are fixed, y is the variable, the function $\varphi(x,y)$ is known, and $f_+(y,\lambda)$ is an unknown function. Since $V_+(x,\lambda)$ is a Hilbert–Schmidt operator, then $V_+(x,\lambda)$ is completely continuous in the space $L_2(-1,1)$. Therefore, to prove the unique solvability of equation (5.116), it is sufficient to show that the homogeneous equation $(I+V_+(x,\lambda))T_xf_+(y,\lambda)=0$ has only the trivial solution $f_+(y,\lambda)\equiv 0$.

Indeed, suppose that this is not like that, i.e., for some λ_0 and at some point x_0 there exists a nonzero function $f_+(y,\lambda_0)$ such that $(I+V_+(x_0,\lambda_0))T_{x_0}f_+(y,\lambda_0)=0$.

We set $\varphi(x,y,\lambda_0)=(I+V_+(x,\lambda_0))T_xf_+(y,\lambda_0)$. Then according to Lemma 5.9 the function $\varphi(x,y,\lambda_0)$ is a solution of the transport equation (5.105), satisfying the condition: $\varphi(x_0,y,\lambda_0)=0$. By the uniqueness of the solution of the Cauchy problem for the transport equation (5.105), we obtain that $\varphi(x,y,\lambda_0)\equiv 0$ for any x and $y\in(-1,1)$. Consequently, the incident wave $A(y,\lambda_0)$, generated by the solution $\varphi(x,y,\lambda_0)$, is also identically equal to zero. Due to this fact, from (5.149) we obtain the Volterra homogeneous equation for the function $f_+(y, \lambda_0)$, which, as shown in the proof of Lemma 5.10, has in $L_2(-1, 1)$ only the trivial solution $f_+(y, \lambda) \equiv 0$. Thus, the unique solvability of the operator equation (5.116) is proved, whence it follows that there exists the operator $(I + V_+(x, \lambda))^{-1}$. The second statement of the lemma is proved similarly. □

Theorem 5.4. *Let $\varphi(x, y)$ be a bounded solution of the transport equation* (5.105) *for a real parameter λ. Then, there exist unique bounded functions $f_+(y, \lambda)$, $f_-(y, \lambda)$ such that*

$$\varphi(x, y) = (I + V_+(x, \lambda))T_x f_+(y, \lambda), \tag{5.167}$$

$$\varphi(x, y) = (I + V_-(x, \lambda))T_x f_-(y, \lambda), \tag{5.168}$$

where the operators $V_\pm(x, \lambda)$ are defined by Lemma 5.9, and the operator T_x is defined by formula (5.115).

Conversely, for any bounded functions $f_+(y, \lambda)$, $f_-(y, \lambda)$, the formulas (5.167) *and* (5.168) *give bounded solutions of the transport equation* (5.105).

Proof. Let $f_+(y, \lambda)$, be any bounded function, then as shown by Lemma 5.9, the formula (5.167) yields a bounded solution of equation (5.105).

Now let $\varphi(x, y)$ be a bounded solution of (5.105). By virtue of Lemma 5.12, we take the function $f_+(y, \lambda) = (I + V_+(0, \lambda))^{-1}\varphi(0, y)$ and construct the function $\tilde{\varphi}(x, y) = (I + V_+(x, \lambda))T_x f_+(y, \lambda)$. By Lemma 5.9 the function, $\tilde{\varphi}(x, y)$ satisfies equation (5.105). The difference $u(x, y) = \varphi(x, y) - \tilde{\varphi}(x, y)$ is equal to zero at $x = 0$, then by virtue of the uniqueness of the solution of the Cauchy problem for equation (5.105), this difference is identically equal to zero, i.e., $\varphi(x, y) \equiv \tilde{\varphi}(x, y)$. This means that the bounded function $f_+(y, \lambda)$ is uniquely found from a given solution $\varphi(x, y)$ of equation (5.105). The second statement of the theorem is proved similarly. □

Now we consider the homogeneous operator equations:

$$(I + V_+(x, \lambda))T_x f_+(y, \lambda) = 0, \ \text{Im}\lambda > 0, \tag{5.169}$$

$$(I + V_-(x, \lambda))T_x f_-(y, \lambda) = 0, \ \text{Im}\lambda < 0. \tag{5.170}$$

Definition 5.3. *(1) A complex number λ_0, $Im\lambda_0 \neq 0$, for which the homogeneous equation* (5.169) *or* (5.170) *has a nontrivial solution, is called a singular number of equation* (5.105).

(2) A complex number λ_0, $Im\lambda_0 \neq 0$, is called generalized eigenvalue for equation (5.105), *if this equation has a nontrivial generalized solution $\varphi(x, y, \lambda_0)$ such that for almost all $y \in (-1, 1)$: $\varphi(x, y, \lambda_0)e^{i\lambda_0 y_1 x} \to 0$ for $x \to \pm\infty$ for a certain choice of a number y_1 : $-1 < y_1 < 1$.*

Lemma 5.13. *If a complex number λ_0,* $\text{Im}\lambda_0 \neq 0$, *is a singular number of equation* (5.105), *then it is a generalized eigenvalue for this equation.*

Proof. Let λ_0, $\mathrm{Im}\lambda_0 > 0$, be a singular number of equation (5.105), then the homogeneous operator equation (5.169) for $\lambda = \lambda_0$ has a nontrivial solution $f_+(y, \lambda_0)$. Then from (5.169) we have

$$\begin{aligned}
&\theta(y - y_1) f_+(y, \lambda_0) e^{-i\lambda_0 y x} + \\
&\int_{y_1}^{1} f_+(y_0, \lambda_0) e^{-i\lambda_0 y_0 x} dy_0 \int_0^{\infty} H_+(x, s, y, y_0) e^{i\lambda_0 |y - y_0| s} ds = \\
&- [1 - \theta(y - y_1)] f_+(y, \lambda_0) e^{-i\lambda_0 y x} - \\
&\int_{-1}^{y_1} f_+(y_0, \lambda_0) e^{-i\lambda_0 y_0 x} dy_0 \int_0^{\infty} H_+(x, s, y, y_0) e^{i\lambda_0 |y - y_0| s} ds, \qquad (5.171)
\end{aligned}$$

where $\theta(y)$ is the Heaviside function:

$$\theta(y) = \begin{cases} 1 \text{ for } y > 0 \\ 0 \text{ for } y \leq 0, \end{cases}$$

and y_1 is any number from the interval $(-1, 1)$.

Denoting the left and right sides of equality (5.171) by $\Phi_1(x, y, y_1)$ and $-\Phi_2(x, y, y_1)$, respectively, we put

$$\Phi(x, y, y_1) = \Phi_1(x, y, y_1) = -\Phi_2(x, y, y_1).$$

Then according to formulas (5.113), (5.121) and (5.143), we write the function $\Phi(x, y, y_1)$ in the form:

$$\Phi(x, y, y_1) = \int_{y_1}^{1} f_+(y_0, \lambda_0) \chi_1(x, y, y_0) dy_0 = -\int_{-1}^{y_1} f_+(y_0, \lambda_0) \chi_1(x, y, y_0) dy_0,$$

where $\chi_1(x, y, y_0)$ is the solution of equation (5.105).

Hence, the function $\Phi(x, y, y_1)$ is a bounded solution of equation (5.105). Furthermore, it follows from (5.171) that for $y \in (y_1, 1)$:

$$\Phi(x, y, y_1) e^{i\lambda_0 y_1 x} = \Phi_1(x, y, y_1) e^{i\lambda_0 y_1 x} \to 0 \text{ as } x \to -\infty,$$

and for $y \in (-1, y_1)$:

$$\Phi(x, y, y_1) e^{i\lambda_0 y_1 x} = -\Phi_2(x, y, y_1) e^{i\lambda_0 y_1 x} \to 0 \text{ as } x \to +\infty,$$

i.e., for almost all $y \in (-1, 1)$: $\Phi(x, y, y_1) e^{i\lambda_0 y_1 x} \to 0$ as $x \to \pm\infty$.

We show that for a certain choice of a number $y_1 \in (-1, 1)$ the solution $\Phi(x, y, y_1)$ of (5.105) is nontrivial. Suppose the contrary that $\Phi(x, y, y_1) \equiv 0$ for any number $y_1 \in (-1, 1)$. Then from (5.171) we have the homogeneous operator equation:

$$\Phi_{x,\lambda_0}(y, y_1) + B\Phi_{x,\lambda_0}(y, y_1) = 0, \; y_1 < y < 1, \; \mathrm{Im}\lambda_0 > 0. \qquad (5.172)$$

where $\Phi_{x,\lambda_0}(y,y_1)=\theta(y-y_1)f_+(y,\lambda_0)e^{-i\lambda_0 yx}$,

$$B\Phi_{x,\lambda_0}(y,y_1)=\int\limits_{y_1}^{1}\Phi_{x,\lambda_0}(y_0,y_1)dy_0\int\limits_{0}^{\infty}H_+(x,s;y,y_0)e^{i\lambda_0|y-y_0|s}ds. \quad (5.173)$$

In equation (5.172) y and y_1 are variables, x and λ_0 are fixed. By using the Volterra property of this equation with respect to the variable y_1, we prove that the homogeneous equation (5.172) has only the trivial solution. Indeed, putting:

$$||\Phi_{x,\lambda_0}||_{y_1}=\sup_{y_1<y<1}|f_+(y,\lambda_0)e^{-i\lambda_0 yx}|, \quad (5.174)$$

with due regard for the estimate (5.120) and the norm (5.174), from (5.173) we obtain the estimate for any $y_1\in(-1,1)$:

$$|B\Phi_{x,\lambda_0}(y,y_1)|\leq||\Phi_{x,\lambda_0}||_{y_1}\int\limits_{y_1}^{1}dy_0\int\limits_{0}^{\infty}|H_+(x,s;y,y_0)|ds\leq||\Phi_{x,\lambda_0}||_{y_1}\times$$

$$\left\{\tilde{C}\int\limits_{y}^{1}dy_0\int\limits_{0}^{\infty}\frac{(y_0-y)ds}{(1+(y_0-y)|x-s|)^{1+\varepsilon}}+\tilde{C}\int\limits_{y_1}^{y}dy_0\int\limits_{0}^{\infty}\frac{(y-y_0)ds}{(1+(y-y_0)|x+s|)^{1+\varepsilon}}\right\}$$

$$\leq C_1||\Phi_{x,\lambda_0}||_{y_1}\int\limits_{y_1}^{1}dy_0\int\limits_{-\infty}^{\infty}\frac{dp}{(1+|p|)^{1+\varepsilon}}=C_2||\Phi_{x,\lambda_0}||_{y_1}(1-y_1), \quad (5.175)$$

where $\tilde{C}$, C_1 and C_2 are constants.

Taking a number T such that $1>T>1-\frac{1}{2C_2}$, i.e., $1-T<\frac{1}{2C_2}$, then from (5.175) we have:

$$|B\Phi_{x,\lambda_0}(y,T)|\leq\frac{1}{2}||\Phi_{x,\lambda_0}||_T,$$

therefore $||B||_T\leq\frac{1}{2}$, but from (5.172) it follows that

$$||\Phi_{x,\lambda_0}||_T\leq||B||_T\cdot||\Phi_{x,\lambda_0}||_T\leq\frac{1}{2}||\Phi_{x,\lambda_0}||_T,\ \text{i.e.,}\ ||\Phi_{x,\lambda_0}||_T=0. \quad (5.176)$$

Owing to estimates (5.176) and (5.120), from (5.173) we have estimate:

$$|B\Phi_{x,\lambda_0}(y,y_1)|\leq||\Phi_{x,\lambda_0}||_{y_1}\int\limits_{y_1}^{T}dy_0\int\limits_{0}^{\infty}|H_+(x,s;y,y_0)|ds\leq$$

$$C_3||\Phi_{x,\lambda_0}||_{y_1}\cdot(T-y_1),\quad C_3\ \text{is a constant.}$$

Taking a number Δ such that $\Delta = T - y_1$ and $0 < \Delta < \frac{1}{2C_3}$, then from the last inequality we get

$$|B\Phi_{x,\lambda_0}(y, T - \Delta)| \leq C_3 ||\Phi_{x,\lambda_0}||_{T-\Delta} \cdot \Delta \leq \frac{1}{2} ||\Phi_{x,\lambda_0}||_{T-\Delta},$$

whence the norm $||B||_{T-\Delta}$ satisfies the estimate: $||B||_{T-\Delta} < \frac{1}{2}$.

Then from (5.172) it follows that

$$||\Phi_{x,\lambda_0}||_{T-\Delta} \leq ||B||_{T-\Delta} \cdot ||\Phi_{x,\lambda_0}||_{T-\Delta} < \frac{1}{2} ||\Phi_{x,\lambda_0}||_{T-\Delta}, \text{ i.e., } ||\Phi_{x,\lambda_0}||_{T-\Delta} = 0.$$

Continuing this process n times, we obtain that $||\Phi_{x,\lambda_0}||_{T-n\Delta} = 0$, i.e., $\Phi_{x,\lambda_0}(y, y_1) \equiv 0$, whence it follows from (5.174) that $f_+(y, \lambda_0) \equiv 0$, which contradicts the supposition that the equation (5.169) has the nontrivial solution $f_+(y, \lambda_0)$, $\text{Im}\lambda_0 > 0$. Hence, $\Phi(x, y, y_1)$ is the generalized eigenfunction, and λ_0, $\text{Im}\lambda_0 > 0$ is the corresponding generalized eigenvalue for equation (5.105).

Similarly, we prove that if λ_0, $\text{Im}\lambda_0 < 0$ is a singular number of (5.105), then it is a generalized eigenvalue for this equation. This completes the proof. □

Definition 5.4. *To each given "+" preimage $f_+(y, \lambda)$ for a real λ, according to Theorem 5.4, there corresponds a unique solution $\varphi(x, y)$ of transport equation* (5.105)*. According to this Theorem, to the found solution $\varphi(x, y)$ there corresponds a unique "−" preimage $f_-(y, \lambda)$. The one-to-one correspondence between "+" and "−" preimages $f_+(y, \lambda)$ and $f_-(y, \lambda)$ defines the operator function $\tilde{S}(\lambda)$ for the transport equation* (5.105)*:*

$$\tilde{S}(\lambda) f_+(y, \lambda) = f_-(y, \lambda). \tag{5.177}$$

$\tilde{S}(\lambda)$ is called the intermediate operator function and will be considered in the space $L_2(-1, 1)$.

Lemma 5.14. *The intermediate operator function $\tilde{S}(\lambda)$ for the one-dimensional transport equation* (5.105) *with a potential $c(x, y, \tau)$ satisfying condition* (5.106)*, admits the analytic factorization:*

$$\tilde{S}(\lambda) = T_{-x}(I + V_-(x, \lambda))^{-1}(I + V_+(x, \lambda))T_x, \tag{5.178}$$

and the bilateral factorization, i.e., the right and left Volterra factorizations

$$\tilde{S}(\lambda) = T_{-x}(I + K_2^+(x, \lambda))^{-1}(I + K_1^-(x, \lambda))T_x, \tag{5.179}$$

$$\tilde{S}(\lambda) = T_{-x}(I + K_2^-(x, \lambda))^{-1}(I + K_1^+(x, \lambda))T_x, \tag{5.180}$$

where the operator T_x is defined by (5.115)*, the operators $V_\pm(x, \lambda)$ and $K_j^\mp(x, \lambda)$, $j = 1, 2$ are defined by Lemmas 5.9 and 5.10. The kernels of the operators $V_+(x, \lambda)$ and $V_-(x, \lambda)$ admit an analytic continuation into the upper half-plane* $\text{Im}\lambda \geq 0$ *and the lower half-plane* $\text{Im}\lambda \leq 0$*, respectively.*

Proof. Comparing (5.167) with (5.168) and using definition (5.177), we have

$$(I + V_-(x,\lambda))T_x\tilde{S}f_+(y,\lambda) = (I + V_+(x,\lambda))T_x f_+(y,\lambda).$$

From the last equality, making use of Lemma 5.12 and (5.115), we obtain the analytic factorization (5.178) of $\tilde{S}(\lambda)$. Further, comparing (5.147) with (5.148) and (5.145) with (5.146), using (5.115) and (5.177), we get the right and left factorizations (5.179) and (5.180), respectively of $\tilde{S}(\lambda)$. □

The results obtained in this Section show that there exist the relations between the solution $\varphi(x,y)$ of the transport equation (5.105) and the incident $A(y,\lambda)$, scattered $B(y,\lambda)$ waves, and between $\varphi(x,y)$ and the preimages $f_+(y,\lambda)$ and $f_-(y,\lambda)$, and between $A(y,\lambda)$, $B(y,\lambda)$ and $f_+(y,\lambda)$, $f_-(y,\lambda)$. These relations can be described by the following compact schema.

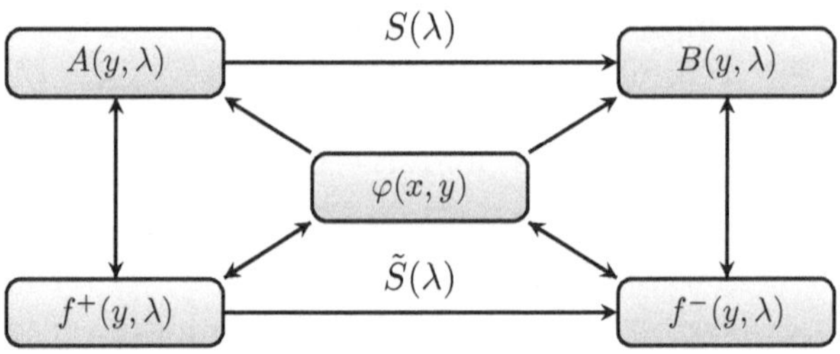

5.2.5 The inverse scattering problem

By Lemma 5.12 and estimates (5.120), we can set for real parameter λ :

$$I + R_+(x,\lambda) = (I + V_+(x,\lambda))^{-1}, \quad I + R_-(x,\lambda) = (I + V_-(x,\lambda))^{-1}, \quad (5.181)$$

where the operators $V_+(x,\lambda)$, $V_-(x,\lambda)$, $R_+(x,\lambda)$ and $R_-(x,\lambda)$, are defined by the formulas:

$$\begin{aligned} V_+(x,\lambda)a(y,\lambda) &= \int_{-1}^{1} a(y_0,\lambda)dy_0 \int_0^{\infty} H_+(x,s;y,y_0)e^{i\lambda|y-y_0|s}ds \\ &= \int_{-1}^{1} a(y_0,\lambda)dy_0 \frac{1}{|y-y_0|}\int_0^{\infty} H_+\left(x,\frac{\xi}{|y-y_0|};y,y_0\right)e^{i\lambda\xi}d\xi, \\ V_-(x,\lambda)a(y,\lambda) &= \int_{-1}^{1} a(y_0,\lambda)dy_0 \int_{-\infty}^{0} H_-(x,s;y,y_0)e^{i\lambda|y-y_0|s}ds \\ &= \int_{-1}^{1} a(y_0,\lambda)dy_0 \frac{1}{|y-y_0|}\int_{-\infty}^{0} H_-\left(x,\frac{\xi}{|y-y_0|};y,y_0\right)e^{i\lambda\xi}d\xi, \end{aligned} \quad (5.182)$$

$$
\begin{aligned}
R_+(x,\lambda)a(y,\lambda) &= \int_{-1}^{1} a(y_0,\lambda)dy_0 \int_{0}^{\infty} R_{+1}(x,\xi;y,y_0)e^{i\lambda\xi}d\xi,\\
R_-(x,\lambda)a(y,\lambda) &= \int_{-1}^{1} a(y_0,\lambda)dy_0 \int_{-\infty}^{0} R_{-1}(x,\xi;y,y_0)e^{i\lambda\xi}d\xi,
\end{aligned} \tag{5.183}
$$

where $a(y,\lambda)$ is any bounded function, the functions $R_{\pm1}(x,\xi;y,y_0)$ are absolutely integrable with respect to ξ, and the integrals:

$$
\int_{0}^{\infty} R_{+1}(x,\xi;y,y_0)e^{i\lambda\xi}d\xi \quad \text{and} \quad \int_{-\infty}^{0} R_{-1}(x,\xi;y,y_0)e^{i\lambda\xi}d\xi,
$$

are bounded functions of y_0. They are the kernels of the Hilbert-Schmidt operators $R_+(x,\lambda)$ and $R_-(x,\lambda)$, respectively.

We introduce the intermediate operators $F(x,\lambda)$ and $G(x,\lambda)$ for the transport equation (5.105) by the equalities:

$$
F(x,\lambda) = T_x\tilde{S}(\lambda)T_{-x} - I, \quad G(x,\lambda) = T_x\tilde{S}^{-1}(\lambda)T_{-x} - I. \tag{5.184}
$$

Lemma 5.15. *The intermediate operators $F(x,\lambda)$ and $G(x,\lambda)$ for equation* (5.105) *with a potential satisfying condition* (5.106) *are defined by the equalities:*

$$
\begin{aligned}
F(x,\lambda)a(y,\lambda) &= \int_{-1}^{1} a(y_0,\lambda)dy_0 \int_{-\infty}^{\infty} f(x,\xi;y,y_0)e^{i\lambda\xi}d\xi,\\
G(x,\lambda)a(y,\lambda) &= \int_{-1}^{1} a(y_0,\lambda)dy_0 \int_{-\infty}^{\infty} g(x,\xi;y,y_0)e^{i\lambda\xi}d\xi,
\end{aligned}
$$

where $a(y,\lambda)$ is any bounded function, the functions $f(x,\xi;y,y_0)$ and $g(x,\xi;y,y_0)$ are absolutely integrable with respect to ξ and are represented in the form:

$$
\begin{aligned}
|y-y_0|f(x,\xi;y,y_0) &= H_+\left(x,\frac{\xi}{|y-y_0|};y,y_0\right) + |y-y_0|R_{-1}(x,\xi;y,y_0)\\
&+ |y-y_0| \times \int_{-1}^{1} d\tau \int_{-\infty}^{\min\left(0,\frac{\xi}{|\tau-y_0|}\right)} R_{-1}(x,|\tau-y_0|p;y,\tau)H_+\left(x,\frac{\xi}{|\tau-y_0|}-p;\tau,y_0\right)dp,
\end{aligned} \tag{5.185}
$$

$$|y-y_0|g(x,\xi;y,y_0) = H_-\left(x,\frac{\xi}{|y-y_0|};y,y_0\right) + |y-y_0|R_{+1}(x,\xi;y,y_0)$$

$$+|y-y_0|\times\int_{-1}^{1}d\tau\int_{\max\left(0,\frac{\xi}{|\tau-y_0|}\right)}^{\infty}R_{+1}(x,|\tau-y_0|p;y,\tau)H_-\left(x,\frac{\xi}{|\tau-y_0|}-p;\tau,y_0\right)dp. \tag{5.186}$$

Proof. From factorization (5.178) with due regard for (5.181)–(5.183) we get the intermediate operators $F(x,\lambda)$ and $G(x,\lambda)$:

$$F(x,\lambda) = R_-(x,\lambda) + V_+(x,\lambda) + R_-(x,\lambda)V_+(x,\lambda), \tag{5.187}$$
$$G(x,\lambda) = R_+(x,\lambda) + V_-(x,\lambda) + R_+(x,\lambda)V_-(x,\lambda). \tag{5.188}$$

Using formulas (5.182) and (5.183), we have:

$$R_-(x,\lambda)V_+(x,\lambda)a(y,\lambda) =$$
$$\int_{-1}^{1}d\tau\int_{-\infty}^{0}R_{-1}(x,\xi;y,\tau)e^{i\lambda\xi}d\xi\int_{-1}^{1}a(y_0,\lambda)dy_0\int_{0}^{\infty}H_+(x,s_1;\tau,y_0)e^{i\lambda|\tau-y_0|s_1}ds_1$$
$$=\int_{-1}^{1}a(y_0,\lambda)dy_0\int_{-\infty}^{\infty}e^{i\lambda s}ds\int_{-1}^{1}d\tau\int_{-\infty}^{\min\left(0,\frac{s}{|\tau-y_0|}\right)}R_{-1}(x,|\tau-y_0|p;y,\tau)\times$$
$$H_+\left(x,\frac{s}{|\tau-y_0|}-p;\tau,y_0\right)dp,$$
$$R_+(x,\lambda)V_-(x,\lambda)a(y,\lambda) =$$
$$\int_{-1}^{1}d\tau\int_{0}^{\infty}R_{+1}(x,\xi;y,\tau)e^{i\lambda\xi}d\xi\int_{-1}^{1}a(y_0,\lambda)dy_0\int_{-\infty}^{0}H_-(x,s;\tau,y_0)e^{i\lambda|\tau-y_0|s}ds$$

$$=\int_{-1}^{1}a(y_0,\lambda)dy_0\int_{-\infty}^{\infty}e^{i\lambda s}ds\int_{-1}^{1}d\tau\int_{\max\left(0,\frac{s}{|\tau-y_0|}\right)}^{\infty}R_{+1}(x,|\tau-y_0|p;y,\tau)\times$$
$$H_-\left(x,\frac{s}{|\tau-y_0|}-p;\tau,y_0\right)dp. \tag{5.189}$$

Owing to arbitrariness of a bounded function $a(y,\lambda)$ and to uniqueness of the Fourier transform, using (5.182)–(5.183) and (5.189), from (5.187) and (5.188) we get the equalities (5.185) and (5.186), respectively. The functions in the right-hand side of these equalities are absolutely integrable with respect to ξ, therefore the functions $f(x,\xi;y,y_0)$ and $g(x,\xi;y,y_0)$ are absolutely integrable with respect to ξ. The lemma is proved. □

We now derive the fundamental equations connecting the functions $H_{\pm}(x, s; y, y_0)$ with the intermediate operator $\tilde{S}(\lambda)$. Using the definition of the intermediate operators (5.184), from factorization (5.178) we have

$$V_-(x,\lambda) + F(x,\lambda) + V_-(x,\lambda)F(x,\lambda) = V_+(x,\lambda), \tag{5.190}$$

$$V_+(x,\lambda) + G(x,\lambda) + V_+(x,\lambda)G(x,\lambda) = V_-(x,\lambda). \tag{5.191}$$

According to the definitions of the operators $V_-(x,\lambda), V_+(x,\lambda), F(x,\lambda)$ and $G(x,\lambda)$, we obtain the equalities:

$$V_-(x,\lambda)F(x,\lambda)a(y,\lambda) =$$

$$\int_{-1}^{1} d\tau \int_{-\infty}^{0} H_-(x,\xi_1;y,\tau)e^{i\lambda|y-\tau|\xi_1}d\xi_1 \int_{-1}^{1} a(y_0,\lambda)dy_0 \int_{-\infty}^{\infty} f(x,\xi;\tau,y_0)e^{i\lambda\xi}d\xi =$$

$$\int_{-1}^{1} a(y_0,\lambda)dy_0 \int_{-\infty}^{\infty} e^{i\lambda s}ds \int_{-1}^{1} \frac{d\tau}{|y-\tau|} \int_{s}^{\infty} H_-\left(x, \frac{s-\xi}{|y-\tau|}; y,\tau\right) f(x,\xi;\tau,y_0)d\xi,$$

$$V_+(x,\lambda)G(x,\lambda)a(y,\lambda) =$$

$$\int_{-1}^{1} a(y_0,\lambda)dy_0 \int_{-\infty}^{\infty} e^{i\lambda s}ds \int_{-1}^{1} \frac{d\tau}{|y-\tau|} \int_{-\infty}^{s} H_+\left(x, \frac{s-\xi}{|y-\tau|}; y,\tau\right) g(x,\xi;\tau,y_0)d\xi.$$

Making use of these equalities, owing to arbitrariness of a bounded function $a(y,\lambda)$ and to uniqueness of the Fourier transform, from (5.190) and (5.191) we have

$$H_+\left(x, \frac{s}{|y-y_0|}; y, y_0\right) = H_-\left(x, \frac{s}{|y-y_0|}; y, y_0\right) + |y-y_0| f(x,s;y,y_0)$$

$$+ |y-y_0| \int_{-1}^{1} \frac{d\tau}{|y-\tau|} \int_{s}^{\infty} H_-\left(x, \frac{s-\xi}{|y-\tau|}; y,\tau\right) f(x,\xi;\tau,y_0)d\xi, \tag{5.192}$$

$$H_-\left(x, \frac{s}{|y-y_0|}; y, y_0\right) = H_+\left(x, \frac{s}{|y-y_0|}; y, y_0\right) + |y-y_0| g(x,s;y,y_0)$$

$$+ |y-y_0| \int_{-1}^{1} \frac{d\tau}{|y-\tau|} \int_{-\infty}^{s} H_+\left(x, \frac{s-\xi}{|y-\tau|}; y,\tau\right) g(x,\xi;\tau,y_0)d\xi. \tag{5.193}$$

Taking into account that $H_+\left(x, \frac{s}{|y-y_0|}; y, y_0\right) = 0$ for $s < 0$ and $H_-\left(x, \frac{s}{|y-y_0|}; y, y_0\right) = 0$ for $s > 0$, from (5.192) and (5.193) we derive fundamental equations, which connect functions $H_{\pm}(x, s; y, y_0)$ with the intermediate operator $\tilde{S}(\lambda)$:

$$H_-\left(x, \frac{s}{|y-y_0|}; y, y_0\right) + |y-y_0| f(x,s;y,y_0)$$
$$+ |y-y_0| \int\limits_{-1}^{1} \frac{d\tau}{|y-\tau|} \int\limits_{s}^{\infty} H_-\left(x, \frac{s-\xi}{|y-\tau|}; y, \tau\right) f(x,\xi;\tau,y_0) d\xi = 0, \quad s<0, \tag{5.194}$$

$$H_+\left(x, \frac{s}{|y-y_0|}; y, y_0\right) + |y-y_0| g(x,s;y,y_0)$$
$$+ |y-y_0| \int\limits_{-1}^{1} \frac{d\tau}{|y-\tau|} \int\limits_{-\infty}^{s} H_+\left(x, \frac{s-\xi}{|y-\tau|}; y, \tau\right) g(x,\xi;\tau,y_0) d\xi = 0, \quad s>0. \tag{5.195}$$

Lemma 5.16. *Suppose that the transport equation* (5.105) *has no generalized eigenvalues. Then, for each fixed* x *the fundamental equations* (5.194) *and* (5.195) *have a unique solution in the space* $L_1(-\infty,\infty;-1,1)$.

Proof. By using Lemma 5.15 and the Fredholm's Theorems [101, 116], we can verify that equations (5.194) and (5.195) are Fredholm integral equations of the second kind in the space $L_1(-\infty,\infty;-1,1)$. Therefore, the existence of solution of equations (5.194) and (5.195) is deduced from the uniqueness of the solution of these equations. Hence, it is sufficient to show that the homogeneous equations:

$$h_-\left(\frac{s}{|y-y_0|}; y, y_0\right) +$$
$$|y-y_0| \int\limits_{-1}^{1} \frac{d\tau}{|\tau-y|} \int\limits_{s}^{\infty} h_-\left(\frac{s-\xi}{|y-\tau|}; y, \tau\right) f(\xi,\tau,y_0) d\xi = 0, \quad s<0, \tag{5.196}$$

$$h_+\left(\frac{s}{|y-y_0|}; y, y_0\right) +$$
$$|y-y_0| \int\limits_{-1}^{1} \frac{d\tau}{|y-\tau|} \int\limits_{-\infty}^{s} h_+\left(\frac{s-\xi}{|y-\tau|}; y, \tau\right) g(\xi,\tau,y_0) d\xi = 0, \quad s>0, \tag{5.197}$$

where $f(\xi;\tau,y_0) = f(0,\xi,\tau,y_0)$ and $g(\xi;\tau,y_0) = g(0,\xi;\tau,y_0)$, have in $L_1(-\infty,\infty;-1,1)$ only the trivial solution $h_\pm(s;y,y_0) \equiv 0$.

From (5.196) and (5.197) it follows that for $y=y_0$, $h_-(-\infty;y_0,y_0) = h_+(\infty;y_0,y_0) \equiv 0$, i.e., the equations (5.196) and (5.197) for $y=y_0$ have only the trivial solution. We now consider equations (5.196) and (5.197) for $y \neq y_0$, and rewrite them in the form:

$$h_-\left(\frac{s}{|y-y_0|}y, y_0\right)+$$

$$|y-y_0|\int_{-1}^{1}\frac{d\tau}{|\tau-y|}\int_{-\infty}^{0}h_-\left(\frac{z}{|y-\tau|}y,\tau\right)f(s-z,\tau,y_0)dz=0,\ \text{for } y\neq y_0, s<0, \tag{5.198}$$

$$h_+\left(\frac{s}{|y-y_0|}y, y_0\right)+$$

$$|y-y_0|\int_{-1}^{1}\frac{d\tau}{|y-\tau|}\int_{0}^{\infty}h_+\left(\frac{z}{|y-\tau|}y,\tau\right)g(s-z;\tau,y_0)dz=0,\ \text{for } y\neq y_0, s>0. \tag{5.199}$$

In equation (5.199), instead of $g(s-z;\tau,y_0)$, substituting the expression (5.186) for $x=0$ and $\xi=s-z$, yields for $s>0$ and $y\neq y_0$:

$$h_+\left(\frac{s}{|y-y_0|};y,y_0\right)+$$

$$|y-y_0|\int_{-1}^{1}\frac{1}{|y-\tau|}d\tau\int_{0}^{s}h_+\left(\frac{z}{|y-\tau|};y,\tau\right)R_{+1}(s-z;\tau,y_0)dz+$$

$$|y-y_0|\int_{-1}^{1}\frac{1}{|y-\tau|}d\tau\int_{s}^{\infty}h_+\left(\frac{z}{|y-\tau|};y,\tau\right)\frac{1}{|\tau-y_0|}.H_-\left(\frac{s-z}{|\tau-y_0|};\tau,y_0\right)dz+$$

$$|y-y_0|\int_{-1}^{1}\frac{1}{|y-\tau|}d\tau\int_{0}^{\infty}h_+\left(\frac{z}{|y-\tau|};y,\tau\right)dz\times$$

$$\int_{-1}^{1}d\tau_1\int_{\max\left(0,\frac{s-z}{\tau_1-y_0}\right)}^{\infty}R_{+1}\left(|\tau_1-y_0|p_1;\tau,\tau_1\right)H_-\left(\frac{s-z}{|\tau_1-y_0|}-p_1;\tau_1,y_0\right)dp_1=0,$$

where $H_\pm(0,s;y,y_0)=H_\pm(s;y,y_0)$, $R_\pm(0,s;y,y_0)=R_\pm(s;y,y_0)$.

Upon changing the variables and the order of integration in the double and fourfold integrals of the last equality, we obtain for $y \neq y_0$ and $s > 0$:

$$|y-y_0|\left\{\frac{1}{|y-y_0|}h_+\left(\frac{s}{|y-y_0|};y,y_0\right)+\right.$$
$$\left.\int_{-1}^{1}\frac{1}{|y-\tau|}d\tau\int_0^s h_+\left(\frac{z}{|y-\tau|};y,\tau\right)R_{+1}(s-z;\tau,y_0)dz\right\}+$$
$$|y-y_0|\int_{-1}^{1}\frac{1}{|y-\tau|}d\tau\int_s^\infty h_+\left(\frac{p}{|y-\tau|};y,\tau\right)\frac{1}{|y_0-\tau|}H_-\left(\frac{s-p}{|y_0-\tau|};\tau,y_0\right)dp+$$
$$|y-y_0|\int_{-1}^{1}\frac{1}{|y_0-\tau|}d\tau\int_s^\infty H_-\left(\frac{s-p}{|y_0-\tau|};\tau,y_0\right)dp\int_{-1}^{1}\frac{1}{|y-\tau_1|}d\tau_1\times$$
$$\int_0^p h_+\left(\frac{z}{|y-\tau_1|};y,\tau_1\right)R_{+1}(p-z;\tau_1,\tau)dz=0. \tag{5.200}$$

The equality (5.200) can be written in the form:

$$Q_+(s;y,y_0)+\int_{-1}^{1}\frac{1}{|y_0-\tau|}d\tau\int_s^\infty Q_+(p;y,\tau)H_-\left(\frac{s-p}{|y_0-\tau|};\tau,y_0\right)dp=0,$$
$$y\neq y_0,\quad s>0, \tag{5.201}$$

where

$$Q_+(s;y,y_0)=\frac{1}{|y-y_0|}h_+\left(\frac{s}{|y-y_0|};y,y_0\right)+$$
$$\int_{-1}^{1}\frac{1}{|y-\tau|}d\tau\int_0^s h_+\left(\frac{z}{|y-\tau|};y,\tau\right)R_{+1}(s-z;\tau,y_0)dz,\ y\neq y_0,\ s>0. \tag{5.202}$$

If we set $Q_+(s,y,y_0)=0$ for $s<0$, $y\neq y_0$, and

$$U_-(s;y,y_0)=\begin{cases}\displaystyle\int_{-1}^{1}\frac{1}{|y_0-\tau|}d\tau\int_s^\infty Q_+(p;y,\tau)H_-\left(\frac{s-p}{|y_0-\tau|};\tau,y_0\right)dp\\ \quad\text{for } s<0 \text{ and } y\neq y_0,\\ 0\quad\text{for } s>0 \text{ and } y\neq y_0,\end{cases}$$

then from (5.201) we have

$$Q_+(s;y,y_0) + \int_{-1}^{1} \frac{1}{|y_0-\tau|} d\tau \int_{s}^{\infty} Q_+(p;y,\tau) H_-\left(\frac{s-p}{|y_0-\tau|};\tau,y_0\right) dp$$
$$= U_-(s;y,y_0) \text{ for } -\infty < s < \infty \text{ and } y \neq y_0. \quad (5.203)$$

We write the equality (5.203) in the operator form:

$$\tilde{Q}_+(\lambda)\,(I + V_-(0,\lambda)) = \tilde{U}_-(\lambda), \quad (5.204)$$

where

$$\tilde{Q}_+(\lambda)a(y,\lambda) = \int_{-1}^{1} a(y_0,\lambda)dy_0 \int_{0}^{\infty} Q_+(s;y,y_0)e^{i\lambda s}ds,$$

$$V_-(0,\lambda)a(y,\lambda) = \int_{-1}^{1} a(y_0,\lambda)dy_0 \int_{-\infty}^{0} H_-(s;y,y_0)e^{i\lambda|y_0-y|s}ds,$$

$$\tilde{U}_-(\lambda)a(y,\lambda) = \int_{-1}^{1} a(y_0,\lambda)dy_0 \int_{-\infty}^{0} U_-(s;y,y_0)e^{i\lambda s}ds,$$

and $a(y,\lambda)$ is any bounded function.

If the equation (5.105) has no generalized eigenvalues, then according to Lemma 5.13, it has no singular numbers. In view of this fact and Lemma 5.12 there exists the bounded operator $(I + V_-(0,\lambda))^{-1}$, $\mathrm{Im}\lambda \leq 0$. Using this fact and equality (5.204), we set $N(\lambda) = \tilde{Q}_+(\lambda) = \tilde{U}_-(\lambda)(I + V_-(0,\lambda))^{-1}$. Then the operator function $N(\lambda)$ admits an analytic continuation into the entire plane and is uniformly bounded with respect to λ. Therefore, by Liouville's Theorem [114], $N(\lambda) = constant$, but for any bounded function $a(y,\lambda)$:

$$\lim_{\mathrm{Im}\lambda\to\infty} \tilde{Q}_+(\lambda)a(y,\lambda) = \lim_{\mathrm{Im}\lambda\to\infty} \int_{-1}^{1} a(y_0,\lambda)dy_0 \int_{0}^{\infty} Q_+(s;y,y_0)e^{i\lambda s}ds = 0,$$

consequently, $N(\lambda) = \tilde{Q}_+(\lambda) \equiv 0$. Owing to the arbitrariness of $a(y,\lambda)$ and uniqueness of the Fourier transform, we obtain that $Q_+(s;y,y_0) \equiv 0$, $s > 0$, $y \neq y_0$. Using this fact, from (5.202), we have a homogeneous Volterra equation for $h_+\left(\frac{s}{|y-y_0|};y,y_0\right)$, which, as it is known, has in $L_1(0,\infty;-1,1)$ only the zero solution $h_+(s;y,y_0) \equiv 0$ for $y \neq y_0$ and $s > 0$.

Using expression (5.185) and the equation (5.194), in a similar manner one proves that the equation (5.198) has in $L_1(-\infty,0;-1,1)$ only the zero solution $h_-(s;y,y_0) \equiv 0$ for $y \neq y_0$ and $s < 0$. This completes the proof. □

We turn to solving the ISP, which consists of the reconstruction of the potential $c(x, y, y_0)$ from the known transition operator $S(\lambda)$. Suppose that in the space $L_2(-1, 1)$ there is a known transition operator $S(\lambda)$ for equation (5.105), which has no generalized eigenvalues. According to Lemma 5.11, we find in a unique manner the Volterra integral Hilbert-Schmidt operators $K_1^-(x, \lambda)$ and $K_2^+(x, \lambda)$ from the known transition operator $S(\lambda)$. According to the factorization (5.179), we construct the intermediate operator $T_x\tilde{S}(\lambda)T_{-x}$. Thus, the intermediate operators $F(x, \lambda) = T_x\tilde{S}(\lambda)T_{-x} - I$ and $G(x, \lambda) = T_x\tilde{S}^{-1}(\lambda)T_{-x} - I$ are found. Knowing these intermediate operators, we consider the fundamental equations (5.194) and (5.195). According to Lemma 5.16, these equations have a unique solution $H_-\left(x, \frac{s}{|y-y_0|}; y, y_0\right)$, $s < 0$; $H_+\left(x, \frac{s}{|y-y_0|}; y, y_0\right)$, $s > 0$, $y \neq y_0$, therefore the potential $c(x, y, y_0)$ in equation (5.105) is uniquely recovered by means of formulas (5.118) and (5.119). The obtained results of the inverse problem are stated in the following theorem.

Theorem 5.5. *Suppose that we are given a transition operator $S(\lambda)$ for the one-dimensional transport equation* (5.105)*, which has no generalized eigenvalues. Then, the functions $f(x, s; y, y_0)$ and $g(x, s; y, y_0)$ in the fundamental equations* (5.194) *and* (5.195) *are constructed uniquely from the intermediate operator $\tilde{S}(\lambda)$ which is calculated by Lemma 5.11 from the known transition operator $S(\lambda)$. By Lemma 5.16 for fixed x and $y \neq y_0$, the fundamental equations* (5.194) *and* (5.195) *have a unique solution $H_-\left(x, \frac{s}{|y-y_0|}; y, y_0\right)$, $s < 0$ and $H_+\left(x, \frac{s}{|y-y_0|}; y, y_0\right)$, $s > 0$, in the space $L_1(-\infty, 0; -1, 1)$ and in $L_1(0, \infty; -1, 1)$, respectively. The potential $c(x, y, y_0)$ in the transport equation* (5.105) *is uniquely reconstructed from the found solutions of equations* (5.194) *and* (5.195) *by means of the formulas* (5.118) *and* (5.119).

Remark 5.1. (1) The functions f and g in systems of fundamental equations (1.42)–(1.43), (1.44)–(1.45) in Chapter 1 and (3.38), (3.39) in Chapter 3 are constructed from the known scattering function of the respective scattering problem. The Jost solutions of the system of n first-order ODEs (5.5) and of the transport equation (5.105) do not admit any analytic continuation from the real axis of the parameter λ. This fact leads to the situation that the functions f and g in systems of fundamental equations (5.93), (5.94) and (5.194), (5.195) are constructed from the intermediate operators, which are constructed from the known transition operators for system (5.5) and for the transport equation (5.105), respectively.

(2) The results of the direct and ISP for the transport equation (5.105) are generalized from the corresponding results of the direct and ISP for the system of n first-order ODEs (5.5). The deep analogy between results obtained in Sections 5.1 and 5.2 of this chapter can be listed in the table of comparison:

The system (5.5): $-i\frac{d}{dx}u(x,\lambda)+C(x)u(x,\lambda)=\lambda J_n u(x,\lambda)$, $J_n=\operatorname{diag}(a_1,...,a_n), a_1>...>a_n$, $-\infty<x<\infty$, $C(x)=\left[c_{jk}(x)\right], c_{jj}(x)=0, j,k=1,2,...,n$, $\lvert c_{jk}(x)\rvert \le \frac{\tilde{C}}{(1+\lvert x\rvert)^{2(1+\varepsilon)}}$.	The transport equation (5.105): $i\frac{\partial}{\partial x}\varphi(x,y)+\int\limits_{-1}^{1}c(x,y,\tau)\varphi(x,\tau)d\tau=$ $\lambda y\varphi(x,y)$, $-\infty<x<\infty; y,\tau\in(-1,1)$, $\lvert c(x,y,\tau)\rvert \le \frac{\tilde{C}\lvert y-\tau\rvert}{(1+\lvert x\rvert)^{2(1+\varepsilon)}}$.
The representations of solutions (5.44) of system (5.5) in the form of the transformation operators. The transition matrix $S(\lambda)$ for system (5.5) admits the right and left factorizations (5.68) by the triangular matrix multipliers.	The representations of solutions (5.116), (5.117) of equation (5.105) in the form of the Hilbert-Schmidt operators. The transition operator $S(\lambda)$ for equation (5.105) admits the right and left Volterra factorizations (5.153) and (5.154) by the Volterra integral Hilbert–Schmidt operators.
The intermediate matrix $\tilde{S}(\lambda)$ admits the analytic factorization (5.51) and the right and left factorizations (5.74) by the triangular matrix multipliers.	The intermediate operator $\tilde{S}(\lambda)$ admits the analytic factorization (5.178) and the right and left Volterra factorizations (5.179), (5.180) by Volterra integral Hilbert-Schmidt operators with variable lower and upper limits.
The relation between the transition and intermediate matrices. The relation between singular number and the generalized eigenvalue of system (5.5).	The relation between the transition and intermediate operators. The relation between singular number and the generalized eigenvalue of equation (5.105).
The triangular matrix multipliers of the bilateral factorization of the transition matrix $S(\lambda)$ are found uniquely from the known elements of $S(\lambda)$. $\tilde{S}(\lambda)$ is constructed from the found multipliers.	The Volterra integral Hilbert-Schmidt operators with variable lower and upper limits are found uniquely from the bilateral factorization of the known transition operator $S(\lambda)$. $\tilde{S}(\lambda)$ is constructed from the found operators.
Theorem 5.2 for the reconstruction of the matrix potential $C(x)$ in terms of the solution of the fundamental equations in the inverse problem.	Theorem 5.5 for the reconstruction of the potential $c(x,y,\tau)$ in terms of the solution of the fundamental equations in the inverse problem.

5.3 Integration of the continual system of non-linear interaction waves

5.3.1 The generalized Lax equation

Let the potential in the one-dimensional transport equation (5.105) depend on additional parameter t (time). We consider the time-dependent transport equation (5.105) in the operator form:

$$i\frac{\partial\varphi}{\partial x}+U\varphi=\lambda J\varphi, \tag{5.205}$$

where λ is the spectral parameter,

$$U\varphi(x,y;t)=\int\limits_{-1}^{1}u(x,y,\tau;t)\varphi(x,\tau;t)d\tau,\ J\varphi(x,y;t)=y\varphi(x,y;t),$$
$$-\infty<x<\infty,\ -1<y<1. \tag{5.206}$$

The time-dependence of the solution $\varphi(x, y; t)$ of equation (5.205) is given by the operator equation:

$$i\frac{\partial\varphi}{\partial t} + V\varphi = \lambda T\varphi, \tag{5.207}$$

where

$$V\varphi(x, y; t) = \int_{-1}^{1} v(x, y, \tau; t)\varphi(x, \tau; t)d\tau, \; T\varphi(x, y; t) = \alpha(y)\varphi(x, y; t), \tag{5.208}$$

wherein $\alpha(y)$ is any continuous function and $\alpha(y) \neq 0$ for $y \in (-1, 1)$.

The potentials $u(x, y, \tau; t)$ and $v(x, y, \tau; t)$ in (5.206) and (5.208) are complex-valued continuous functions possessing the continuous partial derivatives with respect to t and x, and satisfying the estimate for any $t \in (-\infty, \infty)$:

$$|u(x, y, \tau; t)| \leq \tilde{C}\frac{|y-\tau|}{(1+|x|)^{2(1+\varepsilon)}}, \; |v(x, y, \tau; t)| \leq \tilde{C}\frac{|\alpha(y)-\alpha(\tau)|}{(1+|x|)^{2(1+\varepsilon)}}, \tag{5.209}$$

where $\tilde{C}$ is a constant, $\varepsilon > 0$.

The pair of equations (5.205) and (5.207) is related to the Lax equation [44, 156]:

$$\frac{\partial L}{\partial t} = [L, B], \tag{5.210}$$

where

$$L = J^{-1}(i\frac{\partial}{\partial x} + U), \quad B = T^{-1}(i\frac{\partial}{\partial x} + V). \tag{5.211}$$

The Lax equation (5.210) is deduced from the generalized Lax equation [94]:

$$[P, L] = PL - LP = 0, \tag{5.212}$$

where $P = \frac{\partial}{\partial t} - B$.

In fact,

$$PL - LP = \left(\frac{\partial}{\partial t} - B\right)L - L\left(\frac{\partial}{\partial t} - B\right) =$$
$$\frac{\partial}{\partial t}.L - L.\frac{\partial}{\partial t} + LB - BL = -\frac{\partial L}{\partial t} + [L, B] = 0.$$

Here the operator L is determined by the transport equation (5.205). The equation (5.207) governs the time-evolution of eigenfunctions of the operator L.

The ISP for the equation (5.205) has been studied in Section 5.2. By using the results of this ISP, we can prove that the generalized Lax equation (5.212) enables us to apply the ISM to integration of the continual system of non-linear interaction waves:

$$\frac{\partial q(x,y,\tau;t)}{\partial t} + w(y,\tau)\frac{\partial q(x,y,\tau;t)}{\partial x} + \\ i\int_{-1}^{1} \beta(y,\tau,\tau_1)q(x,y,\tau_1;t)q(x,\tau_1,\tau;t)d\tau_1 = 0, \tag{5.213}$$

where $x \in (-\infty, \infty)$; $y, \tau \in (-1,1)$, $w(y,\tau) = \frac{\alpha(\tau)-\alpha(y)}{y-\tau}$, $\beta(y,\tau,\tau_1) = \alpha(\tau_1) - \alpha(\tau) + \frac{(\tau-\tau_1)}{y-\tau}(\alpha(y)-\alpha(\tau))$, $\alpha(y)$ is any continuous function, $\alpha(y) \neq 0$ for any $y \in (-1,1)$.

The derivation of system (5.213) will be presented below in the proof of the Lemma 5.17.

To integrate the continual system (5.213), we need to derive the time-evolution of the transition operator $S(\lambda;t)$ instead of the time-evolution of the operator L. This assertion will be proved by the following Lemmas 5.17 and 5.18 and Theorem 5.6.

Lemma 5.17. *The operator equations* (5.205) *and* (5.207) *are compatible, i.e.,* $\varphi_{xt} = \varphi_{tx}$*, in the class of isospectral potentials, which are continuous functions possessing the first-order partial continuous derivatives with respect to* t *and to* x*, then the potentials* $u(x,y,\tau;t)$ *and* $v(x,y,\tau;t)$ *in these equations vary with time as the solution of the continual system of non-linear interaction waves* (5.213)*. Conversely, if* $u(x,y,\tau;t)$ *and* $v(x,y,\tau;t)$ *satisfy the continual system* (5.213)*, then* $\lambda_t = 0$ *and the linear eigenvalue equations* (5.205) *and* (5.207) *are compatible.*

Proof. Let $\lambda_t = 0$, then differentiating (5.205) with respect to t and (5.207) with respect to x gives

$$\varphi_{xt} = iU_t\varphi + (U - \lambda J)(\lambda T - V)\varphi, \tag{5.214}$$

$$\varphi_{tx} = iV_x\varphi + (V - \lambda T)(\lambda J - U)\varphi. \tag{5.215}$$

By the supposition that the functions $u_t(x,y,\tau;t)$ and $v_x(x,y,\tau;t)$ are continuous, the functions φ_{xt} and φ_{tx} are also continuous, therefore the eigenvalue operator equations (5.205) and (5.207) are compatible, i.e., $\varphi_{xt} = \varphi_{tx}$, then the last equality can be written in the operator form:

$$\{U_t - V_x + i(UV - VU) + i\lambda(TU - UT + VJ - JV)\}\varphi = 0. \tag{5.216}$$

Making use of (5.206) and (5.208), we write out equation (5.216) as follows:

$$\int_{-1}^{1} \varphi(x,\tau;t)d\tau \Big\{ u_t(x,y,\tau;t) - v_x(x,y,\tau;t) +$$

$$+ i\int_{-1}^{1} \big(u(x,y,\tau_1;t)v(x,\tau_1,\tau;t) - v(x,y,\tau_1;t)u(x,\tau_1,\tau;t)\big)d\tau_1 +$$

$$+ i\lambda u(x,y,\tau;t)\big(\alpha(y)-\alpha(\tau)\big) + i\lambda v(x,y,\tau;t)(\tau - y)\Big\} = 0. \tag{5.217}$$

Since the potentials $u(x,y,\tau;t)$ and $v(x,y,\tau;t)$ in (5.205) and (5.207) satisfy estimate (5.209), then

$$u(x,y,y;t) = v(x,y,y;t) = 0. \tag{5.218}$$

Owing to the condition (5.218), the operators U and V defined by (5.206) and (5.208) are expressible through the coefficient $q(x,y,\tau;t)$ in the continual system (5.213):

$$U\varphi = [J,Q]\varphi = (JQ - QJ)\varphi = \int_{-1}^{1} q(x,y,\tau;t)(y-\tau)\varphi(x,\tau;t)d\tau, \tag{5.219}$$

$$V\varphi = [T,Q]\varphi = (TQ - QT)\varphi = \int_{-1}^{1} q(x,y,\tau;t)(\alpha(y)-\alpha(\tau))\varphi(x,\tau;t)d\tau. \tag{5.220}$$

Due to (5.219) and (5.220), the third term in the left-hand side of (5.217) is equal to zero. Indeed,

$$\int_{-1}^{1} \varphi(x,\tau;t)u(x,y,\tau;t)\{\alpha(y)-\alpha(\tau)\}d\tau + \int_{-1}^{1} \varphi(x,\tau;t)v(x,y,\tau;t)(\tau-y)d\tau$$

$$= \int_{-1}^{1} \varphi(x,\tau;t)q(x,y,\tau;t)(y-\tau)\{\alpha(y)-\alpha(\tau)\}d\tau$$

$$+ \int_{-1}^{1} \varphi(x,\tau;t)q(x,y,\tau;t)\{\alpha(y)-\alpha(\tau)\}(\tau-y)d\tau = 0, \text{ i.e.,}$$

$$\{TU - UT + VJ - JV\}\varphi = 0. \tag{5.221}$$

With due regard for (5.221), from (5.216) and (5.217) we obtain the Lax equation for the Lax compatibility condition for equations (5.214) and (5.215) in the operator form:

$$U_t - V_x + i[U,V] = 0, \tag{5.222}$$

and in the kernel form:

$$u_t(x,y,\tau;t) - v_x(x,y,\tau;t) + \\ + i\int_{-1}^{1}\{u(x,y,\tau_1;t)v(x,\tau_1,\tau;t) - v(x,y,\tau_1;t)u(x,\tau_1,\tau;t)\}d\tau_1 = 0. \tag{5.223}$$

Owing to (5.219) and (5.220) the potentials $u(x,y,\tau;t)$ and $v(x,y,\tau;t)$ in (5.205) and (5.207) are expressible through the coefficient $q(x,y,\tau;t)$ in the continual system (5.213). In fact, by virtue of (5.219) and (5.220), the Lax compatibility condition (5.223) of the equalities (5.214) and (5.215) is written in term of the coefficient $q(x,y,\tau;t)$:

$$(y-\tau)q_t(x,y,\tau;t) - [\alpha(y)-\alpha(\tau)]\,q_x(x,y,\tau;t) \\ + i\int_{-1}^{1}\Big\{(y-\tau_1)q(x,y,\tau_1;t)\times[\alpha(\tau_1)-\alpha(\tau)]\,q(x,\tau_1,\tau;t) \\ - [\alpha(y)-\alpha(\tau_1)]\,q(x,y,\tau_1;t)\times(\tau_1-\tau)q(x,\tau_1,\tau;t)\Big\}d\tau_1 = 0.$$

After some elementary transformations, from the last equality we derive the continual system (5.213) for the coefficient $q(x,y,\tau;t)$. Thus, the Lax compatibility condition in the operator form (5.222) is equivalent to the continual system (5.213) for the kernel coefficient $q(x,y,\tau;t)$. Hence, this system admits its Lax generalized equation (5.212) by two linear operator equations (5.205) and (5.207).

Conversely, if the coefficient $q(x,y,\tau;t)$ satisfies the continual system (5.213), then $\lambda_t = 0$ and the operator equations (5.205) and (5.207) are compatible. Indeed, let $\lambda = \lambda_0$ be a point of the spectrum of L for $t = t_0$ and let $\varphi(x,y,\lambda_0;t_0)$ be a corresponding eigenfunction. We define the function $r(x,y,\lambda_0;t)$ by the equality:

$$r(x,y,\lambda_0;t) = \frac{\partial}{\partial x}\varphi(x,y,\lambda_0;t) - M\varphi(x,y,\lambda_0;t), \tag{5.224}$$

where $M = i(U - \lambda_0 J)$.

Putting: $\dfrac{\partial\varphi}{\partial t} = i(V - \lambda_0 T)\varphi = N\varphi$ and differentiating (5.224) with respect to t gives

$$\frac{\partial r}{\partial t} = \frac{\partial}{\partial x}\frac{\partial\varphi}{\partial t} - M\frac{\partial\varphi}{\partial t} - \frac{\partial M}{\partial t}\varphi = \frac{\partial N}{\partial x}\varphi + N\frac{\partial\varphi}{\partial x} - MN\varphi - \frac{\partial M}{\partial t}\varphi,$$

with the initial condition: $r(x,y,\lambda_0;t_0) = 0$.

Using (5.224) to get rid of $\frac{\partial \varphi}{\partial x}$ in the last equality, we obtain the equality:

$$\frac{\partial r}{\partial t} - i(V - \lambda_0 T)r = \{i(V_x - U_t) + NM - MN\}\varphi =$$
$$\Big\{i(V_x - U_t) + UV - VU + \lambda_0(VJ - JV + TU - UT)\Big\}.$$

By supposition, the coefficient $q(x, y, \tau; t)$ satisfies the system (5.213), which is equivalent to the operator equation (5.216). Then, due to (5.216) the right-hand side of the last equality is equal to zero. Consequently, we obtain the equation:

$$\frac{\partial r}{\partial t} - i(V - \lambda_0 T)r = 0. \tag{5.225}$$

Since the potentials $u(x, y, \tau; t)$ and $v(x, y, \tau; t)$ are continuous functions of t, then by the unique solvability of the Cauchy problem for the equation (5.225) with the initial condition:

$$r(x, y, \lambda_0; t_0) = 0,$$

we find that $r(x, y, \lambda_0; t) = 0$ for any $t \in (-\infty, \infty)$. Hence, the point $\lambda = \lambda_0$ belongs to the spectrum for all values of t-time and $\varphi(x, y, \lambda_0; t)$ is the corresponding eigenfunction of the operator L. Thus, the spectrum of L is invariant with t, i.e., $\lambda_t = 0$. Based on this fact, from (5.205) and (5.207) we find the operator equations (5.214) and (5.215), respectively. The right-hand sides of the equalities (5.214) and (5.215) are continuous functions, therefore the functions φ_{xt} and φ_{tx} are continuous, and $\varphi_{xt} = \varphi_{tx}$. Hence, the Lax compatibility condition of the operator equations (5.205) and (5.207) is satisfied. The lemma is proved. □

According to Definition 5.2, the one-to-one correspondence between the incident wave $A(y, \lambda; t)$ and the scattered wave $B(y, \lambda; t)$ for the solution of transport equation (5.205) defines the transition operator $S(\lambda; t)$. This operator $S(\lambda; t)$ for the time-dependent transport equation (5.205) is defined by the relation equality (5.110):

$$S(\lambda; t)A(y, \lambda; t) = B(y, \lambda; t), \tag{5.226}$$

where

$$\lim_{x\to-\infty} \varphi(x, y; t)e^{i\lambda yx} = A(y, \lambda; t), \;\; B(y, \lambda; t) = \lim_{x\to\infty} \varphi(x, y; t)e^{i\lambda yx},$$

and $\varphi(x, y; t)$ is a bounded solution of the time-dependent transport equation (5.205).

The time-dependent transition operator $S(\lambda; t)$ is defined in space $L_2(-1, 1; -\infty, \infty)$. We now introduce the time-dependent transition operators:

$$\tilde{\mathcal{F}}(\lambda; t) = S(\lambda; t) - I, \;\; \tilde{\mathcal{G}}(\lambda; t) = S^{-1}(\lambda; t) - I.$$

5.3.2 The time-evolution of the operators $\tilde{\mathcal{F}}(\lambda;t)$ and $\tilde{\mathcal{G}}(\lambda;t)$

Lemma 5.18. *Let φ be a solution of the operator equation: $L\varphi = \lambda\varphi$, and the operators L and P satisfy the generalized Lax equation (5.212), then the function $\psi = P\varphi$ also satisfies the equation:*

$$L\psi = \lambda\psi, \tag{5.227}$$

and the functions $\left(\frac{\partial}{\partial t} - \lambda T^{-1}J\right)A(y,\lambda;t)$ and $\left(\frac{\partial}{\partial t} - \lambda T^{-1}J\right)B(y,\lambda;t)$ give the incident and scattered waves, respectively for the solution ψ of the operator equation (5.227).

Proof. Applying the generalized Lax equation (5.212) to the function $\psi = P\varphi$, we obtain that the function ψ satisfies the Lax generalized operator equation (5.227):

$$L\psi = LP\varphi = PL\varphi = \lambda P\varphi = \lambda\psi.$$

Making use of equation (5.205), estimate (5.209) and definition (5.226), we have:

$$\lim_{x\to-\infty}\left\{i\frac{\partial\varphi}{\partial x} + U\varphi\right\}e^{i\lambda yx} = \lim_{x\to-\infty}\{\lambda J\varphi\}e^{i\lambda yx} = \lambda JA(y,\lambda;t),$$

$$\lim_{x\to\infty}\left\{i\frac{\partial\varphi}{\partial x} + U\varphi\right\}e^{i\lambda yx} = \lim_{x\to\infty}\{\lambda J\varphi\}e^{i\lambda yx} = \lambda JB(y,\lambda;t),$$

Using the last equalities and (5.209), (5.211), we determine the incident and scattered waves for the solution ψ of the operator Lax equation (5.227):

$$\begin{aligned}\lim_{x\to-\infty}\psi(x,y,\lambda;t)e^{i\lambda yx} &= \lim_{x\to-\infty}\{P\varphi(x,y,\lambda;t)\}e^{i\lambda yx}\\ &= \lim_{x\to-\infty}\left\{\left[\frac{\partial}{\partial t} - B\right]\varphi(x,y,\lambda;t)\right\}e^{i\lambda yx}\\ &= \lim_{x\to-\infty}\left\{\left[\frac{\partial}{\partial t} - T^{-1}\Big(i\frac{\partial}{\partial x} + V\Big)\right]\varphi(x,y,\lambda;t)\right\}e^{i\lambda yx}\\ &= \left(\frac{\partial}{\partial t} - \lambda T^{-1}\cdot J\right)A(y,\lambda;t),\end{aligned} \tag{5.228}$$

$$\begin{aligned}\lim_{x\to\infty}\psi(x,y,\lambda;t)e^{i\lambda yx} &= \lim_{x\to\infty}\left\{\left[\frac{\partial}{\partial t} - T^{-1}\Big(i\frac{\partial}{\partial x} + V\Big)\right]\varphi(x,y,\lambda;t)\right\}e^{i\lambda yx}\\ &= \Big(\frac{\partial}{\partial t} - \lambda T^{-1}\cdot J\Big)B(y,\lambda;t).\end{aligned} \tag{5.229}$$

Thus, the lemma is proved. □

The functions (5.228) and (5.229) are the incident and scattered waves, respectively for the solution ψ of equation (5.227), therefore by the definition (5.226), we obtain:

$$\begin{aligned} S\cdot\Big(\frac{\partial}{\partial t}-\lambda T^{-1}\cdot J\Big)A(y,\lambda;t) &= \Big(\frac{\partial}{\partial t}-\lambda T^{-1}\cdot J\Big)B(y,\lambda;t) \\ &= \Big(\frac{\partial}{\partial t}-\lambda T^{-1}\cdot J\Big)\cdot SA(y,\lambda;t), \\ S^{-1}\cdot\Big(\frac{\partial}{\partial t}-\lambda T^{-1}\cdot J\Big)B(y,\lambda;t) &= \Big(\frac{\partial}{\partial t}-\lambda T^{-1}\cdot J\Big)A(y,\lambda;t) \\ &= \Big(\frac{\partial}{\partial t}-\lambda T^{-1}\cdot J\Big)\cdot S^{-1}B(y,\lambda;t). \end{aligned}$$

The evolution equations for the time-dependent transition operators $S(\lambda;t)$ and $S^{-1}(\lambda;t)$ are derived from the last equalities:

$$\frac{\partial}{\partial t}\cdot S - S\cdot\frac{\partial}{\partial t} = \frac{\partial S}{\partial t} = \lambda T^{-1}\cdot J\cdot S - \lambda S\cdot T^{-1}\cdot J = \lambda[T^{-1}\cdot J, S], \quad (5.230)$$

$$\frac{\partial}{\partial t}\cdot S^{-1} - S^{-1}\cdot\frac{\partial}{\partial t} = \frac{\partial S^{-1}}{\partial t} = \lambda T^{-1}\cdot J\cdot S^{-1} - \lambda S^{-1}\cdot T^{-1}\cdot J = \lambda[T^{-1}\cdot J, S^{-1}]. \quad (5.231)$$

Since $[I, T^{-1}J] = [T^{-1}J, I] = 0$, then from equalities (5.230) and (5.231) we derive the time-evolution equations for the transition operators $\tilde{\mathcal{F}}(\lambda;t) = S(\lambda;t) - I$ and $\tilde{\mathcal{G}}(\lambda;t) = S^{-1}(\lambda;t) - I$:

$$\frac{\partial\tilde{\mathcal{F}}}{\partial t} = \lambda[T^{-1}\cdot J, \tilde{\mathcal{F}}], \quad (5.232)$$

$$\frac{\partial\tilde{\mathcal{G}}}{\partial t} = \lambda[T^{-1}\cdot J, \tilde{\mathcal{G}}]. \quad (5.233)$$

Remark 5.2. By the generalized Lax equation (5.212), instead of the time-evolution of the operator L, the time-evolution equations (5.232) and (5.233) for the transition operators $\tilde{\mathcal{F}}(\lambda;t)$ and $\tilde{\mathcal{G}}(\lambda;t)$ are derived. Hence, the time-evolution of the operator L for the transport equation (5.205) corresponds to the time-evolution of the transition operators $\tilde{\mathcal{F}}(\lambda;t)$ and $\tilde{\mathcal{G}}(\lambda;t)$. These operators evolve by time in accordance with the evolution equations (5.232) and (5.233). This enables us to solve the Cauchy problem for the continual system (5.213).

Remark 5.3. The potentials $u(x,y,\tau;t)$ and $v(x,y,\tau;t)$ in (5.206) and (5.208) relate to the coefficient $q(x,y,\tau;t)$ in the continual system (5.213) by the commutators (5.219), and (5.220):

$$u(x,y,\tau;t) = (y-\tau)q(x,y,\tau;t), \quad v(x,y,\tau;t) = (\alpha(y)-\alpha(\tau))\,q(x,y,\tau;t). \quad (5.234)$$

It follows from the estimate (5.209) and relation equalities (5.234) that the coefficient $q(x, y, \tau; t)$ in the continual system (5.213) satisfies the estimate:

$$|q(x, y, \tau; t)| \leq \frac{\tilde{C}}{(1 + |x|)^{2(1+\varepsilon)}}, \; \tilde{C} \text{ is a constant, } \varepsilon > 0. \qquad (5.235)$$

5.3.3 The Cauchy problem for the continual system (5.213)

We consider the Cauchy initial-value problem for the continual system of non-linear interaction waves (5.213) with the known initial condition:

$$q(x, y, \tau; t) \mid_{t=0} = q(x, y, \tau). \qquad (5.236)$$

According to Lemma 5.8, we define the transition operator $S(\lambda)$ for the scattering problem for transport equation (5.105) with the known initial potential (5.236) satisfying estimate (5.235). The time-dependence of the transition operator $S(\lambda; t)$ for the time-dependent transport equation (5.205) obeys evolution equations (5.232) and (5.233). Further, by Lemma 5.11 we construct Volterra integral Hilbert-Schmidt operators $K_2^+(x, \lambda; t)$ and $K_1^-(x, \lambda; t)$ from the found operator $S(\lambda; t)$. Then the time-dependent intermediate operator $\tilde{S}(\lambda; t)$ is constructed by the factorization (5.179). Therefore, the intermediate operators $F(x, \lambda; t)$ and $G(x, \lambda; t)$ are defined by formulas (5.184), and the functions $f(x, s; y, y_0; t)$ and $g(x, s; y, y_0; t)$ are the kernels of the operators $F(x, \lambda; t)$ and $G(x, \lambda; t)$, respectively. These kernels are constructed by Lemma 5.15. Then by Theorem 5.5 the potential is uniquely reconstructed in terms of solutions of fundamental equations (5.194) and (5.195). Thus, the Cauchy problem (5.213), (5.236) is solved.

Theorem 5.6. *Consider the Cauchy initial-value problem for the continual system of non-linear interaction waves* (5.213) *with the known initial condition* (5.236). *Suppose that the associated transport equation* (5.205) *has no generalized eigenvalues, and the potential $q(x, y, \tau; t)$ in this equation is isospectral and belongs to the class continuous functions satisfying estimate* (5.235) *and possessing the first partial continuous derivatives with respect to t and x. Then,*

(1) The continual system (5.213) *is equivalent to the Lax compatibility condition (5.222) for the transport equation* (5.205) *and equation* (5.207) *governing the time-evolution of the solution of equation* (5.205), *wherein the potentials $u(x, y, \tau; t)$ and $v(x, y, \tau; t)$ satisfy the corresponding estimate* (5.209);

(2) The continual system of non-linear interaction waves (5.213) *is integrated by the ISM. The scheme for solving this Cauchy problem consists of the following steps:*

- *Determining the transition operator $S(\lambda)$ from the known initial condition* (5.236),

- *Using the generalized Lax equation* (5.212) *for deriving time-evolution equations* (5.232) *and* (5.233) *for transition operators* $\tilde{\mathcal{F}}(\lambda;t)$ *and* $\tilde{\mathcal{G}}(\lambda;t)$,
- *Constructing the time-dependent intermediate operators* (5.187) *and* (5.188) *from the known transition operator* $S(\lambda;t)$ *by Lemmas 5.11, 5.14 and 5.15,*
- *Finding uniquely the solution* $q(x,y,\tau;t)$ *of the Cauchy problem* (5.213), (5.236). *The found solution is expressed through solutions* $H_-\left(x,\frac{s}{|y-y_0|};y,y_0;t\right)$, $s<0$ *and* $H_+\left(x,\frac{s}{|y-y_0|};y,y_0;t\right)$, $s>0$, $y\neq y_0$, *of fundamental equations* (5.194) *and* (5.195), *respectively by means of the equalities* (5.118) *and* (5.119)*:*

$$i(y-y_0)q(x,y,y_0;t)=H_-(x,0,y,y_0;t)=-H_+(x,0,y,y_0;t)\ \textit{for}\ y>y_0,$$
$$i(y_0-y)q(x,y,y_0;t)=-H_-(x,0,y,y_0;t)=H_+(x,0,y,y_0;t)\ \textit{for}\ y<y_0.$$

There exits the one-to-one correspondence relation between the transition operator $S(\lambda;t)$ *for the transport equation* (5.205) *and the solution of the Cauchy problem for the continual system* (5.213) *with the known initial condition* (5.236).

It should be noted that since the ISP for the transport equation (5.105) associated with the Cauchy problem (5.213), (5.236) were solved well, then with the help of the generalized Lax equation (5.212), the application of the ISM to the integration of the continuous system of non-linear interaction waves (5.213) is effective. It is embedded in the schema of the ISP.

Chapter 6

Some Problems for The KdV Equation and Associated Inverse Scattering

In this chapter we apply known results written in the monograph [86] about the inverse scattering problem for the Schrödinger equation on the half-line to solving the initial-boundary value problem (IBVP) for the Korteweg-de Vries (KdV) equation:

$$p_t - 6pp_x + p_{xxx} = 0, \;\; x \geq 0, \; t \geq 0$$

with the known initial and boundary conditions:

$$p(x, 0) = p(x), \; p(x) \text{ is a real-valued rapidly decreasing function,}$$
$$p(0, t) = f(t), \; f(t) \text{ is an arbitrary integrable real-valued function.}$$

The chapter is based on the works [86, 139, 144].

The Korteweg-de Vries (KdV) equation is the generic equation for the study of weakly non-linear long waves. It arises in physical systems, which involve a balance between weak non-linearity and weak dispersion at leading order (Whitham [154, 155]).

Interest in solitons was revived in connection with plasma studies. In 1958, Sagdeev postulated that solitons could propagate in plasma similar to solitons on a liquid surface. Gardner and Morikawa established a direct analogy between the equations for plasma in strong magnetic field and equations for shallow water [46]. From this point onwards the KdV gained general recognition in physics, and soon attempts were made to use it in various wave topic [156].

The KdV equation arises in many physical situations, such as surface water waves, internal waves in density-stratified fluid, plasma waves, Rossby waves and magma flow. The KdV equation is integrable and can be solved on the infinite line using the so-called inverse scattering method (ISM) (Gardner et al., 1967 [44]). The discovery to the ISM was preceded by numerical simulation of the KdV. As early as 1954, Fermi, Pasta and Ulam with the help of a new computer detected anomalously slow stochastization of a non-linear oscillator system. In 1964 Kruskal and Zabuskii, using numerical simulation, concluded that solitons in the KdV formalism suffer elastic collision. These results gave an impetus to new analytical studies. Finally, in 1967, the progress was crowned with the discovery of the ISM [44].

DOI: 10.1201/9781003370543-6

The inverse scattering solution shows that when a larger KdV solitary wave overtakes a smaller one, both the solitary waves retain their original shapes, with the only memory of the collision being a phase shift. Due to this special, particle-like property, amongst others, the solitary-wave solution of the KdV equation is termed a soliton. The explicit solution for interacting KdV solitons was developed, using the ISM and Hirota method [58, 149]. The numerous results of studies of the KdV and the generalizations were achieved in [25, 45, 87, 88, 89, 108]. The Cauchy initial-value problems for the KdV equation on the half-line was considered in [134, 136, 139, 141].

A number of physical applications exists for the KdV equation on the positive quarter-plane, such as the generation of waves in a shallow channel by a wave-making device or the critical withdrawal of a stratified fluid from a reservoir [16].

Solutions of the KdV equation on the infinite line, such as the soliton, cnoidal wave, mean height variation and undular bore solution are used to find approximate solutions of the IBVPs for the KdV equation on the positive quarter-plane and on the negative quarter-plane [82, 83]. The form of the solution is found to depend markedly on the specific boundary and initial values. Many of solutions of the IBVPs are based on the modulation theory for the KdV equation [154, 155].

However, the rather ingenious techniques developed for models for the unbounded space cannot be applied in the presence of boundaries occurring at final distances with arbitrary boundary conditions or when external forcings exist, a situation that certainly is of great practical importance. In spite of the above success, the question of extending the ISM to solve IBVPs for non-linear evolution equations (NLEEs) remains difficult [35, 40, 41]. The simplest such problem arises if an equation is formulated on a half-infinite line. Different approaches to study of IBVPs for NLEEs have been developed during recent years [37, 38]. In [24, 106] an "elbow scattering" in the (x, t) plane was introduced to deal with the half-line problem for the KdV equation, leading to a Gelfand–Levitan–Marchenko formulation.

In this chapter, we prove that the IBVP for the KdV equation on the half-line $0 \leq x < \infty$ is associated with the Sturm–Liouville scattering problem in the interval $(0, b)$ (b is a large positive number) generated by the linear Schrödinger equation with the zero boundary conditions at $x = 0$ and at $x = b$. The time-dependence of the scattering data of the associated scattering problem is determined by unknown boundary values (BVs) evaluated at $x = 0$ of the Jost solution of the Schrödinger equation. We overcome this difficulty by deriving the asymptotic form of the normalization eigenfunction of the considered Sturm–Liouville problem. This fact enables us to prove that the scattering phase of the BV is approximately time-independent, and the square of time-dependent moduli of unknown BVs can be calculated approximately in terms of the known initial and boundary conditions. Then the potential in the time-dependent Schrödinger equation is uniquely reconstructed in terms of the solution of the fundamental equation in the inverse problem.

The chapter is organized as follows. In Section 6.1 we recall the known results of the direct and inverse scattering problem (ISP) on the half-line associated with the IBVP for the KdV equation on the positive quarter-plane. The considered IBVP is solved in Section 6.2. In Section 6.3 we present explicit solutions of the Cauchy problem for the KdV equation for $(x,t) \in [0,\infty) \times (-\infty,\infty)$ in the class of non-scattering potentials.

6.1 The direct and ISP

The IBVP for the KdV equation on the positive quarter-plane is associated with the scattering problem generated by the linear Schrödinger equation on the half-line [86]:

$$-y_{xx} + p(x)y = \rho^2 y, \quad 0 \leq x < \infty, \tag{6.1}$$

and by the zero boundary condition at the origin $x = 0$:

$$y(\rho, 0) = 0, \tag{6.2}$$

where the potential $p(x)$ is real-valued function satisfying the inequality:

$$\int_0^\infty x|p(x)|dx < \infty. \tag{6.3}$$

In this section we recall the known results of the direct and ISP (6.1)–(6.2) from the Marchenko's monograph [86]. The proofs of the lemmas and theorems presented in this section can be found in [86], so we omit them.

Due to condition (6.3), the equation (6.1), as x tends to infinite, reduces to the simpler equation: $-y_{xx} + \rho^2 y = 0$. This enables a complete investigation of properties of solutions of equation (6.1).

We introduce the following notations for convenience:

$$\sigma_1(x) = \int_x^\infty |p(\xi)|d\xi, \quad \sigma_2(x) = \int_x^\infty \sigma_1(\xi)d\xi. \tag{6.4}$$

It is easy to verify that $\sigma_2(0) = \int_0^\infty d\xi \int_\xi^\infty |p(\xi_1)|d\xi_1 = \int_0^\infty \xi_1|p(\xi_1)|d\xi_1 < \infty$, i.e., the condition (6.3) is equivalent to the inequality $\sigma_2(0) < \infty$.

6.1.1 The Jost solution and the analytic solution

Lemma 6.1. *For any ρ from the closed upper half-plane $\operatorname{Im}\rho \geq 0$, the equation (6.1) has a solution that can be represented in the form:*

$$e(\rho, x) = e^{i\rho x} + \int_x^\infty K(x,\xi) e^{i\rho\xi} d\xi, \quad \operatorname{Im}\rho \geq 0, \tag{6.5}$$

where the kernel $K(x,\xi)$ satisfies the inequality:

$$|K(x,\xi)| \leq \frac{1}{2}\sigma_1\left(\frac{x+\xi}{2}\right) e^{\sigma_2(x)-\sigma_2\left(\frac{x+\xi}{2}\right)}. \tag{6.6}$$

In addition,

$$K(x,x) = \frac{1}{2}\int_x^\infty p(\xi)d\xi. \tag{6.7}$$

Consider the integral equation:

$$e(\rho, x) = e^{i\rho x} + \int_x^\infty \frac{\sin\rho(\xi - x)}{\rho} p(\xi) e(\rho,\xi) d\xi. \tag{6.8}$$

The integral equation (6.8) is equivalent to the differential equation (6.1) with the boundary condition:

$$\lim_{x\to\infty} e^{-i\rho x} e(\rho, x) = 1. \tag{6.9}$$

Substituting the representation of solution (6.5) into equation (6.8), we obtain the integral equation for the kernel $K(x,\xi)$:

$$K(x,\xi) = \frac{1}{2}\int_{\frac{x+\xi}{2}}^\infty p(s)ds + \frac{1}{2}\int_x^\infty p(s) \int_{\xi-(s-x)}^{\xi+(s-x)} K(s,u)du ds, \tag{6.10}$$

and the condition: $K(x,\xi) = 0$ for $\xi < x$.

Upon changing the variables in the integral (6.10): $u+s = 2\alpha, u-s = 2\beta$, we are led to the equation for $K(x,\xi)$:

$$K(x,\xi) = \frac{1}{2}\int_{\frac{x+\xi}{2}}^\infty p(s)ds + \int_{\frac{x+\xi}{2}}^\infty d\alpha \int_0^{\frac{\xi-x}{2}} p(\alpha-\beta) K(\alpha-\beta,\alpha+\beta) d\beta, \tag{6.11}$$

where the identity $K(x,\xi) = 0$ for $\xi < x$, is automatically accounted for.

The solution $e(\rho, x)$ of equation (6.1) with the boundary condition at infinity (6.9) is represented in the form (6.5) and is called the Jost solution. The operator $\mathbf{I}+\mathbf{K}$ determined via the formula:

$$(\mathbf{I}+\mathbf{K})f = f(x) + \int_x^\infty K(x,\xi)f(\xi)d\xi,$$

$$f \in L_j(0,\infty), j = 1,2,\infty, \|f\|_{L_\infty}(0,\infty) = \text{vrai} \sup_{0<\xi<\infty} |f(\xi)|,$$

is called the transformation operator usually used in scattering problems [7]. This operator transforms the solution $e^{i\rho x}$ of the Schrödinger equation with zero potential into the solution of this equation with the potential $p(x)$.

Remark 6.1. The estimate (6.6) implies that the transformation operator $\mathbf{I}+\mathbf{K}$ defines a one-to-one mapping of each space $L_j(0,\infty)$, $j = 1,2,\infty$, onto itself. Hence, there exists the inverse operator $(\mathbf{I}+\mathbf{K})^{-1} = \mathbf{I}+\mathbf{L}$, which has the same form:

$$(\mathbf{I}+\mathbf{L})f = f(x) + \int_x^\infty L(x,\xi)f(\xi)d\xi,$$

where the kernel $L(x,\xi)$ satisfies the equation:

$$L(x,y) + K(x,y) + \int_x^y L(x,\xi)K(\xi,y)d\xi = 0,$$

and the estimate:

$$|L(x,y)| \le \frac{1}{2}\sigma_1\left(\frac{x+y}{2}\right)\exp\left\{\sigma_2\left(\frac{x+y}{2}\right) - \sigma_2(y)\right\}.$$

Lemma 6.2. *The function $K(x,\xi)$ possesses first-order partial derivatives with respect to both variables, and satisfies the estimate:*

$$\left|\frac{\partial K(x_1,x_2)}{\partial x_j} + \frac{1}{4}p\left(\frac{x_1+x_2}{2}\right)\right| \le \frac{1}{2}\sigma_1(x_1)\sigma_1\left(\frac{x_1+x_2}{2}\right)e^{\sigma_2(x_1)-\sigma_2\left(\frac{x_1+x_2}{2}\right)}. \tag{6.12}$$

Remark 6.2. If the potential function $p(x)$ is differentiable, then the kernel $K(x,\xi)$ is the twice differentiable function satisfying the equation:

$$\frac{\partial^2 K(x,\xi)}{\partial x^2} - \frac{\partial^2 K(x,\xi)}{\partial \xi^2} - p(x)K(x,\xi) = 0, \tag{6.13}$$

and the conditions:

$$\frac{d}{dx}K(x,x) = -\frac{1}{2}p(x), \tag{6.14}$$

$$\lim_{x+\xi\to\infty}\frac{\partial K(x,\xi)}{\partial x} = \lim_{x+\xi\to\infty}\frac{\partial K(x,\xi)}{\partial \xi} = 0. \tag{6.15}$$

Thus, in order for the function $K(x,\xi)$ to be the kernel of the transformation operator, it is necessary and sufficient that it satisfies the equalities (6.13), (6.14) and (6.15).

Lemma 6.3. *The Jost solution* $e(\rho,x)$ *is an analytic function of* ρ *in the upper half-plane* $\mathrm{Im}\rho > 0$ *and is continuous on the real axis. For any* ρ *in the closed upper half-plane* $\mathrm{Im}\rho \geq 0$, *the Jost solution satisfies the following estimates:*

$$|e(\rho,x)| \leq \exp\{-\mathrm{Im}\rho x + \sigma_2(x)\}, \tag{6.16}$$

$$|e(\rho,x) - e^{i\rho x}| \leq \left\{\sigma_2(x) - \sigma_2\left(x + \frac{1}{|\rho|}\right)\right\}\exp\{-\mathrm{Im}\rho x + \sigma_2(x)\}, \tag{6.17}$$

$$|e_x(\rho,x) - i\rho e^{i\rho x}| \leq \sigma_1(x)\exp\{-\mathrm{Im}\rho x + \sigma_2(x)\}. \tag{6.18}$$

Without further mention we shall write k for ρ whenever ρ is real. For real $k \neq 0$ both functions $e(k,x)$ and $e(-k,x)$ are well-defined and satisfy one and the same equation (6.1). Consequently, their Wronskian $W\{e(k,x), e(-k,x)\}$ does not depend on x. Then due to (6.17) and (6.18):

$$W\{e(k,x), e(-k,x)\} = e_x(k,x)e(-k,x) - e(k,x)e_x(-k,x) = 2ik \text{ for } k \neq 0. \tag{6.19}$$

Hence, the solutions $e(k,x)$ and $e(-k,x)$ form a fundamental system of solutions of the equation (6.1). From the reality of the potential $p(x)$ it follows that the kernel $K(x,\xi)$ given by (6.11) is a real-valued function, therefore

$$e(k,x) = \overline{e(-k,x)}, \quad e_x(k,x) = \overline{e_x(-k,x)}. \tag{6.20}$$

The BVs of $e(\rho,x)$ and $e_x(\rho,x)$ at $x = 0$ are denoted by $e(\rho)$ and $e_x(\rho)$, respectively. It follows from (6.3), (6.5) and (6.8) that these BVs satisfy the estimates :

$$e(\rho) = 1 + O\left(\frac{1}{\rho}\right), \quad e_x(\rho) = i\rho + o(1) \text{ as } |\rho| \to \infty, \quad \mathrm{Im}\rho \geq 0. \tag{6.21}$$

Lemma 6.4. *For any value of* ρ, *the equation* (6.1) *has a solution* $\omega(\rho,x)$ *which satisfies the boundary conditions as* $x \to 0$:

$$\omega(\rho,x) = x(1 + o(1)), \quad \omega_x(\rho,x) = 1 + o(1). \tag{6.22}$$

This solution is an entire function of ρ *and for* $\mathrm{Im}\rho \geq 0$ *satisfies the inequality:*

$$|(\rho\omega(\rho,x) - \sin\rho x)\, e^{i\rho x}| \leq \left\{\sigma_2(0) - \sigma_2\left(\frac{1}{|\rho|}\right)\right\}\exp\left\{\int_0^x t|p(t)|dt\right\}. \tag{6.23}$$

The analytic solution $\omega(k,x)$ of the boundary problem (6.1)–(6.2) is a linear combination of solutions of the fundamental system of solutions of equation (6.1) and has the representation:

$$\omega(k,x)=\frac{e(-k,0)}{2ik}e(k,x)-\frac{e(k,0)}{2ik}e(-k,x).$$

Owing to (6.19) $e(k,0)=e(k)\neq 0$ for all $k\neq 0$. Indeed, if $e(k)=0$ for $k\neq 0$, then due to (6.20), from the above representation of $\omega(k,x)$ it would imply that $\omega(k,x)\equiv 0$, and this is impossible. The above results can be formulated as follows.

Lemma 6.5. *The identity*

$$-\frac{2ik\omega(k,x)}{e(k)}=e(-k,x)-S(k)e(k,x), \tag{6.24}$$

holds for all real $k\neq 0$, where the function $S(k)$ is continuous for $k\neq 0$ and

$$S(k)=\frac{e(-k)}{e(k)}=\overline{S(-k)}=[S(-k)]^{-1},\ \ |S(k)|=1, \tag{6.25}$$

$$S(k)=1+o(1)\ \textit{for large}\ |k|. \tag{6.26}$$

We call $S(k)$ the scattering function for the scattering problem (6.1)–(6.2). Note that $e(k)\neq 0$ for all $k\neq 0$, therefore the point $k=0$ is only possible real zero of the function $e(k)$. It follows from (6.21) that $e(\rho)\to 1$ as $|\rho|\to\infty$. Hence, the zeros of $e(\rho)$ form at most a countable set having 0 as the only possible limit point.

Lemma 6.6. *The function $e(\rho)$ may have only a finite number of zeros in the upper half-plane* Im >0. *All these zeros are simple and lie on the imaginary axis. In addition, the function $\rho[e(\rho)]^{-1}$ is bounded in a neighborhood of the point $\rho=0$.*

The function in the left-hand side of (6.24) is the meromorphic function in the upper half-plane Im$\rho>0$ with poles lying at the zeros of the function $e(\rho)$. We call these zeros the singular numbers of the problem (6.1)–(6.2).

Denote $\dot{e}(\rho,x)=\frac{\partial}{\partial\rho}e(\rho,x)$. Differentiating equation (6.1) with respect to ρ, and multiplying the obtained equation by $e(\rho,x)$, we receive

$$\{e(\rho,x)\dot{e}_x(\rho,x)-e_x(\rho,x)\dot{e}(\rho,x)\}_{x=a}^{b}+2\rho\int_a^b e^2(\rho,x)dx=0. \tag{6.27}$$

Let $\rho_j=i\mu_j,\ \ \mu_j>0,\ \ j=1,...,N$, be the zeros of $e(\rho)$, then from (6.5) and (6.10) it follows that the function $e(i\mu_j,x)$ is real-valued. Taking this fact and

(6.16), (6.18) into account, from (6.27) we find the square of norm m_j^{-2} of the Jost solution $e(i\mu_j, x)$ in $L_2[0, \infty)$:

$$m_j^{-2} = \int_0^\infty |e(i\mu_j, x)|^2 dx = -\frac{e_x(i\mu_j, 0)\dot{e}(i\mu_j, 0)}{2i\mu_j} > 0. \tag{6.28}$$

Since $m_j^{-2} > 0$, then $\dot{e}(i\mu_j, 0) \neq 0$, i.e., the simplicity of the zeros of $e(\rho)$ is shown.

Lemma 6.7. *The function* $1-S(k)$ *is the Fourier transformation of a function* $F_S(x)$ *of the form:*

$$F_S(x) = F_S^{(1)}(x) + F_S^{(2)}(x),$$

where $F_S^{(1)}(x) \in L_1(-\infty, \infty)$, *whereas* $F_S^{(2)}(x) \in L_2(-\infty, \infty)$ *and* $\sup\limits_{-\infty<x<\infty} |F_S^{(2)}(x)| < \infty.$

6.1.2 The Parseval's equality and the fundamental equation

Let $i\mu_j$, $j = 1, ..., N$ be the zeros of the function $e(\rho, 0)$, numbered in the order of increase of their moduli ($0 < \mu_1 < \mu_2 < ... < \mu_N$), then

$$W\left\{\omega\left(i\mu_j, x\right), e\left(i\mu_j, x\right)\right\} = e\left(i\mu_j, 0\right) = 0, j = 1, \ldots, N,$$

whence it follows that $e\left(i\mu_j, x\right) = e_x\left(i\mu_j, 0\right)\omega\left(i\mu_j, x\right), j = 1, \ldots, N$. Using the last equalities and formulas (6.24), (6.28), we obtain that the functions

$$\begin{aligned} u(k, x) &= e(-k, x) - S(k)e(k, x), \quad k \in [0, \infty), \\ u(i\mu_j, x) &= m_j e(i\mu_j, x), \quad j = 1, ..., N, \end{aligned}$$

are bounded solutions of the boundary problem (6.1)–(6.2). They form a complete set of normalized eigenfunctions of this problem in the sense that, i.e., for every pair of functions $f(x)$, $g(x)$ from the space $L_2(0, \infty)$ the Parseval's equality hold [86]:

$$(f, g) = \sum_{j=1}^{N} u(i\mu_j, f)\overline{u(i\mu_j, g)} + \frac{1}{2\pi}\int_0^\infty u(k, f)\overline{u(k, g)}dk \tag{6.29}$$

where (f, g) designates the inner product in $L_2[0, \infty)$, and

$$u(k, f) = \int_0^\infty f(x)u(k, x)dx.$$

From estimate (6.6) it follows that the transformation operator $\mathbf{I} + \mathbf{K}$ and its adjoint operator $\mathbf{I} + \mathbf{K}^*$ are bounded in $L_2[0, \infty)$. Therefore, if $f(x) \in$

$L_2[0,\infty)$, then the function $f^*(x) = [(\mathbf{I}+\mathbf{K}^*)f](x)$ also belongs to this space and according to Lemma 6.1:

$$\int_0^\infty f(x)e(-\rho,x)dx = \int_0^\infty f(x)\left\{e^{-i\rho x} + \int_x^\infty K(x,\xi)e^{-i\rho\xi}d\xi\right\}dx =$$

$$\int_0^\infty \left\{f(x) + \int_0^x f(\xi)K(\xi,x)d\xi\right\}e^{-i\rho x}dx = \int_0^\infty f^*(x)e^{-i\rho x}dx = \tilde{f^*}(\rho),$$

where $\tilde{f^*}(\rho)$ designates the ordinary Fourier transformation of the function $f^*(x)$. Consequently,

$$u(k,f) = \tilde{f^*}(k) - S(k)\tilde{f^*}(-k),$$

$$u(i\mu_j,f) = m_j\int_0^\infty f^*(x)e^{-\mu_j x}dx.$$

For brevity we denote the expression of the right-hand side of the formula (6.29) by J. Substituting the expressions for $u(k,f)$ and $u(i\mu_j,f)$ into the right-hand side of (6.29) with due regard for (6.25), we obtain

$$J = (f^*,g^*) + \int_0^\infty\int_0^\infty F(x+y)f^*(x)\overline{g^*(y)}dxdy, \tag{6.30}$$

where the function $F(x)$ is expressible in the form:

$$F(x) = \sum_{j=1}^N m_j^2 e^{-\mu_j x} + F_S(x), \tag{6.31}$$

in which the function $F_S(x)$ possesses the properties written in Lemma 6.7, and

$$F_S(x) = \frac{1}{2\pi}\int_{-\infty}^\infty \{1-S(k)\}e^{ikx}dk. \tag{6.32}$$

Since $\overline{S(k)} = S(-k)$, then the function $F(x)$ is real. Therefore, the operator $\mathbf{F}$ defined by the formula

$$\mathbf{F}[f] = \int_0^\infty F(x+y)f(y)dy,$$

is bounded and self-adjoint in $L_2[0,\infty)$. Using this operator and bearing in mind that $f^* = (\mathbf{I}+\mathbf{K}^*)f$, we write the formula (6.30) as follows

$$\begin{aligned} J = (\{\mathbf{I}+\mathbf{F}\}f^*,g^*) &= (\{\mathbf{I}+\mathbf{F}\}\{\mathbf{I}+\mathbf{K}^*\}f,\{\mathbf{I}+\mathbf{K}^*\}g) \\ &= (\{\mathbf{I}+\mathbf{K}\}\{\mathbf{I}+\mathbf{F}\}\{\mathbf{I}+\mathbf{K}^*\}f,g). \end{aligned}$$

Hence, for the Parseval's equality (6.29) to hold, it is necessary and sufficient that the operators $\mathbf{K}, \mathbf{F}$ and $\mathbf{K}^*$ be related to each other through the identity:

$$\begin{aligned}&(\mathbf{I}+\mathbf{K})(\mathbf{I}+\mathbf{F})(\mathbf{I}+\mathbf{K}^*) = \mathbf{I}, \text{ or equivalently,}\\ &\mathbf{K}+\mathbf{K}^*+\mathbf{K}\mathbf{K}^*+\mathbf{F}+\mathbf{K}\mathbf{F}+\mathbf{F}\mathbf{K}^*+\mathbf{K}\mathbf{F}\mathbf{K}^* = 0.\end{aligned} \tag{6.33}$$

The left-hand side of (6.33) is the integral operator with the kernel:

$$\begin{aligned}\Phi(x,y) = {} & K(x,y) + K(y,x) + F(x+y)\\ & + \int_x^\infty K(x,t)F(t+y)dt + \int_y^\infty F(x+\xi)K(y,\xi)d\xi\\ & + \int_0^\infty K(x,\xi)K(y,\xi)d\xi + \int_y^\infty\int_x^\infty K(x,t)F(t+\xi)K(y,\xi)dtd\xi.\end{aligned} \tag{6.34}$$

It is easily seen that $\Phi(x,y) = \Phi(y,x)$, the identities (6.33) are equivalent to the equality $\Phi(x,y) = 0$ for $y > x$.

We define the function $\varphi_x(y)$ on the half-line (x,∞) by the formula:

$$\varphi_x(y) = K(x,y) + F(x+y) + \int_x^\infty K(x,t)F(y+t)dt.$$

Since $K(y,x) = 0$ for $y > x$, then it follows from (6.34) that for $y > x$:

$$\Phi(x,y) = \varphi_x(y) + \int_y^\infty K(y,\xi)\varphi_x(\xi)d\xi.$$

By Remark 6.1, the operator $\mathbf{I}+\mathbf{K}$ is invertible. Therefore, the identity $\Phi(x,y) = 0$ for $y > x$, and hence the identity (6.33) is equivalent to the equality:

$$\varphi_x(y) = F(x+y) + K(x,y) + \int_x^\infty K(x,t)F(y+t)dt = 0, \quad 0 \le x < y < \infty,$$

This equality is equivalent to the Parseval's equality (6.29). Hence, for the Parseval's equality (6.29) to hold, it is necessary and sufficient that for every $x \in [0,\infty)$, the kernel $K(x,y)$ of the transformation operator, as the function of the variable $y \in [x,\infty)$, satisfies the integral fundamental equation:

$$F(x+y) + K(x,y) + \int_x^\infty K(x,t)F(y+t)dt = 0, \tag{6.35}$$

in which the function $F(x)$ is constructed by the formula (6.31)–(6.32) from the scattering data set:

$$s = \left\{S(k), -\infty < k < \infty, \lambda_j = (i\mu_j)^2, \mu_j > 0, \ m_j, j = 1, ..., N\right\}. \qquad (6.36)$$

The characteristics of the data set (6.36) completely determine the spectrum of the scattering problem (6.1)–(6.2). The quantities of the set (6.36) are called the scattering data for the problem (6.1)–(6.2). The equation (6.35) connecting the scattering data set (6.36) with the kernel $K(x, y)$ of the transformation operator, enables one to reconstruct the potential $p(x)$ by the formula (6.7). This equation is called the fundamental equation.

To derive the fundamental equation (6.35), we use the identity (6.24) with due regard for the formula (6.5) and estimate (6.23).

Theorem 6.1. *The identity* (6.33) *is equivalent to the Parseval's equality* (6.29). *For the identity* (6.33) *to hold, it is necessary and sufficient that for every* $x \in [0, \infty)$ *the kernel* $K(x, y)$ *of the transformation operator as the function of the variable* $y \in [x, \infty)$ *satisfies the fundamental integral equation* (6.35). *The function* $F(x)$ *in* (6.35) *is determined from the scattering data set* (6.36) *by the formula* (6.31)–(6.32).

6.1.3 The necessary conditions of the scattering data

We now use the fundamental equation (6.35) to extract more information on the function $F(x)$. First of all, we note that from the continuity of the kernel $K(x, y)$ it follows that the function $F(x)$ is continuous on $[0, \infty)$.

Next, it follows from the differentiability of the function $K(x, y)$ and the estimate (6.12) for $|K_x(x, y)|$ that the derivative $F'(x)$ exists for $x > 0$, and hence the fundamental equation (6.35) can be differentiated with respect to x. By using this fact and the formula (6.7), estimates (6.6), (6.12), we can prove that the function $x|F'(x)|$ is summable on $[0, \infty)$, and therefore so is $x|F_S'(x)|$. Consequently, the functions $x|F'(x)|$ and $x|F_S'(x)|$ satisfy the same condition:

$$\int_0^\infty x|F'(x)|dx < \infty, \quad \int_0^\infty x|F_S'(x)|dx < \infty, \qquad (6.37)$$

as the potential function $p(x)$.

The scattering function $S(k)$ for the scattering problem (6.1)–(6.2) is defined by the formula: $S(k) = \frac{e(-k)}{e(k)}$. $S(k)$ is continuous for all real $k \neq 0$. This function is continuous at zero in the case, when $e(0) = e(0, 0) \neq 0$, in addition $S(0) = 1$. If $e(0, 0) = 0$, then using equation (6.35) and formulas (6.25), (6.5), we can prove that the scattering function $S(k)$ is continuous at zero, in addition $S(0) = -1$. Hence,

$$S(0) = \begin{cases} 1 & \text{if } e(0,0) \neq 0, \\ -1 & \text{if } e(0,0) = 0. \end{cases} \qquad (6.38)$$

By Lemma 6.3, the function $e(\rho, 0)$ is analytic in the upper half-plane $\mathrm{Im}\rho > 0$ and is continuous in the closed half-plane $\mathrm{Im}\rho \geq 0$. Furthermore, it satisfies the estimate (6.21) and the relation:

$$e(k, 0) = \overline{e(-k, 0)} \text{ for real } k \neq 0. \tag{6.39}$$

Putting $e(k) = |e(k)|e^{i\delta(k)}$, we call the argument $\delta(k)$ the scattering phase of the scattering function $S(k)$. It follows from (6.25) and (6.39) that

$$\delta(k) = -\delta(-k), \;\; \ln S(k) = -2i\delta(k) \text{ for all real } k \neq 0. \tag{6.40}$$

We now apply the argument principle to the function $e(\rho)$ for establishing the relationship between the total increment of the phase $\delta(k)$ and the number N of negative eigenvalues of the problem (6.1)–(6.2). The increment of the phase $\delta(k)$ as k runs over the real axis from $-\infty$ to $+\infty$, bypassing the point $k = 0$ along a half-circle with sufficiently small radius ε in the upper half-plane, is equal to the number of negative eigenvalues N multiplied by 2π:

$$\begin{aligned} 2\pi N &= \{\delta(-\varepsilon) - \delta(-\infty)\} + \{\delta(+\varepsilon) - \delta(-\varepsilon)\} + \{\delta(+\infty) - \delta(+\varepsilon)\} \\ &= 2\{\delta(+\infty) - \delta(+\varepsilon)\} + \{\delta(+\varepsilon) - \delta(-\varepsilon)\}. \end{aligned} \tag{6.41}$$

If $e(0, 0) \neq 0$, then

$$\lim_{\varepsilon \to 0} \{\delta(+\varepsilon) - \delta(-\varepsilon)\} = 0.$$

If $e(0, 0) = 0$, then due to (6.38):

$$\lim_{\varepsilon \to 0} \{\delta(+\varepsilon) - \delta(-\varepsilon)\} = -\pi.$$

Then from (6.41) we obtain an important relation between the increase in the phase $\delta(k)$ and the number N of zeros of $e(\rho)$ lying in the upper half-plane:

$$\frac{2\{\delta(+\infty) - \delta(+0)\}}{2\pi} = \begin{cases} N & \text{if } e(0,0) \neq 0, \\ N + \frac{1}{2} & \text{if } e(0,0) = 0. \end{cases} \tag{6.42}$$

Using (6.38), (6.40), we rewrite the relationship (6.42) as follows:

$$N = \frac{1}{2\pi i} \{\ln S(+0) - \ln S(+\infty)\} - \frac{1 - S(0)}{4}. \tag{6.43}$$

The theorem on necessary conditions of the scattering data (6.36) can be stated as follows.

Theorem 6.2. *The scattering data set* (6.36) *of the boundary problem* (6.1)–(6.2) *with the real potential* $p(x)$ *satisfying estimate* (6.3) *always satisfy the following conditions:*

(I). The scattering function $S(k)$ is continuous on the whole line and possesses the properties (6.25) *and estimate* (6.26). *The function* $1 - S(k)$ *is the Fourier transformation of the function* (6.32). *This function is represented in the form:* $F_S(x) = F_S^{(1)}(x) + F_S^{(2)}(x)$, *where* $F_S^{(1)}(x) \in L_1(-\infty, \infty), F_S^{(2)}(x)$ *is bounded and* $F_S^{(2)}(x) \in L_2(-\infty, \infty)$. *On the positive half-line the function* $F_S(x)$ *has the derivative* $F_S^{'}(x)$ *which satisfies the condition* (6.37)*;*

(II). The full increment of the phase $\delta(k)$ of the scattering function $S(k)$ is related with the number N of negative eigenvalues of the scattering problem (6.1)–(6.2) *by the formula* (6.43).

6.1.4 The necessary and sufficient conditions of a given data set

We consider the operators $\mathbf{F}_{S,a}^{+}, \mathbf{F}_{a}^{+}, (a \geq 0)$ acting in the spaces $L_j(0, \infty)$, $j = 1, 2$ by the rules:

$$\mathbf{F}_{S,a}^{+} f = \int_0^{\infty} F_S(\xi + y + 2a) f(\xi) d\xi,$$

$$\mathbf{F}_{a}^{+} f = \int_0^{\infty} F(\xi + y + 2a) f(\xi) d\xi.$$

These operators appear in the fundamental equation (6.35).

Lemma 6.8. *The operators* $\mathbf{F}_{S,a}^{+}$ *and* $\mathbf{F}_{a}^{+}$ *are compact (completely continuous) in each of the spaces* $L_j(0, \infty), j = 1, 2$ *for every choice of* $a \geq 0$.

Theorem 6.3. *Let the given quantities of a data set of the type* (6.36) *satisfy the conditions (I), then the fundamental equation* (6.35) *has for every* $x > 0$ *a unique solution* $K(x, y) \in L_1(x, \infty)$, *and the function*

$$e(\rho, x) = e^{i\rho x} + \int_x^{\infty} K(x, y) e^{i\rho y} dy, \quad (Im\rho \geq 0) \tag{6.44}$$

satisfies the equation:

$$y_{xx} - p(x) y + \rho^2 y = 0, \quad 0 < x < \infty, \tag{6.45}$$

where

$$p(x) = -2 \frac{d}{dx} K(x, x), \tag{6.46}$$

and for all $\varepsilon > 0$:

$$\int_{\varepsilon}^{\infty} x|p(x)|dx < \infty. \tag{6.47}$$

If, in addition, the operator $\mathbf{I} + \mathbf{F}_a^+$ is also invertible for $a = 0$, then the inequality (6.47) *remains valid for $\varepsilon = 0$, i.e., the function $p(x)$ satisfies the required condition* (6.3).

Theorem 6.4. *For the given quantities of the data set* (6.36) *to be the scattering data of a boundary problem for an equation of the form* (6.1) *on a half-line with the boundary condition* (6.2) *and with a real potential subject to the condition* (6.3), *it is necessary and sufficient that the given quantities of the set* (6.36) *satisfy the conditions (I) and (II).*

Remark 6.3. (1) By Theorem 6.3, the condition (I) guarantees the solvability of the fundamental equation (6.35) for $x > 0$. In this case the function (6.44) satisfies the Schrödinger equation (6.45) with the potential $p(x)$ reconstructed by formula (6.46), and $p(x)$ satisfies estimate (6.47) for $\varepsilon > 0$;

(2) If the conditions (I) and (II) are satisfied, then

(a) The fundamental equation (6.35) has a unique solution both in $L_2[x, \infty)$ and $L_1[x, \infty)$ for each $x \geq 0$, i.e., the homogeneous equation: $f + F_0^+ f = 0$ has no nonzero solution in the corresponding space;

(b) The given quantities of set (6.36) are the scattering data of the boundary problem (6.1)–(6.2) with the reconstructed potential $p(x)$ satisfying inequality (6.47) for $\varepsilon = 0$, i.e., the given quantities satisfy the conditions:

(i) $e(i\mu_j, 0) = 0, \quad j = 1, 2, ..., N,$

(ii) $e(-k, 0) - S(k)e(k, 0) = 0, \quad -\infty < k < \infty.$

6.2 The IBVP for the KdV equation

In this Section we apply the results presented in Section 6.1 on the direct and ISP (6.1)–(6.2) to solving the IBVP for the KdV equation on the positive quarter-plane [144]:

$$p_t - 6pp_x + p_{xxx} = 0, \quad x \geq 0, \quad t \geq 0 \tag{6.48}$$

with the known initial and boundary conditions:

$$p(x,0) = p(x), \tag{6.49}$$
$$p(0,t) = f(t), \tag{6.50}$$

where $f(t)$ is a real-valued continuous function and $p(x,t)$ is a real-valued function satisfying the inequality:

$$\int_0^{\infty} x|p(x,t)|dx < \infty \text{ for any } t \geq 0. \tag{6.51}$$

Let the potential in equation (6.1) depend on an additional parameter t (time). We consider the scattering problem for the time-dependent Schrödinger equation:

$$-y_{xx} + p(x,t)y = \lambda y, \quad x \geq 0, \quad t \geq 0, \quad \lambda = \rho^2 \tag{6.52}$$

with the zero boundary condition at the origin $x = 0$:

$$y(\rho, 0; t) = 0, \quad t \geq 0, \tag{6.53}$$

where the potential $p(x,t)$ is a real-valued function subject to the condition (6.51).

By $L(t)$ we mean the energy operator in $L_2(0,\infty)$ defined by the Schrödinger equation (6.52) with the boundary condition (6.53). The time-dependence of the solution of equation (6.52) is given by the time-evolution equation:

$$\begin{aligned} y_t &= -4y_{xxx} + 6py_x + 3p_x y - 4i\rho^3 y \\ &= (4\rho^2 + 2p)y_x - (4i\rho^3 + p_x)y. \end{aligned} \tag{6.54}$$

The crucial step of the transform technique is to determine from the time-evolution equation (6.54) how the scattering data evolve. This problem will be dealt with in the following subsections.

6.2.1 The Lax compatibility condition

Denote by P the class of real-valued continuous functions satisfying estimate (6.51) and possessing the third-order partial continuous derivative with respect to x, and first-order partial continuous derivative with respect to t. Suppose that the equations (6.52) and (6.54) are considered in the class of isospectral potentials belonging to the class P. Then the compatibility condition of the equations (6.52) and (6.54) for all ρ^2 is equivalent to the KdV equation (6.48). In fact, to verify that the KdV equation (6.48) is the Lax compatibility condition of equations (6.52) and (6.54), it is more convenient to proceed as follows. Transform the second-order equation (6.52) into a system of two first-order equations with the help of an unknown function y_1, such that

$$y_x = i\rho y + y_1. \tag{6.55}$$

Then, from (6.52) we have:

$$y_{1x} = -i\rho y_1 + py. \tag{6.56}$$

The system (6.55) and (6.56) is evidently identical to equation (6.52). Using equations (6.55) and (6.56), we eliminate all x-derivatives from equation (6.54), thus:

$$y_t = 4i\rho^3 y + 4\rho^2 y_1 + 2i\rho p y - p_x y + 2py_1. \tag{6.57}$$

From the equation (6.57), with the help of (6.55) and (6.56), we derive a similar equation for y_1:

$$\begin{aligned} y_{1t} =& -4i\rho^3 y_1 + 4\rho^2 p y + 2i\rho p_x y - 2i\rho p y_1 \\ &+ (2p^2 - p_{xx})y + p_x y_1. \end{aligned} \tag{6.58}$$

By forming the column $\begin{pmatrix} y \\ y_1 \end{pmatrix} = \varphi$ from y and y_1, we rewrite the systems (6.55), (6.56) and (6.57), (6.58) in a compact form:

$$\varphi_x = U(\rho)\varphi, \tag{6.59}$$

$$\varphi_t = V(\rho)\varphi, \tag{6.60}$$

where the matrices U and V depend on an arbitrary complex-valued parameter ρ:

$$U(\rho) = i\rho \begin{pmatrix} 1 & 0 \\ 0 & -1 \end{pmatrix} + \begin{pmatrix} 0 & 1 \\ p & 0 \end{pmatrix}, \tag{6.61}$$

$$V(\rho) = 4i\rho^3 \begin{pmatrix} 1 & 0 \\ 0 & -1 \end{pmatrix} + 4\rho^2 \begin{pmatrix} 0 & 1 \\ p & 0 \end{pmatrix} + 2i\rho \begin{pmatrix} p & 0 \\ p_x & -p \end{pmatrix} + \begin{pmatrix} -p_x & 2p \\ 2p^2 - p_{xx} & p_x \end{pmatrix}. \tag{6.62}$$

The matrices U and V depend on x and t through the function $p(x,t)$. The equations (6.59) and (6.60) form an overdefined system for the vector φ. Differentiating (6.59) with the respect to t, and (6.60) with the respect to x, then equating them, we obtain the Lax compatibility condition of systems (6.59) and (6.60):

$$\frac{\partial U(\rho)}{\partial t} - \frac{\partial V(\rho)}{\partial x} + [U(\rho), V(\rho)] = \begin{pmatrix} 0 & 0 \\ 0 & 0 \end{pmatrix}. \tag{6.63}$$

Substituting (6.61) and (6.62) into (6.63), we find that the left-hand side of the equality (6.63) takes the form:

$$\frac{\partial U(\rho)}{\partial t} - \frac{\partial V(\rho)}{\partial x} + [U(\rho), V(\rho)] = \begin{pmatrix} 0 & 0 \\ p_t - 6pp_x + p_{xxx} & 0 \end{pmatrix}. \tag{6.64}$$

Comparing the right-hand side of equality (6.63) with the right-hand side of equality (6.64), we find that the matrix on the right-hand side of equality (6.64) should vanish. This means that the function $p(x,t)$ should satisfy the KdV equation (6.48). Thus, the compatibility condition (6.63) for equations (6.59) and (6.60) is the new commutative representation of the KdV equation.

Conversely, in the class of potentials belonging to the class P, if the condition (6.63) for compatibility condition of the equations (6.59) and (6.60) is satisfied for all ρ^2, then the spectrum of the operator $L(t)$ is invariant with time, i.e, $\dot{\lambda} = \dfrac{\partial \lambda}{\partial t} = 0$. Indeed, let $\lambda = \lambda_0$ be a point of the spectrum of the operator $L(t)$ for $t = t_0 > 0$, and let $\varphi(\lambda_0, x; t_0)$ be a corresponding eigenfunction. We define the function $r(\lambda_0, x; t)$ by the equality:

$$r(\lambda_0, x; t) = \frac{\partial}{\partial x}\varphi(\lambda_0, x; t) - U\varphi(\lambda_0, x; t), \quad \lambda_0 = \rho_0^2.$$

Differentiating this equality with respect to t, using (6.59), (6.60) and (6.63), yields:

$$\frac{\partial r}{\partial t} = \frac{\partial}{\partial x}\frac{\partial \varphi}{\partial t} - \frac{\partial U}{\partial t}\varphi - U\frac{\partial \varphi}{\partial t} = \left\{\frac{\partial V}{\partial x} - \frac{\partial U}{\partial t} + VU - UV\right\}\varphi = 0.$$

Consequently, we obtain the Cauchy problem for the equation:

$$\frac{\partial r}{\partial t} = 0 \text{ with the initial condition: } r(\lambda_0, x; t_0) = 0.$$

Since $p(x,t)$ in equation (6.59) is a continuous function, then by the unique solvability of the obtained Cauchy problem, we find that $r(\lambda_0, x; t) \equiv 0$ for all $t > 0$, and $\varphi(\lambda_0, x; t)$ is the corresponding eigenfunction of the operator $L(t)$. Thus, the spectrum of $L(t)$ is invariant with time, i.e., $\dot{\lambda} = 0$. The following lemma is proved.

Lemma 6.9. *In the class of isospectral potentials which are continuous functions belonging to the class P, the equations* (6.59) *and* (6.60) *are compatible for all ρ^2, and the matrix* (6.63) *should vanish, i.e., the potential $p(x,t)$ should satisfy the KdV equation* (6.48). *Conversely, if the Lax compatibility condition* (6.63) *is satisfied for all ρ^2, then the spectrum of $L(t)$ is invariant with t, i.e., $\dot{\lambda} = 0$.*

Hence, the KdV equation (6.48) on the half-line can be written as the Lax compatibility condition of two linear eigenvalue equations (6.52) and (6.54), i.e., the KdV equation (6.48) admits its Lax pair by these equations. The equations (6.52) and (6.54) constitute the Lax pair for the KdV equation (6.48) and are usually referred as the x part and the t part, respectively of the Lax pair.

Let the time-dependence of the solution of equation (6.52) is given by the evolution equation:

$$y_t(\rho, x; t) = 2[2\rho^2 + p(x,t)]y_x(\rho, x; t) - [\gamma + p_x(x,t)]y(\rho, x; t), x \geq 0, y \geq 0, \tag{6.65}$$

where γ is an arbitrary constant.

In the class of isospectral potentials belonging to the class P, it is easy to verify that the evolution equation (6.54) is obtained from the evolution equation (6.65) for the solution of equation (6.52), hence

$$y_t = -4y_{xxx} + 6py_x + 3p_x y - \gamma y.$$

Therefore, the equations (6.52) and (6.65) are compatible for all ρ^2 and the potential $p(x,t)$ should satisfy the KdV equation (6.48). Conversely, if $p(x,t)$ satisfies the KdV equation (6.48), then $\dot{\lambda} = 0$. □

6.2.2 The time-dependent Jost solution

The Jost solution $e(\rho, x; t)$ of the Schrödinger equation (6.52) with the boundary condition at infinity:

$$\lim_{x\to\infty} e^{-i\rho x} e(\rho, x; t) = 1 \tag{6.66}$$

is represented in the form (6.5):

$$e(\rho, x; t) = e^{i\rho x} + \int_x^\infty K(x, \xi; t) e^{i\rho\xi} d\xi, \quad \text{Im}\rho \geq 0, \tag{6.67}$$

where t enters (6.67) as a parameter and the kernel $K(x, \xi; t)$ is a real-valued function satisfying equation (6.11) and estimate (6.6).

For real $k \neq 0$ the Jost solutions $e(k, x; t)$ and $e(-k, x; t)$ of the problem (6.52), (6.66) form a fundamental system of solutions of this problem and evolve according to the evolution equation (6.65) with $\gamma = 4ik^3$:

$$e_t(k, x; t) = 2[2k^2 + p(x,t)]e_x(k, x; t) - [4ik^3 + p_x(x,t)]e(k, x; t), \tag{6.68}$$

$$e_t(-k, x; t) = 2[2k^2 + p(x,t)]e_x(-k, x; t) + [4ik^3 - p_x(x,t)]e(-k, x; t). \tag{6.69}$$

Owing to estimates (6.17), (6.18) and (6.51), we obtain that $e_t(\rho, x; t) = o(1)$ for $\text{Im}\rho \geq 0$ as $x \to \infty$. Consequently, the asymptotic behavior of the Jost solution $e(\rho, x; t)$, $\text{Im}\rho \geq 0$ at infinity does not depend on time.

The boundary value (BV) $e(\rho, 0; t)$ is denoted by $e(\rho; t)$. The following lemma on the properties of the Jost solution and its BVs is deduced from (6.19)–(6.21) and Lemmas 6.3 and 6.6.

Lemma 6.10. *The Jost solution $e(\rho, x; t)$ of the problem* (6.52), (6.66) *with the real-valued potential $p(x, t)$ satisfying estimate* (6.51) *is analytic in ρ in the upper half-plane $Im\rho > 0$ and continuous on the real axis. This solution possesses the properties:*

$$e(k, x; t) = \overline{e(-k, x; t)} \text{ for any } (x, t) \in [0, \infty) \times [0, \infty), \quad (6.70)$$
$$e(k; t) \neq 0 \text{ for all } k \neq 0 \text{ and } t \geq 0,$$

and obeys the asymptotic estimate as $x \to \infty$:

$$e(\rho, x; t) = e^{i\rho x} + O\left(\frac{1}{\rho}\right), \quad e_x(\rho, x; t) = i\rho e^{i\rho x} + o(1),$$
$$e_t(\rho, x; t) = o(1), \quad Im\rho \geq 0. \quad (6.71)$$

The BV $e(\rho; t)$ at the origin $x = 0$ of the Jost solution behaves asymptotically like $e(\rho; t) = 1 + O\left(\frac{1}{\rho}\right)$ for large $|\rho|, Im\rho \geq 0$, and has a finite number of simple zeros on the imaginary axis at the points: $\rho = i\mu_j, \mu_j > 0, j = 1, ..., N$. The function $\rho[e(\rho; t)]^{-1}$ is bounded in a neighborhood of the point $\rho = 0$.

6.2.3 The normalization eigenfunction

The normalization eigenfunction of the problem (6.52)–(6.53) is defined by the equality:

$$\Omega(\rho, x; t) = -\frac{2i\rho\omega(\rho, x; t)}{e(\rho; t)} \qquad \text{Im}\rho \geq 0, \ \ x \geq 0, \ \ t \geq 0, \quad (6.72)$$

where $\omega(\rho, x; t)$ is an entire function of ρ satisfying equation (6.52), boundary conditions (6.22) and estimate (6.23).

Due to (6.22) the normalization eigenfunction (6.72) satisfies the boundary conditions at the origin $x = 0$:

$$\Omega(\rho, 0; t) = 0, \ \ \Omega_x(\rho, 0; t) = -\frac{2i\rho}{e(\rho; t)}, \ \ \text{Im}\rho \geq 0. \quad (6.73)$$

From the first condition of (6.73) it follows that

$$\Omega_t(\rho, 0; t) = 0 \ \text{ for any } \ t \geq 0, \ \ \text{Im}\rho \geq 0. \quad (6.74)$$

Consider first the normalization eigenfunction $\Omega(k, x; t)$ corresponding to the continuous spectrum of the scattering problem (6.52), (6.53). We put

$$e(k; t) = |e(k; t)|e^{i\delta(k;t)}, \ \ |e(k; t)| \neq 0 \ \text{ for } \ k \neq 0, \ \ t \geq 0. \quad (6.75)$$

The function $\delta(k;t)$ is called the scattering phase or phase shift of the Jost solution of equation (6.52). Due to (6.70) and (6.75) the scattering phase possesses the property:

$$\delta(k;t) = -\delta(-k;t). \tag{6.76}$$

According to (6.24), (6.25) and (6.76), the normalization eigenfunction $\Omega(k,x;t)$ corresponding to the continuous spectrum of the problem (6.52)–(6.53) is represented in the form:

$$\Omega(k,x;t) = e(-k,x;t) - S(k;t)e(k,x;t) \quad \text{for } k \neq 0, \ \ t \geq 0, \tag{6.77}$$

where

$$S(k;t) = \frac{e(-k;t)}{e(k;t)} = \overline{S(-k;t)} = [S(-k;t)]^{-1} = e^{-2i\delta(k;t)}. \tag{6.78}$$

By virtue of (6.70) the representation (6.77) of $\Omega(k,x;t)$ is rewritten as follows:

$$\Omega(k,x;t) = -\frac{2i}{e(k;t)}\mathrm{Im}\{e(-k;t)e(k,x;t)\} \ \text{for } k \neq 0 \ \text{and } x \geq 0, \ \ t \geq 0. \tag{6.79}$$

Substituting formula (6.67) with $\mathrm{Im}\rho = 0$ into (6.79), and taking formula (6.75) into account, we obtain the important formula for the representation of the normalization eigenfunction corresponding to the continuous spectrum of the scattering problem (6.52)–(6.53):

$$\Omega(k,x;t) = -\,2ie^{i\delta(k;t)}\left\{\sin(kx - \delta(k;t)) + \int_x^\infty K(x,\xi;t)\sin(k\xi - \delta(k;t))d\xi\right\}$$
$$\text{for } k \neq 0 \ \text{and } x \geq 0, \ \ t \geq 0, \tag{6.80}$$

where by Lemmas 6.1 and 6.2 the kernel $K(x,\xi;t)$ satisfies equation (6.11) and estimates (6.6), (6.12). Therefore, the improper integral in (6.80) is convergent.

We now derive the evolution equation for the normalization eigenfunction $\Omega(k,x;t)$. Differentiating (6.77) with respect to t yields

$$\Omega_t(k,x;t) = e_t(-k,x;t) - S_t(k;t)e(k,x;t) - S(k;t)e_t(k,x;t). \tag{6.81}$$

Substituting (6.68) and (6.69) into (6.81), then using (6.77), we get

$$\Omega_t(k,x;t) = \left[4ik^3 - p_x(x,t)\right]\Omega(k,x;t) + 2[2k^2 + p(x,t)]\Omega_x(k,x;t) +$$
$$[8ik^3 S(k;t) - S_t(k;t)]e(k,x;t). \tag{6.82}$$

Owing to conditions (6.73) and (6.74), the passage in equation (6.82) to the limit as $x \to 0$ leads to the evolution equation for the scattering function $S(k;t)$:

$$S_t(k;t) = 8ik^3 S(k;t) - 4ik\frac{2k^2 + p(0,t)}{e^2(k;t)} \quad \text{for} \quad k \neq 0, \quad t \geq 0. \tag{6.83}$$

Substituting (6.83) into (6.82), we derive the evolution equation for the normalization eigenfunction $\Omega(k,x;t)$ corresponding to the continuous spectrum of problem (6.52)–(6.53):

$$\begin{aligned} \Omega_t(k,x;t) =& [4ik^3 - p_x(x,t)]\Omega(k,x;t) + 2[2k^2 + p(x,t)]\Omega_x(k,x;t) + \\ & 4ik\frac{2k^2 + p(0,t)}{e^2(k;t)} e(k,x;t) \quad \text{for} \quad k \neq 0 \quad \text{and} \quad x \geq 0, \quad t \geq 0. \end{aligned} \tag{6.84}$$

Since $\Omega(k,x;t)$ and $\Omega_x(k,x;t)$ in the right-hand side of equality (6.84) satisfy the boundary conditions (6.73), then $\Omega_t(k,x;t)$ on the left-hand side of this equality obeys the condition (6.74). We thus arrive the following lemma. □

Lemma 6.11. *The normalization eigenfunction corresponding to the continuous spectrum of the scattering problem* (6.52)–(6.53) *with the real-valued potential satisfying estimate* (6.51) *is represented in the form* (6.80) *and evolves according to the evolution equation* (6.84). *The scattering function* $S(k;t)$ *for this scattering problem obeys the evolution equation* (6.83).

Remark 6.4. If the third term in the right-hand side of the equation (6.84) is canceled, then this equation becomes the known evolution equation for the eigenfunction of the Cauchy initial-value problem for the KdV equation on the full-line: $-\infty < x < \infty$ [156].

Remark 6.5. The BVs $e(\rho;t)$ evaluated at $x = 0$ of the Jost solution give the complete characterization of the spectrum of the operator $L(t)$. The time-dependence of the scattering function is determined by the evolution equation (6.83), which contains the unknown BVs $e(k;t)$. Therefore, the main difficulty in solving the IBVP (6.48)–(6.49)–(6.50) lies in the determination of these unknown BVs in terms of the known conditions (6.49) and (6.50). We deal with this difficulty in the next two subsections.

6.2.4 The Sturm–Liouville scattering problem

We overcome the difficulty mentioned in Remark 6.5 by considering the Sturm–Liouville scattering problem for equation (6.52) on the interval $(0,b)$, (b is a large positive number), with the boundary conditions (6.73) at $x = 0$ and the zero boundary condition at $x = b$:

$$\Omega(k,b;t) = 0 \quad \text{for} \quad k \neq 0 \quad \text{and} \quad t \geq 0. \tag{6.85}$$

We regard the Sturm–Liouville problem for equation (6.52) with boundary conditions (6.73) and (6.85) as a linear problem associated with the IBVP (6.48)–(6.49)–(6.50). Theorem 6.6 at the end of this section proves that the potential in equation (6.52) is uniquely reconstructed from the scattering data of the Sturm-Lioville (6.52), (6.73), (6.85) and is the solution of the IBVP (6.48)–(6.49)–(6.50). In order to prove this theorem, we need to derive the time-dependence of the scattering data. The normalization eigenfunction $\Omega(k,x;t)$ of equation (6.52) with boundary condition (6.53) is represented by the formula (6.80). Suppose that the given eigenfunction $\Omega(k,x;t)$ of equation (6.52) with boundary conditions (6.73) at $x=0$ satisfies one more zero boundary condition (6.85) at $x=b$. Then the equation (6.85) is determined as the zero boundary condition at $x=b$ of the normalization eigenfunction $\Omega(k,x;t)$ of the Sturm–Liouville problem (6.52), (6.73), (6.85), wherein t enters equation (6.85) as a parameter and $\Omega(k,b;t)$ is the function of k. Therefore, the positive eigenvalues of this spectral Sturm–Liouville problem in essence are the roots of equation (6.85). Since the improper integral in (6.80) is regularly convergent, then the asymptotic form of the normalization eigenfunction (6.80) as $b\to\infty$ can be written as follows:

$$\Omega(k,b;t) = -2ie^{i\delta(k;t)}\sin(kb-\delta(k;t)) + o(1) = 0 \quad (b\to\infty),$$

whence it follows that

$$\sin(kb-\delta(k;t)) = o(1) \;\; (b\to\infty) \;\; \text{for} \;\; k\neq 0, \;\; t\geq 0. \tag{6.86}$$

The roots of equation (6.86) are determined approximately by the asymptotic equation:

$$kb-\delta(k;t) = \arcsin(o(1)) = j\pi + o(1) \;\; (b\to\infty) \;\; \text{for} \;\; k\neq 0, \;\; t\geq 0, \quad j=0,1,2,... \tag{6.87}$$

It follows from (6.87) that the scattering phase $\delta(k;t)$ for the Sturm–Liouville problem (6.52), (6.73), (6.85) is approximately time-independent, i.e.,

$$\delta_t(k;t) = o(1) \;\; (b\to\infty) \;\; \text{for} \;\; k\neq 0, \;\; t\geq 0. \tag{6.88}$$

The asymptotic relation between $\delta(k;t)$ and $\delta(k)$ is deduced from (6.87) and (6.88):

$$\delta(k;t) = \delta(k) + o(1) = kb - j\pi + o(1) \;\; (b\to\infty) \;\; \text{for} \;\; k\neq 0, \;\; t\geq 0, \quad j=0,1,2,... \tag{6.89}$$

Owing to (6.89) the scattering function $S(k;t)$ given by (6.78) approximately tends to $S(k)=S(k;0)$ as $b\to\infty$:

$$S(k;t) = e^{-2i\delta(k,t)} = e^{-2i[\delta(k)+o(1)]} = S(k)[1+o(1)], (b\to\infty) \text{ for } k\neq 0, t\geq 0, \tag{6.90}$$

where $\delta(k) = kb - j\pi, j=0,1,2,...$

Thus, the scattering function for the considered Sturm–Liouville problem is approximately time-independent. The positive eigenvalues of this Sturm–Liouville problem are found approximately by (6.87) and (6.89):

$$\lambda_j = k_j^2 = \left(\frac{\delta(k)+j\pi}{b}\right)^2 + \left(\frac{o(1)}{b}\right)^2 \quad (b\to\infty) \text{ for } k\neq 0,\ j=0,1,2,... \tag{6.91}$$

It follows from (6.42) and (6.89) that the full increase in the scattering phase $\delta(k;t)$ of the scattering function $S(k;t)$ is related with the number N of negative eigenvalues of the problem (6.52), (6.73) by the asymptotic formula:

$$\frac{2\{\delta(+\infty)-\delta(+0)\}}{2\pi} + o(1) = \begin{cases} N+o(1) & \text{if } e(0,0)\neq 0, \\ N+\frac{1}{2}+o(1) & \text{if } e(0,0)=0. \end{cases} \quad (b\to\infty). \tag{6.92}$$

Using (6.43), we write the asymptotic formula (6.92) as follows:

$$N+o(1) = \frac{1}{2\pi i}\{\ln S(+0) - \ln S(+\infty)\} - \frac{1-S(0)}{4} + o(1) \quad (b\to\infty). \tag{6.93}$$

6.2.5 Calculation of unknown BVs

By using (6.70) and (6.78), we derive from (6.83) the evolution equation for the scattering phase:

$$\delta_t(k;t) = -4k^3 + 2k\frac{2k^2+p(0,t)}{|e(k;t)|^2} \quad \text{for } k\neq 0,\ t\geq 0.$$

From this evolution equation, using (6.88), we get the asymptotic formula for the relation between $|e(k;t)|^2$ and $p(0,t)$:

$$\frac{2k^2+p(0,t)}{|e(k;t)|^2} - 2k^2 = o(1) \quad (b\to\infty) \text{ for } k\neq 0,\ t\geq 0.$$

Since $e(k;t)\neq 0$ for $k\neq 0$ and $t\geq 0$, then the square of the time-dependent modulus $|e(k;t)|^2$ of the unknown BV $e(k;t)$ is calculated from the known boundary condition $p(0,t)$ by the formula:

$$|e(k;t)|^2 = 1 + \frac{p(0,t)}{2k^2} + o(1) \quad (b\to\infty) \text{ for } k\neq 0,\ t\geq 0. \tag{6.94}$$

Owing to (6.89) and (6.94), we obtain the asymptotic formulas for finding the unknown time-dependent BVs:

$$e(k,t) = |e(k;t)|e^{i\delta(k;t)} = |e(k;t)|e^{i[\delta(k)+o(1)]}, \quad (b\to\infty), \tag{6.95}$$

$$|e(k;t)| = \sqrt{1+\frac{p(0,t)}{2k^2}+o(1)} \quad (b\to\infty) \text{ for } k\neq 0,\ t\geq 0,$$

subject to $1+\frac{p(0,t)}{k^2}\geq 0$.

In the formulas (6.95) the scattering phase $\delta(k)$ is calculated from the known initial condition (6.49), therefore the unknown BV $e(k;t)$ is determined by (6.95) in terms of the known initial and boundary conditions (6.49) and (6.50).

Remark 6.6. If the given potential function $p(x,t)$ in equation (6.52) is decreasing so fast that

$$p(x,t) \equiv 0 \text{ for } x \geq b \text{ and } \int_0^b x|p(x,t)|dx < \infty \text{ for any } t \geq 0, \tag{6.96}$$

then it follows from (6.10) that

$$K(x,\xi;t) \equiv 0 \text{ for } \xi \geq x \geq b \text{ and } t \geq 0.$$

Consequently, under the conditions (6.96), the zero boundary condition (6.85) at $x = b$ of the normalization eigenfunction (6.80) is represented by the exact formula:

$$\Omega(k,b;t) = -2ie^{-i\delta(k;t)} \sin\left(kb - \delta(k;t)\right) = 0 \text{ for } k \neq 0, \ t \geq 0. \tag{6.97}$$

The roots of equation (6.97) are determined by

$$kb - \delta(k;t) = j_1\pi \text{ for } \quad k \neq 0, \ t \geq 0, \ j_1 = 0,1,2,....$$

Hence, if the potential $p(x,t)$ satisfies conditions (6.96), then the scattering phase is time-independent, i.e., $\delta_t(k;t) = 0$, and

$$\delta(k;t) = \delta(k) = kb - j_1\pi \text{ for } \quad k \neq 0, \ t \geq 0, \ j_1 = 0,1,2,....$$

Due to this fact, the exact formulas for calculating the unknown BV for the case (6.96) are deduced from (6.95)

$$e(k;t) = |e(k;t)|e^{i\delta(k)}, \ |e(k;t)| = \sqrt{1 + \frac{p(0,t)}{2k^2}} \text{ for } k \neq 0, \ t \geq 0. \tag{6.98}$$

6.2.6 The time-dependent scattering data

The BV $e(k;t)$ calculated by formulas (6.95) enables us to construct the time-dependent scattering data of the Sturm-Liouville scattering problem (6.52), (6.73), (6.85). In fact, the Jost solution and its BV have a bounded analytic continuation $e(\rho,x;t)$ and $e(\rho;t)$, respectively, into the upper half-plane $\text{Im}\rho > 0$. The properties of the analytic continuation $e(\rho.t)$ are assembled in Lemma 6.10. Knowing the calculated function $e(\rho;t)$, $\text{Im}\rho > 0$, we consider the normalization eigenfunction (6.72) corresponding to the discrete spectrum of the Sturm–Liouville problem (6.52), (6.73), (6.85):

$$\Omega(\rho,x;t) = -\frac{2i\rho\omega(\rho,x;t)}{e(\rho;t)} \ \text{Im}\rho > 0, \ x \geq 0, \ t \geq 0.$$

The time-dependence of the Jost solution $e(\rho, x; t)$, $\text{Im}\rho > 0$ is given by the evolution equation (6.65).

$$e_t(\rho, x; t) = 2[2\rho^2 + p(x,t)]e_x(\rho, x; t) - [\gamma + p_x(x,t)]e(\rho, x; t), x \geq 0, t \geq 0. \tag{6.99}$$

By Lemma 6.10 the BV $e(\rho; t)$ has the simple zeros on the imaginary axis at the points: $\rho = i\mu_j, \;\; \mu_j > 0$:

$$e(i\mu_j, 0; t) = e(i\mu_j; t) = 0, \;\; j = 1, ..., N.$$

According to (6.28), the normalization multipliers $m_j(t)$ are calculated by the formula:

$$m_j^2(t) = \frac{2\mu_j}{ie_x(i\mu_j; t)\dot{e}(i\mu_j; t)}, \;\; j = 1, ..., N, \tag{6.100}$$

where $e_x(i\mu_j; t) = e_x(i\mu_j, 0; t), \;\; \dot{e}(i\mu_j; t) = \frac{\partial}{\partial\rho}e(\rho, 0; t)|_{\rho=i\mu_j}, \;\; j = 1, ..., N.$

By using evolution equation (6.99) for $e(x, \rho; t)$, we get rid of $e_x(i\mu_j; t)$ in the formula (6.100). Indeed, in equation (6.99) for $e(x, \rho; t)$ putting: $x = 0$ and $\rho = i\mu_j$, then with due regard for that $e(i\mu_j, t) = 0$, we obtain that $e_x(i\mu_j; t)$ is expressible in terms through the known data by the formula:

$$e_x(i\mu_j; t) = \frac{e_t(i\mu_j; t)}{2[p(0,t) - 2\mu_j^2]}, \;\; j = 1, ..., N.$$

Substituting this formula for $e_x(i\mu_j; t)$ into (6.100), we obtain the representation of the square of normalization multipliers:

$$m_j^2(t) = 4\mu_j \frac{p(0,t) - 2\mu_j^2}{i\dot{e}(i\mu_j; t)e_t(i\mu_j; t)}, \;\; j = 1, ..., N. \tag{6.101}$$

where $m_j^2(0) = m_j^2$ and m_j^{-2} is defined by the formula (6.28) from the known initial condition (6.49). The functions $\dot{e}(i\mu_j; t)$ and $e_t(i\mu_j; t)$ are determined by the bounded analytic continuation $e(\rho; t)$ into the upper half-plane $\text{Im}\rho > 0$. This continuation is obtained from the calculated BV (6.95).

The scattering function $S(k; t)$ defined by (6.90) is approximately time-independent. This function and the simple zeros $i\mu_j, \mu_j > 0, j = 1, ..., N$, of $e(\rho, t)$, the normalization multipliers $m_j(t)$ characterize the Sturm–Liouville scattering problem (6.52), (6.73), (6.85) on the continuous and discrete spectrum, respectively. Thereby, the time-dependent scattering data set $s(t)$ for this problem is constructed:

$$s(t) = \{S(k)[1 + o(1)], (b \to \infty), -\infty < k < \infty; i\mu_j, \mu_j > 0; m_j(t), j = 1, ..., N\}, \tag{6.102}$$

where the quantities of the constructed set $s(t)$ for every $t \geq 0$ satisfy the conditions (I) and (II) enumerated in the Theorem 6.4.

Thus, the following theorem on the characterizations of the considered Sturm-Liouville problem is proved.

Theorem 6.5. *The IBVP* (6.48)–(6.49)–(6.50) *is associated with the Sturm–Liouville scattering problem in the interval* $(0, b)$ *(b is a large positive number) generated by the Schrödinger equation* (6.52) *with an isospectral potential* $p(x, t)$ *belonging to the class* P *and with the zero boundary conditions* (6.73) *at* $x = 0$ *and* (6.85) *at* $x = b$. *This associated scattering problem possesses the following characteristics:*

(1) The characterization of the scattering problem on the continuous spectrum is described by the approximately time-independent scattering phase and scattering function, which are represented by the asymptotic formulas (6.89) *and* (6.90), *respectively;*

(2) The full increase in the scattering phase $\delta(k; t)$ *of the scattering function* $S(k; t)$ *is related to the number* N *of the negative eigenvalues of the scattering problem* (6.52), (6.73) *by the asymptotic formula* (6.93);

(3) The unknown time-dependent BV $e(k; t)$ *evaluated at* $x = 0$ *of the Jost solution* $e(k, x; t)$ *of problem* (6.52), (6.66) *is found by the asymptotic formulas* (6.95) *in terms of the known initial and boundary conditions* (6.49) *and* (6.50) *of the considered IBVP* (6.48)–(6.49)–(6.50);

(4) The characterization of the scattering problem on the discrete spectrum is given by the time-dependent normalization multipliers $m_j(t)$, *which are calculated by the formulas* (6.101) *in terms of the known boundary condition* (6.50) *and the bounded analytic continuation* $e(\rho; t)$ *into the upper half-plane* $Im\rho > 0$. *This continuation is obtained from the calculated BV* $e(k; t)$.

6.2.7 The IBVP (6.48)–(6.49)–(6.50)

The objective of this section is to solve the IBVP (6.48)–(6.49)–(6.50). We consider this problem in the class of isospectral potentials which are real-valued continuous functions belonging to the class P.

As shown in Section 6.1, we construct from the known initial condition (6.49) a data set of the form (6.36) of the quantities satisfying the conditions (I) and (II) enumerated in the Theorem 6.2. Further, a time-dependent data set of the form (6.102) is constructed in accordance with the Theorem 6.5 from the data set (6.36). Then the function $F(x, t)$ is made up from the time-dependent data set (6.102) by the formula:

$$F(x; t) = F_S(x; t) + \sum_{j=1}^{N} m_j^2(t) e^{-\mu_j x}, \tag{6.103}$$

where $m_j^2(t)$ are calculated by (6.101) and $F_S(x)$ is constructed from the scattering function (6.90) by the formula:

$$F_S(x;t) = \frac{1}{2\pi}\int_{-\infty}^{\infty} \{1 - S(k)[1+o(1)]\} e^{ikx} dk, \quad (b \to \infty). \tag{6.104}$$

Thus, Theorems 6.4 and 6.5 reduce the problem of solving the IBVP (6.48)–(6.49)–(6.50) to that of solving the fundamental equation of the form (6.35):

$$F(x+y;t) + K(x,y;t) + \int_x^{\infty} K(x,\xi;t)F(\xi+y;t)d\xi = 0 \text{ for } 0 \le x < y < \infty,$$

$$t \ge 0, \tag{6.105}$$

where both x and t enter equation (6.105) merely as parameters, $F(x;t)$, $F_S(x;t)$ are known real functions constructed from the given data set (6.102), $K(x,y;t)$ is an unknown function of y for every $(x,t) \in [0,\infty] \times [0,\infty)$.

Since for every fixed $t \ge 0$ the functions $S(k;t)$ and $F_S(x;t)$ satisfy the conditions (I) and (II), then the functions $F(x;t)$ and $F_S(x;t)$ have derivatives with respect to x on the positive half-line and satisfy the inequality (6.37):

$$\int_0^{\infty} x|F'(x;t)|dx < \infty, \quad \int_0^{\infty} x|F_S'(x;t)|dx < \infty. \tag{6.106}$$

Thus, the known function $F(x,t)$ in equation (6.105) possesses the properties similar to the properties of the potential. Hence the function $F(x,t)$ belongs to the class P. Further, using the fundamental equation (6.105), estimates (6.106), by Theorems 6.4 and 6.5 and Remark 6.3, we can show that:

(1) The fundamental equation (6.105) has for every $(x,t) \in [0,\infty] \times [0,\infty)$ a unique solution $K(x,y;t)$ belonging to the class P;

(2) The function constructed from the solution $K(x,y;t)$ of equation (6.105):

$$e(\rho,x;t) = e^{i\rho x} + \int_x^{\infty} K(x,y;t)e^{i\rho y}dy, \quad \text{Im}\rho \ge 0, \tag{6.107}$$

satisfies an equation of the form (6.52) with the reconstructed potential:

$$p(x,t) = -2\frac{d}{dx}K(x,x;t), \quad x \ge 0, \quad t \ge 0, \tag{6.108}$$

where $p(x,t)$ is a real-valued function belonging to the class P and satisfying estimate (6.51);

(3) The quantities of the data set of the type (6.102) constructed from the given initial and boundary conditions (6.49) and (6.50) satisfy the conditions (I) and (II) enumerated in Theorem 6.2. Therefore they are the scattering data of the Sturm–Liouville scattering problem for an equation of the form (6.52) with the recovered potential (6.108), i.e., the given quantities of the set (6.102) satisfy the conditions:

$$e(-k,0;t) - S(k)[1+o(1)]e(k,0;t) = 0 \ \ (b \to \infty), -\infty < k < \infty, \ \ t \geq 0,$$
$$e(i\rho_j,0;t) = 0, \ \ t \geq 0, \ \ \rho_j > 0, \ \ j = 1,...,N;$$

(4) A correspondence between the data set of the type (6.102) and the reconstructed potential (6.108) is one-to-one. The equations (6.59) and (6.60) with the constructed potential (6.108) are compatible, because the potential (6.108) belongs to the class P. Then by Lemma 6.9 the reconstructed potential (6.108) is the solution of the IBVP (6.48)–(6.49)–(6.50).

Thus, the theorem on the solution of the considered IBVP is proved.

Theorem 6.6. *Consider the IBVP* (6.48)–(6.49)–(6.50) *in the class of isospectral potentials belonging to the class P. Then*

(1) The solution $p(x,t)$ *of the IBVP* (6.48)–(6.49)–(6.50) *is uniquely reconstructed by formula* (6.108) *in terms of the solution* $K(x,y;t)$ *of the fundamental equation* (6.105). *In this equation the functions* (6.103) *and* (6.104) *are made up of the data* (6.102). *The obtained solution* (6.108) *of the IBVP is a real-valued function belonging to the class P;*

(2) The data set $s(t)$ *of the type* (6.102) *constructed from the known initial and boundary conditions* (6.49) *and* (6.50) *is the scattering data of the Sturm–Liouville scattering problem for the Schrödinger equation* (6.52) *with the recovered potential* (6.108) *and with the zero boundary conditions* (6.73) *and* (6.85) *at* $x = 0$ *and at* $x = b$, *respectively;*

(3) Every scattering data set of the type (6.102) *corresponds with a unique solution of the IBVP* (6.48)–(6.49)–(6.50) *and evolves from the continuous and discrete spectrum of the associated Sturm–Liouville scattering problem.*

Remark 6.7. The potential $p(x,t)$ in the Schrödinger equation (6.52) depends on an additional parameter t and satisfies constraint (6.51). If this parameter is taken at any $t \in (-\infty,\infty)$, and the potential $p(x,t)$ is required to satisfy the constraint (6.51) for any $t \in (-\infty,\infty)$, then Theorems 6.5 and 6.6 remain valid for any $t \in (-\infty,\infty)$; i.e., the IBVP (6.48)–(6.49)–(6.50) can be solved for the region: $(x,t) \in [0,\infty) \times (-\infty,\infty)$.

6.3 Exact soliton-solutions of the Cauchy problem for the KdV equation

In this Section we apply results of the direct and ISP (6.52)–(6.53) for $(x,t) \in [0,\infty) \times (-\infty,\infty)$ to finding exact soliton-solutions of the KdV equation in the class of non-scattering potentials. With the aim to find exact soliton-solutions of the KdV equation on a half-line, we consider in the space $L_2[0,\infty)$ the scattering problem for the Schrödinger equation:

$$-y_{xx} + p(x,t)y = \rho^2 y, \quad (x,t) \in [0,\infty) \times (-\infty,\infty), \tag{6.109}$$

with the boundary condition:

$$y(\rho, 0; t) = 0, \tag{6.110}$$

where the potential $p(x,t)$ is a real-valued function satisfying the condition for any $t \in (-\infty,\infty)$ and some $\varepsilon > 0$:

$$\int\limits_0^\infty e^{\varepsilon x}|p(x,t)|dx < \infty. \tag{6.111}$$

The problem (6.109)–(6.110) with a potential which is not assumed to be real-valued function satisfying condition (6.111) was solved in [79].

6.3.1 The direct and inverse problem (6.109)–(6.110)

The scattering problem (6.109)–(6.110) is associated with the Cauchy problem for the KdV equation on a half-line. Since the potential $p(x,t)$ satisfies the condition (6.111), then it satisfies the condition (6.51). Hence, the problem (6.109)–(6.110) with the real-valued potential satisfying condition (6.111) is a particular case of the problem (6.52)–(6.53) considered in Section 6.2. We omit the proof of lemmas and theorems of the direct and inverse problem (6.109)–(6.110), because it can be deduced from the proof of corresponding lemmas and theorems of the direct and inverse scattering problem (6.52)–(6.53) in Section 6.2.

In Subsections 6.3.1 and 6.3.2, t is a merely parameter, therefore we omit it for convenient presentation. The equation (6.109) with the potential $p(x)$ satisfying the condition (6.111) has a solution $e(\rho,x)$, which for each $x \geq 0$ is a holomorphic function of ρ when $\text{Im}\rho > -\frac{\varepsilon}{2}$ and satisfies the asymptotic condition as $x \to \infty$:

$$e(\rho,x) = e^{i\rho x}[1+o(1)], \quad e'_x(\rho,x) = e^{i\rho x}[i\rho + o(1)]. \tag{6.112}$$

For each $\rho_0 > 0$ the equation (6.109) has a solution $e_1(\rho, x)$, which for each $x \geq 0$ is a holomorphic function of ρ in the domain $|\rho| > \rho_0$, Im$\rho > 0$, and satisfies the asymptotic condition as $x \to \infty$ [7, 79]:

$$e_1(\rho, x) = e^{-i\rho x}[1 + o(1)], \;\; e'_{1x}(\rho, x) = e^{-i\rho x}[-i\rho + o(1)] \qquad (6.113)$$

uniformly in ρ in the domain $|\rho| > \rho_0 > 0$.

It follows from the asymptotic relations (6.112) and (6.113) that the functions $e(\rho, x), e(-\rho, x)$ and $e(\rho, x), e_1(\rho, x)$ form the fundamental systems of solutions of equation (6.109) and their Wronskians are equal to:

$$W[e(\rho, x), e(-\rho, x)] = -2i\rho \;\text{ for }\; |\text{Im}\rho| < \frac{\varepsilon}{2}, \qquad (6.114)$$

$$W[e(\rho, x), e_1(\rho, x)] = -2i\rho \;\text{ for }\; |\rho| > \rho_0, \;\; \text{Im}\rho > 0, \;\; \rho_0 > 0. \qquad (6.115)$$

The properties of the solutions $e(\rho, x)$ and $e_1(\rho, x)$ were studied in detail, in particular in the book [7]. It was proved in [7] that

$$e(\rho, x) = e^{i\rho x} + \int_x^\infty K(x, \xi) e^{i\rho\xi} d\xi, \qquad (6.116)$$

where the kernel $K(x, \xi)$ has first-order continuous partial derivatives with respect to x and ξ.

From the estimates obtained in [7] and condition (6.111) we can easily deduce that for $0 \leq x \leq \xi < \infty$:

$$\begin{aligned} &|K(x, \xi)| < C \exp\{-\varepsilon\frac{x + \xi}{2}\}, \\ &|K'_x(x, \xi)|, \; |K'_\xi(x, \xi)| \leq \frac{1}{4}|p(\frac{x + \xi}{2})| + C \exp\{-\varepsilon(\frac{3}{2}x + \xi)\}, \end{aligned} \qquad (6.117)$$

where C is a certain number.

We denote by $\omega(\rho, x)$ the solution of the eigenvalue problem generated by the equation:

$$-\frac{d^2}{dx^2}\omega(\rho, x) + p(x)\omega(\rho, x) = \rho^2\omega(\rho, x), x \in [0, \infty), \qquad (6.118)$$

with the initial conditions:

$$\omega(\rho, 0) = 0, \;\; \frac{d}{dx}\omega(\rho, 0) = 1. \qquad (6.119)$$

By virtue of Wronskians (6.114) and (6.115), the solution of the problem (6.118)–(6.119) is represented in the form:

$$\omega(\rho, x) = \frac{e(-\rho)e(\rho, x) - e(\rho)e(-\rho, x)}{2i\rho} \;\text{ for }\; |\text{Im}\rho| < \frac{\varepsilon}{2}, \qquad (6.120)$$

$$\omega(\rho, x) = \frac{e_1(\rho)e(\rho, x) - e(\rho)e_1(\rho, x)}{2i\rho} \;\text{ for }\; |\rho| > \rho_0, \;\; \rho > 0, \qquad (6.121)$$

where $e(\rho) = e(\rho, 0)$ and $e_1(\rho) = e_1(\rho, 0)$.

Differentiating the equality (6.121) with respect to x, using the initial conditions (6.119), we have:

$$e_1(\rho) = \frac{2i\rho + e(\rho)e'_{1x}(\rho)}{e'_x(\rho)} \text{ for } |\rho| > \rho_0, \ \rho > 0. \tag{6.122}$$

By L we mean the operator generated in the space $L_2[0,\infty)$ by the equation (6.109) and boundary condition (6.110), where the potential $p(x)$ in equation (6.109) is a real-valued function satisfying condition (6.111). Consider an eigenfunction $\Omega(\rho, x)$ of the operator L normalized in the following way:

$$\Omega(\rho, x) = \frac{2i\rho\omega(\rho, x)}{e(\rho)}, \ \text{Im}\rho > -\frac{\varepsilon}{2}. \tag{6.123}$$

By (6.120) and (6.121), the normalized eigenfunction $\Omega(\rho, x)$ is represented in the form:

$$\Omega(\rho, x) = S(\rho)e(\rho, x) - e(-\rho, x) \text{ for } |\text{Im}\rho| < \frac{\varepsilon}{2},$$
$$\Omega(\rho, x) = S_1(\rho)e(\rho, x) - e_1(\rho, x) \text{ for } |\rho| > \rho_0,$$

where

$$S(\rho) = \frac{e(-\rho)}{e(\rho)} \text{for } |\text{Im}\rho| < \frac{\varepsilon}{2},$$
$$\tag{6.124}$$
$$S_1(\rho) = \frac{e_1(\rho)}{e(\rho)} \text{for } |\rho| > \rho_0, \ \text{Im}\rho > 0.$$

The functions $S(\rho)$ and $S_1(\rho)$ are called the scattering function and the reflection coefficient of the operator L, respectively.

Since the potential in the equation (6.118) is a real-valued function satisfying estimate (6.111), then by Lemma 6.6 all the zeros ρ_j of the function $e(\rho)$ are simple and lie on the imaginary axis, i.e., $\rho_j = i\mu_j, \ \mu_j > \varepsilon_0 > 0, \ j = 1, ..., N$. By virtue of this fact, using the expression (6.122) and (6.28), we calculate [114]:

$$f_j(x) = i\text{Res}|_{\rho=i\mu_j}\left\{\frac{e_1(\rho)}{e(\rho)}e^{i\rho x}\right\} = i\text{Res}|_{\rho=i\mu_j}\left\{\frac{2i\rho + e(\rho)e'_{1x}(\rho)}{e(\rho)e'_x(\rho)}e^{i\rho x}\right\} =$$
$$i\text{Res}|_{i\mu_j}\left\{\frac{2i\rho e^{i\rho x}}{e(\rho)e'_x(\rho)}\right\} = i\frac{2i(i\mu_j)e^{-\mu_j x}}{e'_\rho(i\mu_j)e'_x(i\mu_j)}, \tag{6.125}$$

$$e^{\mu_j x}f_j(x) = -\frac{2i\mu_j}{e'_\rho(i\mu_j)e'_x(i\mu_j)} = \left(\int_0^\infty |e(i\mu_j, x)|^2 dx\right)^{-1} = m_j^2 > 0, \ j = 1, ..., N. \tag{6.126}$$

m_j^2 are called normalization multipliers of the operator L corresponding to the simple poles $i\mu_j$, $j = 1, ..., N$.

We introduce the function :

$$F_S(x) = \frac{1}{2\pi} \int\limits_{-\infty+i\eta}^{+\infty+i\eta} [S(\rho) - 1] e^{i\rho x} d\rho, \tag{6.127}$$

where η is a number satisfying the inequality: $0 < \eta < \varepsilon_0$, ϵ_0 is defined by (1.23).

The integral (6.127) is applied to analytic function $S(\rho) - 1$ in the strip $0 < |\mathrm{Im}\rho| < \epsilon_0$, therefore its value will not depend on η.

The function $F_S(x)$ like the scattering function $S(\rho)$, is a spectral characteristic of the operator L on its continuous spectrum. The functions $f_j(x)$ defined by (6.125) characterize the operator L on its point spectrum.

The scattering function $S(\rho)$, the singular numbers $i\mu_j, .., i\mu_N$ and the normalization multipliers $m_1^2, ..., m_N^2$ are called the scattering data of the operator L. The scattering data are not independent of each other. By Theorem 6.4 the scattering data uniquely determine the operator L. To reconstruct this operator from the scattering data, we construct the function (6.31):

$$F(x) = F_S(x) + \sum_{j=1}^{N} f_j(x) = F_S(x) + \sum_{j=1}^{N} m_j^2 e^{-\mu_j x}. \tag{6.128}$$

The kernel $K(x, y)$ from (6.116) satisfies the fundamental equation (6.35):

$$F(x+y) + K(x,y) + \int\limits_{x}^{\infty} K(x,\xi) F(y+\xi) d\xi = 0 \quad 0 \le x < y < \infty. \tag{6.129}$$

By Theorems 6.3 and 6.4 and Remark 6.3, the equation (6.129) has a unique solution $K(x, y)$, and the potential $p(x)$ is recovered through the found solution by the equality:

$$p(x) = -2\frac{d}{dx}K(x,x), \quad x \ge 0, \tag{6.130}$$

where the reconstructed function (6.130) is real-valued and satisfies the same estimate (6.111), as the estimate for the potential in the Schrödinger equation (6.109).

6.3.2 Non-scattering potentials

There exists a remarkable class of potentials, for which the equation (6.129) can be solved exactly. These are non-scattering potentials on the half-line, for which the inverse Fourier transform $F_S(x)$ defined by (6.127) in the sense of generalized functions is equal to zero. Hence, in the class of non-scattering potentials the functions $F_S(x)$ and $F(x)$ defined by (6.127) and (6.128), respectively, are

$$F_S(x) = 0, \quad F(x) = \sum_{j=1}^{N} m_j^2 e^{-\mu_j x}. \tag{6.131}$$

Under the condition (6.131), the equation (6.129) can be solved exactly. Indeed, the solution $K(x, y)$ of this equation is to be sought in the form:

$$K(x, y) = \sum_{j=1}^{N} K_j(x) e^{-\mu_j y}, \quad 0 \le x \le y, \quad \mu_j > \varepsilon_0, \quad j = 1, ..., N. \tag{6.132}$$

Substituting (6.132) into the equation (6.129), after some simple transformations, we obtain a system of linear algebraic equations for $K_j(x)$:

$$K_j(x) + f_j(x) \sum_{n=1}^{N} \frac{e^{-\mu_n x}}{\mu_n + \mu_j} K_n(x) = -f_j(x), \quad j = 1, ..., N. \tag{6.133}$$

Let $D(x)$ denote an $N \times N$ square matrix consisting of the elements:

$$D_{jn}(x) = \delta_{jn} + f_j(x) \frac{e^{-\mu_n x}}{\mu_n + \mu_j}, \quad j, \ n = 1, ..., N. \tag{6.134}$$

From linear algebra, we know that the solution of the system (6.133) is

$$K_j(x) = \frac{\det D^{(j)}(x)}{\det D(x)}, \quad j = 1, ..., N,$$

where $D^{(j)}(x)$ stands for the matrix obtained from the matrix $D(x)$ on substituting the elements in its j-th column by the elements $-f_n(x)$:

$$D_{nj}^{(j)}(x) = -f_n(x) = -m_n^2 e^{-\mu_n x}, \quad n = 1, ..., N.$$

Since the potential $p(x)$ is determined by $K(x, x)$, then we calculate it with the help of (6.132):

$$K(x, x) = \Big(\det D(x)\Big)^{-1} \sum_{j=1}^{N} \det D^{(j)}(x) e^{-\mu_j x}, \quad x \ge 0.$$

Using the rule of differentiation of determinants, we find that the numerator in this expression is equal to the derivative of $\det D(x)$, because $K(x, x) = \frac{d}{dx} \ln \det D(x)$. Hence, the formula (6.130) for the potential $p(x)$ in the class of non-scattering potentials is written in a compact form:

$$p(x) = -2 \frac{d^2}{dx^2} \ln \det D(x), x \ge 0. \tag{6.135}$$

The expression (6.135) completely describes the whole family of non-scattering potentials.

6.3.3 The time-dependence of the reflection coefficient

It is known [156] that the KdV equation:

$$p_t - 6pp_x + p_{xxx} = 0, \quad (x,t) \in [0,\infty) \times (-\infty,\infty) \tag{6.136}$$

is identical to the equation defined by the Lax representation:

$$\dot{L} = [L, A] = LA - AL, \tag{6.137}$$

where $L = -\frac{d^2}{dx^2} + p$, $A = 4\frac{d^3}{dx^3} - 3p_x - 6p\frac{d}{dx} - \gamma$, $\dot{L}$ is derivative of $L(t)$ with respect to t, and γ is some constant, which will be determined by Lemma 6.13 below.

Indeed, $\dot{L}$ is the operator of multiplication by p_t, and the commutator $[L, A]$ is also, as simple calculations show, an operator of multiplication by $6pp_x - p_{xxx}$, so that the equation (6.137) coincides with the KdV equation (6.136).

The potential $p(x,t)$ in the operator $L(t)$ is called isospectral if the spectrum of $L(t)$ is invariant with t, i.e., $\dot{\lambda} = 0$.

Lemma 6.12. *The potential $p(x,t)$ in the operator $L(t)$ is isospectral if and only if $p(x,t)$ satisfies the Lax representation* (6.137).

Proof. The Lax pair for the KdV equation (6.136) consists of the operator $L(t)$ for the spectral problem and the operator A governing the time-dependence of eigenfunctions. Namely,

$$L\Omega = \lambda\Omega, \quad \lambda = \rho^2, \quad \mathrm{Im}\rho > -\frac{\varepsilon}{2}, \tag{6.138}$$

$$\dot{\Omega} = -A\Omega, \quad (x,t) \in [0,\infty) \times (-\infty,\infty), \tag{6.139}$$

where Ω is the normalized eigenfunction defined by (6.123).

Differentiating equation (6.138) with respect to t and using (6.139), we have:

$$\dot{L}\Omega + L\dot{\Omega} = \dot{L}\Omega - LA\Omega = \dot{\lambda}\Omega + \lambda\dot{\Omega} = \dot{\lambda}\Omega - \lambda A\Omega = \dot{\lambda}\Omega - AL\Omega,$$

$$\left\{\dot{L} - (LA - AL)\right\}\Omega = \dot{\lambda}\Omega. \tag{6.140}$$

It follows from (6.140) that the Lax representation (6.137) for the nontrivial eigenfunction Ω holds if and only if $\dot{\lambda} = 0$, which is that we wished to prove. □

Lemma 6.13. *If the potential $p(x,t)$ in the operator $L(t)$ satisfies the KdV equation* (6.136), *then the time-dependence of the normalization eigenfunction* (6.123) *is defined by the evolution equation:*

$$\dot{\Omega} = \left(4i\rho^3 - 4\frac{d^3}{dx^3} + 6p\frac{d}{dx} + 3p_x\right)\Omega, \quad Im\rho > -\frac{\varepsilon}{2}, (x,t) \in [0,\infty) \times (-\infty,\infty) \tag{6.141}$$

and the reflection coefficient $S_1(\rho;t)$ *evolves according to the equation:*

$$S_1(\rho;t) = \frac{e_1(\rho;t)}{e(\rho;t)} = \frac{e_1(\rho)}{e(\rho)} e^{8i\rho^3 t} \ \textit{for}\ |\rho| > \rho_0, \ \ Im\rho > 0, \ \ t \in (-\infty, \infty). \tag{6.142}$$

Proof. Let the potential $p(x,t)$ in $L(t)$ satisfy the KdV equation (6.136), then the time-dependence of the normalized eigenfunction (6.123) is given by the evolution equation (6.139). Using (6.137), we write the equation (6.139) in the form:

$$\dot{\Omega} = (\gamma - p_x)\Omega + (4\lambda + 2p)\Omega_x, \ \ \lambda = \rho^2. \tag{6.143}$$

Due to (6.112) and (6.113) the normalization eigenfunction (6.123) obeys the asymptotic condition as $x \to \infty$:

$$\Omega(\rho, x;t) = \frac{e_1(\rho;t)}{e(\rho;t)} e^{i\rho x} - e^{-i\rho x} + o(1) \ \text{ for } \ |\rho| > \rho_0 > 0, \text{Im}\rho > 0. \tag{6.144}$$

Since $p(x,t)$ is a solution of the KdV equation (6.136), then by Lemma 6.12 the potential $p(x,t)$ is an isospectral potential. Using this fact and (6.111), (6.144), in (6.143) letting x tend to ∞, we find

$$\begin{aligned} &\frac{d}{dt}\left(\frac{e_1(\rho;t)}{e(\rho;t)}\right) e^{i\rho x} = \\ &\gamma\left(\frac{e_1(\rho;t)}{e(\rho;t)} e^{i\rho x} - e^{-i\rho x}\right) + 4i\rho^3\left(\frac{e_1(\rho;t)}{e(\rho;t)} e^{i\rho x} + e^{-i\rho x}\right), \end{aligned}$$

whence, it follows that for $|\rho| > \rho_0 > 0$, $\text{Im}\rho > 0$:

$$\begin{aligned} &-\gamma e^{-i\rho x} + 4i\rho^3 e^{-i\rho x} = 0, \\ &\frac{d}{dt}\left(\frac{e_1(\rho;t)}{e(\rho;t)}\right) e^{i\rho x} = (\gamma + 4i\rho^3)\frac{e_1(\rho;t)}{e(\rho;t)} e^{i\rho x}. \end{aligned} \tag{6.145}$$

Hence, $\gamma = 4i\rho^3$, and the time-dependence of the functions Ω and S_1 defined by evolution equations (6.141) and (6.142) are deduced from (6.139) and (6.145), respectively. The lemma is proved. □

The Lemma 6.13 enables us to find the time-dependent potential $p(x,t)$ belonging to the class of non-scattering potentials. In fact, the time-dependent matrix $D(x;t)$ is obtained from the matrix $D(x)$ given by (6.134) with the help of the following substitution:

$$D_{jn}(x) = \delta_{jn} + f_j(x)\frac{e^{-\mu_n x}}{\mu_n + \mu_j} \to D_{jn}(x;t) = \delta_{jn} + f_j(x;t)\frac{e^{-\mu_n x}}{\mu_n + \mu_j},$$

where $f_j(x,t)$ is calculated by the formulas (6.125), (6.126) and (6.142):

$$f_j(x;t) = i\mathrm{Res}|_{\rho=i\mu_j}\left\{S_1(\rho;t)e^{i\rho x}\right\} = i\mathrm{Res}|_{\rho=i\mu_j}\left\{S_1(\rho)e^{8i\rho^3 t+i\rho x}\right\}$$
$$= \frac{2\mu_j}{ie'_\rho(i\mu_j)e'_x(i\mu_j)}e^{-\mu_j x+8\mu_j^3 t} = m_j^2 e^{-\mu_j x+8\mu_j^3 t} = f_j(x)e^{8\mu_j^3 t}, \qquad (6.146)$$

in addition $f_j(x,0) = f_j(x),\ j = 1, ..., N,\ S_1(\rho;0) = S_1(\rho),\ (x,t) \in [0,\infty) \times (-\infty,\infty)$.

The formulas (6.135) and (6.146) give exact soliton-solutions of the KdV equation (6.136) in the class of non-scattering potentials:

$$p(x,t) = -2\frac{d^2}{dx^2}\ln \det D(x;t), \qquad (6.147)$$

where

$$D_{jn}(x,t) = \delta_{jn} + \frac{2\mu_j}{i(\mu_j+\mu_n)e'_\rho(i\mu_j)e'_x(i\mu_j)}e^{-(\mu_j+\mu_n)x+8\mu_j^3 t},\quad j,\ n = 1, ..., N. \qquad (6.148)$$

The soliton-solution (6.147) of the KdV equation (6.136) is constructed from the non-scattering data s of the associated scattering problem (6.109)–(6.110):

$$s = \{S(\rho;t) \equiv 1 \text{ for } |\mathrm{Im}\rho| < \frac{\varepsilon_0}{2}, \rho_j = i\mu_j, \mu_j > \varepsilon_0 > 0, m_j^2 > 0, j = 1, ..., N\}. \qquad (6.149)$$

The non-scattering data (6.149) are formulated from the known initial condition $p(x) = p(x,0)$ of the Cauchy problem for the KdV equation (6.136) considered in the class of non-scattering potentials.

Theorem 6.7. *Let the function $p(x)$ in the operator L be an isospectral non-scattering potential which is a real-valued continuous function satisfying the estimate* (6.111). *Then the normalization multipliers m_j^2 are defined by formula* (6.126), *and the time-dependence of the normalization eigenfunction and of the reflection coefficient is defined by formulas* (6.141) *and* (6.142), *respectively. By these formulas, the non-scattering potential* (6.147) *constructed from the given non-scattering data* (6.149) *describes the whole family of non-scattering potentials which are soliton-solutions of the Cauchy problem for the KdV equation* (6.136) *with the known initial condition $p(x)$.*

6.3.4 Some examples

Example 1. Let the scattering data (6.149) consist of only one simple pole $\rho = i\mu,\ \ \mu > 0$. We construct the non-scattering potential from this known data by the formulas (6.147)–(6.148). As a result, the non-scattering potential is presented in the form:

$$p(x,t) = -8\mu^2 \frac{ie'_\rho(i\mu)e'_x(i\mu)\exp\Big\{-2\mu x + 8\mu^3 t\Big\}}{\Big[ie'_\rho(i\mu)e'_x(i\mu) + \exp\Big\{-2\mu x + 8\mu^3 t\Big\}\Big]^2}, \tag{6.150}$$

$$(x,t) \in [0,\infty) \times (-\infty,\infty).$$

Owing to (6.126), we can put

$$\varphi = \frac{1}{2\mu}\ln[ie'_\rho(i\mu)e'_x(i\mu)]^{-1}. \tag{6.151}$$

Using (6.151), we transform the non-scattering potential defined by (6.150) to the form:

$$p(x,t) = -\frac{2\mu^2}{ch^2\mu(x - 4\mu^2 t - \varphi)}, \quad (x,t) \in [0,\infty) \times (-\infty,\infty). \tag{6.152}$$

The explicit expression (6.152) gives an exact solitary pulse solution of the KdV equation (6.136). This soliton-solution is a non-scattering potential containing one bound state. The energy of this state, μ^2 determines both the amplitude and the speed v of the soliton: $v = 4\mu^2$. The quantity φ defined by (6.151) represents the coordinate of the center of the soliton at time $t = 0$. φ is called the phase of the soliton.

Now we clarify how the solitary pulse (called the soliton) emerges on the half-line. If t is very large negative, then from (6.150) it follows that for $x \geq 0$:

$$p(x,t) = o(1) \ \text{ as } \ t \to -\infty.$$

Suppose that we are observing the solitary pulse at the origin $x = 0$, and that at the moment $t = t_1 = -\dfrac{\varphi}{4\mu^2}$ the pulse $p(x,t)$ takes the form: $p(x,t_1) = -\dfrac{2\mu^2}{ch^2(\mu x)}$. There exists a positive number δ such that the pulse $p(x,t)$ emerges in the neighbourhood of the point $x = 0$ at the moment $t = t_1$. In Figure 6.1 the sketch of this pulse is depicted at four values of time t_0, t_1, $t_2 \in (t_1-\delta,\ t_1+\delta)$ and at the value $t_3 \bar{\in} (t_1-\delta, t_1+\delta)$ such that $t_1-\delta < t_0 < t_1 < t_2 < t_1+\delta < t_3$. The full shape of the pulse is depicted at $t = t_3$. After the moment $t = t_3$ the pulse travels with the speed $4\mu^2$ to the right of the axis of x.

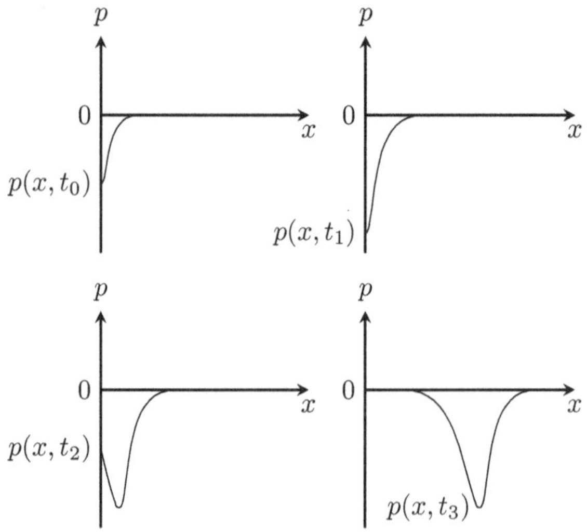

FIGURE 6.1: The solitary pulse soliton-solution of the KdV equation (6.136) with one bound state at four successive moment times: $t = t_0,\ t_1,\ t_2$ and t_3.

Example 2. Let the non-scattering data (6.149) consist of two simple poles $\rho_1 = i\mu_1$ and $\rho_2 = i\mu_2$, $\mu_1 > \mu_2 > 0$. In this case the elements $D_{jn}(x,t)$ of the matrix $D(x,t)$ are calculated by formulas (6.146) and (6.148):

$$\begin{aligned} D_{11}(x,t) &= 1 + \frac{m_1^2}{2\mu_1} e^{-2\mu_1 x + 8\mu_1^3 t}, \quad D_{12}(x,t) = \frac{m_1^2}{\mu_1+\mu_2} e^{-(\mu_1+\mu_2)x + 8\mu_1^3 t}, \\ D_{21}(x,t) &= \frac{m_2^2}{\mu_1+\mu_2} e^{-(\mu_1+\mu_2)x + 8\mu_2^3 t}, \quad D_{22}(x,t) = 1 + \frac{m_2^2}{2\mu_2} e^{-2\mu_2 x + 8\mu_2^3 t}, \end{aligned} \tag{6.153}$$

where m_1^2 and m_2^2 are defined by the formula (6.109).

Putting

$$\xi = \mu_1 x - 4\mu_1^3 t + \xi_0, \quad \eta = \mu_2 x - 4\mu_2^3 t + \eta_0, \tag{6.154}$$

$$\xi_0 = \frac{1}{2}\ln\left(\frac{2\mu_1}{m_1^2}\frac{\mu_1+\mu_2}{\mu_1-\mu_2}\right), \quad \eta_0 = \frac{1}{2}\ln\left(\frac{2\mu_2}{m_2^2}\frac{\mu_1+\mu_2}{\mu_1-\mu_2}\right), \tag{6.155}$$

and using (6.153), we calculate the determinant D of the matrix $D(x,t)$:

$$D = \det D(x,t) = 1 + \frac{\mu_1+\mu_2}{\mu_1-\mu_2} e_1 + \frac{\mu_1+\mu_2}{\mu_1-\mu_2} e_2 + e_1 e_2, \tag{6.156}$$

where

$$e_1 = e^{-2\xi} = \frac{m_1^2}{2\mu_1}.\frac{\mu_1 - \mu_2}{\mu_1 + \mu_2}e^{-2\mu_1 x + 8\mu_1^3 t}, e_2 = e^{-2\eta} = \frac{m_2^2}{2\mu_2}.\frac{\mu_1 - \mu_2}{\mu_1 + \mu_2}e^{-2\mu_2 x + 8\mu_2^3 t}. \tag{6.157}$$

The non-scattering real-valued potential $p(x,t)$ is calculated by formulas (6.147)–(6.148)

$$p(x,t) = -2\frac{d^2}{dx^2}\ln\det D(x,t) = -2\frac{DD_{xx} - (D_x)^2}{D^2}, (x,t) \in [0,\infty) \times (-\infty,\infty) \tag{6.158}$$

where the first and the second partial derivatives D_x and D_{xx} of the determinant D are found from (6.156), using (6.157):

$$\begin{aligned} D_x &= -2\mu_1 A_{12} e_1 - 2\mu_2 A_{12} e_2 - 2(\mu_1 + \mu_2)e_1 e_2, \\ D_{xx} &= 4\mu_1^2 A_{12} e_1 + 4\mu_2^2 A_{12} e_2 + 4(\mu_1 + \mu_2)^2 e_1 e_2, \\ A_{12} &= \frac{\mu_1 + \mu_2}{\mu_1 - \mu_2}. \end{aligned} \tag{6.159}$$

If t is very large negative, then using (6.156), (6.157) and (6.159), from (6.158) we obtain for every fixed $x \geq 0$:

$$p(x,t) = o(1) \text{ as } t \to -\infty.$$

It is easy to verify that

$$\begin{aligned} D &= 1 + A_{12}e^{-2\xi} + A_{12}e^{-2\eta} + e^{-2\xi-2\eta} \\ &= \frac{e^{-\xi-\eta}}{\mu_1 - \mu_2}\left\{(\mu_1 - \mu_2)e^{\xi+\eta} + (\mu_1 + \mu_2)e^{-\xi+\eta} + (\mu_1 + \mu_2)e^{\xi-\eta} + (\mu_1 - \mu_2)e^{-\xi-\eta}\right\} \\ &= \frac{e^{-\xi-\eta}}{\mu_1 - \mu_2}\left\{\mu_1(e^{\xi} + e^{-\xi})(e^{\eta} + e^{-\eta}) - \mu_2(e^{\xi} - e^{-\xi})(e^{\eta} - e^{-\eta})\right\} \\ &= \frac{e^{-\eta}(e^{\eta} + e^{-\eta})e^{-\xi}(e^{\xi} - e^{-\xi})}{\mu_1 - \mu_2}\left\{\mu_1\frac{(e^{\xi} + e^{-\xi})}{(e^{\xi} - e^{-\xi})} - \mu_2\frac{(e^{\eta} - e^{-\eta})}{(e^{\eta} + e^{-\eta})}\right\} \\ &= \frac{(1 + e^{-2\eta})(1 - e^{-2\xi})}{\mu_1 - \mu_2}\left(\mu_1 \operatorname{cth}\xi - \mu_2 \operatorname{th}\eta\right). \end{aligned}$$

Hence,

$$D^2 = \frac{(1 + e_2)^2(1 - e_1)^2}{(\mu_1 - \mu_2)^2}\left(\mu_1 \operatorname{cth}\xi - \mu_2 \operatorname{th}\eta\right)^2. \tag{6.160}$$

Using (6.156), (6.157) and (6.159), we calculate:

$$\begin{aligned}DD_{xx} - D_x^2 &= \Big(1 + A_{12}e_1 + A_{12}e_2 + e_1e_2\Big)\times\\ &\Big(4\mu_1^2A_{12}e_1 + 4\mu_2^2A_{12}e_2 + 4(\mu_1+\mu_2)^2e_1e_2\Big)-\\ &4\mu_1^2A_{12}^2e_1^2 - 4\mu_2^2A_{12}^2e_2^2 - 4(\mu_1+\mu_2)^2e_1^2e_2^2-\\ &-8\mu_1(\mu_1+\mu_2)A_{12}e_1^2e_2 - 8\mu_2(\mu_1+\mu_2)A_{12}e_1e_2^2 - 8\mu_1\mu_2A_{12}^2e_1e_2\\ &= 4\mu_1^2A_{12}e_1 + 4\mu_2^2A_{12}e_2 + \Big[4(\mu_1+\mu_2)^2 + 4(\mu_1^2+\mu_2^2-2\mu_1\mu_2)A_{12}^2\Big]e_1e_2+\\ &\Big[4(\mu_1+\mu_2)^2 + 4\mu_1^2 - 8\mu_1(\mu_1+\mu_2)\Big]A_{12}e_1^2e_2+\\ &\Big[4(\mu_1+\mu_2)^2 + 4\mu_2^2 - 8\mu_2(\mu_1+\mu_2)\Big]A_{12}e_1e_2^2\\ &= 4\mu_1^2A_{12}e_1 + 4\mu_2^2A_{12}e_2 + 8(\mu_1+\mu_2)^2e_1e_2 + 4\mu_2^2A_{12}e_1^2e_2 + 4\mu_1^2A_{12}e_1e_2^2\\ &= 4\mu_1^2A_{12}e_1 + 4\mu_2^2A_{12}e_2 + 8(\mu_1^2-\mu_2^2)A_{12}e_1e_2 + 4\mu_2^2A_{12}e_1^2e_2 + 4\mu_1^2A_{12}e_1e_2^2\\ &= 4\mu_1^2A_{12}e_1(1+2e_2+e_2^2) + 4\mu_2^2A_{12}e_2(1-2e_1+e_1^2)\\ &= 4\mu_1^2A_{12}e_1(1+e_2)^2 + 4\mu_2^2A_{12}e_2(1-e_1)^2\\ &= A_{12}(1+e_2)^2(1-e_1)^2\left\{\mu_1^2\frac{4e_1}{(1-e_1)^2} + \mu_2^2\frac{4e_2}{(1+e_2)^2}\right\}\\ &= \frac{\mu_1+\mu_2}{\mu_1-\mu_2}(1+e_2)^2(1-e_1)^2\left\{\mu_1^2\frac{4}{(e^{\xi}-e^{-\xi})^2} + \mu_2^2\frac{4}{(e^{\eta}+e^{-\eta})^2}\right\}. \quad (6.161)\end{aligned}$$

The explicit soliton-solution $p(x,t)$ of the KdV equation (6.109) with two bound states is obtained from (6.158), using (6.159), (6.160) and (6.161):

$$p(x,t) = -2(\mu_1^2-\mu_2^2)\frac{\mu_1^2\ \text{cosech}^2\xi + \mu_2^2\ \text{sech}^2\eta}{\Big(\mu_1\text{cth}\xi - \mu_2\ \text{th}\eta\Big)^2}, (x,t)\in[0,\infty)\times(-\infty,\infty)\,, \tag{6.162}$$

where ξ and η are determined by (6.154) and (6.155), $\mu_1 > \mu_2 > 0$.

Thus, $p(x,t)$ represents the non-linear superposition of two forms, one traveling with speed $4\mu_1^2$, the other traveling with speed $4\mu_2^2$.

We suppose that $\mu_1 > \mu_2$, then from (6.154)–(6.155) it follows that for every x:

$$\xi - \eta = (\mu_1-\mu_2)\Big[x - 4(\mu_1^2+\mu_1\mu_2+\mu_2^2)t\Big] + \xi_0 - \eta_0 \to \infty \ \text{ as } \ t\to-\infty,$$

i.e., for t much less than zero and for those values of x, where ξ is about one and η is very negative, we have

$$\operatorname{sech}\eta \to 0, \quad \text{and} \quad \operatorname{th}\eta \to -1, \quad \text{therefore it follows from (6.162) that}$$

$$p(x,t) \to -2(\mu_1^2 - \mu_2^2)\frac{\mu_1^2 \operatorname{cosech}^2\xi}{(\mu_1 \operatorname{cth}\xi + \mu_2)^2} = -2\mu_1^2 \operatorname{sech}^2(\xi + \Delta),$$

where $\Delta = \frac{1}{2} ln\left(\frac{\mu_1+\mu_2}{\mu_1-\mu_2}\right)$.

While in the region of x, where η is about one and ξ is very positive, we obtain:

$$\xi - \eta \to \infty \quad \text{as} \quad t \to -\infty, \quad \text{and} \quad \operatorname{cosech}\xi \to 0, \quad \operatorname{cth}\xi \to 1, \quad \text{therefore,}$$

$$p(x,t) \to -2(\mu_1^2 - \mu_2^2)\frac{\mu_2^2 \operatorname{sech}^2\eta}{(\mu_1 - \mu_2 \operatorname{th}\eta)^2} = -2\mu_2^2 \operatorname{sech}^2(\eta - \Delta).$$

That is, for very large negative t the solution looks like two solitary pulses, the large one to the left of the small one.

After a long time, when t is large positive, it follows from (6.154)–(6.155) that for every $x \geq 0$:

$$\eta - \xi = (\mu_2 - \mu_1)[x - 4(\mu_1^2 + \mu_1\mu_2 + \mu_2^2)t] + \eta_0 - \xi_0 \to \infty \quad \text{as} \quad t \to \infty,$$

i.e., for large positive t and for those values of x, where η is order one and ξ very large negative, then

$$\operatorname{cosech}\xi \to 0, \quad \text{and} \quad \operatorname{cth}\xi \to -1 \quad \text{as} \quad \xi \to -\infty, \quad \text{therefore,}$$

$$p(x,t) \to -2(\mu_1^2 - \mu_2^2)\frac{\mu_2^2 \operatorname{sech}^2\eta}{(\mu_1 + \mu_2 \operatorname{th}\eta)^2} = -2\mu_2^2 \operatorname{sech}^2(\eta + \Delta).$$

While if ξ is order one, and η very large positive, then

$$\operatorname{sech}\eta \to 0, \quad \text{and} \quad \operatorname{th}\eta \to 1, \quad \text{as} \quad \eta \to \infty, \quad \text{therefore,}$$

$$p(x,t) \to -2(\mu_1^2 - \mu_2^2)\frac{\mu_1^2 \operatorname{cosech}^2\xi}{(\mu_1 \operatorname{cth}\xi - \mu_2)^2} = -2\mu_1^2 \operatorname{sech}^2(\xi - \Delta).$$

That is after a long time the large solitary pulse is to the right of the small solitary pulse. They have coalesced and reemerged with their shapes unscathed. The only remnant of the interaction is the phase shift $\Delta = \frac{1}{2} ln\left(\frac{\mu_1+\mu_2}{\mu_1-\mu_2}\right)$. That is the large pulse is moved forward by an amount $2\Delta/\mu_1$ relative to where it would have been in the absence of an interaction, and the small pulse is retarded by an amount $2\Delta/\mu_2$ relative to where it would have been in an unperturbed situation. In Figure 6.2 we show a sketch of this soliton-solution of the KdV equation (6.136) with two bound states at four successive times.

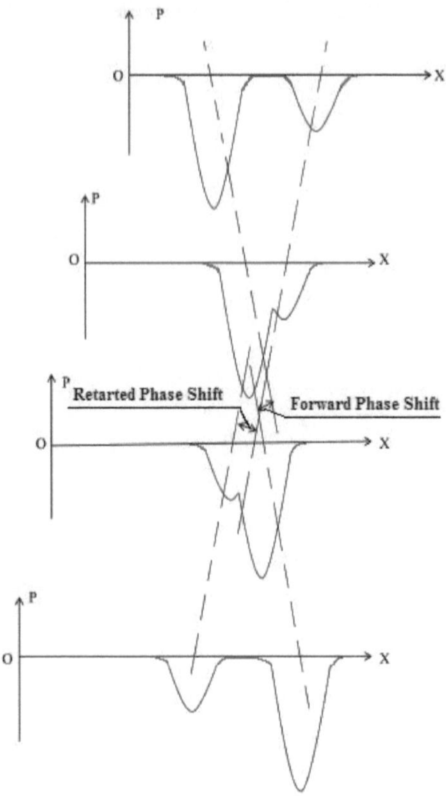

FIGURE 6.2: Two soliton-solutions of the KdV equation (6.136) with two bound states at the four successive moments of time $t = t_0, t_1, t_2$ and t_3.

In general, the non-scattering solution with N bound states has a similar behavior. In this case the non-scattering data (6.149) consist of N simple poles: $\rho_j = i\mu_j,\ \mu_j > 0,\ j = 1, ..., N$. With N bound states the solution resembles the superposition of N solitary pulses whose speeds and amplitudes are determined by the positive values $\mu_j,\ j = 1, ..., N$. The solitary pulses emerge unscathed from interaction except for a phase shift given by the sum of phase shifts from all possible pairwise interactions.

Chapter 7

Inverse Scattering and Its Application to The KdV Equation with Dominant Surface Tension

The results of inverse scattering problem associated with the initial-boundary value problem (IBVP) for the Korteweg-de Vries (KdV) equation with dominant surface tension are formulated. The necessary and sufficient conditions for given functions to be the left- and right-reflection coefficients of the scattering problem are established. The time-dependence t, $t > 0$ of each element of the scattering matrix $s(k,t)$ is found in respective sector of the k-spectral plane by expansion formulas which are constructed from the known initial and boundary conditions of the IBVP. Knowing the right-reflection coefficient calculated from the elements of $s(k,t)$, we solve the Gelfand–Levitan–Marchenko (GLM) equation in the inverse problem. Then the solution of the IBVP is expressible through the solution of the GLM equation. The asymptotic behavior at infinity of time of the solution of the IBVP is shown.

The chapter is based on the published papers [119, 146, 147].

In this chapter we study the IBVP for the KdV equation on the first quadrant:

$$p_t - p_{xxx} + 6pp_x = 0, \quad x > 0, \quad t > 0, \tag{7.1}$$

$$p_{|x=0} = 0, \quad p_{xx|x=0} = 0, \tag{7.2}$$

$$p_{|t=0} = p(x), \quad p(x,t) \to 0, \text{ as } x \to \infty \tag{7.3}$$

where the boundary conditions (7.2) are consistent at the corner point, i.e., $p(0,0) = p_{xx}(0,0) = 0$, and the function $p(x)$ that determines the initial condition (7.3) is required to satisfy the following conditions, which will be referred to as Condition I:

Condition I: The function $p(x)$ is real-valued infinitely smooth and tends to zero at infinity in the Schwartz sense [116], i.e., $p(x)$ and all its derivatives decrease faster than any positive power of x^{-1}. At $x = 0$ $p(x)$ vanishes together with all derivatives, and $\int\limits_0^\infty p(x)dx \neq 0$.

DOI: 10.1201/9781003370543-7

The ISM is the best-known single development This method has allowed a detailed investigation of the initial value problems (IVPs) for integrable non-linear evolution equations (NLEEs) within different classes of the initial data. However, it is difficult to transfer to an IBVP for the KdV equation (7.1) with arbitrary boundary and initial conditions. The boundary condition (7.2) is a special type that agrees with the main integrability criteria and is called integrable.

Applying the transformation: $x \to -x$ and $p \to -p$ to the IBVP (7.1)–(7.2)–(7.3) gives the following IBVP:

$$p_t + p_{xxx} + 6pp_x = 0, \quad x < 0, \quad t > 0, \tag{7.4}$$
$$p_{|x=0} = 0, \quad p_{xx|x=0} = 0,$$
$$p_{|t=0} = p(x), \quad p(x)_{|x\to\infty} \to 0.$$

Thus, by this transformation the IBVP for the KdV (7.1) with the negative coefficient of dispersive term p_{xxx} on the positive quarter-plane is transformed to the IBVP for the usual KdV equation (7.4) on the negative quarter-plane for $x < 0$ and $t > 0$.

The IBVP (7.1)–(7.2)–(7.3) is of interest for a number of reasons. Firstly, its mathematical formulation is different from the IBVP (6.48)–(6.49)–(6.50) considered in Section 6.2 of Chapter 6. Two boundary conditions (7.2) are required at $x = 0$, in contrast to the one boundary condition needed at $x = 0$ for the IBVP (6.48)–(6.49)–(6.50). This leads to solution for the IBVP (7.1)–(7.2)–(7.3) being different to those for the IBVP (6.48)–(6.49)–(6.50). Secondly, it is related to the KdV equation (7.1) with dominant surface tension when the coefficient of p_{xxx} has a negative coefficient. One physical example for which the equation (7.1) is appropriate is weakly non-linear long waves propagating on a fluid with sufficiently large surface tension. The coefficient of dispersive term becomes negative, if the Bond number which measures the amplitude of the surface tension, is greater than one-third (Kichenassamy and Olver 1992 [60]).

A rigorous methodology for the analysis of IBVPs on the half-line, $0 < x < \infty$, $t > 0$ is applied to the NLS, to the sG equation in laboratory coordinates and to the KdV equation with dominant surface tension [38]. The R-H approach is used in [38, 51, 52, 53] for deriving the time-dependence of the scattering matrix of the SP associated with the IBVP (7.1)–(7.2)–(7.3). The construction of the broad classes of explicit solutions of the IBVP (7.1)–(7.2)–(7.3) is presented in [6].

In this chapter the IBVP (7.1)–(7.2)–(7.3) is described differently from the method cited in [38, 51, 53]. Indeed, we reduce the problem of solving the considered IBVP to that of solving two linear inverse scattering problems (SPs). The first SP is associated with the KdV equation (7.1), the second SP is self-adjoint. The properties of the scattering matrix $s(k)$ and the left- and right-reflection coefficients for the first SP are studied in Section 7.1. In this Section the necessary and sufficient conditions for given functions to be the left- and right-reflection coefficients are established. The system of evolution equations for the scattering matrix $s(k,t)$ is derived in Section 7.2. By using the linear change of dependent variables, we lead the system of evolution equations governing the time-dependence of eigenfunctions for the boundary condition (7.2) to the system of equations with the self-adjoint potential matrix. The Theorem 7.2 in Section 7.3 proves that the scattering function of the self-adjoint SP is expressible through elements of the given scattering matrix $s(k) = s(k,0)$ of the first SP. Knowing the scattering function, we solve the inverse SP for finding the unknown potential self-adjoint matrix. Consequently, the unknown object $p_x(0,t)$ entering the coefficients of the system of evolution equations for $s(k,t)$ is found from an element of the self-adjoint matrix. In Section 7.4 the time-dependence of elements of the matrix $s(k,t)$ is defined in every respective sector of the k-spectral plane. Knowing the right reflection coefficient constructed from the found elements of $s(k,t)$, we solve the linear Gelfand–Levitan–Marchenko (GLM) equation. The solution of the IBVP is expressible through the found solution of the GLM equation.

7.1 The direct and inverse SP

The IBVP (7.1)–(7.2)–(7.3) is associated with the SP on a half-line for a system of equations:

$$\frac{d}{dx}\begin{pmatrix} y \\ y_x \end{pmatrix} = \begin{pmatrix} 0 & 1 \\ p(x)-\lambda & 0 \end{pmatrix}\begin{pmatrix} y \\ y_x \end{pmatrix}, \quad 0 \le x < \infty \tag{7.5}$$

with the boundary conditions as $x \to 0$:

$$(y(k,x), y_x(k,x)) = (1, -ik) + o(1), \quad \lambda = k^2. \tag{7.6}$$

The SP for the Schrödinger equation on the whole line is well studied, therefore it is convenient to reduce the SP (7.5)–(7.6) on the half-line to a problem on the whole line by continuing the potential $p(x)$ from the positive half-line to the whole line. The potential is continued trivially by setting $p(x) \equiv 0$ for all $x < 0$. Accordingly, we write system (7.5) in the operator form:

$$y_x = Uy,\ U = \begin{pmatrix} 0 & 1 \\ p-\lambda & 0 \end{pmatrix},\ y = (y_1(k,x), y_2(k,x)),\ \lambda = k^2,\ -\infty < x < \infty, \tag{7.7}$$

where the real-valued function $p(x)$ satisfies Condition I, $p(x) \equiv 0$ for $x < 0$, $y_2 = y_{1x}$ and y_1 satisfies the Schrödinger equation:

$$-y_{xx} + p(x)y = k^2 y, \quad -\infty < x < \infty. \tag{7.8}$$

The formulas (7.9)–(7.27) presented below are deduced from the known facts of the scattering theory for equation (7.8) (see [76, 86]). We construct the matrix solutions of system (7.5):

$$E = \begin{pmatrix} e(k,x) & e(-k,x) \\ e_x(k,x) & e_x(-k,x) \end{pmatrix}, \quad W = \begin{pmatrix} \omega(-k,x) & \omega(k,x) \\ \omega_x(-k,x) & \omega_x(k,x) \end{pmatrix},$$

with the conditions for all real $k = \sqrt{\lambda}$:

$$E = T(k) \begin{pmatrix} e^{ikx} & 0 \\ 0 & e^{-ikx} \end{pmatrix} + o(1) \quad (x \to \infty), \tag{7.9}$$

$$W = T(k) + o(1) \quad (x \to 0), \quad T(k) = \begin{pmatrix} 1 & 1 \\ ik & -ik \end{pmatrix}. \tag{7.10}$$

The function $e(k,x)$ satisfies the integral equation:

$$e(k,x) = e^{ikx} + \int_x^\infty \frac{\sin k(s-x)}{k} p(s) e(k,s) ds, \tag{7.11}$$

and can be represented as:

$$e(k,x) = e^{ikx} + \int_x^\infty K(x,\xi) e^{ik\xi} d\xi, \quad Imk \geq 0, \tag{7.12}$$

where the function $K(x,\xi)$ is the kernel of the transformation operator and is determined from the integral equation of the Volterra type:

$$K(x,\xi) = \frac{1}{2} \int_{\frac{x+\xi}{2}}^\infty p(s) ds + \int_{\frac{x+\xi}{2}}^\infty d\alpha \int_0^{\frac{\xi-x}{2}} p(\alpha-\beta) K(\alpha-\beta, \alpha+\beta) d\beta, \tag{7.13}$$

with $K(x,\xi) = 0$ for $\xi < x$. Consequently, $K(x,\xi)$ satisfies the estimate:

$$|K(x,\xi)| \leq C \int_{\frac{x+\xi}{2}}^\infty |p(s)| ds, \tag{7.14}$$

and the conditions:

$$-2\frac{d}{dx} K(x,x) = p(x), \quad \left(\frac{d}{dx} K(x,x) \right)_{x=0} = -\frac{1}{2} p(0) = 0. \tag{7.15}$$

Remark 7.1. From (7.12) it follows that there exists a linear operator which transforms the solution of the Schrödinger equation (7.8) with zero potential (e^{ikx}) into the solution of this equation with the potential $p(x)$.

The function $\omega(k,x)$ satisfies the integral equation:

$$\omega(k,x) = e^{-ikx} + \int_0^x \frac{\sin k(x-s)}{k} p(s)\omega(k,s)ds, \quad \mathrm{Im}k \geq 0,$$

and can be represented as [53]:

$$\omega(k,x) = \begin{cases} e^{-ikx} + \int\limits_{-\infty}^{x} K^-(x,\xi)e^{-ik\xi}d\xi & \text{for } x > 0, \\ e^{-ikx} & \text{for } x \leq 0, \end{cases} \tag{7.16}$$

where $K^-(x,\xi) \equiv 0$ for $\xi \leq x \leq 0$, and $2\frac{d}{dx}K^-(x,x) = \begin{cases} p^-(x) & \text{for } x > 0, \\ 0 & \text{for } x \leq 0. \end{cases}$

In view of the reality of $p(x)$, the functions $K(x,\xi)$ and $K^-(x,\xi)$ are real-valued and therefore

$$e(k,x) = \overline{e(-k,x)}, e_x(k,x) = \overline{e_x(-k,x)}, \omega(k,x) = \overline{\omega(-k,x)}, \omega_x(k,x) = \overline{\omega_x(-k,x)}. \tag{7.17}$$

The solutions $e(k,x)$ and $\omega(k,x)$ of system (7.7) admit from the real line an analytical continuation into the upper half-plane $\mathrm{Im}k > 0$. Since the Wronskians of the solutions do not depend on x, then

$$\begin{aligned} &W\{\omega(-k,x), \omega(k,x)\} = W\{e(k,x), e(-k,x)\} \\ &= e(k,x)e_x(-k,x) - e_x(k,x)e(-k,x) = -2ik \quad \text{for real } k \neq 0. \end{aligned}$$

Hence, $(e(k,x), e(-k,x))$ and $(\omega(-k,x), \omega(k,x))$ for real $k \neq 0$ constitute fundamental systems of solutions and are bases of solutions of system (7.7), therefore

$$\begin{pmatrix} e(k,x) & e(-k,x) \\ e_x(k,x) & e_x(-k,x) \end{pmatrix} = \begin{pmatrix} \omega(-k,x) & \omega(k,x) \\ \omega_x(-k,x) & \omega_x(k,x) \end{pmatrix} \begin{pmatrix} s_{11}(k) & s_{12}(k) \\ s_{21}(k) & s_{22}(k) \end{pmatrix}, \quad k \neq 0. \tag{7.18}$$

The properties of the coefficients $s_{ij}(k)$ of the matrix $s(k)$ are clarified in the following lemma.

Lemma 7.1 ([76, 86]). *For real values of the parameter k the entries $s_{ij}(k)$ of $s(k)$ possess the properties:*

(1) The involutions: $s_{11}(-k) = \overline{s_{11}(k)}$, $s_{21}(-k) = \overline{s_{21}(k)}$, $s_{22}(k) = s_{11}(-k)$, $s_{12}(k) = s_{21}(-k)$,

(2) The constraint: $\det s(k) = 1 = |s_{11}(k)|^2 - |s_{12}(k)|^2, |t(k)|^2 + |\tilde{r}(k)|^2 = 1, t(k) = s_{11}^{-1}(k)$,

$$\tilde{r}(k) = \frac{s_{21}(k)}{s_{11}(k)}, \ \tilde{R}(k) = -\frac{s_{12}(k)}{s_{11}(k)} = -\frac{s_{11}(-k)}{s_{11}(k)}\tilde{r}(-k), \ s_{11}(k) \neq 0.$$

The functions $s_{11}(k)$ and $s_{21}(k)$ are called the refraction and reflection coefficients, respectively. The functions $\tilde{R}(k)$ and $\tilde{r}(k)$ are called the right- and left-reflection coefficients, respectively for the waves incident on the potential $p(x)$ from the right.
Using Lemma 7.1, from (7.18) we obtain

$$\begin{pmatrix} \omega(-k,x) & \omega(k,x) \\ \omega_x(-k,x) & \omega_x(k,x) \end{pmatrix} = \begin{pmatrix} e(k,x) & e(-k,x) \\ e_x(k,x) & e_x(-k,x) \end{pmatrix} \begin{pmatrix} s_{11}(-k) & -s_{21}(-k) \\ -s_{21}(k) & s_{11}(k) \end{pmatrix}, k \neq 0. \tag{7.19}$$

The matrix $s(k) = \begin{pmatrix} s_{11}(k) & s_{12}(k) \\ s_{21}(k) & s_{22}(k) \end{pmatrix}$ thus defined is called the scattering matrix of the SP (7.7), (7.10). Using (7.10), (7.18), we determine the entries $s_{ij}(k)$ of $s(k)$ in terms of boundary values (BVs) $(e(k), e_x(k)) = (e(k,0), e_x(k,0))$ for $k \neq 0$:

$$\begin{aligned} s_{11}(k) &= \frac{1}{2}e(k) + \frac{1}{2ik}e_x(k), \quad s_{12}(k) = \frac{1}{2}e(-k) + \frac{1}{2ik}e_x(-k), \\ s_{21}(k) &= \frac{1}{2}e(k) - \frac{1}{2ik}e_x(k), \quad s_{22}(k) = \frac{1}{2}e(-k) - \frac{1}{2ik}e_x(-k), \end{aligned} \tag{7.20}$$

where due to (7.10) and (7.17) the BVs $(e(k), e_x(k))$ are different from zero for all real $k \neq 0$.

Substituting (7.11) for $x = 0$ into (7.20) with due regard for (7.12), we obtain:

$$\begin{aligned} 2ik\left(s_{11}(k) - 1\right) &= 2ik\left(s_{22}(-k) - 1\right) = -\int_0^\infty e^{-iks}p(s)e(k,s)ds = \\ & -\int_0^\infty p(\xi)d\xi - \int_0^\infty e^{ik\xi}d\xi \int_0^\infty p(s)K(s,\xi+s)ds, \quad \operatorname{Im} k \geq 0, \end{aligned} \tag{7.21}$$

$$\begin{aligned} 2iks_{21}(k) =& 2iks_{12}(-k) = \int_0^\infty e^{iks}p(s)e(k,s)ds = \\ & \frac{1}{2}\int_0^\infty e^{ik\xi}p\left(\frac{\xi}{2}\right)d\xi + \int_0^\infty e^{ik\xi}d\xi \int_0^{\frac{\xi}{2}} p(s)K(s,\xi-s)ds, \quad \operatorname{Im} k \geq 0. \end{aligned} \tag{7.22}$$

Owing to Condition I and (7.14) the functions $s_{ij}(k)$ have the asymptotic behavior:

$$s_{11}(k) - 1 = s_{22}(-k) - 1 = O\left(\frac{1}{|k|}\right) \quad \text{as } |k| \to \infty, \tag{7.23}$$

$$s_{21}(k) = s_{12}(-k) = o\left(\frac{1}{|k|}\right) \quad \text{as } |k| \to \infty. \tag{7.24}$$

The integral representations for $s_{ij}(k)$ are obtained from (7.21) and (7.22):

$$2iks_{12}(-k) = 2iks_{21}(k) = \int_0^\infty e^{ik\xi} A(\xi) d\xi \ \ \mathrm{Im}k \geq 0, \tag{7.25}$$

$$2ik\left(s_{11}(k) - 1\right) = 2ik\left(s_{22}(-k) - 1\right) = -\int_0^\infty p(\xi)d\xi - \int_0^\infty e^{ik\xi} B(\xi) d\xi \ \ \mathrm{Im}k \geq 0, \tag{7.26}$$

where

$$A(\xi) = \frac{1}{2} p\left(\frac{\xi}{2}\right) + \int_0^{\frac{\xi}{2}} p(s) K(s, \xi - s) ds, \ \ B(\xi) = \int_0^\infty p(s) K(s, \xi + s) ds. \tag{7.27}$$

We prove now the following crucial lemma.

Lemma 7.2. *The coefficients $s_{21}(k)$ and $s_{11}(k)$ of the scattering matrix $s(k)$ of the SP (7.7), (7.10) are infinitely differentiable for functions $k \neq 0$, $Imk \geq 0$. Their derivatives satisfy the estimates as $k \to \infty$:*

$$|s_{21}^{(m)}(k)| \leq C_{m,j} |k|^{-j} \quad m, \ j = 0, 1, 2, ... \tag{7.28}$$

$$|s_{11}(k) - 1| \leq C|k|^{-1}, |s_{11}^{(m)}(k)| \leq C_m |k|^{-1} \ \ m = 1, 2, ... \tag{7.29}$$

The functions $2iks_{11}(k)$ and $2iks_{21}(k)$ are continuous in the closed half-plane $Imk \geq 0$.

Proof. Owing to Condition I and (7.14) of the potential p, the function $K(x, \xi)$ from equation (7.13) is real-valued and infinitely differentiable with respect to each variable of x and ξ. Furthermore, $K(x, \xi)$ and all its derivatives decrease faster than any positive power of x^{-1} and ξ^{-1}. Therefore, the functions $A(\xi)$ and $B(\xi)$ defined by (7.27) are infinitely differentiable and decrease faster than any positive power of ξ^{-1}:

$$|A(\xi)| \leq C_A |\xi|^{-j}, \ \ |B(\xi)| \leq C_B |\xi|^{-j}, \ \ j = 0, 1, 2, ... \tag{7.30}$$

Using Condition I of p and the smoothness and estimates of K, from (7.27) we get:

$$|A^{(m)}(\xi)| \leq C_{Am}|\xi|^{-j}, \quad m, j = 0, 1, 2, ... \tag{7.31}$$

Since $p(0) = 0$, then $A(0) = 0$. It can be proved by induction that

$$|A^{(m)}(\xi)|_{\xi=0} = 0, \quad m = 1, 2, ... \tag{7.32}$$

Due to (7.30) the functions $s_{21}(k)$ and $s_{11}(k)$ defined by (7.25) and (7.26), respectively are infinitely differentiable for $k \neq 0$. Then, from (7.25)–(7.27) it follows that the functions $2iks_{11}(k)$ and $2iks_{21}(k)$ are continuous in the closed half-plane $Imk \geq 0$. The estimate (7.28) can be proved by induction with respect to m. Indeed, by using equality (7.30)–(7.32) and integrating the Fourier integral (7.25) by parts j times, we get

$$2ik(ik)^j s_{21}(k) = \int_0^\infty e^{ik\xi} A^{(j)}(\xi) d\xi,$$

whence it follows that $|s_{21}(k)| \leq C_{0,j}|k|^{-j}$, $j = 0, 1, 2, ...$ and $m = 0$. Thus, the estimate (7.28) is true for $j = 0, 1, 2, ...$ and $m = 0$. Supposing that the estimate (7.28) is proved for $m - 1$:

$$\left| s_{21}^{(m-1)}(k) \right| \leq C_{m-1,j}|k|^{-j}.$$

Differentiating (7.25) m times and using the Leibniz's Rule, we obtain:

$$(2iks_{21}(k))^{(m)} = 2iks_{21}^{(m)}(k) + 2ims_{21}^{(m-1)}(k) = \int_0^\infty e^{ik\xi}(i\xi)^m A(\xi) d\xi. \tag{7.33}$$

Integrating the integral in the right-hand side of (7.33) by parts $j - 1$ times, using (7.24), (7.30)–(7.32), we calculate:

$$\int_0^\infty e^{ik\xi}(i\xi)^m A(\xi)\, d\xi = \frac{1}{ik} e^{ik\xi}[(i\xi)^m A(\xi)]_{\xi=0,\infty} - \frac{1}{ik}\int_0^\infty e^{ik\xi}[(i\xi)^m A(\xi)]' d\xi$$

$$= \ldots = \frac{1}{(ik)^{j-1}} e^{ik\xi}\left[(i\xi)^m A(\xi)\right]_{\xi=0,\infty}^{(j-2)} - \frac{1}{(ik)^{j-1}} \int_0^\infty e^{ik\xi}[(i\xi)^m A(\xi)]^{(j-1)} d\xi$$

$$= -\frac{1}{(ik)^{j-1}} \int_0^\infty e^{ik\xi}[(i\xi)^m A(\xi)]^{(j-1)} d\xi.$$

Using the right-hand side of the above equality and the induction hypothesis for $s_{21}^{(m-1)}(k)$, from (7.33) we obtain the estimate (7.28) for $s_{21}^{(m)}(k)$.

The first inequality of (7.29) is deduced from estimate (7.23). We prove the second inequality of (7.29). Differentiating (7.26) m times, yields:

$$(2iks_{11}(k))^{(m)} = 2iks_{11}^{(m)}(k) + 2ims_{11}^{(m-1)}(k) = -\int_0^\infty e^{ik\xi}(i\xi)^m B(\xi)d\xi.$$

Using (7.30) and the induction hypothesis for $s_{11}^{(m-1)}(k)$, from the latest equality we obtain the second estimate of (7.29). The number $B(0)$ is nonzero. Indeed, using (7.13) and Condition I of p, we calculate:

$$B(0) = \int_0^\infty p(s)K(s,s)ds = \frac{1}{2}\int_0^\infty p(s)ds\int_s^\infty p(\xi)d\xi = \frac{1}{2}\left\{\int_0^\infty p(s)ds\}\right\}^2 -$$

$$\frac{1}{2}\int_0^\infty p(s)ds\int_0^s p(\xi)d\xi = \frac{1}{2}\left\{\int_0^\infty p(s)ds\right\}^2 - B(0).$$

Hence, $B(0) = \frac{1}{4}\left\{\int_0^\infty p(s)ds\right\}^2 \neq 0$. On account of this fact, the function $s_{11}^{(m)}(k)$ obeys the estimate (7.29) as the first power of k^{-1}. Lemma 7.2 is proved. □

The following lemma is deduced from Lemmas 7.1 and 7.2 and properties of the solutions of problems (7.7), (7.9) and (7.7), (7.10).

Lemma 7.3 ([76, 86]). *The analytic continuation of the function $s_{11}(k)$ with respect to k from the real axis into the upper half-plane Im$k > 0$ can have a finite number of simple zeros on the positive imaginary axis at $k_j = i\mu_j$, $\mu_j > 0$, $j = 1, ..., N$.*

Suppose that in equation (7.8) the potential function $p(x)$ is to be subjected to the following restriction, which will be referred to as Condition II. *Condition II.* The potential function $p(x)$ is to be subject to the condition that the equation (7.8) must not have a discrete spectrum. Then by Lemma 7.3, $s_{11}(k) \neq 0$ for all k, Im$k > 0$.

The scattering matrix $s(k)$ gives complete information about the continuous spectrum of the Schrödinger operator. By Condition II, Lemma 7.3 and the dispersion relation we can show that essentially, all information about $s(k)$ is contained in the right-reflection coefficient $\tilde{R}(k)$. Denote by $\mathcal{S}$ the class of all real-valued functions satisfying Conditions I and II. From (7.18) and (7.19) it follows that the functions:

$$[s_{11}(k)]^{-1}e(k,x) = \omega(-k,x) + \tilde{r}(k)\omega(k,x), \tag{7.34}$$

$$[s_{11}(k)]^{-1}\omega(k,x) = e(-k,x) + \tilde{R}(k)e(k,x), \tag{7.35}$$

are bounded solutions of system (7.5) with the potential $p(x)$ belonging to the class $\mathcal{S}$. The behavior of the function $ks_{11}(k)$ as $k \to 0$ is closely related to that of functions $\tilde{R}(k)$ and $\tilde{r}(k)$ as $k \to 0$. In fact, upon rewriting (7.34) and (7.35) and letting $k \to 0$, gives:

$$0 = e(0,x) \lim_{k\to 0} ks_{11}(k) \left\{\tilde{R}(k) + 1\right\} = \omega(0,x) \lim_{k\to 0} ks_{11}(k) \left\{\tilde{r}(k) + 1\right\}.$$

Consequently,

$$\lim_{k\to 0} ks_{11}(k) \left\{\tilde{R}(k) + 1\right\} = \lim_{k\to 0} ks_{11}(k) \left\{\tilde{r}(k) + 1\right\} = 0. \tag{7.36}$$

From condition (7.36) we have the estimate: $k\left\{s_{11}(k) + s_{21}(k)\right\} = o(1)$ as $k \to 0$.

To fulfill this estimate, the following condition must be satisfied:

$$s_{11}(k) + s_{21}(k) = O(1) \ \text{ as } \ k \to 0. \tag{7.37}$$

Conversely, if the condition (7.37) is fulfilled, then the condition (7.36) is satisfied.

Lemma 7.4. *The left-reflection coefficient $\tilde{r}(k)$ of the SP for system (7.5) with the potential $p(x)$ belonging to the class $\mathcal{S}$ and boundary condition (7.6) obeys the following conditions:*

(1) For all k, $Imk \geq 0$, the function $\tilde{r}(k)$ is completely continuous and infinitely differentiable. $\tilde{r}(k)$ and all its derivatives decrease faster than any positive power of k^{-1}, and

$$\overline{\tilde{r}(k)} = \tilde{r}(-k), \ \ |\tilde{r}(k)| < 1 \ \text{ for real } \ k \neq 0. \tag{7.38}$$

If the residues of the functions $s_{11}(k)$ and $s_{21}(k)$ at $k = 0$ are different from zero, then $\tilde{r}(0) = -1$.

(2) The function $\tilde{r}(k)$ admits the Fourier integral representation:

$$\tilde{r}(k) = \int_0^\infty e^{ikx} r(x) dx \ \text{ for all } k, \ \ Imk \geq 0, \tag{7.39}$$

where $r(x)$ is a completely continuous and rapidly decreasing function, which is defined by the inverse Fourier transform:

$$r(x) = \frac{1}{2\pi} \int_{-\infty}^{\infty} e^{-ikx} \tilde{r}(k) dk, \ \text{ for } \ x > 0. \tag{7.40}$$

The function $r(x)$ is infinitely differentiable:

$$r^{(m)}(x) = \frac{1}{2\pi} \int_{-\infty}^{\infty} (-ik)^m e^{-ikx} \tilde{r}(k) dk \ \text{ for } x > 0, \ m = 1, 2, ...$$

In addition $r(x)$ and $r^{(m)}(x)$ are real-valued functions vanishing at $x = 0$.

Proof. The validity of assertion (1) for $k \neq 0$, Im$k \geq 0$ follows from Conditions I and II of $p(x)$ and Lemmas 7.1–7.3. We prove the smoothness of $\tilde{r}(k)$ and its derivatives at $k = 0$. From (7.20) we have the relations:

$$2iks_{21}(k) = ike(k) - e_x(k), \quad 2iks_{11}(k) = ike(k) + e_x(k).$$

These relations make clear that the entries $s_{ij}(k), i, j = 1, 2$ of $s(k)$ have generally a simple pole at $k = 0$. The residues of functions $2iks_{11}(k)$ and $2iks_{21}(k)$ at $k = 0$ are defined by:

$$Res[2iks_{11}(k), 0] = -Res[2iks_{21}(k), 0] = e_x(0). \tag{7.41}$$

If $e_x(0) \neq 0$, then the function $\tilde{r}(k)$ and its derivatives are continuous and smooth at $k = 0$:

$$\lim_{k\to 0} \tilde{r}(k) = \lim_{k\to 0} \frac{2iks_{21}(k)}{2iks_{11}(k)} = \lim_{k\to 0} \frac{-e_x(k)}{e_x(k)} = \frac{-e_x(0)}{e_x(0)} = -1.$$

If $e_x(0) = 0$, i.e., the residues (7.41) are zero, then the functions $s_{11}(k)$ and $s_{21}(k)$ are continuous and analytic at $k = 0$. The constraint condition $|s_{11}(k)|^2 - |s_{21}(k)|^2 = 1$ implies that $|s_{11}(k)| > 1$ for $k \in \mathbb{R}$. Therefore, the function $\tilde{r}(k)$ and its derivatives are continuous and smooth at $k = 0$.

By Lemma 7.2 the functions $\tilde{r}(k)$ and $(-ik)^m \tilde{r}(k)$ are analytic in the upper half-plane $Imk \geq 0$ and rapidly decrease at infinity for any nonnegative integer m. On account of this fact, the function $\tilde{r}(k)$ admits the Fourier integral representation (7.39) of a function $r(x)$ for $x > 0$. Since the Fourier transform (7.39) and its inverse Fourier transform (7.40) map $\mathcal{S}$ onto $\mathcal{S}$ mutually continuously one-to-one [116], then $r(x)$ and its derivatives $r^{(m)}(x)$ for $x > 0$ defined by (7.40) are rapidly decreasing functions. Due to (7.38) these functions are real-valued. We have from (7.40):

$$r^{(m)}(0) = \frac{1}{2\pi} \int_{-\infty}^{\infty} (-ik)^m \tilde{r}(k) dk; \quad m = 0, 1, 2, ...$$

where the integrand function $G(k) = (-ik)^m \tilde{r}(k)$ and its derivatives $G^{(j)}(k)$ are analytic in the upper half-plane Im$k \geq 0$ and rapidly decrease at infinity for nonnegative integer m, and $G(\infty) = G^{(j)}(\infty) = 0$, $j = 1, 2, ...$ Hence, the function $G(k)$ is analytic in the closed upper half-plane Im$k \geq 0$, therefore, by Cauchy's Theorem the right-hand side of the last formula vanishes, i.e., $r^{(m)}(0) = 0, m = 0, 1, 2, ...,$ [114]. Lemma 7.4 is proved. □

The further conditions of $\tilde{R}(k)$ are deduced from Lemmas 7.1–7.4.

Lemma 7.5. *The right reflection coefficient* $\tilde{R}(k)$ *of the SP for system* (7.5) *with the potential* $p(x)$ *belonging to the class* $\mathcal{S}$ *and boundary condition* (7.6) *obeys the following conditions:*

(1) The function $\tilde{R}(k)$ *is completely continuous and infinitely differentiable for all real* $k \in (-\infty, \infty)$. $\tilde{R}(k)$ *and all its derivatives decrease faster than any positive power of* k^{-1}, *and*

$$\overline{\tilde{R}(k)} = \tilde{R}(-k), \quad |\tilde{R}(k)| = |\tilde{r}(k)| < 1 \text{ for real } k \neq 0; \qquad (7.42)$$

If the residues of the functions $s_{11}(k)$ *and* $s_{12}(k)$ *at* $k = 0$ *are different from zero, then* $\tilde{R}(0) = -1$;

(2) For real k *the function* $\tilde{R}(k)$ *admits the Fourier transform:*

$$\tilde{R}(k) = \int_{-\infty}^{\infty} e^{-ikx} R(x) dx, \qquad (7.43)$$

where $R(x)$ *is a real completely continuous and rapidly decreasing function, which is defined by the inverse Fourier transform:*

$$R(x) = \frac{1}{2\pi} \int_{-\infty}^{\infty} e^{ikx} \tilde{R}(k) dk. \qquad (7.44)$$

The function $R(x)$ *is infinitely differentiable.*

$$R^{(m)}(x) = \frac{1}{2\pi} \int_{-\infty}^{\infty} (ik)^m e^{ikx} \tilde{R}(k) dk, \ m = 1, 2, ...$$

where the Fourier transform (7.43) and its inverse Fourier transform (7.44) map $\mathcal{S}$ *onto* $\mathcal{S}$ *mutually continuously one-to-one [116]. Due to this fact and (7.42) the functions* $R(x)$ *and* $R^{(m)}(x)$, $m = 1, 2, ...$ *are real-valued and rapidly decreasing.*

To recover the SP (7.5)–(7.6) from the right-reflection coefficient $\tilde{R}(k)$, we derive the fundamental integral equation connecting the given $\tilde{R}(k)$ with the kernels of the transformation operator. The following equation is derived from (7.35), (see [76, 86]):

$$R(x+y) + K(x,y) + \int_{x}^{\infty} R(y+\xi) K(x,\xi) d\xi = 0 \ \text{ for } y > x, \qquad (7.45)$$

The equation (7.45) is derived for finding the potential function in equation (7.8) which is supposed that has only a purely continuous spectrum.

In equation (7.45) x is a parameter, $R(x+y)$ is a known function satisfying conditions enumerated in Lemma 7.5, and $K(x,y)$ is an unknown function of y for every $x \in [0, \infty)$. Owing to conditions (7.42)–(7.44) of the function $R(x+y)$, the equation (7.45) has a unique solution $K(x,y)$ either in $L_1[x,\infty)$ or $L_2[x,\infty)$.

Remark 7.2. The equation of the inverse problem in the form of the integral equations (7.45) is derived. The form of equations of the inverse problem is usually for the Schrödinger (Sturm–Liouville) operator is so-called the Gelfand–Levitan–Marchenko equation.

Analogously, the following fundamental integral equation is derived from (7.34):

$$r(x+y) + K^{-}(x,y) + \int_{-\infty}^{x} r(y+\xi)K^{-}(x,\xi)d\xi = 0 \text{ for } y < x \tag{7.46}$$

Owing to conditions (7.38)–(7.40) of the known function $r(x+y)$ the equation (7.46) has a unique solution in either $L_1(-\infty, x]$ or $L_2(-\infty, x]$.

We use the equation (7.45) to extract information on the solution $K(x,y)$, $y > x$ from the conditions of the known function $R(x)$ in this equation. In fact, the function $K(x,y), y > x$ satisfies conditions, which are analogous to the conditions of the function $R(x)$. As has been proved in [76, 86], the solution $K(x,y)$ of equation (7.45) is the kernel of the transformation operator and the function constructed from $K(x,y)$:

$$e(k,x) = e^{ikx} + \int_{x}^{\infty} K(x,\xi)e^{ik\xi}d\xi, \quad Imk \geq 0, \tag{7.47}$$

satisfies the Schrödinger equation (7.8):

$$-e''(k,x) + p(x)e(k,x) = k^2 e(k,x), \tag{7.48}$$

with the potential $p(x)$ constructed from the solution of equation (7.45) by the formula:

$$p(x) = -2\frac{d}{dx}K(x,x) \text{ for } x > 0. \tag{7.49}$$

Substituting (7.47) into (7.11), we obtain integral equation (7.13) with the constructed potential (7.49). Since the solution of (7.13) is unique, then $K(x,y)$ satisfies (7.15).

By an argument analogous to the previous one, we can prove that the function constructed from the solution $K^{-}(x,y)$ of equation (7.46):

$$\omega(k,x) = e^{-ikx} + \int_{-\infty}^{x} K^{-}(x,\xi)e^{-ik\xi}d\xi, \quad \mathrm{Im}k \geq 0, \tag{7.50}$$

satisfies the Schrödinger equation (7.8):

$$-\omega''(k,x) + p^{-}(x)\omega(k,x) = k^2\omega(k,x) \tag{7.51}$$

with the potential $p^{-}(x)$ constructed by $p^{-}(x) = 2\frac{d}{dx}K^{-}(x,x)$ for $x > 0$.

The principal mathematical issue in inverse problems lies in the proof of the fact that under conditions (properties) of functions $s_{11}(k)$, $s_{21}(k)$ enumerated in Lemmas 7.1 and 7.2, the procedure actually leads to the same differential equation, i.e., $p^-(x) \equiv p(x)$. This leads to establishing the necessary and sufficient conditions of functions $\tilde{r}(k)$ and $\tilde{R}(k)$ to be the left- and right-reflection coefficients of the considered problem.

Theorem 7.1. *Suppose that the functions $s_{11}(k)$ and $s_{21}(k)$, $-\infty < k < \infty$ satisfy the conditions enumerated in Lemmas 7.1 and 7.2, condition (7.37) and the function $s_{11}(k)$ admits an analytical continuation into the upper half-plane $Imk > 0$ and has no zeros there. Then*

(1) The functions $e(k,x)$ and $\omega(k,x)$ constructed from the solutions $K(x,y) \in L_j[x,\infty)$ and $K^-(x,y) \in L_j[-\infty,x], j = 1,2$ of equations (7.45) and (7.46), respectively satisfy the same Shrödinger equation (7.8) with the constructed potential:

$$p^-(x) \equiv p(x) = -2\frac{d}{dx}K(x,x), \tag{7.52}$$

$$p(0) = -2\frac{d}{dx}\Big(K(x,x)\Big)_{x=0} = 0. \tag{7.53}$$

(2) The conditions of functions $s_{11}(k)$ and $s_{21}(k)$ are both necessary and sufficient for the ratios of the type: $\tilde{r}(k) = \dfrac{s_{21}(k)}{s_{11}(k)}$. $Imk \geq 0$ and $\tilde{R}(k) = -\dfrac{s_{12}(k)}{s_{11}(k)}$, $-\infty < k < \infty$ to be the left-reflection and right-reflection coefficients of the SP for one and the same system (7.5) with boundary condition (7.6) and constructed real-valued potential (7.52) belonging to the class $\mathcal{S}$. The Schrödinger equation (7.8) is restored precisely from $\tilde{R}(k)$.

Proof. The functions given by (7.47) and (7.50) admit analytic continuations into the upper half-plane Im$k > 0$. Extend the domain of the function $K^-(x,y)$ by setting: $K^-(x,y) = 0$ for $y > x$. Further, we put:

$$\tilde{\Phi}(x,y) = K^-(x,y) + r(x+y) + \int_{-\infty}^{x} r(y+\xi)K^-(x,\xi)d\xi \text{ for all real } y. \tag{7.54}$$

Due to equation (7.46), $\tilde{\Phi}(x,y) = 0$ for $y < x$. For $y > x$:

$$\tilde{\Phi}(x,y) = r(x+y) + \int_{-\infty}^{x} r(y+\xi)K^-(x,\xi)d\xi, \quad \int_{-\infty}^{\infty} |\tilde{\Phi}(x,y)|dy = \int_{x}^{\infty} |\tilde{\Phi}(x,y)|dy < \infty.$$

Multiply both sides of equality (7.54) by e^{iky}, then integrate with respect to y and apply the inverse Fourier formula to $r(x+y)$:

$$\int_x^\infty \tilde{\Phi}(x,y)e^{iky}dy = \int_{-\infty}^x K^-(x,y)e^{iky}dy + \frac{s_{21}(k)}{s_{11}(k)}e^{-ikx} + \frac{s_{21}(k)}{s_{11}(k)}\int_{-\infty}^x K^-(x,\xi)e^{-ik\xi}d\xi.$$

Adding e^{ikx} to the right- and left-hand sides of the last equality and using (7.50), gives:

$$e^{ikx} + \int_x^\infty \tilde{\Phi}(x,y)e^{iky}dy = \omega(-k,x) + \frac{s_{21}(k)}{s_{11}(k)}\omega(k,x). \tag{7.55}$$

Multiply (7.55) by $s_{11}(k)$, then

$$s_{11}(k)\omega(-k,x) + s_{21}(k)\omega(k,x) = e^*(k,x), \tag{7.56}$$

$$e^*(k,x) = s_{11}(k)\left\{e^{ikx} + \int_x^\infty \tilde{\Phi}(x,y)e^{iky}dy\right\}. \tag{7.57}$$

Replacing k by $-k$ in (7.56), yields:

$$s_{11}(-k)\omega(k,x) + s_{21}(-k)\omega(-k,x) = e^*(-k,x). \tag{7.58}$$

Solving the system (7.56), (7.58) for $\omega(k,x)$, using properties of $s_{11}(k)$ and $s_{21}(k)$, gives:

$$\omega(k,x) = s_{11}(k)e^*(-k,x) - s_{21}(-k)e^*(k,x). \tag{7.59}$$

In order to prove the identity (7.52), we need to prove that

$$e^*(k,x) \equiv e(k,x). \tag{7.60}$$

Indeed, if identity (7.60) will be proved, then due to (7.19) and (7.48), it follows from (7.59):

$$-\omega''(k,x) = -s_{11}(k)e''(-k,x) + s_{21}(-k)e''(k,x) = [k^2 - p(x)]\omega(k,x).$$

Comparing the last equalities with (7.51), we obtain identity (7.52).

To prove identity (7.60), certain properties of the function $e^*(k,x)$ should be established.

(a) The function $e^*(k,x)$ defined by formula (7.57) admits an analytic continuation into the upper half-plane $\mathrm{Im}k > 0$. Using estimate (7.29) for large k, from (7.57) we obtain the following estimate:

$$|e^*(k,x) - e^{ikx}| = O\left(\frac{e^{-x\mathrm{Im}k}}{|k|}\right) \quad \text{as } k \to \infty. \tag{7.61}$$

(b) By Lemma 7.2 the function $ke^*(k,x)$ is continuous in the closed upper half-plane Im$k \geq 0$. In the neighbourhood of the point $k = 0$, this function satisfies uniformly the estimate:

$$ke^*(k,x) = o(1) \text{ as } k \to 0. \tag{7.62}$$

Indeed, to prove estimate (7.62) two cases may arise.

(1) The function $s_{11}(k)$ is bounded in neighborhood of point $k = 0$, in which case the function $e^*(k,x)$ is also bounded in a neighborhood of this point, therefore, $ke^*(k,x) \to 0$ as $k \to 0$. Hence, the estimate (7.62) follows from the continuity of $ke^*(k,x)$.

(2) The function $s_{11}(k)$ is not bounded in a neighborhood of the point $k = 0$, in which case there exists a sequence $k_n \to 0$ such that $s_{11}(k_n) \to \infty$. It follows from condition (7.36) that $\lim_{n\to\infty} k_n s_{11}(k_n) = O(1)$, $\lim_{n\to\infty} \dfrac{s_{21}(k_n)}{s_{11}(k_n)} = -1$. Putting $k = k_n$ in (7.55), passing to the limit as $k_n \to 0$, gives:$1 + \int\limits_x^\infty \tilde{\Phi}(x,y)dy = \omega(0,x) - \omega(0,x) = 0$. Therefore, $\lim_{n\to\infty} k_n e^*(k_n,x) = 0$, which proves estimate (7.62) in this case too, due to the continuity of the function $ke^*(k,x)$.

(c)

$$e^*(k,x) - e^{ikx} \in L_2(-\infty,\infty). \tag{7.63}$$

Since the function $s_{11}(k)$ satisfies the first estimate of (7.29) for large k, therefore it suffices to show that $e^*(k,x)$ is square integrable in a neighborhood of the point $k = 0$. To prove this assertion, we need to show that $e^*(k,x)$ is a bounded function in a neighborhood of the point $k = 0$. We write equality (7.56) in the form:

$$e^*(k,x) = [s_{11}(k) + s_{21}(k)]\omega(-k,x) + s_{21}(k)[\omega(k,x) - \omega(-k,x)].$$

Taking into account that: $ks_{21}(k) = o(1)$ as $k \to 0$, we get the estimate:

$$s_{21}(k)[\omega(k,x) - \omega(-k,x)] = -2is_{21}(k)\sin kx - 2i\int\limits_{-\infty}^x K^-(x,\xi)s_{21}(k)\sin k\xi d\xi$$

$$= -2i\frac{\sin kx}{k}[ks_{21}(k)] - 2i\int\limits_{-\infty}^x K^-(x,\xi)\frac{\sin k\xi}{k}[ks_{21}(k)]d\xi$$

$$= O(x) + O\Bigg(\int\limits_{-\infty}^x |\xi||K^-(x,\xi)||d\xi|\Bigg) = O(1) \quad \text{as } k \to 0.$$

From this estimate and assumption (7.37), it follows that the function $e^*(k,x)$ is bounded in a neighborhood of the point $k = 0$. Thus, the assertion (c) is proved.

Now we can prove identity (7.60). Consider a function $[e^*(k,x) - e^{ikx}]e^{-iky}$ for $y < x$, which is analytic in the upper half-plane Im$k > 0$. We integrate this function along the contour represented in Figure 7.1. Due to properties (a) and (b) of the function $e^*(k,x)$, the integrals of the considered function along the small and large semicircles tend to zero as $\rho \to 0$ and $R \to \infty$. Hence, by also the property (c) and Cauchy's Theorem [114]:

$$\lim_{R\to\infty} \int_{-R}^{R} [e^*(k,x) - e^{ikx}]e^{-iky} dk = 0 \text{ for } y < x.$$

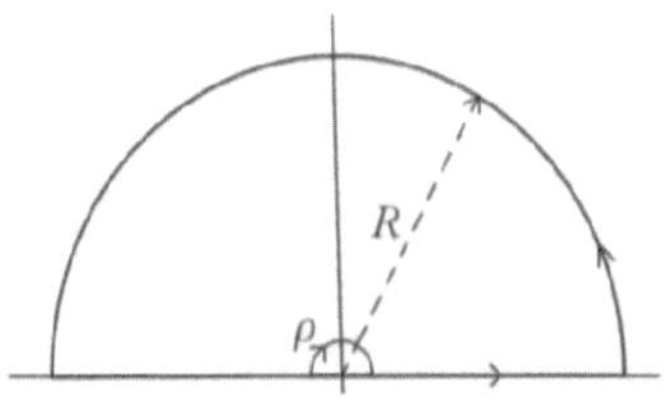

FIGURE 7.1: The path of integration.

Due to the property (c), there exists a function $K^*(x,y)$ so that: $\frac{1}{2\pi}\int_{x}^{\infty} [e^*(k,x) - e^{ikx}]e^{-iky} dk = K^*(x,y)$ for $y > x$. Hence, $K^*(x,y)$, $y > x$ is the inverse Fourier transformation of the function $e^*(k.x) - e^{ikx}$, which belongs to the space $L_2(-\infty,\infty)$ and

$$e^*(k,x) - e^{ikx} = \int_{x}^{\infty} K^*(x,y)e^{iky} dy. \tag{7.64}$$

The Fourier transform (7.64) and its inverse Fourier transform map $L_2[x,\infty)$ onto $L_2[x,\infty)$ mutually continuously one-to-one, therefore $K^*(x,y) \in L_2(x,\infty)$. Dividing equality (7.59) by $s_{11}(k)$ and using (7.64), we have:

$$\frac{1}{s_{11}(k)}\omega(k,x) - e^{-ikx} = \int_{x}^{\infty} K^*(x,\xi)e^{-ik\xi}d\xi - \frac{s_{21}(-k)}{s_{11}(k)}\left[e^{ikx} + \int_{x}^{\infty} K^*(x,\xi)e^{ik\xi}d\xi\right]. \tag{7.65}$$

The function $s_{11}(k)$ is analytical in the upper half-plane $\mathrm{Im}k > 0$ and has no zero there. Due to (7.23), (7.24), (7.59) and (7.61), we have:

$$\begin{aligned} h(k,x) &= \left[\frac{1}{s_{11}(k)}\omega(k,x) - e^{-ikx}\right]e^{iky} \\ &= \left[e^{*}(-k,x) - \frac{s_{21}(-k)}{s_{11}(k)}e^{*}(k,x) - e^{-ikx}\right]e^{iky} \\ &= O\left(\frac{e^{-(y-x)\mathrm{Im}k}}{|k|}\right), y > x \end{aligned}$$

as $|k| \to \infty$.

Multiply both sides of identity (7.65) by $\frac{1}{2\pi}e^{iky}$ for $y > x$, and integrate with respect to k, then due to the analytical property of the function $h(k,x)$ and its estimate, the left-hand side of the obtained equality vanishes when $y > x$ and $|k| \to \infty$, while by virtue of the Fourier integral theorem [61], the right-hand side of this equality gives:

$$R(x+y) + K^{*}(x,y) + \int_{x}^{\infty} R(y+\xi)K^{*}(x,\xi)d\xi.$$

Therefore, the function $K^{*}(x,y)$ satisfies the integral fundamental equation (7.45). From the uniqueness of a solution of equation (7.45), it follows that $K^{*}(x,y) = K(x,y)$. Then by (7.47) and (7.64) $e^{*}(k,x) \equiv e(k,x)$. Since the solution of equation (7.45) is the kernel of the transformation operator, then it satisfies equation (7.13). Due to the uniqueness of solution of equation (7.13), the solution $K(x,y)$ of equation (7.45) is related to the potential by formula (7.52) and satisfies condition (7.53). Therefore, the restored potential (7.52) belongs to the class $\mathcal{S}$. Thus, the Schrödinger equation (7.8) is restored with the potential (7.52) satisfying condition (7.53). In addition, the negative spectrum of the restored equation is absent. Thus, the first assertion is proved.

We proceed to prove the second assertion. Let $s_{11}(k)$ and $s_{21}(k)$, $-\infty < k < \infty$ be any given functions satisfying conditions enumerated in Theorem 7.1. We prove that these given functions are the scattering data of the SP (7.5)–(7.6) with the restored potential belonging to the class $\mathcal{S}$. In fact, consider the SP with the restored potential (7.52) satisfying condition (7.53). Let $\tilde{s}_{11}(k)$ and $\tilde{s}_{21}(k)$ be scattering data, and let $\tilde{K}(x,y)$, $y > x$ be the kernel of the transformation operator of this considered SP. Then the function $\tilde{K}(x,y)$ satisfies the equation of the form (7.45) for $y > x$:

$$\hat{R}(x+y) + \tilde{K}(x,y) + \int_{x}^{\infty} \hat{R}(y+\xi)\tilde{K}(x,\xi)d\xi = 0, \tag{7.66}$$

where $\hat{R}(x+y)$ is the known function determined by the formula:

$$\hat{R}(x+y) = -\frac{1}{2\pi}\int_{-\infty}^{\infty}\frac{\tilde{s}_{21}(-k)}{\tilde{s}_{11}(k)}e^{ik(x+y)}dk.$$

Since the solutions $K(x,y)$ and $\tilde{K}(x,y)$ of equations (7.45) and (7.66), respectively, satisfy the same integral equation (7.13) with the potential (7.52), then owing to the uniqueness of solution of equation (7.13), we obtain the identity: $K(x,y) \equiv \tilde{K}(x,y)$ for $y > x$. Taking this identity into account and subtracting equation (7.66) from equation (7.45), we have the homogeneous equation:

$$R(x+y) - \hat{R}(x+y) + \int_{x}^{\infty}[R(y+\xi) - \hat{R}(y+\xi)]K(x,\xi)d\xi = 0 \quad \text{for} \quad y > x. \tag{7.67}$$

For a sufficiently large positive $x = x_0$, the integral operator in homogeneous equation (7.67) is a contracting operator in the space of functions bounded on the interval (x_0, ∞). Hence, for $x \geq x_0$ and $y \geq x$: $\hat{R}(x+y) \equiv R(x+y)$, from which it follows that, $\hat{R}(x) = R(x)$ for $x \geq 2x_0$. Therefore, the integral equation (7.67) takes the form:

$$R(x+y) - \hat{R}(x+y) + \int_{x}^{2x_0-y}[R(y+\xi) - \hat{R}(y+\xi)]K(x,\xi)d\xi = 0 \quad \text{for} \quad y > x.$$

For fixed x_0 and y, the above equation is Volterra homogeneous, therefore $R(x+y) - \hat{R}(x+y) \equiv 0$ for any x and $y \geq x$, i.e., $\hat{R}(x)$ is identical to $R(x)$, or, in expanded form:

$$\frac{1}{2\pi}\int_{-\infty}^{\infty} -\frac{s_{21}(-k)}{s_{11}(k)}e^{ikx}dk = \frac{1}{2\pi}\int_{-\infty}^{\infty} -\frac{\tilde{s}_{21}(-k)}{\tilde{s}_{11}(k)}e^{ikx}dk.$$

Then owing to uniqueness of expansion of a function in the Fourier integral: $\dfrac{\tilde{s}_{21}(-k)}{\tilde{s}_{11}(k)} \equiv \dfrac{s_{21}(-k)}{s_{11}(k)}$.

Since the functions $s_{11}(k)$ and $\tilde{s}_{11}(k)$ have no zeros in the upper half-plane $\text{Im}k > 0$, then by the dispersion relation, the functions $s_{11}(k)$ and $\tilde{s}_{11}(k)$ are uniquely restorable from the modulus of the right-reflection coefficient $|\tilde{R}(k)| = \left|\dfrac{s_{21}(-k)}{s_{11}(k)}\right|$, whence it follows that:

$$s_{11}(k) \equiv \tilde{s}_{11}(k), \quad s_{21}(-k) \equiv \tilde{s}_{21}(-k), \quad s_{12}(k) = -\tilde{R}(k)s_{11}(k),$$
$$\tilde{r}(k) = -\frac{s_{11}(-k)}{s_{11}(k)}\tilde{R}(-k).$$

Thus, the ratios: $\tilde{R}(k) = -\dfrac{s_{12}(k)}{s_{11}(k)}$ and $\tilde{r}(k) = \dfrac{s_{21}(k)}{s_{11}(k)}$ satisfy conditions enumerated in Lemmas 7.4 and 7.5, respectively. Therefore, they are right- and left-reflection coefficients of the considered SP, and the equation (7.8) is restored precisely from $\tilde{R}(k)$. Thus, the sufficiency of conditions of right-reflection and left-reflection coefficients $\tilde{R}(k)$ and $\tilde{r}(k)$ is proved. The necessity of conditions of these coefficients has been established by Lemmas 7.1, 7.2, 7.4 and 7.5. The proof of Theorem 7.1 is completed. □

7.2 The system of evolution equations for the scattering matrix

The KdV equation (7.1) with the negative coefficient of the dispersive term is derived from the Lax condition for compatibility of two systems of linear equations of x and t:

$$-y_{xx} + p(x,t)y = \lambda y, \quad 0 \le x < \infty, \tag{7.68}$$

$$y_t = p_x(x,t) - 2(p(x,t) + 2\lambda)y_x. \tag{7.69}$$

For convenience, we pass the scalar pair of equations (7.68) and (7.69) to the matrix equations:

$$Y_x = UY, \;\; Y = (y_1, y_2), \tag{7.70}$$

$$Y_t = VY, \tag{7.71}$$

where U and V are the given matrices having the form:

$$U = \begin{pmatrix} 0 & 1 \\ p - \lambda & 0 \end{pmatrix}, \; V = \begin{pmatrix} p_x & -4\lambda - 2p \\ p_{xx} - (4\lambda + 2p)(p - \lambda) & -p_x \end{pmatrix}, \; k^2 = \lambda.$$

The equations (7.68) and (7.69) are equivalent to systems (7.70) and (7.71), respectively. The potential $p(x,t)$ for every $t > 0$ belongs to the class $\mathcal{S}$, therefore the systems (7.70) and (7.71) are compatible, i.e.,

$$y_{2xt} = y_{2tx} \text{ or } y_{1xxt} = y_{1xtx}.$$

The compatible equalities will be referred to as the Lax compatibility condition for systems (7.68) and (7.69). It is easy to verify that this compatibility condition is equivalent to the KdV equation (7.1). The boundary conditions (7.2) are satisfied if and only if the system (7.71) along the axis $x = 0$ takes the form:

$$Y_t = \begin{pmatrix} p_x(0,t) & -4\lambda \\ 4\lambda^2 & -p_x(0,t) \end{pmatrix} Y. \tag{7.72}$$

It follows from (7.70) that $y_2 = y_{1x}$ and y_1 satisfies the Schrödinger equation:

$$-y_{xx} + p(x,t)y = k^2 y, \quad 0 \le x < \infty, \tag{7.73}$$

where the initial function $p(x,0) = p(x)$ belongs to the class $\mathcal{S}$.

The matrix solutions E and W of problems for system (7.70) with condition (7.9) and for system (7.70) with condition (7.10) are denoted as follows:

$$E(k,x,t) = \left(E^+(k,x,t),\ E^-(k,x,t)\right) = \begin{pmatrix} e(k,x,t) & e(-k,x,t) \\ e_x(k,x,t) & e_x(-k,x,t) \end{pmatrix},$$

$$W(k,x,t) = \left(W^-(k,x,t),\ W^+(k,x,t)\right) = \begin{pmatrix} \omega(-k,x,t) & \omega(k,x,t) \\ \omega_x(-k,x,t) & \omega_x(k,x,t) \end{pmatrix}.$$

These matrix functions E and W have the asymptotic behavior for all real $k = \sqrt{\lambda}$:

$$E(k,x,t) = T(k)e^{ikx\sigma_3} + o(1), (x \to \infty), e^{ikx\sigma_3} = \begin{pmatrix} e^{ikx} & 0 \\ 0 & e^{-ikx} \end{pmatrix}, \tag{7.74}$$

$$W(k,x,t) = T(k) + o(1), \quad (x \to 0), \quad T(k) = \begin{pmatrix} 1 & 1 \\ ik & -ik \end{pmatrix}. \tag{7.75}$$

Here and every where in what follows, $\sigma_i, i = 1,2,3$ denote the Pauli matrices:

$$\sigma_1 = \begin{pmatrix} 0 & 1 \\ 1 & 0 \end{pmatrix}, \sigma_2 = \begin{pmatrix} 0 & i \\ -i & 0 \end{pmatrix}, \sigma_3 = \begin{pmatrix} 1 & 0 \\ 0 & -1 \end{pmatrix}.$$

Note that the Jost solutions E^+ and E^- of the problem (7.70), (7.74) are defined and analytic only on the half-planes Im$k \ge 0$ and Im$k \le 0$, respectively. They have the asymptotic behavior:

$$E^+(k,x,t) = \begin{pmatrix} 1 \\ ik \end{pmatrix} e^{ikx}\left(1 + O(k^{-1})\right), \quad (k \to \infty), \quad \text{Im}k \ge 0,$$

$$E^-(k,x,t) = \begin{pmatrix} 1 \\ -ik \end{pmatrix} e^{-ikx}\left(1 + O(k^{-1})\right), \quad (k \to \infty), \quad \text{Im}k \le 0.$$

Whereas the matrix solution W of the problem (7.70), (7.75) is an entire analytical function and its column solutions W^+ and W^- have the asymptotic behavior:

$$W^+(k,x,t) = \begin{pmatrix} 1 \\ -ik \end{pmatrix} e^{-ikx}\left(1 + O(k^{-1})\right), \quad (k \to \infty), \quad \text{Im}k \ge 0,$$

$$W^-(k,x,t) = \begin{pmatrix} 1 \\ ik \end{pmatrix} e^{ikx}\left(1 + O(k^{-1})\right), \quad (k \to \infty), \quad \text{Im}k \le 0.$$

By Lemma 7.2, the columns $s_1(k) = (s_{11}(k), s_{21}(k))$ and $s_2(k) = (s_{12}(k), s_{22}(k))$ are defined and analytic in the upper Im$k \ge 0$ and lower

$\mathrm{Im}k \le 0$ half-planes for $k \ne 0$, respectively. We compose the matrices Φ^+ and Φ^- by the columns of matrices E and W according to the rule.

$$\Phi^+(k,x) = \left(s_{11}^{-1}(k)E^+(k,x), W^+(k,x)\right) e^{-ikx\sigma_3}, \quad \mathrm{Im}k > 0, \tag{7.76}$$

$$\Phi^-(k,x) = \left(W^-(k,x), s_{22}^{-1}(k)E^-(k,x)\right) e^{-ikx\sigma_3}, \quad \mathrm{Im}k < 0, \tag{7.77}$$

where $E^\pm(k,x) = E^\pm(k,x,0), \;\; W^\pm(k,x) = W^\pm(k,x,0)$.

The matrix functions Φ^+ and Φ^- are analytic in the upper $\mathrm{Im}k > 0$ and lower $\mathrm{Im}k < 0$ half-planes of k, respectively. Using the representation (7.18), we write the matrices Φ^+ and Φ^- in the form:

$$\Phi^+(k,x) = \left(W^-(k,x), W^+(k,x)\right) \begin{pmatrix} 1 & 0 \\ s_{11}^{-1}(k)s_{21}(k) & 1 \end{pmatrix} e^{-ikx\sigma_3},$$

$$\Phi^-(k,x) = \left(W^-(k,x), W^+(k,x)\right) \begin{pmatrix} 1 & s_{22}^{-1}(k)s_{12}(k) \\ 0 & 1 \end{pmatrix} e^{-ikx\sigma_3}.$$

Due to (7.75), the matrices Φ^+ and Φ^- at $x = 0$ are expressible in terms of entries of $s(k)$:

$$T^{-1}(k)\Phi^+(k,0) = \begin{pmatrix} 1 & 0 \\ s_{11}^{-1}(k)s_{21}(k) & 1 \end{pmatrix}, \quad \mathrm{Im}k > 0,$$

$$T^{-1}(k)\Phi^-(k,0) = \begin{pmatrix} 1 & s_{22}^{-1}(k)s_{12}(k) \\ 0 & 1 \end{pmatrix}, \quad \mathrm{Im}k < 0.$$

In the generic case the functions $s_{11}(k)$ and $s_{22}(k)$ have a pole at the point $k = 0$, and in this case $s_{11}^{-1}(0) = 0, \;\; s_{22}^{-1}(0) = 0$, and $s_{11}^{-1}(0)E^+(0,x) = \begin{pmatrix} 0 \\ 0 \end{pmatrix}$, $s_{22}^{-1}(0)E^-(0,x) = \begin{pmatrix} 0 \\ 0 \end{pmatrix}$. Then the matrices $\Phi^+(0,x)$ and $\Phi^-(0,x)$ satisfy following conditions:

$$\Phi^+(0,x)\begin{pmatrix} 1 \\ 0 \end{pmatrix} = \begin{pmatrix} 0 \\ 0 \end{pmatrix}, \quad \Phi^-(0,x)\begin{pmatrix} 0 \\ 1 \end{pmatrix} = \begin{pmatrix} 0 \\ 0 \end{pmatrix}.$$

In the special case when the scattering matrix $s(k)$ is bounded at $k = 0$ these conditions are not valid.

Definition 7.1. *A non-singular matrix solution $Y(k,x)$ of system (7.70) such that the function $Y(k,x)e^{-ikx\sigma_3}$ is bounded for all $x \ge 0$, is called a solution of the SP on the half-axis $x \ge 0$ for the fixed point k. Such a formulation of the SP on the whole axis was given by Shabat [110].*

The constructed functions $Y(k,x) = \Phi^+(k,x)e^{ikx\sigma_3}$ and $Y(k,x) = \Phi^-(k,x)e^{ikx\sigma_3}$ are some bounded solutions of system (7.70) defined on the half-planes $\mathrm{Im}k > 0$ and $\mathrm{Im}k < 0$, respectively, excluding the points k, where $s_{11}(k) = 0$ $\mathrm{Im}k > 0$ and $s_{22}(k) = 0$ $\mathrm{Im}k < 0$. We note that solution of this problem contains some arbitrary matrix-valued multiplier, in contrast to the similar problem on the whole axis, that must be bounded at $+\infty$ and at $-\infty$. Due to this fact, any other solution $(\tilde{Y}^1(k,x), \tilde{Y}^2(k,x))$, $\mathrm{Im}k > 0$ of system (7.70) is different from the solution $Y(k,x) = (s_{11}^{-1}(k)E^+(k,x), W^+(k,x))$, $\mathrm{Im}k > 0$ by a non-singular matrix $a(k) = \begin{pmatrix} a_{11}(k) & a_{12}(k) \\ a_{21}(k) & a_{22}(k) \end{pmatrix}$:

$$(\tilde{Y}^1(k,x), \tilde{Y}^2(k,x) = (s_{11}^{-1}(k)E^+(k,x), W^+(k,x))a(k), \quad \mathrm{Im}k > 0.$$

By Definition 7.1 the function $(\tilde{Y}^1(k,x), \tilde{Y}^2(k,x))e^{-ikx\sigma_3}$ is bounded, whence it follows that the function:

$$\tilde{Y}^1(k,x) = s_{11}^{-1}(k)e(k,x)a_{11}(k)e^{-ikx} + \omega(k,x)a_{21}(k)e^{-ikx}, \quad \mathrm{Im}k > 0$$

is bounded, whence it follows that the element $a_{21}(k)$ of the matrix $a(k)$ must be equal to zero. Indeed, if $a_{21}(k) \neq 0$, then the solution $(\omega(k,x)a_{21}(k)e^{-ikx})e^{ikx}$, $\mathrm{Im}k > 0$, exponentially decreases as $x \to \infty$. Since the function $E^+(k,x), \mathrm{Im}k > 0$ satisfies condition (7.74), and all solutions of system (7.70) that decrease exponentially as $x \to \infty$, are proportional to the function $E^+(k,x)$, $\mathrm{Im}k > 0$, then the function $\omega(k,x)$, $\mathrm{Im}k > 0$ is proportional to the function $s_{11}^{-1}(k)e(k,x), \mathrm{Im}k > 0$ and the matrix $(s_{11}^{-1}(k)E^+(k,x), W^+(k,x))$ is singular. This conflicts with the solution of system (7.70):

$$Y(k,x) = (s_{11}^{-1}(k)E^+(k,x), W^+(k,x)), \mathrm{Im}k > 0,$$

that is the non-singular matrix solution constructed by the Definition 7.1.

Analogously prove that any other solution $(\tilde{Y}^1(k,x), \tilde{Y}^2(k,x)$, $\mathrm{Im}k < 0$ of system (7.70) is different from the solution $Y(k,x) = (W^-(k,x), s_{22}^{-1}(k) E^-(k,x))$, $\mathrm{Im}k < 0$ by a non-singular lower-triangular matrix $a(k) = \begin{pmatrix} a_{11}(k) & 0 \\ a_{21}(k) & a_{22}(k) \end{pmatrix}$:

$$(\tilde{Y}^1(k,x), \tilde{Y}^2(k,x)) = (W^-(k,x), s_{22}^{-1}(k)E^-(k,x)) \begin{pmatrix} a_{11}(k) & 0 \\ a_{21}(k) & a_{22}(k) \end{pmatrix}, \mathrm{Im}k < 0.$$

The time-dependence of the Jost solution $(e(k,x,t), e_x(k,x,t))$ and the analytic solution $(\omega(k,x,t), \omega_x(k,x,t))$ are defined by the evolution equation (7.71) with the boundary conditions (7.74) and (7.75), respectively:

$$\begin{pmatrix} e_t(k,x,t) \\ e_{xt}(k,x,t) \end{pmatrix} = V \begin{pmatrix} e(k,x,t) \\ e_x(k,x,t) \end{pmatrix} + 4ik^3 \begin{pmatrix} e(k,x,t) \\ e_x(k,x,t) \end{pmatrix}, \tag{7.78}$$

$$\begin{pmatrix} \omega_t(k,x,t) \\ \omega_{xt}(k,x,t) \end{pmatrix} = V \begin{pmatrix} \omega(k,x,t) \\ \omega_x(k,x,t) \end{pmatrix} + a(k,t) \begin{pmatrix} \omega(k,x,t) \\ \omega_x(k,x,t) \end{pmatrix} + b(k,t) \begin{pmatrix} \omega(-k,x,t) \\ \omega_x(-k,x,t) \end{pmatrix}, \tag{7.79}$$

where $a(k,t)$ and $b(k,t)$ are determined by the boundary conditions (7.75) at $x=0$.

Indeed, using (7.71), (7.75), from (7.79) we obtain the system of equations for determining $a(k,t)$ and $b(k,t)$:

$$\begin{aligned}
&\omega_t(k,0,t)+[2p(0,t)+4k^2]\omega_x(k,0,t)-p_x(0,t)\omega(k,0,t)\\
&=a(k,t)\omega(k,0,t)+b(k,t)\omega(-k,0,t), \qquad (7.80)\\
&\omega_{xt}(k,0,t)+\left\{[2p(0,t)+4k^2][p(0,t)-k^2]-p_{xx}(0,t)\right\}\omega(k,0,t)\\
&+p_x(0,t)\omega_x(k,0,t)=a(k,t)\omega_x(k,0,t)+b(k,t)\omega_x(-k,0,t).
\end{aligned}$$

Due to the boundary condition (7.75), the system (7.80) is reduced to the following system:

$$\begin{aligned}
&-ik[2p(0,t)+4k^2]-p_x(0,t)=a(k,t)+b(k,t),\\
&[2p(0,t)+4k^2][p(0,t)-k^2]-ikp_x(0,t)-p_{xx}(0,t)=-ik[a(k,t)-b(k,t)].
\end{aligned}$$

The quantities $a(k,t)$ and $b(k,t)$ in (7.79) are calculated from this system:

$$a(k,t)=-4ik^3+\frac{p_{xx}(0,t)-2p^2(0,t)}{2ik}, \qquad (7.81)$$

$$b(k,t)=-2ikp(0,t)-p_x(0,t)+\frac{2p^2(0,t)-p_{xx}(0,t)}{2ik}. \qquad (7.82)$$

Now, we derive the system of evolution equations governing the time-dependence of the scattering matrix $s(k,t)$. Upon differentiating equality (7.18) with respect to t, we have the equality:

$$E_t(k,x,t)=W_t(k,x,t)s(k,t)+W(k,x,t)s_t(k,t), \qquad (7.83)$$

where $s(k,t)$ is defined by (7.18) as the ratio of two fundamental solutions:

$$s(k,t)=W^{-1}(k,x,t)E(k,x,t). \qquad (7.84)$$

Substituting (7.78), (7.79) and (7.84) into equality (7.83), we get the equality:

$$\begin{aligned}
&W(k,x,t)s_t(k,t)=E_t(k,x,t)-W_t(k,x,t)W^{-1}(k,x,t)E(k,x,t)=\\
&V(k,x,t)\left\{E(k,x,t)-W(k,x,t)W^{-1}(k,x,t)E(k,x,t)\right\}+E(k,x,t)4ik^3\sigma_3\\
&-W(k,x,t)\begin{pmatrix} a(-k,t) & b(k,t)\\ b(-k,t) & a(k,t)\end{pmatrix}W^{-1}(k,x,t)E(k,x,t),
\end{aligned}$$

where $W_t=VW+W\begin{pmatrix} a(-k,t) & b(k,t)\\ b(-k,t) & a(k,t)\end{pmatrix}$, $E_t=VE+E4ik^3\sigma_3$.
Using (7.84), from the last equality we obtain

$$s_t(k,t)=s(k,t)4ik^3\sigma_3-\begin{pmatrix} a(-k,t) & b(k,t)\\ b(-k,t) & a(k,t)\end{pmatrix}s(k,t).$$

Substituting the calculated quantities (7.81) and (7.82) into the right-hand side of the last equality gives:

$$s_t(k,t) = 4ik^3[s,\sigma_3] + \left\{ p_x(0,t)\sigma_1 + \frac{2p^2(0,t) - p_{xx}(0,t)}{2k} i\sigma_3 + \right.$$
$$\left. \frac{p_{xx}(0,t) - 2p^2(0,t) - 4k^2 p(0,t)}{2k}\sigma_2 \right\} s(k,t). \tag{7.85}$$

Owing to boundary conditions (7.2), the derived system (7.85) reduces to the system of linear differential equations governing the time-dependence of the scattering matrix $s(k,t)$:

$$s_t(k,t) = 4ik^3[s,\sigma_3] + p_x(0,t)\sigma_1 s(k,t). \tag{7.86}$$

Thus, the time-dependence of $s(k,t)$ defined by system (7.86) implicitly depends on the time t. This is the main difference between the IBVP (7.1)–(7.2)–(7.3) and the Cauchy problem, therein lies the difficulty in passing from the Cauchy problem to this IBVP. Moreover, the system (7.86) is undetermined, because the function $p_x(0,t)$ entering the coefficients of this system is unknown. In Section 7.3 we shall prove that the unknown object $p_x(0,t)$ can be expressed through entries of the given scattering matrix $s(k)$.

7.3 The self-adjoint problem

7.3.1 The linear change of dependent variables

The system (7.72) describes the time-evolution of the eigenfunction for the boundary point $x = 0$. Using the linear change of dependent variables: $Y(k,t) = J(k)\tilde{y}(\mu,t)$, we reduce system (7.72) to the form:

$$\tilde{y}_t(k,t) = \left(J^{-1}V_0 J\right)\tilde{y}(k,t), \tag{7.87}$$

where $V_0 = p_x(0,t)\begin{pmatrix} 1 & 0 \\ 0 & -1 \end{pmatrix} - 4k^2\begin{pmatrix} 0 & 1 \\ 0 & 0 \end{pmatrix} + 4k^4\begin{pmatrix} 0 & 0 \\ 1 & 0 \end{pmatrix}$, and J is a matrix, which is to be taken so that $J^{-1}\sigma_3 J = \sigma_1$.

From the equality: $J\sigma_1 = \sigma_3 J$, it follows that the matrix J takes the form: $J = \begin{pmatrix} \alpha & \alpha \\ -\beta & \beta \end{pmatrix}$, α and β are arbitrary constants and $\alpha \neq 0$, $\beta \neq 0$. Then

$$J^{-1}\sigma_3 J = \sigma_1, J^{-1}\begin{pmatrix} 0 & 1 \\ 0 & 0 \end{pmatrix} J = \frac{1}{2\alpha\beta}\begin{pmatrix} \beta & -\alpha \\ \beta & \alpha \end{pmatrix}\begin{pmatrix} 0 & 1 \\ 0 & 0 \end{pmatrix}\begin{pmatrix} \alpha & \alpha \\ -\beta & \beta \end{pmatrix}$$
$$= \frac{\beta}{2\alpha}\begin{pmatrix} -1 & 1 \\ -1 & 1 \end{pmatrix} = \frac{\beta}{2\alpha}(-\sigma_3 - i\sigma_2),$$
$$J^{-1}\begin{pmatrix} 0 & 0 \\ 1 & 0 \end{pmatrix} J = \frac{\alpha}{2\beta}\begin{pmatrix} -1 & -1 \\ 1 & 1 \end{pmatrix} = \frac{\alpha}{2\beta}(-\sigma_3 + i\sigma_2).$$

Substituting the last equalities into the matrix $J^{-1}V_0J$ gives:

$$J^{-1}V_0J = p_x(0,t)\sigma_1 - 4k^2\frac{\beta}{2\alpha}(-\sigma_3 - i\sigma_2) + 4k^4\frac{\alpha}{2\beta}(-\sigma_3 + i\sigma_2) =$$
$$p_x(0,t)\sigma_1 + (4k^2\frac{\beta}{2\alpha} - 4k^4\frac{\alpha}{2\beta})\sigma_3 + (4k^2\frac{\beta}{2\alpha} + 4k^4\frac{\alpha}{2\beta})i\sigma_2,$$

Thus, after the change of variables J, the system (7.87) takes the form:

$$\tilde{y}_t(k,t) = \left\{p_x(0,t)\sigma_1 + 2k^2\left(\frac{\beta}{\alpha} - k^2\frac{\alpha}{\beta}\right)\sigma_3 + 2k^2\left(\frac{\beta}{\alpha} + k^2\frac{\alpha}{\beta}\right)i\sigma_2\right\}\tilde{y}(k,t).$$

Taking $\alpha = 1$ and $\beta = \pm ik$, which are the roots of the equation: $\beta^2 + k^2 = 0$, we reduce the above system to the form:

$$\tilde{y}_t(\mu,t) = (\pm 4i\mu\sigma_3 + p_x(0,t)\sigma_1)\,\tilde{y}(\mu,t), \quad \mu = k^3. \tag{7.88}$$

Let $\beta = -ik$, then the matrix J in (7.87) coincides with the matrix $T(k)$ defined by (7.75):

$$J(k) = \begin{pmatrix} 1 & 1 \\ ik & -ik \end{pmatrix} = T(k), \;\; Y(k,t) = T(k)\tilde{y}(\mu,t), \;\; \mu = k^3, \; 0 \le t < \infty. \tag{7.89}$$

Thus, by virtue of the linear change of dependent variables given by (7.89), we lead system (7.88) into the system of first-order ordinary differential equations:

$$i\sigma_3\tilde{y}_t(\mu,t) + C(t)\tilde{y}(\mu,t) = 4\mu\tilde{y}(\mu,t), \;\; \mu = k^3, \;\; 0 \le t < \infty, \tag{7.90}$$

where the potential matrix $C(t)$ is self-adjoint:

$$C(t) = \begin{pmatrix} 0 & c_1(t) \\ c_2(t) & 0 \end{pmatrix} = \overline{C(t)}, \tag{7.91}$$

$c_1(t) = -ip_x(0,t), c_2(t) = ip_x(t)$ and the potential $p_x(0,t)$ is a real-valued function.

For convenience we denote $\tilde{y}(\mu,t)$ by $y(\mu,t)$ and consider the self-adjoint problem generated by system (7.90):

$$i\sigma_3 y_t(\mu,t) + C(t)y(\mu,t) = 4\mu y(\mu,t), \;\; \mu = k^3, \;\; 0 \le t < \infty \tag{7.92}$$

with the boundary condition at the origin $t = 0$:

$$y_1(\mu,0) = y_2(\mu,0). \tag{7.93}$$

We also consider the problem for system (7.92) with boundary conditions at infinity:

$$y_1(\mu,t) = A(\mu)e^{-4i\mu t} + o(1), \;\; y_2(\mu,t) = B(\mu)e^{4i\mu t} + o(1), \;\; (t \to \infty), \tag{7.94}$$

where μ is a real number.

Assume that the potential function $p_x(0,t)$ in (7.91) satisfies the estimate:

$$|p_x(0,t)| \le \frac{C}{1+t^{1+\varepsilon}}, \quad t>0, \;\; \varepsilon>0, \tag{7.95}$$

C is a constant.

The problems (7.92), (7.94) and (7.92)–(7.93) with the potential satisfying estimate (7.95) have been solved in [119]. Since the potential matrix (7.91) is a particular case of the potential self-adjoint matrix of the problem investigated in Section 1.2, then the following propositions are deduced from corresponding assertions proved in Section 1.2 without proving. We shall see that owing to the self-adjoint condition of the potential matrix, this problem possesses a series of characters, which are essentially different from characters of the non-self-adjoint problem [97].

7.3.2 The characters of the self-adjoint problem

Proposition 7.1. *The boundary problem (7.92)–(7.93) has a unique bounded solution $(y_1(\mu,t), y_2(\mu,t))$ for real μ such that for any given number $A(\mu)$ there exists a unique number $B(\mu)$ defined from $A(\mu)$ so that the asymptotics (7.94) are satisfied. Furthermore, this solution has the representation:*

$$y_1(\mu,t) = A(\mu)e^{-4i\mu t}\left(1+\int_0^\infty H_{11}(t,t+\xi)e^{-4i\mu\xi}d\xi\right)+ B(\mu)e^{4i\mu t}\int_0^\infty H_{21}(t,t+\xi)e^{4i\mu\xi}d\xi, \tag{7.96}$$

$$y_2(\mu,t) = A(\mu)e^{-4i\mu t}\int_0^\infty H_{21}(t,t+\xi)e^{-4i\mu\xi}d\xi + B(\mu)e^{4i\mu t}\left(1+\int_0^\infty H_{11}(t,t+\xi)e^{4i\mu\xi}\right),$$

where the matrix $H(t,s) = \begin{pmatrix} H_{11}(t,s) & H_{21}(t,s) \\ H_{21}(t,s) & H_{11}(t,s)\end{pmatrix}$, $0\le t\le s$ is an analog of the kernel of the transformation operator. The kernel functions $H_{11}(t,s)$ and $H_{21}(t,s)$ satisfy the estimate:

$$|H_{j1}(t,s)| \le \frac{C}{1+(t+s)^{1+\varepsilon}}, \quad 0\le t\le s, \;\; j=1,2, \tag{7.97}$$

C is a constant.

The estimate (7.95) for $p_x(0,t)$ is a sufficient condition, which ensures the unique solvability of integral equations (1.17)–(1.20) for kernel functions $H_{j1}(t,s)$, $j=1,2$. Using the self-adjoint condition (7.91), wherein the functions $c_1(t)$ and $c_2(t)$ are pure imaginary functions, from the integral equations for the kernels we obtain that the solutions $H_{jk}(t,s), j,k=1,2$ of these equations are real-valued functions, $H_{11}(t,s)=H_{22}(t,s), H_{12}(t,s)=H_{21}(t,s)$. In addition, the kernel function $H_{21}(t,t+\xi)$ is related to the potential function $p_x(0,t)$ by the formula:

$$-2H_{21}(t,t)=p_x(0,t) \text{ for } t>0. \tag{7.98}$$

Proposition 7.2. *There exist the bounded Jost solutions* $e^{(1)}(\mu,t) = \left(e_1^{(1)}(\mu,t), e_2^{(1)}(\mu,t)\right)$ *and* $e^{(2)}(\mu,t) = \left(e_1^{(2)}(\mu,t), e_2^{(2)}(\mu,t)\right)$ *of system* (7.92) *with the boundary at infinity:*

$$\begin{aligned} e^{(1)}(\mu,t)e^{4i\mu t} &= (1+o(1), o(1)), \quad Im\mu \le 0, \quad (t\to\infty), \\ e^{(2)}(\mu,t)e^{-4i\mu t} &= (o(1), 1+o(1)), \quad Im\mu \ge 0, \quad (t\to\infty). \end{aligned}$$

These Jost solutions are represented in the form:

$$\begin{aligned} e_1^{(1)}(\mu,t) &= e^{-4i\mu t}\left(1+\int_0^\infty H_{11}(t,t+\xi)e^{-4i\mu\xi}d\xi\right), \\ e_2^{(1)}(\mu,t) &= e^{-4i\mu t}\int_0^\infty H_{21}(t,t+\xi)e^{-4i\mu\xi}d\xi, \quad Im\lambda \le 0, \\ e^{(2)}(\mu,t) &= \left(\overline{e_2^{(1)}(\mu,t)}, \overline{e_1^{(1)}(\mu,t)}\right), \quad Im\lambda \ge 0. \end{aligned}$$

It is easy to verify that $\det\left(e^{(1)}(\mu,t), e^{(2)}(\mu,t)\right)=1$, therefore for any real μ the solutions $e^{(1)}(\mu,t)$ and $e^{(2)}(\mu,t)$ of system (7.92) constitute a fundamental system of vector-solutions of this system.

Definition 7.2. *The one-to-one correspondence between the numbers* $A(\mu)$ *and* $B(\mu)$ *determines the scattering function* $S(\mu)$*:* $S(\mu)A(\mu)=B(\mu)$, $-\infty<\mu<\infty$ *for the SP* (7.92)–(7.93) *on the half-axis:* $0\le t\le\infty$.

By an argument analogous to the problem (7.92)–(7.93) on the half-axis $0\le t<\infty$, for any $t\ge 0$, we consider the problem generated by system (7.92) on a half-axis $t\le\tau<\infty$ with the boundary condition at $\tau=t$:

$$z_1(\mu,t,t)=z_2(\mu,t,t), \tag{7.99}$$

and the problem for this system with the boundary condition at infinity:

$$\left.\begin{aligned} z_1(\mu,\tau,t) &= A(\mu,t)e^{-4i\mu(\tau-t)}+o(1), \\ z_2(\mu,\tau,t) &= B(\mu,t)e^{4i\mu(\tau-t)}+o(1). \end{aligned}\right. \quad (\tau\to\infty), \quad t\le\tau<\infty.$$

Definition 7.3. *The one-to-one correspondence between the numbers* $A(\mu,t)$ *and* $B(\mu,t)$ *determines the scattering function* $S(\mu,t)$: $S(\mu,t)A(\mu,t) = B(\mu,t)$, $-\infty < \mu < \infty$ *for the SP generated by system* (7.92) *on the half-axis:* $t \le \tau < \infty$ *with condition* (7.99).

By this definition and using (7.96), (7.99), we derive the factorization of $S(\mu,t)$:

$$S(\mu,t) = (1 + H_-(\mu,t))\,(1 + H_+(\mu,t))^{-1}, \quad -\infty < \mu < \infty, \tag{7.100}$$

where $H_-(\mu,t) = \int_0^\infty H(t,\xi)e^{-4i\mu\xi}d\xi$, $H_+(\mu,t) = \int_0^\infty H(t,\xi)e^{4i\mu\xi}d\xi$, $H(t,\xi) = H_{11}(t,t+\xi) - H_{21}(t,t+\xi)$ and $H_{j1}(t,t+\xi)$, $j = 1,2$ are real-valued functions satisfying estimate (7.97).

By virtue of estimate (7.97) and the self-adjoint property of matrix (7.91), the function $H(t,\xi)$ is absolutely integrable with respect to ξ, the numerator $1 + H_-(\mu,t)$ and denominator $1 + H_+(\mu,t)$ of ratio (7.100) are different from zero and analytic in the half-planes $\mathrm{Im}\mu \le 0$ and $\mathrm{Im}\mu \ge 0$, respectively. Then there exists an absolutely integrable with respect to ξ function $K(t,\xi)$ such that

$$\left(1 + \int_0^\infty H(t,\xi)e^{4i\mu\xi}d\xi\right)^{-1} = 1 + \int_0^\infty K(t,\xi)e^{4i\mu\xi}d\xi,$$

$$\left(1 + \int_0^\infty H(t,\xi)e^{-4i\mu\xi}d\xi\right)^{-1} = 1 + \int_0^\infty K(t,\xi)e^{-4i\mu\xi}d\xi.$$

Furthermore, for any $t \ge 0$ the scattering function $S(\mu,t)$ possesses the property:

$$\overline{S(\mu,t)} = S^{-1}(\mu,t) = S(-\mu,t), \quad -\infty < \mu < \infty. \tag{7.101}$$

Proposition 7.3. *For any* $t \ge 0$ *the scattering functions* $S(\mu,t) - 1$ *and* $S^{-1}(\mu,t) - 1$ *for the SP* (7.92), (7.99) *are the Fourier transformations:*

$$S(\mu,t) = 1 + F(\mu,t) = 1 + \int_{-\infty}^{\infty} f(t,\xi)e^{4i\mu\xi}d\xi, \tag{7.102}$$

$$S^{-1}(\mu,t) = 1 + G(\mu,t) = 1 + \int_{-\infty}^{\infty} g(t,\xi)e^{4i\mu\xi}d\xi, \tag{7.103}$$

where $f(t,\xi)$ and $g(t,\xi)$ are real-valued and absolutely integrable with respect to ξ functions defined by the formulas:

$$f(t,\xi) = H(t,-\xi) + K(t,\xi) + \int_0^\infty K(t,\zeta)H(t,\zeta-\xi)d\zeta, \tag{7.104}$$

$$g(t,\xi) = H(t,\xi) + K(t,-\xi) + \int_{-\infty}^0 K(t,-\zeta)H(t,\xi-\zeta)d\zeta. \tag{7.105}$$

From the properties of $S(\mu,t)$ given by (7.101)–(7.105) it follows that

$$S(\mu,0) = S(\mu) = 1+ \int_{-\infty}^{+\infty} f(\xi)e^{4i\mu\xi}d\xi, \quad S^{-1}(\mu,0) = S^{-1}(\mu) = 1+ \int_{-\infty}^{+\infty} g(\xi)e^{4i\mu\xi}d\xi,$$

where $f(\xi) = f(0,\xi)$, $g(\xi) = g(0,\xi)$; $f(-\xi) = g(\xi)$.

Proposition 7.4. *There exists uniquely a solution $\varphi(\mu,t) = (\varphi_1(\mu,t), \varphi_2(\mu,t))$ of system (7.92) with the initial condition: $\varphi_1(\mu,0) = \varphi_2(\mu,0) = 1$. The function $\varphi(\mu,t)$ is entirely analytic of the parameter μ and has the representation:*

$$\begin{aligned} D^{-1}(\mu)\varphi(\mu,t) &= e^{(1)}(\mu,t) + S(\mu)e^{(2)}(\mu,t), \quad Im\mu \geq 0, \\ N^{-1}(\mu)\varphi(\mu,t) &= S^{-1}(\mu)e^{(1)}(\mu,t) + e^{(2)}(\mu,t), \quad Im\mu \leq 0, \end{aligned} \tag{7.106}$$

where $N(\mu) = e_1^{(1)}(\mu,0) - e_2^{(1)}(\mu,0) = 1 + H_-(\mu,0) \neq 0$ *for* $Im\mu \leq 0$,

$$\begin{aligned} &D(\mu) = e_2^{(2)}(\mu,0) - e_1^{(2)}(\mu,0) = \overline{N(\mu)} \neq 0 \ \text{ for } Im\mu \geq 0, \\ &N(\mu) = 1 + o(1) \ \ (|\mu| \to \infty), \ \ D(\mu) = 1 + o(1) \ \ (|\mu| \to \infty), \\ &S(\mu) = \frac{N(\mu)}{D(\mu)}, \ \ S^{-1}(\mu) = \overline{S(\mu)} = S(-\mu), \ \ -\infty < \mu < \infty, \end{aligned} \tag{7.107}$$

here by the uniqueness of solution of boundary problem (7.92)–(7.93) and the self-adjoint property of this problem, the numerator $N(\mu)$ and denominator $D(\mu)$ are different from zero and analytic in the half-planes $Im\mu \leq 0$ and $Im\mu \geq 0$, respectively.

Proposition 7.5. *For any $t \geq 0$ the functions $f(t,\xi)$ and $g(t,\xi)$ defined by (7.104) and (7.105) are closely related to $f(\xi)$ and $g(\xi)$, respectively by the formula:*

$$\begin{aligned} &f(t,\xi) = f(\xi - 2t) \ \text{ for } \xi < 0, \ \ g(t,\xi) = g(\xi + 2t) \ \text{ for } \xi > 0, \\ &f(-\xi - 2t) = g(\xi + 2t) \ \text{ for } \xi > 0. \end{aligned} \tag{7.108}$$

Furthermore, the functions $f(-\xi)$ and $g(\xi)$ for $\xi > 0$ satisfy the estimate of the type (7.95):

$$|f(-\xi)| \leq \frac{\tilde{C}}{1+\xi^{1+\varepsilon}}, \quad |g(\xi)| \leq \frac{\tilde{C}}{1+\xi^{1+\varepsilon}}, \quad \xi > 0, \varepsilon > 0.$$

By Propositions 7.2 and 7.4 the vector-function:

$$\frac{1}{e_2^{(2)}(\mu,0)-e_1^{(2)}(\mu,0)}\begin{pmatrix}\varphi_1(\mu,t)e^{4i\mu t}\\ \varphi_2(\mu,t)e^{4i\mu t}\end{pmatrix}-\begin{pmatrix}1\\ e^{8i\mu t}\end{pmatrix},$$

does not have singular numbers and is analytic in the upper half-plane $\mathrm{Im}\mu \geq 0$. It tends to zero as $|\mu| \to \infty$. Therefore, by an application of the Jordan's lemma and the Fourier integral theorem to this vector-function, using condition (7.91) and Propositions 7.1 and 7.5, we derive the Fredholm system of integral equations connecting the scattering function $S(\mu,t)$ with the real kernel functions $(H_{11}(t,t+\xi), H_{21}(t,t+\xi)) = (H_{22}(t,t+\xi), H_{12}(t,t+\xi))$:

$$\begin{cases} H_{22}(t,t+\xi)+\int\limits_0^\infty H_{21}(t,t+\zeta)g(2t+\xi+\zeta)d\zeta=0,\\ g(2t+\xi)+H_{21}(t,t+\xi)+\int\limits_0^\infty H_{22}(t,t+\zeta)g(2t+\xi+\zeta)d\zeta=0.\end{cases} \tag{7.109}$$

Proposition 7.6. *For every fixed $t \geq 0$ the Fredholm system* (7.109) *has a unique solution $(H_{11}(t,t+\xi),\ H_{21}(t,t+\xi))$ in the space $L_1[0,\infty)$.*

Proposition 7.7. *For the given function $S(\mu)$ to be the scattering function for the self-adjoint problem* (7.92)–(7.93)*, it is necessary and sufficient that there exists a function $S(\mu,t)$ such that $S(\mu) = S(\mu,0)$ and*

(1) the function $S(\mu,t)$ admits the factorization of the form (7.100)*;*

(2) for any $t \geq 0$ the functions $S(\mu,t)-1$ and $S^{-1}(\mu,t)-1$ are the Fourier transformations (7.102) *and* (7.103) *of the functions* (7.104) *and* (7.105)*, respectively;*

(3) for any $t \geq 0$ the functions $f(t,\xi)$ and $g(t,\xi)$ are closely related to the functions $f(-\xi)$ and $g(\xi)$ by the equality (7.108)*, respectively. $f(-\xi)$ and $g(\xi)$ are real-valued functions satisfying the estimate of the type* (7.95).

To solve the inverse SP (7.92)–(7.93) for finding the unknown object $p_x(0,t)$, we need to find the unknown scattering function $S(\mu)$ of this problem first. To find the function $S(\mu)$, in the next subsection we shall express the sought function $S(\mu)$ through known elements of the scattering matrix $s(k)$ of the SP (7.5)–(7.6). This is the key step to recover the potential matrix (7.91), i.e., the system (7.92).

7.3.3 The problem of finding the scattering function $S(\mu,t)$

The matter is that the first SP (7.5), (7.6) and the second SP (7.92)–(7.93) are formulated on two different spectral planes. The first SP is considered on the k-spectral plane, whereas the second SP is considered on the μ-spectral plane, ($\mu = k^3$). To compare functions on the k-spectral plane with those on the μ-spectral plane, we use the conjugation contour for the SP (7.92)–(7.93). This conjugation contour $\mathrm{Im}\mu = \mathrm{Im}k^3 = 0$ splits into a system of rays $\{n_j\}_{j=1}^6$ coming from the origin of the k-plane with the slope angles $\gamma_j = (j-1)\pi/3$ with respect to the positive direction of the axis $\mathrm{Im}k = 0$. We denote by N_j the interior sector confined between the rays n_{j+1} and n_j (see Figure 7.2).

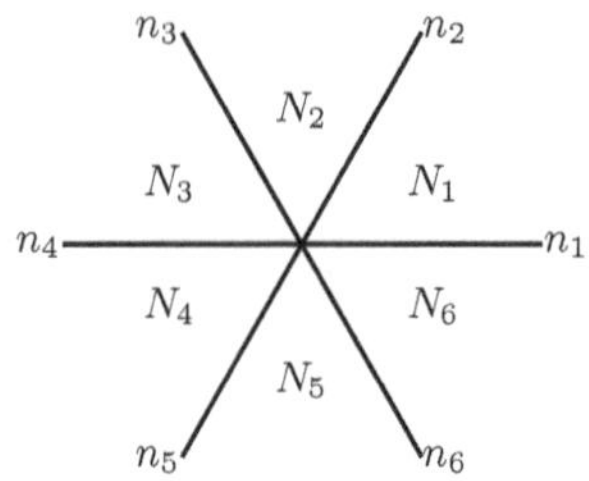

FIGURE 7.2: The conjugation contour.

Taking into account that $\arg\mu = \arg k^3 = 3\arg k$, we obtain:

$$\text{If } k \in N_1 \cup N_3 \cup N_5, \text{ then } \mathrm{Im}\mu > 0. \tag{7.110}$$

$$\text{If } k \in N_2 \cup N_4 \cup N_6, \text{ then } \mathrm{Im}\mu < 0. \tag{7.111}$$

Conversely, if $\mathrm{Im}\mu > 0$, then $\frac{m}{3}2\pi < \arg k < \frac{m}{3}2\pi + \frac{\pi}{3}, m = 0, 1, 2.$

$$\text{if } \mathrm{Im}\mu < 0, \text{ then } \frac{m}{3}2\pi + \frac{\pi}{3} < \arg k < \frac{m}{3}2\pi + \frac{2\pi}{3}, m = 0, 1, 2.$$

By taking the Jost solutions $e^{(1)}(\mu, t)$, $e^{(2)}(\mu, t)$ and analytic solution $\varphi(\mu, t)$ of the SPs (7.92), (7.94) and (7.92)–(7.93), respectively, we compose the matrix functions $\psi^+(\mu, t)$ and $\psi^-(\mu, t)$ according to the rule.

$$\psi^+(\mu, t) = \left(\frac{\varphi(\mu, t)}{D(\mu)}, e^{(2)}(\mu, t) \right) e^{4i\mu t\sigma_3} \text{ for } k \in N_1 \cup N_3 \cup N_5, \tag{7.112}$$

$$\psi^-(\mu, t) = \left(e^{(1)}(\mu, t), \frac{\varphi(\mu, t)}{N(\mu)} \right) e^{4i\mu t\sigma_3} \text{ for } k \in N_2 \cup N_4 \cup N_6. \tag{7.113}$$

Since the matrix functions $\left(\frac{\varphi(\mu,t)}{D(\mu)}, e^{(2)}(\mu, t)\right)$ and $\left(e^{(1)}(\mu, t), \frac{\varphi(\mu,t)}{N(\mu)}\right)$ satisfy system (7.92), then the constructed matrix functions $\psi^+(\mu, t)$ and $\psi^-(\mu, t)$ satisfy system (7.86). Hence, all the matrix functions $s(k, t)$, $\psi^+(\mu, t)$ and $\psi^-(\mu, t)$ satisfy the same system (7.86). We are persuaded that the functions $\psi^+(\mu, t)$ and $\psi^-(\mu, t)$ are the fundamental solutions of system (7.86). Indeed, as was shown by the Propositions 7.2 and 7.4 the functions $\varphi(\mu, t)$ and $e^{(1)}(\mu, t)$, $e^{(2)}(\mu, t)$ satisfy the same system (7.92), and the last two functions are the fundamental solutions of this system. Taking into account that the determinant of the matrix function $\left(e^{(1)}(\mu,t),e^{(2)}(\mu,t)\right)$ does not depend on t, we calculate the estimate

$$\left(e^{(1)}(\mu,t),e^{(2)}(\mu,t)\right)e^{4i\mu\sigma_3}\to\begin{pmatrix}1&0\\0&1\end{pmatrix}\quad\text{as } t\to\infty,$$

Using this estimate and representations (7.112) and (7.113), we easily verify that the determinants of $\psi^+(\mu,t)$ and $\psi^-(\mu,t)$ do not depend on t and $\det(\psi^+(\mu,t))=\det(\psi^-(\mu,t))=1$. Hence, the matrix functions $\psi^+(\mu,t)$ and $\psi^-(\mu,t)$ constructed by formulas (7.112) and (7.113) are two bases of fundamental solutions of system (7.86). Furthermore, they are invertible and are normalized at infinity for any $t>0:\ \psi^+(\infty,t)=\psi^-(\infty,t)=\begin{pmatrix}1&0\\0&1\end{pmatrix}$.

Denote the columns of the scattering matrix $s(k,t)$ by $s_1(k,t)=(s_{11}(k,t),s_{21}(k,t))$ and $s_2(k,t)=(s_{12}(k,t),s_{22}(k,t))$. Both columns are solutions of system (7.86), but they are defined in different half-planes, therefore we consider them separately for convenience. Every column solution $s_j(k,t),\ \ j=1,2$ can be represented in a form of linear combinations of fundamental solutions $\psi^+(\mu,t)$ or $\psi^-(\mu,t)$:

$$s_1(k,t)=\psi^+(\mu,t)\begin{pmatrix}\alpha(k,t)\\\beta(k,t)\end{pmatrix}\quad\text{for } k\in N_1,\tag{7.114}$$

$$s_1(k,t)=\psi^-(\mu,t)\begin{pmatrix}\alpha(k,t)\\\beta(k,t)\end{pmatrix}\quad\text{for } k\in N_2,\tag{7.115}$$

$$s_1(k,t)=\psi^+(\mu,t)\begin{pmatrix}\alpha(k,t)\\\beta(k,t)\end{pmatrix}\quad\text{for } k\in N_3,\tag{7.116}$$

$$s_2(k,t)=\psi^-(\mu,t)\begin{pmatrix}\alpha(k,t)\\\beta(k,t)\end{pmatrix}\quad\text{for } k\in N_4,\tag{7.117}$$

$$s_2(k,t)=\psi^+(\mu,t)\begin{pmatrix}\alpha(k,t)\\\beta(k,t)\end{pmatrix}\quad\text{for } k\in N_5,\tag{7.118}$$

$$s_2(k,t)=\psi^-(\mu,t)\begin{pmatrix}\alpha(k,t)\\\beta(k,t)\end{pmatrix}\quad\text{for } k\in N_6,\tag{7.119}$$

where the coefficients $\alpha(k,t)$ and $\beta(k,t)$ are unknown, which will be defined below.

The functions $s_1(k,t)$ and $s_2(k,t)$ are defined in the half-planes $\mathrm{Im}k\geq 0$ and $\mathrm{Im}k\leq 0$, those are in the closure of the union of the sectors N_1,N_2,N_3 and N_4,N_5,N_6, respectively. Using the consistency condition for systems (7.70) and (7.71) in the quarter of the plane $x\geq 0,\ \ t\geq 0$, we calculate $\alpha(k,0)$ and $\beta(k,0)$. In view of system (7.72) for the eigenfunction along the axis $x=0$ and the change of variables (7.89), this consistency condition means that the functions $T(k)\psi^+(\mu,t),\ T(k)\psi^-(\mu,t)$ defined by (7.89) and (7.112), (7.113) on μ-plane and $\Phi^+(k,0,t),\ \Phi^-(k,0,t)$ defined by (7.76) and (7.77) on k-plane must be consistent at the corner point $(x,t)=(0,0)$ for all values of the spectral parameter k. Making use of this fact, we calculate $\alpha(k,0)$ and $\beta(k,0)$

for $k \in N_j$, $j = 1, 2, \cdots, 6$. Indeed, let $k \in N_1$, then by relation (7.110): $\omega^2 k \in I_5$, where $\omega = e^{2i\frac{\pi}{3}}$ is a cubic root of unity. Conversely, if $\omega^2 k \in N_5$, then $k \in N_1$. Therefore, by the change of variables (7.89) and the consistency condition, we have the consistency equality:

$$T(\omega^2 k)\Psi^+(\mu, 0) = \Phi^-(\omega^2 k, 0, 0) \text{ for } k \in N_1.$$

Using this consistency equality and the representation (7.77) of the function $T^{-1}(\omega^2 k)\Phi^-(\omega^2 k, 0, t)$, from (7.114) we have for $k \in N_1$:

$$\begin{pmatrix} s_{11}(k,0) \\ s_{21}(k,0) \end{pmatrix} = \psi^+(\mu, 0) \begin{pmatrix} \alpha(k,0) \\ \beta(k,0) \end{pmatrix} =$$
$$T^{-1}(\omega^2 k)\Phi^-(\omega^2 k, 0, 0) \begin{pmatrix} \alpha(k,0) \\ \beta(k,0) \end{pmatrix} = \begin{pmatrix} 1 & r_{12}(\omega^2 k) \\ 0 & 1 \end{pmatrix} \begin{pmatrix} \alpha(k,0) \\ \beta(k,0) \end{pmatrix},$$

where $r_{12}(\omega^2 k) = s_{22}^{-1}(\omega^2 k, 0) s_{12}(\omega^2 k, 0)$.

Hence, $\alpha(k, 0)$ and $\beta(k, 0)$ for $k \in N_1$ are calculated in terms of entries of $s(k)$:

$$\alpha(k,0) = s_{11}(k,0) - r_{12}(\omega^2 k) s_{21}(k,0), \;\; \beta(k,0) = s_{21}(k,0) \text{ for } k \in N_1. \tag{7.120}$$

Using (7.89), (7.111) and the consistency condition gives: $T(k)\psi^-(\mu, 0) = \Phi^+(k, 0, 0)$ for $k \in N_2$. Then owing to (7.76), from (7.115) we get:

$$\begin{pmatrix} s_{11}(k,0) \\ s_{21}(k,0) \end{pmatrix} = \psi^-(\mu, 0) \begin{pmatrix} \alpha(k,0) \\ \beta(k,0) \end{pmatrix} = T^{-1}(k)\Phi^+(k, 0, 0) \begin{pmatrix} \alpha(k,0) \\ \beta(k,0) \end{pmatrix}$$
$$= \begin{pmatrix} 1 & 0 \\ r_{21}(k) & 1 \end{pmatrix} \begin{pmatrix} \alpha(k,0) \\ \beta(k,0) \end{pmatrix},$$

where $r_{21}(k) = s_{11}^{-1}(k, 0) s_{21}(k, 0)$, whence it follows that:

$$\alpha(k,0) = s_{11}(k,0), \;\; \beta(k,0) = 0 \text{ for } k \in N_2. \tag{7.121}$$

In view of (7.89), (7.110) and of consistency condition yields: $T(\omega k)\psi^+(\mu, 0) = \Phi^-(\omega k, 0, 0)$ for $k \in N_3$. Then due to (7.77), from (7.116) it follows that:

$$\begin{pmatrix} s_{11}(k,0) \\ s_{21}(k,0) \end{pmatrix} = \psi^+(\mu, 0) \begin{pmatrix} \alpha(k,0) \\ \beta(k,0) \end{pmatrix} = T^{-1}(\omega k)\Phi^-(\omega k, 0, 0) \begin{pmatrix} \alpha(k,0) \\ \beta(k,0) \end{pmatrix}$$
$$= \begin{pmatrix} 1 & r_{12}(\omega k) \\ 0 & 1 \end{pmatrix} \begin{pmatrix} \alpha(k,0) \\ \beta(k,0) \end{pmatrix} \text{ for } k \in N_3,$$

where since $k \in N_3$, then $\omega k \in N_5$. Conversely if $\omega k \in N_5$, then $k \in N_3$. $r_{12}(\omega k) = s_{22}^{-1}(\omega k, 0) s_{12}(\omega k, 0)$.

Hence, the coefficients $\alpha(k, 0)$, $\beta(k, 0)$ for $k \in N_3$ are calculated by:

$$\alpha(k,0) = s_{11}(k,0) - r_{12}(\omega k) s_{21}(k,0), \;\; \beta(k,0) = s_{21}(k,0) \text{ for } k \in N_3. \tag{7.122}$$

By an argument analogous to the previous one, from (7.117), (7.118) and (7.119) we calculate coefficients $\alpha(k,0), \beta(k,0)$ for the cases when $k \in N_4, k \in N_5$ and $k \in N_6$:

$$\begin{pmatrix} s_{12}(k,0) \\ s_{22}(k,0) \end{pmatrix} = \psi^-(\mu,0) \begin{pmatrix} \alpha(k,0) \\ \beta(k,0) \end{pmatrix} = T^{-1}(\omega^2 k)\Phi^+(\omega^2 k,0,0) \begin{pmatrix} \alpha(k,0) \\ \beta(k,0) \end{pmatrix}$$
$$= \begin{pmatrix} 1 & 0 \\ r_{21}(\omega^2 k) & 1 \end{pmatrix} \begin{pmatrix} \alpha(k,0) \\ \beta(k,0) \end{pmatrix}, \quad k \in N_4,$$

where since $k \in N_4$, then $\omega^2 k \in N_2$, and conversely. $r_{21}(\omega^2 k) = s_{11}^{-1}(\omega^2 k,0)s_{21}(\omega^2 k,0)$. Therefore

$$\alpha(k,0) = s_{12}(k,0), \quad \beta(k,0) = s_{22}(k,0) - r_{21}(\omega^2 k)s_{12}(k,0) \quad \text{for } k \in N_4. \tag{7.123}$$

$$\begin{pmatrix} s_{12}(k,0) \\ s_{22}(k,0) \end{pmatrix} = \psi^+(\mu,0) \begin{pmatrix} \alpha(k,0) \\ \beta(k,0) \end{pmatrix} = T^{-1}(k)\Phi^-(k,0,0) \begin{pmatrix} \alpha(k,0) \\ \beta(k,0) \end{pmatrix}$$
$$= \begin{pmatrix} 1 & r_{12}(k) \\ 0 & 1 \end{pmatrix} \begin{pmatrix} \alpha(k,0) \\ \beta(k,0) \end{pmatrix}, \quad k \in N_5,$$

where $r_{12}(k) = s_{22}^{-1}(k,0)s_{12}(k,0)$, and therefore

$$\alpha(k,0) = 0, \quad \beta(k,0) = s_{22}(k,0) \quad \text{for } k \in N_5. \tag{7.124}$$

$$\begin{pmatrix} s_{12}(k,0) \\ s_{22}(k,0) \end{pmatrix} = \psi^-(\mu,0) \begin{pmatrix} \alpha(k,0) \\ \beta(k,0) \end{pmatrix} = T^{-1}(\omega k)\Phi^+(\omega k,0,0) \begin{pmatrix} \alpha(k,0) \\ \beta(k,0) \end{pmatrix}$$
$$= \begin{pmatrix} 1 & 0 \\ r_{21}(\omega k) & 1 \end{pmatrix} \begin{pmatrix} \alpha(k,0) \\ \beta(k,0) \end{pmatrix}, \quad k \in N_6,$$

where since $k \in N_6$, then $\omega k \in N_2$, and conversely. $r_{21}(\omega k) = s_{11}^{-1}(\omega k,0)s_{21}(\omega k,0)$. Therefore

$$\alpha(k,0) = s_{12}(k,0), \quad \beta(k,0) = s_{22}(k,0) - r_{21}(\omega k)s_{12}(k,0) \quad \text{for } k \in N_6. \tag{7.125}$$

Using (7.112), (7.113) with due regard for $\varphi_1(\mu,0) = \varphi_2(\mu,0) = 1$, we write the equalities (7.114)–(7.119) for $t = 0$ as follows:

$$s_1(k,0) = D^{-1}(\mu)\alpha(k,0) \begin{pmatrix} 1 \\ 1 \end{pmatrix} + \beta(k,0)e^{(2)}(\mu,0) \quad \text{for } k \in N_1 \cup N_3, \tag{7.126}$$

$$s_1(k,0) = \alpha(k,0)e^{(1)}(\mu,0) + N^{-1}(\mu)\beta(k,0) \begin{pmatrix} 1 \\ 1 \end{pmatrix} \quad \text{for } k \in N_2, \tag{7.127}$$

$$s_2(k,0) = \alpha(k,0)e^{(1)}(\mu,0) + N^{-1}(\mu)\beta(k,0) \begin{pmatrix} 1 \\ 1 \end{pmatrix} \quad \text{for } k \in N_4 \cup N_6, \tag{7.128}$$

$$s_2(k,0) = D^{-1}(\mu)\alpha(k,0) \begin{pmatrix} 1 \\ 1 \end{pmatrix} + \beta(k,0)e^{(2)}(\mu,0) \quad \text{for } k \in N_5. \tag{7.129}$$

Using formula (7.107) for relationship between the sought quantities $D(\mu)$, $N(\mu)$ and the BVs of the Jost solutions $e^{(1)}(\mu,t)$, $e^{(2)}(\mu,t)$ of the problem (7.92), (7.94), from (7.126), (7.129) and (7.128), (7.127), we get:

$$D(\mu)\beta(k,0)=\begin{cases} s_{21}(k,0)-s_{11}(k,0) & \text{for } k\in N_1\cup N_3 \\ s_{22}(k,0)-s_{12}(k,0) & \text{for } k\in N_5 \end{cases} \tag{7.130}$$

$$N(\mu)\alpha(k,0)=\begin{cases} s_{12}(k,0)-s_{22}(k,0) & \text{for } k\in N_4\cup N_6 \\ s_{11}(k,0)-s_{21}(k,0) & \text{for } k\in N_2 \end{cases} \tag{7.131}$$

where $N(\mu)=e_1^{(1)}(\mu,0)-e_2^{(1)}(\mu,0)$, $D(\mu)=e_2^{(2)}(\mu,0)-e_1^{(2)}(\mu,0)$.

Here on the left-hand sides of (7.130) and (7.131) the coefficients $\alpha(k,0)$, $\beta(k,0)$ are calculated by formulas (7.120)–(7.125):

$$\beta(k,0)=s_{21}(k,0) \text{ for } k\in N_1\cup N_3, \alpha(k,0)=s_{12}(k,0) \text{ for } k\in N_4\cup N_6, \tag{7.132}$$

$$\beta(k,0)=s_{22}(k,0) \text{ for } k\in N_5,\ \ \alpha(k,0)=s_{11}(k,0) \text{ for } k\in N_2. \tag{7.133}$$

Due to (7.24), from (7.132) we have the asymptotics as $|k|\to\infty$:

$$\beta(k,0)=s_{21}(k,0)\to 0 \text{ for } k\in N_1\cup N_3,$$
$$\alpha(k,0)=s_{12}(k,0)\to 0 \text{ for } k\in N_4\cup N_6,$$

whence it follows that the left-hand sides of equality (7.130) for $k\in N_1\cup N_3$ and of equality (7.131) for $k\in N_4\cup N_6$ tend to zero as $|k|\to\infty$, whereas owing to (7.26) the right-hand sides of these equalities tend to -1 as $|k|\to\infty$. Consequently, the equalities (7.130) for $k\in N_1\cup N_3$ and (7.131) for $k\in N_4\cup N_6$ are not valid when $|k|\to\infty$.

Thus, the sought coefficients $\beta(k,0)$and $\alpha(k,0)$ in (7.130) and (7.131) are determined by formulas (7.133). The equalities (7.130) for all $k\in N_5$ and (7.131) for all $k\in N_2$ are derived for establishing the relationship between the sought quantities $N(\mu)$, $D(\mu)$ and entries $s_{ij}(k)$ of the known scattering matrix $s(k)$. From these derived equalities, using Lemma 7.1 about the properties of $s_{ij}(k)$, we obtain the formulas for expressing the sought quantities through entries of the known $s(k)$:

$$D(\mu)=1-\frac{s_{12}(k,0)}{s_{22}(k,0)} \text{ for } k\in N_5,\ \text{Im}\mu>0, \tag{7.134}$$

$$N(\mu)=\overline{D(\mu)}=1-\frac{s_{21}(k,0)}{s_{11}(k,0)} \text{ for } k\in N_2,\ \text{Im}\mu<0. \tag{7.135}$$

Using the continuity of functions $e(k)$, $e_x(k)$, the self-adjoint property of the SP (7.92)-(7.93), the properties of $s_{ij}(k)$ and (7.17), (7.20), (7.23), (7.24), from (7.134) and (7.135) we get:

$$D(\mu) = 1 - \frac{ike(-k) + e_x(-k)}{ike(-k) - e_x(-k)} = 2 + o(1) \text{ as } k \to 0,$$
$$D(\mu) = 1 + o(1) \text{ as } |\mu| \to \infty,$$
$$N(\mu) = 1 - \frac{ike(k) - e_x(k)}{ike(k) + e_x(k)} = 2 + o(1) \text{ as } k \to 0,$$
$$N(\mu) = 1 + o(1) \text{ as } |\mu| \to \infty,$$
$$N(\mu) = D(-\mu) = \overline{D(\mu)}, \mu = k^3.$$

Owing to the sell-adjoint property of the SP (7.92)–(7.93) and properties of entries $s_{ij}(k)$ of the matrix $s(k)$, we have:

$$D(\mu) \neq 0 \text{ for } \mathrm{Im}\mu \geq 0, \; N(\mu) \neq 0 \text{ for } \mathrm{Im}\mu \leq 0,$$
$$S(\mu) = \frac{N(\mu)}{D(\mu)}, \; \overline{S(\mu)} = S^{-1}(\mu) = S(-\mu) \quad -\infty < \mu < \infty. \tag{7.136}$$

The results of the self-adjoint SP (7.92)–(7.93) and the relationship between this SP and the SP (7.70), (7.75) can be formulated as follows.

Theorem 7.2. *(1) The sought quantities $D(\mu)$, $N(\mu)$ and $S(\mu)$ for the self-adjoint SP* (7.92)–(7.93) *are found in terms of entries of the known scattering matrix $s(k)$ of the SP* (7.70), (7.75) *by formulas* (7.134), (7.135) *and* (7.136), *respectively. The found quantities satisfy all the properties assembled in Proposition 7.4;*

(2) By Proposition 7.7 about the description of the scattering function, the function $S(\mu, t)$ of the SP (7.92)–(7.93) *on a half-axis $t \leq \tau < \infty$ is constructed from the calculated scattering function $S(\mu) = S(\mu, 0)$. Then by Proposition 7.6 the system of fundamental equations* (7.109) *has a unique solution $(H_{11}(t, t+\xi)$, $(H_{21}(t, t+\xi))$ in the space $L_1[0, \infty)$, where the function $g(2t+\xi)$ for $t \geq 0$ and $\xi > 0$ in system* (7.109) *is constructed from the found function $S(\mu)$ by Propositions 7.3 and 7.5. Thus, the potential self-adjoint matrix* (7.91) *in system* (7.92) *is found uniquely from known initial and boundary conditions* (7.2) *and* (7.3) *by formula* (7.98) *expressed through the solution of system* (7.109);

(3) The Jost solutions and analytic solution defined by Propositions 7.2 and 7.4 are calculated by solving the self-adjoint problems (7.92), (7.94) *and* (7.92)–(7.93) *with the found potential self-adjoint matrix* (7.91).

7.4 The time-evolution of $s(k,t)$ and solution of the IBVP

7.4.1 The time-evolution of solution $s(k,t)$ of system (7.86)

The differential equations for the column functions $s_1(k,t)$ and $s_2(k,t)$ are derived from system (7.86) for the columns of the scattering matrix $s(k,t)$:

$$\frac{d}{dt}\begin{pmatrix} s_{11}(k,t) \\ s_{21}(k,t) \end{pmatrix} = 8ik^3 \begin{pmatrix} 0 \\ s_{21}(k,t) \end{pmatrix} + p_x(0,t)\begin{pmatrix} s_{21}(k,t) \\ s_{11}(k,t) \end{pmatrix}, \quad \mathrm{Im}k \geq 0, \tag{7.137}$$

$$\frac{d}{dt}\begin{pmatrix} s_{12}(k,t) \\ s_{22}(k,t) \end{pmatrix} = -8ik^3 \begin{pmatrix} s_{12}(k,t) \\ 0 \end{pmatrix} + p_x(0,t)\begin{pmatrix} s_{22}(k,t) \\ s_{12}(k,t) \end{pmatrix}, \quad \mathrm{Im}k \leq 0. \tag{7.138}$$

Further, differentiating equality (7.114) with respect to t and taking into account that the matrix functions $s(k,t)$ and $\psi^+(\mu,t) = \left[\psi^+_{ij}(\mu,t)\right]$ satisfy the same system (7.86), gives:

$$\begin{aligned}\frac{d}{dt}\begin{pmatrix} s_{11}(k,t) \\ s_{21}(k,t) \end{pmatrix} &= \left\{8ik^3\begin{pmatrix} 0 & -\psi^+_{12}(\mu,t) \\ \psi^+_{21}(\mu,t) & 0 \end{pmatrix} + p_x(0,t)\sigma_1\psi^+(\mu,t)\right\} \\ &\times \begin{pmatrix} \alpha(k,t) \\ \beta(k,t) \end{pmatrix} + \psi^+(\mu,t)\frac{d}{dt}\begin{pmatrix} \alpha(k,t) \\ \beta(k,t) \end{pmatrix}, \quad k \in N_1.\end{aligned}$$

By comparison of the last equality with equality (7.137), we get the equation:

$$\psi^+(\mu,t)\frac{d}{dt}\begin{pmatrix} \alpha(k,t) \\ \beta(k,t) \end{pmatrix} = 8ik^3\begin{pmatrix} \psi^+_{12}(\mu,t)\beta(k,t) \\ s_{21}(k,t) - \alpha(k,t)\psi^+_{21}(\mu,t) \end{pmatrix}.$$

From (7.114) we have the equality:

$$s_{21}(k,t) - \alpha(k,t)\psi^+_{21}(\mu,t) = \beta(\mu,t)\psi^+_{22}(\mu,t).$$

Taking this equality into account, from the last equation we derive the evolution equation for $(\alpha(k,t), \beta(k,t))$ for $k \in N_1$:

$$\begin{aligned}\frac{d}{dt}\begin{pmatrix} \alpha(k,t) \\ \beta(k,t) \end{pmatrix} &= 8ik^3\beta(k,t)\left(\psi^+(\mu,t)\right)^{-1}\begin{pmatrix} \psi^+_{12}(\mu,t) \\ \psi^+_{22}(\mu,t) \end{pmatrix} \\ &= 8ik^3\beta(k,t)(\psi^+(\mu,t)^{-1}(\psi^+(\mu,t))\begin{pmatrix} 0 \\ 1 \end{pmatrix} \\ &= 8ik^3\beta(k,t)\begin{pmatrix} 0 \\ 1 \end{pmatrix}, \quad k \in N_1.\end{aligned} \tag{7.139}$$

Using (7.120), from equation (7.139) we obtain the explicit formulas for the coefficient $(\alpha(k,t),\ \beta(k,t))$ for $k \in N_1$:

$$\alpha(k,t) = \alpha(k,0) = s_{11}(k,0) - r_{12}(\omega^2 k)s_{21}(k,0), \quad \beta(k,t) = s_{21}(k,0)e^{8ik^3 t}.$$

Thus, the evolution in time t, $t > 0$ of $s_1(k,t)$ for $k \in N_1$ is derived from (7.114) with this calculated coefficient:

$$s_1(k,t) = \psi^+(\mu,t)\begin{pmatrix} s_{11}(k,0) - r_{12}(\omega^2 k)s_{21}(k,0) \\ s_{21}(k,0)e^{8ik^3t} \end{pmatrix} \quad \text{for } k \in N_1. \tag{7.140}$$

In the same way as in the case of deriving equation (7.139), from (7.115) or (7.116) we derive the same evolution equation for $(\alpha(k,t),\ \beta(k,t))$:

$$\frac{d}{dt}\begin{pmatrix} \alpha(k,t) \\ \beta(k,t) \end{pmatrix} = 8ik^3\beta(k,t)\begin{pmatrix} 0 \\ 1 \end{pmatrix} \quad \text{for } k \in N_2 \cup N_3. \tag{7.141}$$

Using (7.121), from evolution equation (7.141) for $k \in N_2$ we calculate: $\alpha(k,t) = \alpha(k,0) = s_{11}(k,0)$, $\beta(k,t) = \beta(k,0)e^{8ik^3t} = 0$ for $k \in N_2$.

Substituting the calculated coefficients $\alpha(k,t)$ and $\beta(k,t)$ for $k \in N_2$ into (7.115), we derive the evolution in time t, $t > 0$ of $s_1(k,t)$:

$$s_1(k,t) = \psi^-(\mu,t)\begin{pmatrix} s_{11}(k,0) \\ 0 \end{pmatrix} \quad \text{for } k \in N_2. \tag{7.142}$$

Using (7.122), from the evolution equation (7.141) for $k \in N_3$ we get

$$\alpha(k,t) = \alpha(k,0) = s_{11}(k,0) - r_{12}(\omega k)s_{21}(k,0),\ \beta(k,t) = s_{21}(k,0)e^{8ik^3t},\ k \in N_3$$

Then the evolution in time t, $t > 0$ of $s_1(k,t)$ for $k \in N_3$ is derived from (7.116):

$$s_1(k,t) = \psi^+(\mu,t)\begin{pmatrix} s_{11}(k,0) - r_{12}(\omega k)s_{21}(k,0) \\ s_{21}(k,0)e^{8ik^3t} \end{pmatrix} \quad \text{for } k \in N_3. \tag{7.143}$$

In the same way as in the case of deriving the equation for the coefficient $(\alpha(k,t),\ \beta(k,t))$ for $k \in N_1 \cup N_2 \cup N_3$, using (7.138) and (7.117) or (7.118), (7.119), we derive the evolution equation for this coefficient for $k \in N_4 \cup N_5 \cup N_6$:

$$\frac{d}{dt}\begin{pmatrix} \alpha(k,t) \\ \beta(k,t) \end{pmatrix} = -8ik^3\alpha(k,t)\begin{pmatrix} 1 \\ 0 \end{pmatrix},\ k \in N_4 \cup N_5 \cup N_6.$$

From this evolution equation, using (7.123), (7.124) and (7.125), we calculate

$$\begin{aligned} &\alpha(k,t) = \alpha(k,0)e^{-8ik^3t} = s_{12}(k,0)e^{-8ik^3t}, \\ &\beta(k,t) = \beta(k,0) = s_{22}(k,0) - r_{21}(\omega^2 k)s_{12}(k,0), \end{aligned} \qquad k \in N_4,$$

$$\alpha(k,t) = \alpha(k,0)e^{-8ik^3t} = 0,\ \ \beta(k,t) = \beta(k,0) = s_{22}(k,0),\ \ k \in N_5,$$

$$\begin{aligned} &\alpha(k,t) = \alpha(k,0)e^{-8ik^3t} = s_{12}(k,0)e^{-8ik^3t}, \\ &\beta(k,t) = \beta(k,0) = s_{22}(k,0) - r_{21}(\omega k)s_{12}(k,0). \end{aligned} \qquad k \in N_6$$

Substituting the calculated coefficients $\alpha(k,t)$ and $\beta(k,t)$ into (7.117), (7.118) and (7.119), respectively, we derive the time-evolution of $s_2(k,t)$ for $t>0$:

$$s_2(k,t) = \psi^-(\mu,t)\begin{pmatrix} s_{12}(k,0)e^{-8ik^3t} \\ s_{22}(k,0) - r_{21}(\omega^2 k)s_{12}(k,0) \end{pmatrix} \quad \text{for } k \in N_4, \tag{7.144}$$

$$s_2(k,t) = \psi^+(\mu,t)\begin{pmatrix} 0 \\ s_{22}(k,0) \end{pmatrix} \quad \text{for } k \in N_5, \tag{7.145}$$

$$s_2(k,t) = \psi^-(\mu,t)\begin{pmatrix} s_{12}(k,0)e^{-8ik^3t} \\ s_{22}(k,0) - r_{21}(\omega k)s_{12}(k,0) \end{pmatrix} \quad \text{for } k \in N_6. \tag{7.146}$$

Hence, the columns $s_1(k,t)$ and $s_2(k,t)$ of the scattering matrix $s(k,t)$ are defined in the half-planes $\mathrm{Im}k \geq 0$ and $\mathrm{Im}k \leq 0$, respectively, those are in the closure of the union of the sectors N_1, N_2, N_3 and N_4, N_5, N_6, respectively, except at the point $k=0$. The column solutions $s_1(k,t)$ and $s_2(k,t)$ are expressible by the expansion formulas (7.140), (7.142), (7.143) and (7.144), (7.145), (7.146), respectively in their sectors in terms of entries of the matrix $s(k,0)$ and of the fundamental solutions $\psi^{\pm}(\mu,t)$ of system (7.86). By the assertions (2) and (3) of Theorem 7.2 these solutions $\Psi^+(\mu,t)$ and $\Psi^-(\mu,t)$ defined by formulas (7.112) and (7.113), respectively are calculated from the known conditions (7.2) and (7.3).

From (7.110) and (7.111) it follows that for $t>0$:

- if $k \in I_1 \cup I_3 \cup I_5$, then $\mathrm{Im}k^3 > 0$ and $\mathrm{Re}(8ik^3t) < 0$,
- if $k \in I_2 \cup I_4 \cup I_6$, then $\mathrm{Im}k^3 < 0$ and $\mathrm{Re}(-8ik^3t) < 0$.

Due to the inequality $\mathrm{Re}(8ik^3t) < 0$ for $k \in I_1 \cup I_3 \cup I_5$, from (7.112) and (7.140), (7.143) it follows that $s_1(k,t)$ remains a bounded solution of system (7.86) in the upper half-plane $\mathrm{Im}\mu > 0$ for all $t>0$, whereas in view of the inequality $\mathrm{Re}(-8ik^3t) < 0$ for $k \in I_2 \cup I_4 \cup I_6$, from (7.113) and (7.144), (7.146) it is easily seen that $s_2(k,t)$ remains a bounded solution of system (7.86) in the lower half-plane $\mathrm{Im}\mu < 0$ for all $t>0$.

The condition $t \geq 0$ is important precisely here. Indeed, for $t<0$ the function e^{8ik^3t} in (7.140) and (7.143) increases in the domains N_1 and N_3, and the function $s_1(k,t)$ is therefore, no longer bounded at infinity of t. In the same situation, for $t<0$ the function e^{-8ik^3t} in (7.144) and (7.146) increases in the domains N_4 and N_6, and the function $s_2(k,t)$ is therefore, no longer bounded at infinity of t. Thus, the vector functions $s_1(k,t)$ and $s_2(k,t)$ for all $t \geq 0$ remain bounded and analytic in their sectors of the half-planes $\mathrm{Im}\mu \geq 0$ and $\mathrm{Im}\mu \leq 0$, respectively.

7.4.2 The solution of the IBVP (7.1)–(7.3)

We are now in the position to solve the IBVP (7.1)–(7.3). By Theorems 7.1 and 7.2 and results presented in Subsection 7.4.1, this IBVP is reduced to that of solving the GLM time-dependent equation of the form (7.45):

$$R(x+y,t)+K(x,y,t)+\int_x^{\infty} R(y+\xi,t)K(x,\xi,t)d\xi=0 \text{ for } y>x,\ \ t\geq 0, \tag{7.147}$$

where x and t enter equation (7.147) as parameters, $K(x,y,t),\ y>x$ is an unknown function of y for every $(x,t)\in[0,\infty)\times[0,\infty)$, and $R(x+y,t)$ is the function defined by (7.44):

$$R(x+y,t)=\frac{1}{2\pi}\int_{-\infty}^{\infty} e^{ik(x+y)}\tilde{R}(k,t)dk \text{ for } y>x,\ \ t>0, \tag{7.148}$$

$$\tilde{R}(k,t)=-\frac{s_{12}(k,t)}{s_{11}(k,t)},\ \ s_{12}(k,t)=s_{21}(-k,t), \tag{7.149}$$

and for every $t>0$ the functions $s_{11}(k,t)$ and $s_{21}(k,t)$ satisfy the sufficient conditions of Theorem 7.1.

Under the conditions (7.42)–(7.44) of the known function $\tilde{R}(k,t)$, the homogeneous integral equation for every $t\geq 0$:

$$f(y,t)+\int_x^{\infty} R(y+\xi,t)f(\xi,t)d\xi=0 \text{ for } y>x,$$

has only trivial solution, therefore the equation (7.147) has a unique solution in either $L_2[x,\infty)$ or $L_1[x,\infty)$. Using Parseval's relation, condition (7.42) and definition (7.148) gives:

$$\left|\int_{-\infty}^{\infty}\int_{-\infty}^{\infty} f(y,t)f(\xi,t)R(y+\xi)\,d\xi dy\right|=\left|\frac{1}{2\pi}\int_{-\infty}^{\infty}\tilde{R}(k,t)\left[\tilde{f}(-k)\right]^2 dk\right|<\|f\|^2,$$

where $\tilde{f}(k,t)=\int_x^{\infty} e^{-iky}f(y,t)\,dy$.

Thus, we find that $\|R\|_{L_2}<1$. Consequently, the equation (7.147) can be solved by the method of successive approximations, upon setting:

$$K(x,y;t)=\sum_{j=0}^{\infty}K^{(j)}(x,y;t), \tag{7.150}$$

where $K^{(0)}(x,y;t)=-R(x+y,t), K^{(j)}(x,y;t)=-\int_x^{\infty} R(y+\xi,t)K^{(j-1)}(x,\xi,t)d\xi,$

$j=1,2,...$

The solution (7.150) of equation (7.147) may be represented as a convergent Neumann series:

$$K(x,y;t) = -R(x+y,t) + \int_x^\infty R(y+\xi_1;t)R(x+\xi_1,t)d\xi_1 -$$
$$\int_x^\infty R(\xi_2+y,t)d\xi_2 \int_x^\infty R(\xi_1+\xi_2,t)R(x+\xi_1,t)d\xi_1 + \tag{7.151}$$
$$\int_x^\infty R(\xi_3+y,t)d\xi_3 \int_x^\infty R(\xi_2+\xi_3,t)d\xi_2 \int_x^\infty R(\xi_1+\xi_2,t)R(x+\xi_1,t)d\xi_1 - ...,$$

where $R(x+y,t)$ is defined by (7.148) from the known function $\tilde{R}(k,t)$.

The solution $p(x,t)$ of the IBVP (7.1)–(7.3) is constructed by formula (7.15) expressed through the found solution (7.151) of equation (7.147):

$$p(x,t) = -2\frac{d}{dx}K(x,x;t). \tag{7.152}$$

Hence, the solution (7.152) of the KdV equation (7.1) corresponding to solution (7.151) of equation (7.147) is given by:

$$\begin{aligned} p(x,t) &= 2\frac{d}{dx}R(2x,t) + 2R^2(2x,t) \\ &\quad - 4\Big\{ \int_x^\infty R(x+\xi,t)R_x(x+\xi,t)d\xi + R(2x,t)\int_x^\infty R^2(x+\xi,t)d\xi \\ &\quad - \int_x^\infty R_x(x+\xi_2,t)d\xi_2 \int_x^\infty R(\xi_1+\xi_2,t)R(x+\xi_1,t)d\xi_1 \Big\} + ... \end{aligned} \tag{7.153}$$

With sufficient perseverance, one can verify directly that the solution (7.153) satisfies the differential equation (7.1) to any desired order in powers of R. The presentation (7.153) in formality is similar to that of the solution of the KdV equation with the positive coefficient of the dispersive term on the whole-axis, that evolves from a purely continuous spectrum [108].

By Theorem 7.1 the solution of equation (7.147) coincides with the solution of integral equation (7.13) with the constructed potential (7.153), i.e., with the kernel of the transformation operator of the SP (7.5)–(7.6) with the constructed potential (7.153). Hence, the potential function $p(x,t)$ given by formula (7.153) belongs to the class $\mathcal{S}$, and therefore

$$p(0,t) = -2\left(\frac{d}{dx}K(x,x,t)\right)_{x=0} = 0.$$

Consider the asymptotic behavior of $s_1(k,t)$ and $s_2(k,t)$ at infinity of time (t). Substituting (7.112) into (7.140), (7.143) and (7.145), using Propositions 7.2 and 7.4 gives:

$$\begin{pmatrix} s_{11}(k,t) \\ s_{21}(k,t) \end{pmatrix} = \begin{pmatrix} e_1^{(1)}(\mu,t) + S(\mu)e_1^{(2)}(\mu,t) & e_1^{(2)}(\mu,t) \\ e_2^{(1)}(\mu,t) + S(\mu)e_2^{(2)}(\mu,t) & e_2^{(2)}(\mu,t) \end{pmatrix} \begin{pmatrix} e^{4i\mu t} & 0 \\ 0 & e^{-4i\mu t} \end{pmatrix}$$
$$\times \begin{pmatrix} s_{11}(k,0) - r_{12}s_{21}(k,0) \\ s_{21}(k,0)e^{8ik^3t} \end{pmatrix} = \begin{pmatrix} s_{11}(k,0) - r_{12}s_{21}(k,0) + o(1) \\ o(1) \end{pmatrix}$$
$$\text{as } t \to \infty \text{ for } k \in N_1 \cup N_3, \mathrm{Im}k^3 > 0, \tag{7.154}$$
$$\begin{pmatrix} s_{12}(k,t) \\ s_{22}(k,t) \end{pmatrix} = \begin{pmatrix} e_1^{(1)}(\mu,t) + S(\mu)e_1^{(2)}(\mu,t) & e_1^{(2)}(\mu,t) \\ e_2^{(1)}(\mu,t) + S(\mu)e_2^{(2)}(\mu,t) & e_2^{(2)}(\mu,t) \end{pmatrix} \begin{pmatrix} e^{4i\mu t} & 0 \\ 0 & e^{-4i\mu t} \end{pmatrix}$$
$$\times \begin{pmatrix} 0 \\ s_{22}(k,0) \end{pmatrix} = \begin{pmatrix} o(1) \\ s_{22}(k,0) + o(1) \end{pmatrix} \quad \text{as } t \to \infty \text{ for } k \in N_5, \mathrm{Im}k^3 < 0,$$

where $r_{12} = r_{12}(\omega^2 k)$ for $k \in N_1$ and $r_{12} = r_{12}(\omega k)$ for $k \in N_3$.

Substituting (7.113) into (7.142), (7.144) and (7.146), using Proposition 7.2 and 7.4, we obtain:

$$\begin{pmatrix} s_{11}(k,t) \\ s_{21}(k,t) \end{pmatrix} = \begin{pmatrix} e_1^{(1)}(\mu,t) & S^{-1}(\mu)e_1^{(1)}(\mu,t) + e_1^{(2)}(\mu,t) \\ e_2^{(1)}(\mu,t) & S^{-1}(\mu)e_2^{(1)}(\mu,t) + e_2^{(2)}(\mu,t) \end{pmatrix} \begin{pmatrix} e^{4i\mu t} & 0 \\ 0 & e^{-4i\mu t} \end{pmatrix}$$
$$\times \begin{pmatrix} s_{11}(k,0) \\ 0 \end{pmatrix} = \begin{pmatrix} s_{11}(k,0) + o(1) \\ o(1) \end{pmatrix} \quad \text{as } t \to \infty \text{ for } k \in N_2, \ \mathrm{Im}k^3 > 0, \tag{7.155}$$
$$\begin{pmatrix} s_{12}(k,t) \\ s_{22}(k,t) \end{pmatrix} = \begin{pmatrix} e_1^{(1)}(\mu,t) & S^{-1}(\mu)e_1^{(1)}(\mu,t) + e_1^{(2)}(\mu,t) \\ e_2^{(1)}(\mu,t) & S^{-1}(\mu)e_2^{(1)}(\mu,t) + e_2^{(2)}(\mu,t) \end{pmatrix} \begin{pmatrix} e^{4i\mu t} & 0 \\ 0 & e^{-4i\mu t} \end{pmatrix}$$
$$\times \begin{pmatrix} s_{12}(k,0)e^{-8ik^3t} \\ s_{22}(k,0) - r_{21}s_{12}(k,0) \end{pmatrix} = \begin{pmatrix} o(1) \\ s_{22}(k,0) - r_{21}s_{12}(k,0) + o(1) \end{pmatrix}$$
$$\text{as } t \to \infty \text{ for } k \in N_4 \cup N_6, \mathrm{Im}k^3 < 0,$$

where $r_{21} = r_{21}(\omega^2 k)$ for $k \in N_4$ and $r_{21} = r_{21}(\omega k)$ for $k \in N_6$.

Using properties of $s_{11}(k,0)$ and $s_{21}(k,0)$, from (7.154) and (7.155), we obtain the asymptotic behavior of the right- and left-reflection coefficients at infinity of t:

$$\tilde{R}(k,t) = o(1), \ \ \tilde{r}(k,t) = o(1) \ \text{ as } t \to \infty. \tag{7.156}$$

Owing to (7.156), the function $R(x+y,t)$ defined by (7.148) has the asymptotic behavior at infinity of t for any $y > x > 0$:

$$R(x+y,t) = o(1) \ \text{ as } t \to \infty,$$

whence it follows that the asymptotic solution (7.151) of the GLM equation (7.147) at infinity of t is equal to zero for any $y > x > 0$:

$$K(x, y, t) = o(1) \text{ as } t \to \infty.$$

Then the asymptotic solution $p(x, t)$ of the IBVP (7.1)–(7.3) constructed by formula (7.152) at infinity of t is equal to zero for any $x > 0$:

$$p(x, t) = o(1) \text{ as } t \to \infty. \tag{7.157}$$

As has been proved in [108], if the spectrum is purely continuous, then the asymptotic solution p as $t \to \infty$ of the KdV equation with the positive coefficient of the dispersive term on the whole axis is still a slowly varying wave train, oscillating about $p = 0$. Hence, in the case that the spectrum of the associated scattering problem is purely continuous, the asymptotic solution (7.157) at infinity of t of the IBVP (7.1)–(7.3) is different from the asymptotic solution of the KdV equation on the whole axis, (see page 728 [108]).

Thus, the following theorem is proved.

Theorem 7.3. *(1) The time-dependence $t, t \geq 0$ of the column $s_1(k, t) = (s_{11}(k, t),\ s_{21}(k, t))$ is defined for $Imk \geq 0$ in the sectors N_1, N_2 and N_3 by the expansion formulas* (7.140) *for $k \in N_1$,* (7.142) *for $k \in N_2$ and* (7.143) *for $k \in N_3$, respectively. The time-dependence t, $t \geq 0$ of the column $s_2(k, t) = (s_{12}(k, t), s_{22}(k, t))$ is defined for $Imk \leq 0$ in the sectors N_4, N_5 and N_6 by the expansion formulas* (7.144) *for $k \in N_4$,* (7.145) *for $k \in N_5$ and* (7.146) *for $k \in N_6$, respectively.*

For any $t \geq 0$ the solutions $s_1(k, t)$ and $s_2(k, t)$ of system (7.86) *are bounded and analytic functions in every respective sector of the k-spectral plane, wherein they are defined.*

(2) The formulas (7.140), (7.142), (7.143) *and* (7.144), (7.145), (7.146) *for $s_1(k, t)$ and $s_2(k, t)$ are expressible through elements of the scattering matrix $s(k)$ and of fundamental solutions $\psi^+(\mu, t)$, $\psi^-(\mu, t)$, $\mu = k^3$ of system* (7.86)*. The solutions $\psi^+(\mu, t)$ and $\psi^-(\mu, t)$ of system* (7.86) *are constructed by formulas* (7.112) *and* (7.113) *from the solutions of the self-adjoint problems* (7.92), (7.93) *and* (7.92), (7.94)*. By Theorems 7.1 and 7.2 the functions $s(k)$ and $\psi^{\pm}(\mu, t)$ are calculated from the known initial and boundary conditions* (7.2) *and* (7.3).

(3) Let the function $R(x + y, t)$ in equation (7.147) *be defined by the formula* (7.148) *from the known right-refection coefficient $\tilde{R}(k, t)$, which is defined by the ratio* (7.149)*, wherein the functions $s_{11}(k, t)$ and $s_{12}(k, t)$ satisfy the sufficient conditions of Theorem 7.1. By the Lax compatibility condition for systems* (7.70) *and* (7.71)*, the solution $p(x, t)$ given by* (7.153) *of the linear equation* (7.147) *satisfies the KdV equation* (7.1) *with given initial and boundary conditions* (7.2)-(7.3)*. The found solution $p(x, t)$ satisfies the Conditions I and II. For any $x > 0$ the asymptotic behavior of the solution $p(x, t)$ as $t \to \infty$ is equal to zero.*

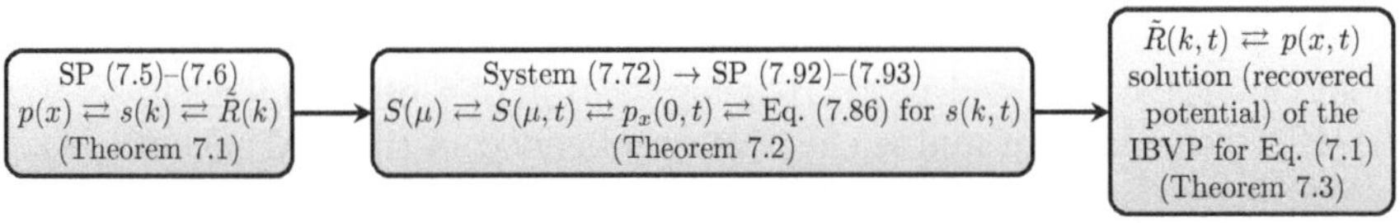

FIGURE 7.3: The schema of the algorithm for finding the solution of the IBVP for the KdV equation (7.1).

The algorithm for finding the solution of the IBVP (7.1)–(7.2)–(7.3) consists of three successive steps:

(1) Solving the SP (7.5)–(7.6) for finding the scattering matrix $s(k) = [s_{ij}(k)]$ from the known initial condition (7.3) satisfying the Conditions I and II;

Deriving the system of evolution equations (7.86) for $s(k,t)$ from the known initial and boundary conditions (7.2) and (7.3). Here the function $p_x(0,t)$ entering the coefficients of system (7.86) is unknown;

(2) Using the linear change of dependent variables (7.89), we reduce the system (7.72) to system (7.92) with the potential self-adjoint matrix (7.91). We find the unknown scattering function $S(\mu,t)$ $(\mu = k^3)$ of the self-adjoint SP (7.92)–(7.93) in terms of known elements $s_{ij}(k)$ of the matrix $s(k)$. Then the unknown object $p_x(0,t)$ is expressed through the solution of system (7.109) by formula (7.98). Further, the fundamental solutions $\Psi^{\pm}(\mu,t)$ of system (7.86) are calculated from solutions of the self-adjoint SP (7.92)–(7.93) and (7.92), (7.94);

(3) We derive the time-evolution of $s(k,t)$ from the calculated $s_{ij}(k)$ and $\Psi^{\pm}(\mu,t)$. Due to this fact, the unique solvability of the GLM time-dependent equation (7.147) is proved. Then the potential $p(x,t)$ is expressed through the found solution of equation (7.147) is restored.

Conclusions

By Theorems 7.1–7.3, the IBVP (7.1)–(7.2)–(7.3) is solved well. These theorems correspond to the successive steps 1, 2 and 3, respectively, of the algorithm for finding the solution of the considered IBVP. The function $\tilde{R}(k,t)$ in equation (7.147) is defined by ratio (7.149), in which $s_{11}(k;t)$ and $s_{12}(k;t)$ for every $t \geq 0$ are any given complex-valued functions satisfying the sufficient conditions of Theorem 7.1. This algorithm can be depicted by the schema consisting of three successive steps corresponding to Theorems 7.1–7.3:

The proved Theorems 7.1–7.3 demonstrate the development of the ISM for solving the considered IBVP. The application of these theorems to solving the IBVP is consistent and is effectively embedded in the ISM schema.

Chapter 8

The Inverse Scattering Problem for The Perturbed String Equation and Its Application to Integration of The Two-Dimensional Generalization from Korteweg-de Vries Equation

The inverse scattering problem for the perturbed string equation in characteristic variables on the whole axis is studied. Using the generalized Lax equation generated by the perturbed string equation, we derive the time-evolution of the scattering operator and the two-dimensional generalization from the one-dimensional Korteweg-de Vries equation. This enables us to solve the system of time-dependent fundamental equations in the inverse problem. Then the solution of the two-dimensional generalization from the Korteweg-de Vries equation is found that is expressed through the found solution of this system.

The chapter is based on the published papers [122, 123, 148].

The non-stationary inverse scattering problems (ISPs) for perturbed string equations on semi-axis and whole axis have been solved in [92] and [122]. In this chapter, we prove that the ISP for the perturbed string equation in characteristic variables on the whole axis is associated with the two-dimensional generalization from the one-dimensional Korteweg-de Vries (KdV) equation. The direct and ISP for this equation are studied in Sections 8.1–8.4 of this chapter. Note that the kernels of transform operators of the considered scattering problem (SP) are different from kernels of transform operators of the SP in [122]. Therefore, the solution of the ISP considered in this chapter is represented in terms that are different from terms used in [122]. In Section 8.5 we derive the two-dimensional generalization from the KdV equation from the generalized Lax equation, which is generated by the associated perturbed string equation [123]. With the help of the generalized Lax equation, the two-dimensional generalization from the KdV equation is derived and the time-dependence of the scattering operator of the considered SP is established. Then the known functions in the system of fundamental equations in inverse problem are constructed from the kernel matrix of the known time-dependent scattering operator. This enables us to solve the system of time-dependent fundamental equations in the inverse problem, and the solution

DOI: 10.1201/9781003370543-8

of the two-dimensional generalization from the KdV equation is expressed through the found solution of this system.

8.1 The scattering problem

First of all, we consider the correctness of setting the scattering problem (SP) for the perturbed string equation in characteristic variables on the whole axis [122, 123]:

$$\frac{\partial^2 u(x,y)}{\partial x \partial y} + C(x,y)\,u(x,y) = \rho(x,y), \quad -\infty < x < \infty, -\infty < y < \infty, \tag{8.1}$$

where $C(x,y)$ and $\rho(x,y)$ are measurable functions of x and y.

Under some conditions of functions $C(x,y)$ and $\rho(x,y)$, we shall find a bounded solution $u(x,y)$ of equation (8.1) so that as $x+y \to -\infty$ $u(x,y)$ tends to zero uniformly with respect to $x-y$:

$$u(x,y) \to 0 \quad \text{as} \quad x+y \to -\infty. \tag{8.2}$$

Hereafter we shall use the following lemma.
Nizhnik's Lemma [93]. *Suppose that we are given the equation:*

$$u(t) = h(t) + (Au)(t), \tag{8.3}$$

in the space $C(B)$ of continuous vector-functions with values belonging to the Banach space B. If with respect to the norm:

$$\|u(t)\|_T = \sup_{t \le T} |u(t)|_B, \tag{8.4}$$

the estimate:

$$\|Au\|_T \le \int_{-\infty}^{T} \alpha(\tau)\|u\|_\tau d\tau, \text{ where } \int_{-\infty}^{\infty} \alpha(\tau)d\tau < \infty \tag{8.5}$$

is valid, then for any right part $h(t) \in C(B)$ there exists a unique solution $u(t)$ of equation (8.3) from the space $C(B)$. The found solution satisfies the estimate:

$$|u(t)|_B \le \|h\|_t \exp\left\{ \int_{-\infty}^{t} \alpha(\tau)d\tau < \infty \right\}.$$

At first we consider the nonperturbed equation:

$$\frac{\partial^2 u(x,y)}{\partial x \partial y} = \rho(x,y), \quad -\infty < x < \infty, -\infty < y < \infty. \tag{8.6}$$

Lemma 8.1. *In the sense of generalized functions, the solution of equation* (8.6) *with the boundary condition* (8.2) *is unique.*

Proof. The solution of the homogeneous ($\rho \equiv 0$) equation corresponding to equation (8.6) has the general solution of the form:

$$u(x,y) = a_1(x) + a_2(y), \tag{8.7}$$

where $a_j(s)$, $j = 1, 2$ are arbitrary uniformly bounded continuous functions from the space $C(\bar{E})$, $\bar{E}$ is the extended axis $\bar{E} = (-\infty, \infty)$.
From condition (8.2) we obtain that the functions:

$$\frac{\partial u(x,y)}{\partial x} = \frac{\partial a_1(x)}{\partial x}, \frac{\partial u(x,y)}{\partial y} = \frac{\partial a_2(y)}{\partial y},$$

as $x + y \to -\infty$ tend to zero uniformly with respect to $x - y$, i.e., $a_j(s) =$ constant, $j = 1, 2$. Then, from condition (8.2) it follows that $u \equiv 0$. This gives the uniqueness of the solution of equation (8.6) with the boundary condition (8.2). □

Lemma 8.2. *Suppose that in equation* (8.1) $C(x,y)$ *and* $\rho(x,y)$ *are continuous functions defined in the space* $C(E^2)$ *and satisfy the condition:*

$$\iint\limits_{E^2} |C(x,y)| dxdy < \infty, \iint\limits_{E^2} |\rho(x,y)| dxdy < \infty. \tag{8.8}$$

Then in the sense of generalized functions there exists a unique solution $u(x,y)$ *of the problem* (8.1)-(8.2).

Proof. Verify directly that the function $u(x,y) = \int\limits_{-\infty}^{x} d\xi \int\limits_{-\infty}^{y} \rho(\xi,\eta)\, d\eta$ satisfies equation (8.6) and condition (8.2). Using this fact and Lemma 8.1, we obtain that in the sense of generalized functions the problem (8.1)–(8.2) is equivalent to the following integral equation in the space $C(E^2)$:

$$u(x,y) = h(x,y) - \int\limits_{-\infty}^{x} d\xi \int\limits_{-\infty}^{y} C(\xi,\eta)\, u(\xi,\eta)\, d\eta,\; h(x,y) = \int\limits_{-\infty}^{x} d\xi \int\limits_{-\infty}^{y} \rho(\xi,\eta)\, d\eta. \tag{8.9}$$

This integral equation can be written in the operator form:

$$u = h + Au, \tag{8.10}$$

where $u = u(x,y)$, $h = h(x,y)$,

$$Au = -\int\limits_{-\infty}^{x} d\xi \int\limits_{-\infty}^{y} C(\xi,\eta)\, u(\xi,\eta)\, d\eta. \tag{8.11}$$

Introduce the norm of $u(x,y)$ from the space $C(E^2)$:

$$\|u\|_T = \sup_{\substack{-\infty\leqslant\eta\leqslant\infty\\ \xi\leqslant T}} |u(\xi,\eta)|. \tag{8.12}$$

Using condition (8.8) and norm (8.12), from (8.11) we get the estimate of the form (8.5):

$$\|Au\|_T \leqslant \int_{-\infty}^{T} d\xi \int_{-\infty}^{\infty} |C(\xi,\eta)|\, d\eta \times \sup_{\substack{-\infty\leqslant\eta\leqslant\infty\\ \xi_1\leqslant\xi}} |u(\xi_1,\eta)| = \int_{-\infty}^{T} \alpha(\xi)\, \|u\|_\xi d\xi,$$

where $\alpha(\xi) = \int_{-\infty}^{\infty} |C(\xi,\eta)|\, d\eta$ and $\int_{-\infty}^{\infty} \alpha(\xi)\, d\xi < \infty$.

Thus, all conditions of the Nizhnik's Lemma [93] for equation (8.10) are fulfilled, therefore there exists a unique solution $u(x,y)$ of the operator equation (8.10) in the space $C(E^2)$. Since the problem (8.1)–(8.2) is equivalent to the integral equation (8.9), then this problem has in the sense of generalized functions a unique solution in the space $C(E^2)$. Lemma 8.2 is proved. □

Consider the perturbed string equation in characteristic variables [122, 123]:

$$\frac{\partial^2 u(x,y)}{\partial x \partial y} + C(x,y)\,u(x,y) = 0, -\infty < x < \infty, -\infty < y < \infty, \tag{8.13}$$

where the potential $C(x,y)$ is a measurable function satisfying condition (8.8).

It is required to find in the sense of generalized functions a bounded solution of equation (8.13), that belongs to the space $C(E^2)$ and has the form:

$$u(x,y) = a_1(x) + a_2(y) + w(x,y), \tag{8.14}$$

where $a_j(s)$, $j = 1,2$, is the profile of the given incident wave and the scattered wave $w(x,y)$ satisfies uniformly with respect to $x - y$ the boundary condition:

$$w(x,y) \to 0 \quad \text{as} \quad x + y \to -\infty. \tag{8.15}$$

Theorem 8.1. *Suppose that $a_j(s), j = 1,2$ are preassigned uniformly bounded differentiable functions and the potential function $C(x,y)$ in equation (8.13) is measurable with respect to x and y and satisfies the condition (8.8). Then there exists a unique solution $u(x,y)$ of the scattering problem (SP) (8.13)–(8.15) in the space $C(E^2)$. This solution is represented in the asymptotic form:*

$$u(x,y) = b_1(x) + b_2(y) + o(1) \quad \textit{as} \quad x + y \to \infty, \tag{8.16}$$

where the reflected waves $b_j(s), j = 1, 2$ *are connected with the solution* $u(x, y)$ *by the equalities:*

$$b_1(x) = a_1(x) - \int_{-\infty}^{x} d\xi \int_{-\infty}^{\infty} C(\xi, \eta) u(\xi, \eta) d\eta, \tag{8.17}$$

$$b_2(y) = a_2(y) + \int_{y}^{\infty} d\eta \int_{-\infty}^{\infty} C(\xi, \eta) u(\xi, \eta) d\xi. \tag{8.18}$$

Proof. We first observe that the function:

$$w(x, y) = u(x, y) - a_1(x) - a_2(y), \tag{8.19}$$

satisfies the equation:

$$\frac{\partial^2 w(x, y)}{\partial x \partial y} = -C(x, y) u(x, y), \tag{8.20}$$

and boundary condition (8.15). Therefore, the function:

$$w(x, y) = -\int_{-\infty}^{x} d\xi \int_{-\infty}^{y} C(\xi, \eta) u(\xi, \eta) d\eta, \tag{8.21}$$

satisfies in the sense of generalized functions equation (8.20) and boundary condition (8.15).

Using relation (8.19) between $w(x, y)$ and $u(x, y)$, we obtain from (8.21) the integral equation of the SP (8.13)-(8.15):

$$u(x, y) = a_1(x) + a_2(y) - \int_{-\infty}^{x} d\xi \int_{-\infty}^{y} C(\xi, \eta) u(\xi, \eta) d\eta, \tag{8.22}$$

where the potential $C(\xi, \eta)$ satisfies the condition (8.8). The equality (8.22) is the integral equation of the SP (8.13)–(8.15).

Using Lemma 8.2 and taking into account that the equation (8.22) is the Volterra integral equation with variable upper limit x, we can prove that the SP (8.13)–(8.15) has in the space $C(E^2)$ a unique solution.

On the other hand, eliminating $a_1(x)$ from equation (8.22), using (8.17), we have

$$u(x, y) = b_1(x) + a_2(y) + \int_{-\infty}^{x} d\xi \int_{y}^{\infty} C(\xi, \eta) u(\xi, \eta) d\eta. \tag{8.23}$$

Eliminating $a_2(y)$ from equation (8.23), using (8.18), we get:

$$u(x, y) = b_1(x) + b_2(y) - \int_{x}^{\infty} d\xi \int_{y}^{\infty} C(\xi, \eta) u(\xi, \eta) d\eta. \tag{8.24}$$

Taking condition (8.8) into account, from (8.24) we obtain the asymptotic (8.16). Theorem 8.1 is proved. □

Further eliminating $a_2(y)$ from equation (8.22), using (8.18), we get:

$$u(x,y) = a_1(x) + b_2(y) + \int_x^\infty d\xi \int_\infty^y C(\xi,\eta)\, u(\xi,\eta)\, d\eta - K, \tag{8.25}$$

where K is the functional defined by the integral:

$$K = K(u) = \int_{-\infty}^{\infty} d\xi \int_{-\infty}^{\infty} C(\xi,\eta)\, u(\xi,\eta)\, d\eta. \tag{8.26}$$

Remark 8.1. Let $b_j(s), j = 1,2$, be preassigned uniformly bounded differentiable functions, then by an argument analogous to the proof of Theorem 8.1, we can prove that there exists a unique solution $u(x,y)$ of equation (8.13), which is represented in the form:

$$u(x,y) = b_1(x) + b_2(y) + v(x,y), \tag{8.27}$$

where the scattered wave $v(x,y)$ satisfies uniformly with respect to $x-y$ the boundary condition:

$$v(x,y) \to 0 \quad \text{as} \quad x+y \to \infty. \tag{8.28}$$

The solution $u(x,y)$ of the SP (8.13), (8.27)–(8.28) is represented uniformly with respect to $x-y$ in the asymptotic form:

$$u(x,y) = a_1(x) + a_2(y) + o(1) \quad \text{as} \quad x+y \to -\infty$$

where the reflected waves $a_j(s), j = 1,2$, are connected with the solution $u(x,y)$ by equalities (8.17) and (8.18).

Definition 8.1. *By Theorem 8.1, to each vector-valued function $a(s) = (a_1(s),\ a_2(s))$ given by the profile of the incident wave, there corresponds the vector-valued function $b(s) = (b_1(s), b_2(s))$ defining the profile of the reflected wave. Thus, an operator S, that transforms $a(s)$ into $b(s)$, is defined. This operator S is called the scattering operator for the SP* (8.13)–(8.15)*:*

$$Sa(s) = b(s), -\infty < s < \infty.$$

According to Remark 8.1, to each vector-valued function $b(s) = (b_1(s), b_2(s))$, there corresponds the vector-valued function $a(s) = (a_1(s), a_2(s))$. Thus, an operator S^{-1}, that transforms $b(s)$ into $a(s)$, is defined:

$$S^{-1}b(s) = a(s), -\infty < s < \infty.$$

The scattering operators S *and* S^{-1} *are matrix operators* $S = (s_{ij})$, $S^{-1} = (q_{ij})$, $i, j = 1, 2$. *The operators* s_{11} *and* s_{22} *are the right and left penetration factors and* s_{21} *and* s_{12} *are the right and left coefficients of reflection. We will consider the scattering operators* S *and* S^{-1} *in space* $C(E; E^2)$ *of uniformly bounded continuous vector-valued functions.*

If the potential $C(x, y) \equiv 0$, *then* $b(s) = a(s)$ *and the operator* S *is equal to the identity operator.*

8.2 Transform operators

To study equations (8.22)–(8.25), we consider the special integral equations:

$$A_1(x, y) = a(x) + \int_{-\infty}^{x} d\xi \int_{y}^{\infty} C(\xi, \eta) A_1(\xi, \eta) d\eta, \tag{8.29}$$

$$A_2(x, y) = a(x) - \int_{-\infty}^{y} d\eta \int_{-\infty}^{x} C(\xi, \eta) A_2(\xi, \eta) d\xi, \tag{8.30}$$

$$A_3(x, y) = a(x) - \int_{y}^{\infty} d\eta \int_{x}^{\infty} C(\xi, \eta) A_3(\xi, \eta) d\xi, \tag{8.31}$$

$$A_4(x, y) = a(x) + \int_{x}^{\infty} d\xi \int_{-\infty}^{y} C(\xi, \eta) A_4(\xi, \eta) d\eta, \tag{8.32}$$

$$B_1(x, y) = b(y) + \int_{-\infty}^{x} d\xi \int_{y}^{\infty} C(\xi, \eta) B_1(\xi, \eta) d\eta, \tag{8.33}$$

$$B_2(x, y) = b(y) - \int_{-\infty}^{y} d\eta \int_{-\infty}^{x} C(\xi, \eta) B_2(\xi, \eta) d\xi, \tag{8.34}$$

$$B_3(x, y) = b(y) - \int_{y}^{\infty} d\eta \int_{x}^{\infty} C(\xi, \eta) B_3(\xi, \eta) d\xi, \tag{8.35}$$

$$B_4(x, y) = b(y) + \int_{x}^{\infty} d\xi \int_{-\infty}^{y} C(\xi, \eta) B_4(\xi, \eta) d\eta, \tag{8.36}$$

where $A_j(x,y)$ and $B_j(x,y)$, $j=1,2,3,4$, are unknown functions, $a(x)$ and $b(y)$ are arbitrarily preassigned uniformly bounded continuous functions defined on the whole axis, and $C(x,y)$ satisfies condition (8.8).

Lemma 8.3. *There exists a unique solution of equations* (8.29), (8.30), (8.31) *and* (8.32) *in the space* $C(E^2)$ *for arbitrary given right-hand side* $a(x) \in C(E)$. *The solution of these equations can be represented in the form:*

$$A_j(x,y) = a(x) + \int_{-\infty}^{x} H_{+j}(x,y,p)a(p)\,dp, \quad j=1,2, \tag{8.37}$$

$$A_k(x,y) = a(x) + \int_{x}^{\infty} H_{-k}(x,y,p)a(p)\,dp, \quad k=3,4, \tag{8.38}$$

where the kernel functions $H_{+j}(x,y,p)$, $p \le x, j=1,2$ *and* $H_{-k}(x,y,p)$, $p \ge x, k=3,4$ *are related to the potential* $C(x,y)$ *by the equalities:*

$$-\frac{\partial}{\partial y}H_{+j}(x,y,x) = \frac{\partial}{\partial y}H_{-k}(x,y,x) = C(x,y), \quad j=1,2;\ k=3,4, \tag{8.39}$$

and satisfy the estimate:

$$\begin{gathered}\int_{-\infty}^{\infty}\left(\sup_{-\infty<x,y<\infty}|H_{+j}(x,y,p)|\right)dp < \infty, \quad j=1,2,\\ \int_{-\infty}^{\infty}\left(\sup_{-\infty<x,y<\infty}|H_{-k}(x,y,p)|\right)dp < \infty, \quad k=3,4.\end{gathered} \tag{8.40}$$

Proof. We write equation (8.29) in the operator form:

$$A_1 = a + \mathcal{A}A_1, \tag{8.41}$$

where

$$A_1 = A_1(x,y),\ a = a(x),\ \mathcal{A}A_1 = \int_{-\infty}^{x} d\xi \int_{y}^{\infty} C(\xi,\eta)\,A_1(\xi,\eta)\,d\eta. \tag{8.42}$$

Introduce the norm:

$$\|A_1\|_T = \sup_{\substack{-\infty<y<\infty\\ x\le T}} |A_1(x,y)|,$$

then from (8.42) with due regard for (8.8), it follows that

$$\|\mathcal{A}A_1\|_T \le \int_{-\infty}^{T} d\xi \int_{-\infty}^{\infty} |C(\xi,\eta)|\,d\eta \times \sup_{\substack{-\infty<\eta<\infty\\ \xi_1\le\xi}} |A_1(\xi_1,\eta)| = \int_{-\infty}^{T} \alpha(\xi)\,\|A_1\|_\xi d\xi, \tag{8.43}$$

where

$$\alpha(\xi) = \int_{-\infty}^{\infty} |C(\xi,\eta)|\, d\eta, \int_{-\infty}^{\infty} |\alpha(\xi)|\, d\xi < \infty. \tag{8.44}$$

Due to (8.43) and (8.44), all conditions of the Nizhnik's Lemma [93] for equation (8.41) are fulfilled, therefore there exists a unique solution $A_1(x,y)$ of the operator equation (8.41) in the space $C(E^2)$. The unique solvability of equations (8.30)–(8.36) in the space $C(E^2)$ is proved analogously.

We continue to find the representations (8.37) and (8.38) of the solution of equations (8.29)–(8.32). Substituting formula (8.37) for $j=1$ into equation (8.29), using arbitrariness of the function $a(x)$, we derive the integral equation for the function $H_{+1}(x,y,p)$ for $p<x$:

$$H_{+1}(x,y,p) = \int_{y}^{\infty} C(p,\tau)\, d\tau + \int_{p}^{x} d\xi \int_{y}^{\infty} C(\xi,\tau)\, H_{+1}(\xi,\tau,p)\, d\tau. \tag{8.45}$$

Analogously, substituting formulas (8.37) for $j=2$ and (8.38) for $k=3,4$ into equations (8.30), (8.31) and (8.32), respectively, taking arbitrariness of the function $a(x)$ into account, we obtain:

$$H_{+2}(x,y,p) = -\int_{-\infty}^{y} C(p,\tau)\, d\tau - \int_{p}^{x} d\xi \int_{-\infty}^{y} C(\xi,\tau)\, H_{+2}(\xi,\tau,p)\, d\tau, p<x \tag{8.46}$$

$$H_{-3}(x,y,p) = -\int_{y}^{\infty} C(p,\tau)\, d\tau - \int_{x}^{p} d\xi \int_{y}^{\infty} C(\xi,\tau)\, H_{-3}(\xi,\tau,p)\, d\tau, p>x \tag{8.47}$$

$$H_{-4}(x,y,p) = \int_{-\infty}^{y} C(p,\tau)\, d\tau + \int_{x}^{p} d\xi \int_{-\infty}^{y} C(\xi,\tau)\, H_{-4}(\xi,\tau,p)\, d\tau, p>x. \tag{8.48}$$

The equalities (8.45)–(8.48) are considered as integral equations for unknown functions $H_{+j}(x,y,p)$, $j=1,2$ and $H_{-k}(x,y,p)$, $k=3,4$, respectively, where p is fixed and $H_{+j}(x,y,p), j=1,2; H_{-k}(x,y,p)$, $k=3,4$, are functions of two variables x and y.

We write the equation (8.45) in the operator form:

$$H_{+}(p) = \beta(p) + \mathcal{B}H_{+}(p), \tag{8.49}$$

where $H_{+}(p) = H_{+1}(x,y,p)\,, \beta(p) = \int_{y}^{\infty} C(p,\tau)\, d\tau,$

$$\mathcal{B}H_{+}(p) = \int_{y}^{\infty} d\tau \int_{p}^{x} C(\xi,\tau)\, H_{+1}(\xi,\tau,p)\, d\xi, \quad p<x. \tag{8.50}$$

We introduce the norm:

$$\|H_{+1}(p)\|_T = \sup_{\substack{-\infty<x<\infty \\ y\geq T}} |H_{+1}(x,y,p)|. \tag{8.51}$$

The operator equation (8.49) can be solved by the method of successive approximations. Indeed, we show that the series:

$$H_{+1}(p) = \beta(p) + \mathcal{B}\beta(p) + \mathcal{B}^2\beta(p) + \cdots + \mathcal{B}^n\beta(p) + \cdots \tag{8.52}$$

converges. Using condition (8.8) and norm (8.51), from (8.50) we obtain the estimate for any y and $x > p$:

$$|\mathcal{B}\beta(p)| = \left| \int_y^\infty d\tau \int_p^x C(\xi,\tau)\, d\xi \int_y^\infty C(p,\tau_1)\, d\tau_1 \right| \leq \int_y^\infty |C(p,\tau)|\, d\tau \int_y^\infty \alpha(\tau_1)\, d\tau_1,$$

where $\alpha(\tau_1) = \int\limits_{-\infty}^{\infty} |C(\xi,\tau_1)|\, d\xi$.

Suppose that the following estimate holds for any y and $x > p$:

$$|\mathcal{B}^n\beta(p)| \leq \int_y^\infty |C(p,\tau)|\, d\tau \int_y^\infty \alpha(\tau_1)\, d\tau_1 \int_{\tau_1}^\infty \alpha(\tau_2)\, d\tau_2 \ldots \int_{\tau_{n-1}}^\infty \alpha(\tau_n)\, d\tau_n. \tag{8.53}$$

Using condition (8.8), definition (8.50) and estimate (8.53), by induction we easily show that

$$|\mathcal{B}^{n+1}\beta(p)| \leq \int_y^\infty |C(p,\tau)|\, d\tau \int_y^\infty \alpha(\tau_1)\, d\tau_1 \int_{\tau_1}^\infty \alpha(\tau_2)\, d\tau_2 \ldots \int_{\tau_n}^\infty \alpha(\tau_{n+1})\, d\tau_{n+1}.$$

Using the obtained estimates, we construct the majorant series:

$$\int_y^\infty |C(p,\tau)|\, d\tau \left\{ 1 + \int_y^\infty \alpha(\tau_1)\, d\tau_1 + \int_y^\infty \alpha(\tau_1)\, d\tau_1 \int_{\tau_1}^\infty \alpha(\tau_2)\, d\tau_2 + \cdots \right.$$
$$\left. + \int_y^\infty \alpha(\tau_1)\, d\tau_1 \int_{\tau_1}^\infty \alpha(\tau_2)\, d\tau_2 \ldots \int_{\tau_n}^\infty \alpha(\tau_{n+1})\, d\tau_{n+1} + \cdots \right\}. \tag{8.54}$$

Denote the expression in the bracket by the function $\psi(y)$, then

$$\frac{\partial\psi(y)}{\partial y} = -\alpha(y)\,\psi(y), \quad \psi(y) = \exp\left\{ \int_y^\infty \alpha(\tau)\, d\tau \right\},$$

whence it follows that the series (8.54) is summable and its sum is equal to:

$$\int_y^{\infty} |C(p,\tau)|\, d\tau \ \exp\left\{ \int_y^{\infty} \alpha(\tau)\, d\tau \right\}.$$

The convergence of series (8.52) is deduced from the convergence of the majorant series (8.54). Hence, $|H_{+1}(x,y,p)| \leq \text{const} \int_{-\infty}^{\infty} |C(p,\tau)|\, d\tau$, then using condition (8.8) and from the last inequality we get the estimate (8.40) for $H_{+1}(x,y,p)$.

Analogously prove that the method of successive approximations applied to equations (8.46)–(8.48) converges, in addition the obtained solutions of these equations satisfy estimate (8.40). The relations (8.39) are found from equations (8.45)–(8.48). □

By an argument analogous to the proof of Lemma 8.3, we prove the following lemma.

Lemma 8.4. *There exists a unique solution of equations* (8.33)–(8.36) *in the space* $C(E^2)$ *for arbitrary right-hand side* $b(y) \in C(E)$*. The solution of these equations can be represented in the form:*

$$\mathcal{B}_{-k}(x,y) = b(y) + \int_y^{\infty} L_{-k}(x,y,p)\, b(p)\, dp, \quad k = 1,3, \tag{8.55}$$

$$\mathcal{B}_{j}(x,y) = b(y) + \int_{-\infty}^{y} L_{+j}(x,y,p)\, b(p)\, dp, \quad j = 2,4, \tag{8.56}$$

where the kernel functions $L_{-k}(x,y,p), p \geq y, k = 1,3$ *and* $L_{+j}(x,y,p), p \leq y, j = 2,4,$ *are related to the potential* $C(x,y)$ *by the equalities:*

$$\frac{\partial}{\partial x} L_{-k}(x,y,y) = -\frac{\partial}{\partial x} L_{+j}(x,y,y) = C(x,y), \quad k = 1,3; j = 2,4, \tag{8.57}$$

and satisfy the estimate:

$$\begin{aligned} &\int_{-\infty}^{\infty} \sup_{-\infty<x,y<\infty} (|L_{-k}(x,y,p)|)\, dp < \infty, \quad k = 1,3, \\ &\int_{-\infty}^{\infty} \sup_{-\infty<x,y<\infty} (|L_{+j}(x,y,p)|)\, dp < \infty, \quad j = 2,4. \end{aligned} \tag{8.58}$$

Here substituting formulas (8.55) for $k = 1, 3$, and (8.56) for $j = 2, 4$ into equation (8.33), (8.34), (8.35) and (8.36), respectively, using arbitrariness of the function $b(y)$, we drive the integral equations:

$$L_{-1}(x, y, p) = \int_{-\infty}^{x} C(\xi, p) d\xi + \int_{y}^{p} d\tau \int_{-\infty}^{x} C(\xi, \tau) L_{-1}(\xi, \tau, p) d\xi, p > y, \quad (8.59)$$

$$L_{-3}(x, y, p) = -\int_{x}^{\infty} C(\xi, p) d\xi - \int_{y}^{p} d\tau \int_{x}^{\infty} C(\xi, \tau) L_{-3}(\xi, \tau, p) d\xi, p > y, \quad (8.60)$$

$$L_{+2}(x, y, p) = -\int_{-\infty}^{x} C(\xi, p) d\xi - \int_{p}^{y} d\tau \int_{-\infty}^{x} C(\xi, \tau) L_{+2}(\xi, \tau, p) d\xi, p < y, \quad (8.61)$$

$$L_{+4}(x, y, p) = \int_{x}^{\infty} C(\xi, p) d\xi + \int_{p}^{y} d\tau \int_{x}^{\infty} C(\xi, \tau) L_{+4}(\xi, \tau, p) d\xi, p < y. \quad (8.62)$$

Due to condition (8.8) the method of successive approximations applied to equation (8.59)–(8.62) converges and solutions of these equations satisfy estimate (8.58). Here $H_{+j}(x, y, p), j = 1, 2; H_{-k}(x, y, p), k = 3, 4$ and $L_{-k}(x, y, p), k = 1, 3;\ L_{+j}(x, y, p),\ j = 2, 4$ are kernels of transform operators.

We use Lemmas 8.3 and 8.4 to find the representation of the solution of the scattering problem (8.13)–(8.15). Since the free term of equation (8.22) is the sum of $a_1(x)$ and $a_2(y)$, then the solution $u(x, y)$ of the SP (8.13)–(8.15) is the sum of the solutions (8.37) for $j = 2$ and (8.56) for $j = 2$ of equations (8.30) and (8.34) with the right part $a_1(x)$ and $a_2(y)$, respectively:

$$u(x, y) = a_1(x) + \int_{-\infty}^{x} H_{+2}(x, y, p) a_1(p) dp + a_2(y) + \int_{-\infty}^{y} L_{+2}(x, y, p) a_2(p) dp. \quad (8.63)$$

The solution of equation (8.23) is the sum of solutions (8.37) for $j = 1$ and (8.55) for $k = 1$ of equations (8.29) and (8.33) with the right part $b_1(x)$ and $a_2(y)$, respectively:

$$u(x, y) = b_1(x) + \int_{-\infty}^{x} H_{+1}(x, y, p) b_1(p) dp + a_2(y) + \int_{y}^{\infty} L_{-1}(x, y, p) a_2(p) dp. \quad (8.64)$$

The solution of equation (8.24) is the sum of solutions (8.38) for $k = 3$ and (8.55) for $k = 3$ of equations (8.31) and (8.35) with the right part $b_1(x)$ and

$b_2(y)$, respectively:

$$u(x,y) = b_1(x) + \int_x^\infty H_{-3}(x,y,p) b_1(p)\,dp + b_2(y) + \int_y^\infty L_{-3}(x,y,p) b_2(p)\,dp. \tag{8.65}$$

The solution of equation (8.25) is the sum of solutions (8.38) for $k = 4$ and (8.56) for $j = 4$ of equations (8.32) and (8.36) with the right part $a_1(x) - K$ and $b_2(y)$ or $a_1(x)$ and $b_2(y) - K$:

$$\begin{aligned} u(x,y) &= [a_1(x) - K] + \int_x^\infty H_{-4}(x,y,p)\,[a_1(p) - K]dp + b_2(y) \\ &\quad + \int_{-\infty}^y L_{+4}(x,y,p)\, b_2(p)dp, \\ &= a_1(x) + \int_x^\infty H_{-4}(x,y,p) a_1(p)\,dp + [b_2(y) - K] \\ &\quad + \int_{-\infty}^y L_{+4}(x,y,p)\,[b_2(p) - K]\,dp, \end{aligned} \tag{8.66}$$

where the functional K is calculated by formula (8.26).

Lemma 8.5. *Suppose that the kernels $K_+(\xi,x), x<\xi$ and $K_-(\xi,x), x>\xi$ of the Volterra integral operators K_+ and K_- with variable upper and lower limit, respectively, satisfy the estimate*

$$\int_{-\infty}^\infty \left(\sup_\xi |K_\pm(\xi,x)|\right) dx < \infty. \tag{8.67}$$

Then there exist in the space $C(E)$ the bounded operators:

$$(I + K_+)^{-1} = I + R_+, \quad (I + K_-)^{-1} = I + R_-,$$

where R_+ and R_- are the Volterra integral operators with variable upper and lower limit, respectively. The kernels of these operators satisfy the estimate:

$$\int_{-\infty}^\infty \left(\sup_\xi |R_\pm(\xi,x)|\right) dx < \infty. \tag{8.68}$$

Proof. Consider the operator equation:

$$R_{+,x}(\xi) + AR_{+,x}(\xi) = -K_{+,x}(\xi), \tag{8.69}$$

where ξ is a variable, x is a parameter, $K_{+,x}(\xi) = K_+(\xi, x)$,

$$AR_{+,x}(\xi) = \int_x^{\xi} K_+(\xi,\eta)\, R_{+,x}(\eta)d\eta. \tag{8.70}$$

In equation (8.69) $K_{+,x}(\xi)$ is the known function satisfying estimate (8.67), $R_{+,x}(\xi)$ is a desired function. We put:

$$\|R_{+,x}(\xi)\|_T = \sup_{-\infty<\xi<T} |R_{+,x}(\xi)|, \tag{8.71}$$

where T is a positive number.

Using estimate (8.67) for $K_+(\xi, x)$ and norm (8.71), from (8.70) we obtain the inequality:

$$\|AR_{+,x}(\xi)\|_T \leq \int_x^T \alpha(\tau)\|R_{+,x}\|_\tau d\tau, \tag{8.72}$$

where $\alpha(\tau) = \sup_{-\infty<t<\infty} |K_+(t,\tau)|, \int_{-\infty}^{\infty} \alpha(\tau)d\tau < \infty.$

Taking estimates (8.67) and (8.72) into account, we easily prove the uniqueness of the solution of equation (8.69) in the space $C(E)$. Using estimate (8.67) and norm (8.71), we find the solution $R_{+,x}(\xi)$ of this equation by the method of successive approximations, in addition:

$$\|R_{+,x}\|_T \leq \|K_{+,x}\|_T \exp\left\{\int_x^T \alpha(\tau)\, d\tau\right\}. \tag{8.73}$$

Using estimate (8.67), from (8.73) we obtain estimate (8.68) for $R_+(\xi, x)$.

The second assertion for the kernel $R_-(\xi, x)$ is proved analogously. □

Using estimates (8.40) and (8.58), by Lemma 8.5 we obtain the existence of the following bounded operators in the space $C(E)$:

$$(I+H_{+j}(y))^{-1} = I+R_{+j}(y), j = 1,2,\ (I+L_{-k}(x))^{-1} = I+P_{-k}(x), k = 1,3, \tag{8.74}$$

$$(I+H_{-k}(y))^{-1} = I+R_{-k}(y), j = 3,4,\ (I+L_{+j}(x))^{-1} = I+P_{+j}(x), j = 2,4, \tag{8.75}$$

where for fixed y $H_{+j}(y)$, $R_{+j}(y), j = 1,2$ and $H_{-k}(y), R_{-k}(y), k = 3,4$, are the Volterra operators with variable upper and lower limits, whose kernels are the functions $H_{+j}(x,y,p)$, $R_{+j}(x,y,p), j = 1,2$ and

$H_{-k}(x,y,p), R_{-k}(x,y,p), k = 3,4$. These kernel functions satisfy estimates (8.40) and

$$\int_{-\infty}^{\infty}\left(\sup_{-\infty<x,y<\infty}|R_{+j}(x,y,p)|\right)dp < \infty, j = 1,2,$$
$$\int_{-\infty}^{\infty}\left(\sup_{-\infty<x,y<\infty}|R_{-k}(x,y,p)|\right)dp < \infty, k = 3,4. \quad (8.76)$$

For fixed x $L_{-k}(x)$, $P_{-k}(x), k = 1,3$ and $L_{+j}(x), P_{+j}(x), j = 2,4$, are the Volterra operators with variable lower and upper limits, whose kernels are the functions $L_{-k}(x,y,p)$, $P_{-k}(x,y,p), k = 1,3$ and $L_{+j}(x,y,p)$, $P_{+j}(x,y,p), j = 2,4$. These kernel functions satisfy estimate (8.58) and

$$\int_{-\infty}^{\infty}\left(\sup_{-\infty<x,y<\infty}|P_{-k}(x,y,p)|\right)dp < \infty, k = 1,3,$$
$$\int_{-\infty}^{\infty}\left(\sup_{-\infty<x,y<\infty}|P_{+j}(x,y,p)|\right)dp < \infty, j = 2,4. \quad (8.77)$$

8.3 Properties of the scattering operator

By Theorem 8.1 and Remark 8.1, the scattering operators for SP (8.13)-(8.15) with the potential $C(x,y)$ satisfying condition (8.8) are the matrices $S = (s_{ij})$ and $S^{-1} = (q_{ij}), i,j = 1,2$.

Lemma 8.6. *The operator coefficients s_{11} and q_{22} are represented in the form:*

$$s_{11} = (I + H_{+1}(y))^{-1}(I + H_{+2}(y)), \quad (8.78)$$
$$q_{22} = (I + L_{-1}(x))^{-1}(I + L_{-3}(x)), \quad (8.79)$$

where the operators $H_{+j}(y), j = 1,2$ and $L_{-k}(x), k = 1,3$ are defined by Lemmas 8.3 and 8.4.

Proof. Comparing (8.63) with (8.64), and (8.64) with (8.65), using (8.74) and Definition 8.1, we get the representations (8.78) and (8.79). □

We introduce the matrix operators F and G defined by the equalities:

$$F = (F_{ij}) = S - I, \quad G = (G_{ij}) = S^{-1} - I, \quad j = 1,2. \quad (8.80)$$

Lemma 8.7. *The operators F_{11} and G_{22} are Volterra operators with variable upper and lower limits of integration, respectively. The kernels $F_{ij}(s,p)$ of operators F_{ij} and kernels $G_{ij}(s,p)$ of operators $G_{ij}, i,j = 1,2$ satisfy the estimates:*

$$\int_{-\infty}^{\infty}\Big(\sup_{-\infty<s<\infty}|F_{ij}(s,p)|\Big)dp < \infty,\ \int_{-\infty}^{\infty}\Big(\sup_{-\infty<s<\infty}|G_{ij}(s,p)|\Big)dp < \infty,\ i,j = 1,2. \tag{8.81}$$

Proof. Comparing (8.63) with (8.64), using (8.74), (8.75) and Definition (8.80), we get:

$$F_{11}(x,p)=\theta(x-p)\left[R_{+1}(x,y,p)+H_{+2}(x,y,p)+\int_p^x R_{+1}(x,y,z)\,H_{+2}(z,y,p)\,dz\right], \tag{8.82}$$

$$F_{12}(x,p) = \theta(y-p)\,L_{+2}(x,y,p) - \theta(p-y)\,L_{-1}(x,y,p) + \int_p^x R_{+1}(x,y,z)\,L_{+2}(x,z,p)\,dz - \int_{-\infty}^{\min(x,p)} R_{+1}(x,y,z)\,L_{-1}(x,z,p)dz. \tag{8.83}$$

where $\theta(x)$ is the Heaviside function.

Further, comparing (8.64) with (8.65), using (8.74), (8.75) and Definition (8.80), we have:

$$G_{21}(y,p) = \theta(p-x)\,H_{-3}(x,y,p) - \theta(x-p)\,H_{+1}(x,y,p) + \int_y^p P_{-1}(x,y,z)\,H_{-3}(z,y,p)\,dz - \int_{\max(y,p)}^{\infty} P_{-1}(x,y,z)\,H_{+1}(z,y,p)dz, \tag{8.84}$$

$$G_{22}(y,p)=\theta(p-y)\left[P_{-1}(x,y,p)+L_{-3}(x,y,p)+\int_y^p P_{-1}(x,y,z)\,L_{-3}(x,z,p)\,dp\right]. \tag{8.85}$$

Using estimate (8.8), by Lemmas 8.3–8.5 we get estimates (8.40), (8.58) and (8.76), (8.77). Taking these estimates into account, from (8.82)–(8.85) we obtain estimates for the kernels (8.82), (8.83) and (8.84), (8.85), i.e., estimates (8.81) for $F_{1j}(x,p)$ and $G_{2j}(y,p), j = 1,2$, whence it follows that the operators F_{11} and G_{22} are Volterra operators with variable upper and lower limits of integration. Further, comparing (8.63) with (8.65), using relations (8.74),

(8.75) and definition (8.80), gives:

$$F_{21}(y,p) = H_{+2}(x,y,p) - H_{-3}(x,y,p) - F_{11}(x,p)$$
$$- \int_{\max(y,p)}^{\infty} P_{-3}(x,y,z)\,H_{+2}(z,y,p)dz - \int_{y}^{p} P_{-3}(x,y,z)\,H_{-3}(z,y,p)\,dz$$
$$- \int_{\max(y,p)}^{\infty} P_{-3}(x,y,z)\,F_{11}(z,p)dz - \int_{\max(x,p)}^{\infty} H_{-3}(x,y,z)\,F_{11}(z,p)dz$$
$$- \int_{y}^{\infty} P_{-3}(x,y,\tau)d\tau \int_{\max(\tau,p)}^{\infty} H_{-3}(\tau,y,z)\,F_{11}(z,p)dz, \tag{8.86}$$

$$F_{22}(y,p) = P_{-3}(x,y,p) + L_{+2}(x,y,p) - F_{12}(x,p)$$
$$- \int_{\max(y,p)}^{\infty} P_{-3}(x,y,z)\,L_{+2}(z,y,p)dz - \int_{y}^{\infty} P_{-3}(x,y,z)\,F_{12}(z,p)\,dz$$
$$- \int_{x}^{\infty} H_{-3}(x,y,z)\,F_{12}(z,p)dz - \int_{y}^{\infty} P_{-3}(x,y,\tau)d\tau \int_{\tau}^{\infty} H_{-3}(\tau,y,z)F_{12}(z,p)\,dz, \tag{8.87}$$

$$G_{11}(x,p) = R_{+2}(x,y,p) + H_{-3}(x,y,p) - G_{21}(y,p)$$
$$- \int_{-\infty}^{\min(x,p)} R_{+2}(x,y,z)\,H_{-3}(z,y,p)dz - \int_{-\infty}^{y} L_{+2}(x,y,z)\,G_{21}(z,p)\,dz$$
$$- \int_{-\infty}^{x} R_{+2}(x,y,z)\,G_{21}(z,p)dz - \int_{-\infty}^{x} R_{+2}(x,y,\tau)d\tau \int_{-\infty}^{\tau} L_{+2}(x,\tau,z)G_{21}(z,p)\,dz, \tag{8.88}$$

$$G_{12}(x,p) = L_{-3}(x,y,p) - L_{+2}(x,y,p) - G_{22}(y,p)$$
$$+ \int_{-\infty}^{\min(x,p)} R_{+2}(x,y,z)\,L_{-3}(x,z,p)dz - \int_{p}^{x} R_{+2}(x,y,z)L_{+2}(x,z,p)\,dz$$
$$- \int_{-\infty}^{\min(p,y)} L_{+2}(x,y,z)\,G_{22}(z,p)\,dz - \int_{-\infty}^{\min(p,x)} R_{+2}(x,y,z)\,G_{22}(z,p)\,dz$$
$$- \int_{-\infty}^{x} R_{+2}(x,y,\tau)d\tau \int_{-\infty}^{\min(\tau,p)} L_{+2}(x,\tau,z)G_{22}(z,p)\,dz. \tag{8.89}$$

Using estimates (8.40), (8.58), (8.76), (8.77) and estimates (8.81) for the kernels $F_{1j}(x,p)$ and $G_{2j}(y,p)\,, j = 1,2$, from (8.86)-(8.89) we get estimates (8.81) for the kernels $F_{2j}(y,p)$ and $G_{1j}(x,p)\,, j = 1,2$. Hence, the estimates (8.81) for $i,j = 1,2$ are proved. □

We introduce functionals P and Q defined by the equalities:

$$Pa = K(u)\,, \quad Qb = K(u)\,, \tag{8.90}$$

where $a = (a_1(x)\,,\, a_2(y))\,,\ b = (b_1(x)\,,\, b_2(y))$ and $K(u)$ is defined by formula (8.26).
Instead of u in two equalities of (8.90), we substitute expression (8.22) into the first equality and expression (8.24) into the second equality of (8.90). As a result, the functionals (8.90) are presented in the form:

$$K(u) = P_1 a_1 + P_2 a_2, \quad K(u) = Q_1 b_1 + Q_2 b_2, \tag{8.91}$$

where

$$P_1 a_1 = \int_{-\infty}^{\infty} p_1(s)\, a_1(s) ds = \int_{-\infty}^{\infty} ds \left\{ \int_{-\infty}^{\infty} C(s,\tau)\, d\tau + \int_{s}^{\infty} d\xi \int_{-\infty}^{\infty} C(\xi,\tau) H_{+2}(\xi,\tau,s)\, d\tau \right\} a_1(s)\,,$$

$$P_2 a_2 = \int_{-\infty}^{\infty} p_2(s)\, a_2(s) ds = \int_{-\infty}^{\infty} ds \left\{ \int_{-\infty}^{\infty} C(\xi,s)\, d\xi + \int_{s}^{\infty} d\tau \int_{-\infty}^{\infty} C(\xi,\tau) L_{+2}(\xi,\tau,s)\, d\xi \right\} a_2(s)\,, \tag{8.92}$$

$$Q_1 b_1 = \int_{-\infty}^{\infty} q_1(s)\, b_1(s) ds = \int_{-\infty}^{\infty} ds \left\{ \int_{-\infty}^{\infty} C(s,\tau)\, d\tau + \int_{-\infty}^{s} d\xi \int_{-\infty}^{\infty} C(\xi,\tau) H_{-3}(\xi,\tau,s)\, d\tau \right\} b_1(s)\,,$$

$$Q_2 b_2 = \int_{-\infty}^{\infty} q_2(s)\, b_2(s) ds = \int_{-\infty}^{\infty} ds \left\{ \int_{-\infty}^{\infty} C(\xi,s)\, d\xi + \int_{-\infty}^{s} d\tau \int_{-\infty}^{\infty} C(\xi,\tau)\, L_{-3}(\xi,\tau,s)\, d\xi \right\} b_2(s)\,. \tag{8.93}$$

We put

$$\widehat{S} = \begin{pmatrix} s_{11} & s_{12} \\ \widehat{s}_{21} & \widehat{s}_{22} \end{pmatrix}, \quad \widehat{S}^{-1} = \begin{pmatrix} \widehat{q}_{11} & \widehat{q}_{12} \\ q_{21} & q_{22} \end{pmatrix},$$

where

$$\widehat{s}_{22} = s_{22} - P_2, \quad \widehat{F}_{22} = F_{22} - P_2, \quad \widehat{s}_{21} = \widehat{F}_{21} = F_{21} - P_1, \tag{8.94}$$

$$\widehat{q}_{11} = q_{11} - Q_1, \quad \widehat{G}_{11} = G_{11} - Q_1, \quad \widehat{q}_{12} = \widehat{G}_{12} = G_{12} - Q_2. \tag{8.95}$$

Lemma 8.8. *Let $S = (s_{ij})$ be the scattering operator for the SP* (8.13)-(8.15) *with the potential $C(x,y)$ satisfying condition* (8.8). *Then the coefficient operators $\widehat{s}_{22}$ and $\widehat{q}_{11}$ are represented in the form:*

$$\widehat{s}_{22} = \left(I + L_{+4}(x)\right)^{-1} \left(I + L_{+2}(x)\right), \tag{8.96}$$

$$\widehat{q}_{11} = \left(I + H_{-4}(y)\right)^{-1} \left(I + H_{-3}(y)\right), \tag{8.97}$$

and the operators $\widehat{F}_{22}$ and $\widehat{G}_{11}$ are Volterra operators with variable upper and lower limits of integration, respectively.
The kernels $\widehat{F}_{2j}(y,s)$ of operators $\widehat{F}_{2j}$ and kernels $\widehat{G}_{1j}(x,s)$ of operators $\widehat{G}_{1j}$ satisfy estimate:

$$\int_{-\infty}^{\infty}\left(\sup_{-\infty<y<\infty}\left|\widehat{F}_{2j}(y,s)\right|\right)ds<\infty,\ \int_{-\infty}^{\infty}\left(\sup_{-\infty<x<\infty}\left|\widehat{G}_{1j}(x,s)\right|\right)ds<\infty,\ j=1,2, \tag{8.98}$$

where the operators $H_{-k}(y), k=3,4$ and $L_{+j}(x), j=2,4$ are defined by Lemmas 8.3 and 8.4.

Proof. Comparing (8.63) with (8.66) and (8.65) with (8.66), using (8.75) and the definitions (8.91), (8.94), (8.95), we obtain the equalities:

$$\begin{aligned}&\left[(I+P_{+4})(H_{+2}-H_{-4})+P_1\right]a_1+\left[(I+P_{+4})(I+L_{+2})+P_2\right]a_2=b_2,\\&\left[(I+R_{-4})(I+H_{-3})+Q_1\right]b_1+\left[(I+R_{-4})(L_{-3}-L_{+4})+Q_2\right]b_2=a_1.\end{aligned}$$

From the last equalities and by the definitions (8.94) and (8.95), we get the representations (8.96) and (8.97). From these obtained representations, we have

$$\widehat{F}_{22}(y,s)=\theta(y-s)\left[P_{+4}(x,y,s)+L_{+2}(x,y,s)+\int_s^y P_{+4}(x,y,z)L_{+2}(x,z,s)dz\right], \tag{8.99}$$

$$\widehat{G}_{11}(x,s)=\theta(s-x)\left[R_{-4}(x,y,s)+H_{-3}(x,y,s)+\int_x^s R_{-4}(x,y,z)H_{-3}(z,y,s)dz\right]. \tag{8.100}$$

From (8.92), (8.93) with due regard for estimates (8.8), (8.40), (8.58), it follows that $p_j(s), q_j(s)\in L_1(-\infty,\infty), j=1,2$. Taking this fact, estimate (8.81) and expressions (8.94), (8.95), (8.99), (8.100), we obtain the estimates (8.98), whence it follows that $\widehat{F}_{22}$ and $\widehat{G}_{11}$ are Volterra operators with variable upper and lower limits of integration, respectively. The lemma is proved. □

Further, from (8.22)–(8.24), using estimate (8.8) and definition (8.26), we get the relations:

$$\begin{aligned}&\lim_{x\to-\infty}a_1(x)=\lim_{x\to-\infty}b_1(x),\quad \lim_{y\to\infty}a_2(y)=\lim_{y\to\infty}b_2(y),\\&K(u)=\lim_{x\to\infty}\{a_1(x)-b_1(x)\}=\lim_{y\to-\infty}\{b_2(y)-a_2(y)\}.\end{aligned} \tag{8.101}$$

Using relations (8.101), from definitions (8.80), (8.91) we obtain that the kernels $p_j(s)$ and $q_j(s)$ of the functionals (8.90) are expressed through kernels of operators S and S^{-1}, respectively:

$$p_1(s) = -\lim_{x\to\infty} F_{11}(x,s) = \lim_{y\to-\infty} F_{21}(y,s), \tag{8.102}$$

$$p_2(s) = -\lim_{x\to\infty} F_{12}(x,s) = \lim_{y\to-\infty} F_{22}(y,s), \tag{8.103}$$

$$q_1(s) = \lim_{x\to\infty} G_{11}(x,s) = -\lim_{y\to-\infty} G_{21}(y,s), \tag{8.104}$$

$$q_2(s) = \lim_{x\to\infty} G_{12}(x,s) = -\lim_{y\to-\infty} G_{22}(y,s). \tag{8.105}$$

8.4 Inverse scattering problem

Comparing (8.64) with (8.66), from the obtained expressions, using definitions (8.80), (8.94) and (8.102), (8.103), we get a system of fundamental equations connecting the kernels of transform operators $H_{+1}(x,y,s), L_{+4}(x,y,s), H_{-4}(x,y,s), L_{-1}(x,y,s)$ with $F_{11}(x,s), F_{12}(x,s)$ and $\widehat{F}_{21}(y,s), \widehat{F}_{22}(y,s)$:

$$\begin{aligned} F_{11}(x,s) + H_{+1}(x,y,s) + \int_s^x H_{+1}(x,y,p) F_{11}(p,s)\,dp - \widehat{F}_{21}(y,s) \\ - \int_{-\infty}^y L_{+4}(x,y,p)\widehat{F}_{21}(p,s)\,dp = H_{-4}(x,y,s), \\ \widehat{F}_{22}(y,s) + L_{+4}(x,y,s) + \int_s^y L_{+4}(x,y,p)\widehat{F}_{22}(p,s)\,dp - F_{12}(x,s) \\ - \int_{-\infty}^x H_{+1}(x,y,p) F_{12}(p,s)\,dp = L_{-1}(x,y,s), \end{aligned} \tag{8.106}$$

where $F_{11}(x,s), F_{12}(x,s)$ and $\widehat{F}_{21}(y,s), \widehat{F}_{22}(y,s)$ are the kernels of coefficient operators of the matrix operator $\widehat{S} - I$. From (8.94) and (8.102), (8.103) it follows that the kernel functions $\widehat{F}_{21}(y,s)$ and $\widehat{F}_{22}(y,s)$ are defined by the formulas:

$$\widehat{F}_{21}(y,s) = F_{21}(y,s) + \lim_{x\to\infty} F_{11}(x,s) = F_{21}(y,s) - \lim_{y\to-\infty} F_{21}(y,s), \tag{8.107}$$

$$\widehat{F}_{22}(y,s) = F_{22}(y,s) + \lim_{x\to\infty} F_{12}(x,s) = F_{22}(y,s) - \lim_{y\to-\infty} F_{22}(y,s). \tag{8.108}$$

Hence, the kernels $\widehat{F}_{21}(y,s)$ and $\widehat{F}_{22}(y,s)$ of the operators $\widehat{F}_{21}$ and $\widehat{F}_{22}$ are determined by the kernels of the coefficient operators of the known matrix operator $S-I$.

Since $H_{-4}(x,y,s) = 0$ for $s<x$, and $L_{-1}(x,y,s) = 0$ for $s<y$, from (8.106) we have:

$$\begin{gathered}
F_{11}(x,s) + H_{+1}(x,y,s) + \int_s^x H_{+1}(x,y,p)\, F_{11}(p,s)\, dp - \widehat{F}_{21}(y,s) \\
- \int_{-\infty}^{y} L_{+4}(x,y,p)\, \widehat{F}_{21}(p,s)\, dp = 0 \text{ for } s<x \\
\widehat{F}_{22}(y,s) + L_{+4}(x,y,s) + \int_s^y L_{+4}(x,y,p)\, \widehat{F}_{22}(p,s)\, dp - F_{12}(x,s) \\
- \int_{-\infty}^{x} H_{+1}(x,y,p)\, F_{12}(p,s)\, dp = 0 \text{ for } s<y.
\end{gathered} \tag{8.109}$$

Lemma 8.9. *Let S be a scattering operator for the problem* (8.13)–(8.15) *with the potential $C(x,y)$ satisfying condition* (8.8)*. Then the system of fundamental equations* (8.109) *has a unique solution in space $L_1(E)$.*

Proof. To prove the lemma, it is sufficient to show that the system of homogeneous equations:

$$\begin{gathered}
h(x,y,s) + \int_s^x h(x,y,\omega)\, F_{11}(\omega,s)\, d\omega - \int_{-\infty}^{y} l(x,y,\omega)\, \widehat{F}_{21}(\omega,s)\, d\omega = 0, \quad s<x \\
l(x,y,s) + \int_s^y l(x,y,\omega)\, \widehat{F}_{22}(\omega,s)\, d\omega - \int_{-\infty}^{x} h(x,y,\omega)\, F_{12}(\omega,s)\, d\omega = 0, \quad s<y
\end{gathered} \tag{8.110}$$

has only trivial solution in $L_1(E)$. To this end, we consider the homogeneous equations:

$$g_1(x,y,s) + \int_{-\infty}^{s} g_1(x,y,\omega)\, H_{-4}(\omega,y,s)\, d\omega = 0, \tag{8.111}$$

$$g_2(x,y,s) + \int_{-\infty}^{s} g_2(x,y,\omega)\, L_{-1}(x,\omega,s)\, d\omega = 0, \tag{8.112}$$

where

$$g_1(x,y,s)=h(x,y,s)+\int_s^x h(x,y,\omega)A_{11}(\omega,y,s)d\omega+\int_s^y l(x,y,\omega)A_{12}(x,\omega,s)d\omega, \tag{8.113}$$

$$g_2(x,y,s)=l(x,y,s)+\int_s^x h(x,y,\omega)A_{21}(\omega,y,s)d\omega+\int_s^y l(x,y,\omega)A_{22}(x,\omega,s)d\omega, \tag{8.114}$$

wherein $A_{j1}(\tau,y,s)$, $A_{j2}(x,\tau,s)$, $j=1,2$, satisfy equations for $s \leqslant \tau$:

$$A_{11}(\tau,y,s)+\int_{-\infty}^{s} A_{11}(\tau,y,\omega)\,H_{-4}(\omega,y,s)\,d\omega = F_{11}(\tau,s)\,, \tag{8.115}$$

$$A_{12}(x,\tau,s)+\int_{-\infty}^{s} A_{12}(x,\tau,\omega)\,H_{-4}(\omega,y,s)\,d\omega = -\widehat{F}_{21}(\tau,s)\,, \tag{8.116}$$

$$A_{21}(\tau,y,s)+\int_{-\infty}^{s} A_{21}(\tau,y,\omega)\,L_{-1}(x,\omega,s)\,d\omega = -F_{12}(\tau,s)\,, \tag{8.117}$$

$$A_{22}(x,\tau,s)+\int_{-\infty}^{s} A_{22}(x,\tau,\omega)\,L_{-1}(x,\omega,s)\,d\omega = \widehat{F}_{22}(\tau,s)\,. \tag{8.118}$$

In equations (8.115)–(8.118) s is variable, x,τ and τ,y are parameters, $A_{j1}(\tau,y,s)$, $A_{j2}(x,\tau,s)$, $s \leqslant \tau, j=1,2$ are desired functions. By Lemmas 8.3, 8.4 and 8.7, 8.8, the kernels $H_{-4}(\omega,y,s)$, $L_{-1}(x,\omega,s)$ and free terms in equations (8.115)–(8.118) are absolutely summable functions with respect to s. Therefore, the equations (8.115)–(8.118) are Volterra equations and have a unique solution in $L_1(E)$. Taking into account that in equations (8.115)–(8.118) $A_{j1}(\tau,y,s)$, $A_{j2}(x,\tau,s)=0$ for $s>\tau, j=1,2$, we can verify that the system (8.110) is equivalent to equations (8.111) and (8.112) with the expressions (8.113) and (8.114), respectively, wherein $A_{j1}(\tau,y,s)$, $A_{j2}(x,\tau,s)$, $s \leqslant \tau, j=1,2$, are solutions of equations (8.115)–(8.118).

In considered equations (8.111) and (8.112) s is variable, x and y are parameters, $g_j(x,y,s)$, $j=1,2$, are desired functions, and $H_{-4}(\omega,y,s)$, $L_{-1}(x,\omega,s)$, $\omega \leqslant s$, are absolutely summable functions with respect to s. The homogeneous equations (8.111) and (8.112) are Volterra equations with respect to s, therefore they have only the trivial solution $g_1 \equiv 0, g_2 \equiv 0$ in space $L_1(E)$. Then from (8.113), (8.114) we obtain the matrix Volterra homogeneous equation:

$$\begin{pmatrix} h(x,y,s) \\ l(x,y,s) \end{pmatrix} + \left(\int\limits_{s}^{x} h(x,y,\omega)\, d\omega, \int\limits_{s}^{y} l(x,y,\omega)\, d\omega \right) \begin{pmatrix} A_{11}(\omega,y,s) \, A_{21}(\omega,y,s) \\ A_{12}(x,\omega,s) \, A_{22}(x,\omega,s) \end{pmatrix} = 0,$$

which has also only the trivial solution $h = l \equiv 0$ in space $L_1(E)$. Lemma 8.9 is proved. □

We turn to solving the ISP, which consists of the regeneration of the potential $C(x,y)$ from the known operator S. Suppose that in the space $C(E, E^2)$ a scattering operator S for a string perturbed equation of the form (8.13) on the whole axis is given. Then the operator $F = S - I$ is known. We construct the kernel functions of the operators $\widehat{F}_{2j}, j = 1,2$ by formulas (8.107)-(8.108) and form the system of fundamental equations (8.109). By Lemma 8.9, this system has a unique solution $(H_{+1}(x,y,s), s < x, L_{+4}(x,y,s), s < y)$, where $H_{+1}(x,y,s), L_{+4}(x,y,s)$ are kernels of transform operators of the SP generated by equation (8.13) on the whole axis with the constructed potential:

$$-\frac{\partial}{\partial y} H_{+1}(x,y,x) = -\frac{\partial}{\partial x} L_{+4}(x,y,y) = C(x,y). \tag{8.119}$$

We formulate the obtained result.

Theorem 8.2. *Suppose that we are given a scattering operator S for a string perturbed equation of the form* (8.13)*, where the potential $C(x,y)$ satisfies condition* (8.8)*. Then the kernel functions of coefficient operators of the matrix $\widehat{S} - I$ in system* (8.109) *are constructed by formulas* (8.80)*,* (8.107) *and* (8.108)*, and for fixed x and y the system of fundamental equations* (8.109) *has a unique solution $(H_{+1}(x,y,s), s < x, \ L_{+4}(x,y,s), s < y)$ in the space $L_1(E)$. The potential $C(x,y)$ is uniquely restored from the known scattering operator S by the formula* (8.119) *expressed through the found solution of system* (8.109).

8.5 Integration of the two-dimensional generalization from the KdV equation

The inverse scattering method (ISM) makes possible a rather complete integration of Cauchy problems for non-linear evolution equations (NLEEs) in one space variable [156]. However, it is known that there is an essential difficulty when the ISM is applied to integration of multidimensional NLEEs. The difficulty arises from the strict restriction on coefficients of multidimensional operators, which define the Lax pair. The multidimensional non-linear evolution equations (NLEEs) were derived from the generalized Lax equation [94, 95, 96]. The inverse scattering transform for multidimensional (2+1) problems was studied by Fokas and Ablowitz in [39]. The inverse scattering transform of a non-linear evolution equation with 2+1 dimensions related to nonstrict hyperbolic systems was considered by Iskenderov and Ismailov in [59].

It is known [156] that the Lax equation for the ordinary KdV equation is:

$$\frac{\partial L}{\partial t} = [L, A], \tag{8.120}$$

where L is the Schrödinger operator: $L = -\frac{\partial}{\partial x^2} + u$, and A is the operator that defines the time-dependence of eigenfunctions of the Schrödinger equation.

We consider the generalized Lax equation

$$LP - QL = 0, \tag{8.121}$$

where

$$P = \frac{\partial}{\partial t} - A, \quad Q = \frac{\partial}{\partial t} - B. \tag{8.122}$$

The generalized Lax equation (8.121) with the operators P and Q defined by (8.122) takes the form:

$$\frac{\partial L}{\partial t} = L \cdot \frac{\partial}{\partial t} - \frac{\partial}{\partial t} \cdot L = LA - BL. \tag{8.123}$$

In particular, if $A = B$, then the equation (8.123) is reduced to the Lax equation (8.120).

For deriving the two-dimensional generalization from the KdV equation, the operators L, A and B in equation (8.123) are chosen as follows:

$$L = \frac{\partial^2}{\partial x \partial y} + u, \tag{8.124}$$

$$-A = \alpha_1 \frac{\partial^3}{\partial x^3} + \alpha_2 \frac{\partial^3}{\partial y^3} + 3v_1 \cdot \frac{\partial}{\partial x} + 3v_2 \cdot \frac{\partial}{\partial y}, \tag{8.125}$$

$$-B = \alpha_1 \frac{\partial^3}{\partial x^3} + \alpha_2 \frac{\partial^3}{\partial y^3} + 3\frac{\partial}{\partial x} \cdot v_1 + 3\frac{\partial}{\partial y} \cdot v_2, \tag{8.126}$$

where $u = u(x, y; t)$ is the perturbed potential in the Schrödinger operator L defined by (8.124), α_1 and α_2 are constants, and the coefficients v_1 and v_2 are related to the potential u by the equalities:

$$\frac{\partial v_1}{\partial y} = \alpha_1 \frac{\partial u}{\partial x}, \quad \frac{\partial v_2}{\partial x} = \alpha_2 \frac{\partial u}{\partial y}. \tag{8.127}$$

For deriving the two-dimensional generalization from the KdV equation, we consider the generalized Lax operator equation (8.121) in the space $C(E^2; t)$:

$$(LP - QL)\varphi = 0, \tag{8.128}$$

where $\varphi = \varphi(x, y; t) \in C(E^2; t)$ and the operators P, Q and L are defined by (8.122) and (8.124)–(8.126), respectively.

Substituting the operators (8.124), (8.125) and (8.126) into the generalized Lax operator equation (8.128), we lead equation (8.128) to the equation in the space $C(E^2; t)$:

$$\frac{\partial u}{\partial t} + \alpha_1 \frac{\partial^3 u}{\partial x^3} + \alpha_2 \frac{\partial^3 u}{\partial y^3} + 3\frac{\partial}{\partial x}(v_1 u) + 3\frac{\partial}{\partial y}(v_2 u) = 0, \tag{8.129}$$

where the coefficients v_1 and v_2 are related to the potential u by equalities (8.127).

Using the relation equalities (8.127) for eliminating the coefficient functions v_1 and v_2, from equation (8.129) we obtain the integral-differential equation for the potential u, i.e., the obtained equation (8.129) is the two-dimensional generalization from the one-dimensional KdV equation for u. In fact, if the functions u, v_1 and v_2 depend on space variables in the manner of $x + y$, then

$$v_j(x, y, t) = \alpha_j u(x, y, t) + c_j, \tag{8.130}$$

where α_j and $c_j, j = 1, 2$ are constants.

After substituting (8.130) into equation (8.129), we lead this equation to the equation:

$$\frac{\partial u}{\partial t} = \alpha\left[u'''_{xxx} + (6u + C)\, u'_x\right], \tag{8.131}$$

where $\alpha = \alpha_1 + \alpha_2$, $C = 3\dfrac{c_1 + c_2}{\alpha_1 + \alpha_2}$.

By using the following substitution:

$$u = -v - \frac{C}{6}, \quad \tau = \alpha t, \quad x = -\xi,$$

we lead equation (8.131) to the ordinary KdV equation:

$$v'_\tau - 6vv'_\xi + v'''_{\xi\xi\xi} = 0.$$

Lemma 8.10. *Consider in the space $C(E^2; t)$ the perturbed string equation in characteristic variables:*

$$\frac{\partial^2 \varphi(x, y; t)}{\partial x \partial y} + u(x, y; t)\, \varphi(x, y; t) = 0, \quad -\infty < x < \infty, -\infty < y < \infty, \tag{8.132}$$

where the perturbed potential $u(x, y; t)$ satisfies equation (8.129) and condition (8.8), i.e., $\iint\limits_{E^2} |u(x, y; t)| dx dy < \infty$.

Let φ be a solution of equation (8.132), then the function $\psi = P\varphi$ also satisfies equation (8.132), where the operator P is defined by (8.122) and (8.125).

Proof. Let φ be a solution of equation (8.132), wherein the potential u satisfies equation (8.129). Then the function $\psi = P\varphi$ satisfies the generalized Lax equation (8.121). Due to this fact, we obtain:

$$L\psi = LP\varphi = QL\varphi = 0,$$

where the operators P, Q and L are defined by (8.122), (8.125), (8.126) and (8.132), respectively.

Hence, the function ψ is a solution of equation (8.132). □

We now consider the problem generated by equation (8.129) in the space $C(E^2; t)$ with the known initial condition:

$$u(x, y; t)|_{t=0} = u_0, \tag{8.133}$$

and given boundary conditions of the coefficient functions v_1 and v_2 at infinity:

$$p(x) = 3v_1(x, -\infty) = 3v_1(x, \infty), \tag{8.134}$$
$$q(y) = 3v_2(-\infty, y) = 3v_2(\infty, y). \tag{8.135}$$

Let φ be the solution of equation (8.132) with the potential coefficient u satisfying condition (8.8) and equation (8.129). Then by Theorem 8.1 and equations (8.22) and (8.24), the solution φ satisfies the asymptotics at infinity:

$$\varphi(x, y) = a_1(x) + a_2(y) + o(1) \ \text{ as } x + y \to -\infty, \tag{8.136}$$
$$\varphi(x, y) = b_1(x) + b_2(y) + o(1) \ \text{ as } x + y \to \infty. \tag{8.137}$$

According to the definition 8.1, the scattering operator S of the SP for equation (8.132) is defined by the formula:

$$Sa(s) = b(s), \quad -\infty < s < \infty, \tag{8.138}$$

where $a(s) = (a_1(s), a_2(s))$, $b(s) = (b_1(s), b_2(s))$, $a_j(s)$ and $b_j(s)$, $j = 1, 2$ are profiles of the incident and reflected waves, respectively. These profiles of waves are connected with the solution $\varphi(x, y)$ of equation (8.132) by the asymptotics (8.136) and (8.137).

Further, by Lemma 8.10 we consider the SP generated by equation (8.132) for the solution $\psi = P\varphi$ of this equation, wherein the operator P is defined by the equalities (8.122) and (8.125):

$$P = \frac{\partial}{\partial t} - A = \frac{\partial}{\partial t} + \alpha_1 \frac{\partial^3}{\partial x^3} + \alpha_2 \frac{\partial^3}{\partial y^3} + 3v_1 \cdot \frac{\partial}{\partial x} + 3v_2 \cdot \frac{\partial}{\partial y}. \tag{8.139}$$

By Theorem 8.1, the solution $\psi = P\varphi$ of equation (8.132) belongs to the space $C(E^2)$ and has the form:

$$\psi(x, y) = \widehat{a}_1(x) + \widehat{a}_2(y) + \widehat{w}(x, y), \tag{8.140}$$

where the scattered wave $\widehat{w}(x,y)$ satisfies uniformly with respect to $x-y$ the boundary condition:

$$\widehat{w}(x,y) \to 0 \quad \text{as} \quad x+y \to -\infty. \tag{8.141}$$

In view of the conditions (8.134) and (8.135), there exist at $-\infty$ and ∞ the limits of the function $3v_1(x,y) \in C(E^2)$ for every x:

$$\lim_{x+y\to-\infty} 3v_1(x,y) = 3v_1(x,-\infty) = \lim_{x+y\to\infty} 3v_1(x,y) = 3v_1(x,\infty) = p(x), \tag{8.142}$$

and there exist at $-\infty$ and ∞ the limits of the function $3v_2(x,y) \in C(E^2)$ for every y:

$$\lim_{x+y\to-\infty} 3v_2(x,y) = 3v_2(-\infty,y) = \lim_{x+y\to\infty} 3v_2(x,y) = 3v_2(\infty,y) = q(y). \tag{8.143}$$

According to (8.22) and (8.24), the solution ψ of the SP (8.132), (8.140)–(8.141) satisfies the asymptotics at infinity:

$$\begin{aligned}\lim_{x+y\to-\infty}\psi(x,y) &= \lim_{x+y\to-\infty} P\varphi(x,y) = \widehat{P}\left[a_1(x)+a_2(y)\right]+o(1)\\ &= \widehat{P}\left[a_1(x)\right]+\widehat{P}\left[a_2(y)\right]+o(1)\\ &= \widehat{a}_1(x)+\widehat{a}_2(y)+o(1) \text{ as } x+y\to-\infty, \end{aligned} \tag{8.144}$$

$$\begin{aligned}\lim_{x+y\to\infty}\psi(x,y) &= \lim_{x+y\to\infty} P\varphi(x,y) = \widehat{P}\left[b_1(x)+b_2(y)\right]+o(1)\\ &= \widehat{P}\left[b_1(x)\right]+\widehat{P}\left[b_2(y)\right]+o(1)\\ &= \widehat{b}_1(x)+\widehat{b}_2(y)+o(1) \text{ as } x+y\to\infty, \end{aligned} \tag{8.145}$$

where

$$\widehat{P} = \frac{\partial}{\partial t} + \alpha_1\frac{\partial^3}{\partial x^3} + p(x)\cdot\frac{\partial}{\partial x} + \alpha_2\frac{\partial^3}{\partial y^3} + q(y)\cdot\frac{\partial}{\partial y}, \tag{8.146}$$

$p(x), q(y)$ are given by (8.142), (8.143), respectively, and $\widehat{a}(s) = (\widehat{a}_1(s), \widehat{a}_2(s))$, $\widehat{b}(s) = \left(\widehat{b}_1(s), \widehat{b}_2(s)\right)$ are profiles of the incident and reflected waves, which are connected with the solution ψ of equation (8.132) by the asymptotics (8.144) and (8.145), respectively.

The unique solvability of the SP (8.132), (8.140), (8.141) is proved analogously to the unique solvability of the SP (8.13)–(8.15). Owing to this fact, the relation between the vector-functions $\widehat{a}(s)$ and $\widehat{b}(s)$ is defined by the same definition (8.138) and Remark 8.1:

$$S\widehat{a}(s) = \widehat{b}(s), \quad S^{-1}\widehat{b}(s) = \widehat{a}(s), \quad -\infty < s < \infty. \tag{8.147}$$

where

$$S = I + F = I + \left[F_{ij}\right], S^{-1} = I + G = I + \left[G_{ij}\right], i,j = 1,2. \tag{8.148}$$

By the asymptotics (8.145) and definitions (8.138), (8.147), we obtain the equalities for arbitrary vector-function $(a_1(x), a_2(y))$ belonging to the space $C(E; E^2)$ of uniformly bounded continuous vector-functions:

$$\begin{aligned}\widehat{b}_1(x) + \widehat{b}_2(y) &= \widehat{P}b_1(x) + \widehat{P}b_2(y)\\ &= \widehat{P}.\left[(s_{11}, s_{12})\begin{pmatrix} a_1(x) \\ a_2(y) \end{pmatrix} + (s_{21}, s_{22})\begin{pmatrix} a_1(x) \\ a_2(y) \end{pmatrix}\right]\\ &= s_{11}\widehat{a}_1(x) + s_{12}\widehat{a}_2(y) + s_{21}\widehat{a}_1(x) + s_{22}\widehat{a}_2(y)\\ &= (s_{11} + s_{21}) \cdot \widehat{P}a_1(x) + (s_{12} + s_{22}) \cdot \widehat{P}a_2(y)\\ &= (s_{11}, s_{12}).\widehat{P}\begin{pmatrix} a_1(x) \\ a_2(y) \end{pmatrix} + (s_{21}, s_{22}).\widehat{P}\begin{pmatrix} a_1(x) \\ a_2(y) \end{pmatrix}. \qquad (8.149)\end{aligned}$$

By the asymptotics (8.144) and definitions (8.138), (8.147), we obtain the equalities for any vector function $(b_1(x), b_2(y)) \in C(E; E^2)$:

$$\begin{aligned}\widehat{a}_1(x) + \widehat{a}_2(y) &= \widehat{P}a_1(x) + \widehat{P}a_2(y)\\ &= \widehat{P}.\left[(q_{11}, q_{12})\begin{pmatrix} b_1(x) \\ b_2(y) \end{pmatrix} + (q_{21}, q_{22})\begin{pmatrix} b_1(x) \\ b_2(y) \end{pmatrix}\right]\\ &= (q_{11}, q_{12}).\widehat{P}\begin{pmatrix} b_1(x) \\ b_2(y) \end{pmatrix} + (q_{21}, q_{22}).\widehat{P}\begin{pmatrix} b_1(x) \\ b_2(y) \end{pmatrix}. \qquad (8.150)\end{aligned}$$

Using the arbitrariness of $(a_1(x), a_2(y)) \in C(E; E^2)$, $(b_1(x), b_2(y)) \in C(E; E^2)$ and the properties of the operator coefficients s_{ij} and $q_{ij}, i, j = 1, 2$, from (8.149) and (8.150) we obtain the commutation equations:

$$\left[s_{ij} \cdot \hat{P} - \hat{P} \cdot s_{ij}\right]\begin{pmatrix} a_1(x) \\ a_2(y) \end{pmatrix} = 0 \text{ for any } (a_1(x), a_2(y)) \in C\left(E; E^2\right), \qquad (8.151)$$

$$\left[q_{ij} \cdot \hat{P} - \hat{P} \cdot q_{ij}\right]\begin{pmatrix} b_1(x) \\ b_2(y) \end{pmatrix} = 0 \text{ for any } (b_1(x), b_2(y)) \in C\left(E; E^2\right). \qquad (8.152)$$

From the commutation equations (8.151) and (8.152), we derive the commutator equations for the operators $\widehat{P}$ and F, and $\widehat{P}$ and G:

$$\left[(I + F) \cdot \widehat{P} - \widehat{P} \cdot (I + F)\right]\begin{pmatrix} a_1(x) \\ a_2(y) \end{pmatrix} = 0 \text{ for any } (a_1(x), a_2(y)) \in C\left(E; E^2\right), \qquad (8.153)$$

$$\left[(I + G) \cdot \widehat{P} - \widehat{P} \cdot (I + G)\right]\begin{pmatrix} b_1(x) \\ b_2(y) \end{pmatrix} = 0 \text{ for any } (b_1(x), b_2(y)) \in C\left(E; E^2\right), \qquad (8.154)$$

where the operators $\widehat{P}$ and F, G are defined by the formulas (8.146) and (8.148), respectively.

Putting:

$$\widehat{P} = \frac{\partial}{\partial t} - R, \quad -R = \alpha_1 \frac{\partial^3}{\partial x^3} + p(x) \cdot \frac{\partial}{\partial x} + \alpha_2 \frac{\partial^3}{\partial y^3} + q(y) \cdot \frac{\partial}{\partial y}, \tag{8.155}$$

then using equalities (8.155), from (8.153) and (8.154) we obtain the commutator equalities for any vector-function $(a_1(x), a_2(y)) \in C(E; E^2)$:

$$(I+F) \cdot \left(\frac{\partial}{\partial t} - R\right) \begin{pmatrix} a_1(x) \\ a_2(y) \end{pmatrix} = \left(\frac{\partial}{\partial t} - R\right) \cdot (I+F) \begin{pmatrix} a_1(x) \\ a_2(y) \end{pmatrix}, \tag{8.156}$$

and for any vector-function $(b_1(x), b_2(y)) \in C(E; E^2)$:

$$(I+G) \cdot \left(\frac{\partial}{\partial t} - R\right) \begin{pmatrix} b_1(x) \\ b_2(y) \end{pmatrix} = \left(\frac{\partial}{\partial t} - R\right) \cdot (I+G) \begin{pmatrix} b_1(x) \\ b_2(y) \end{pmatrix}. \tag{8.157}$$

Owing to the arbitrariness of vector-functions $(a_1(x), a_2(y)) \in C(E; E^2)$ and $(b_1(x), b_2(y)) \in C(E; E^2)$, the time-evolution equations for the scattering operators F and G are derived from the commutator equalities (8.156) and (8.157):

$$\frac{\partial F(x,y,t)}{\partial t} = RF - FR = [R, F], \tag{8.158}$$

$$\frac{\partial G(x,y,t)}{\partial t} = RG - GR = [R, G], \tag{8.159}$$

where the partial derivatives of the operators F and G with respect to t are given by:

$$\frac{\partial F}{\partial t} = \frac{\partial}{\partial t} \cdot F - F \cdot \frac{\partial}{\partial t}, \quad \frac{\partial G}{\partial t} = \frac{\partial}{\partial t} \cdot G - G \cdot \frac{\partial}{\partial t}.$$

Hence, the partial derivatives of operators F and G with respect to t are defined by the commutators $[R, F]$ and $[R, G]$, respectively in the right-hand sides of equalities (8.158) and (8.159).

Thus, with the help of the generalized Lax equation (8.121), we find the asymptotics (8.144) and (8.145) at $-\infty$ and ∞ of the solution ψ of equation (8.132). Then from these asymptotics we derive the time-evolution equations (8.158) and (8.159). Hence, the time-evolution of the operator L corresponds to the time-evolution of the partial derivatives of operators F and G with respect to t. These partial derivatives evolve by time in accordance with the evolution equations (8.158) and (8.159). This enables us to integrate the two-dimensional generalization from the KdV equation (8.129). In fact, we can form the system of time-dependent equations (8.109) from the time-dependent operator F. The known functions in system (8.109) are the time-dependent kernel functions: $F_{11}(x, s; t), F_{12}(x, s; t), \widehat{F}_{21}(y, s; t), \widehat{F}_{22}(y, s; t)$. These functions are constructed by formulas (8.107) and (8.108) from the matrix kernel of the known scattering operator $F = S - I$, which obeys time-evolution equation (8.158). By Theorem 8.2, the system of time-dependent equations (8.109) has a unique solution in space $L_1(E)$. The found solution of system (8.109) is the matrix kernel of the transform operator of the SP generated by the perturbed string equation (8.132) with the restored time-dependent potential (8.119):

$$-\frac{\partial}{\partial y} H_{+1}(x, y, x; t) = -\frac{\partial}{\partial x} L_{+4}(x, y, y; t) = u(x, y; t). \tag{8.160}$$

Thus, the solution of equation (8.129) is found by formula (8.160). The following theorem is proved.

Theorem 8.3. *The problem generated by the two-dimensional generalization from the KdV equation* (8.129) *with the given initial condition* (8.133) *and boundary conditions* (8.134), (8.135) *must be solved by the ISM in the space* $C(E^2; t)$. *The solution of this problem is obtained by formula* (8.160) *expressed through the found solution* $(H_{+1}(x, y, s; t), s < x, L_{+4}(x, y, s; t), s < y)$ *of the system of time-dependent fundamental equations* (8.109). *The known functions in this system are constructed by formulas* (8.80) *and* (8.107), (8.108) *from the matrix kernel of the known time-dependent scattering operator* $F = S - I$, *which obeys the time-evolution equation* (8.158).

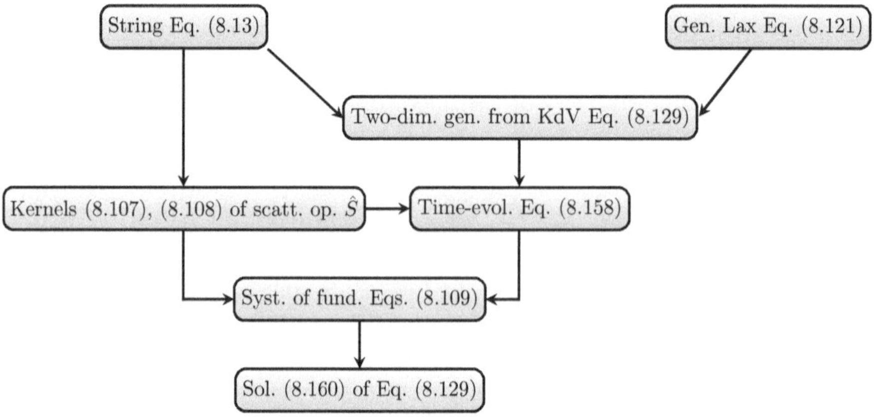

FIGURE 8.1: The main steps in finding the solution of equation (8.129).

One should note that in the proof of Theorem 8.3, we have constructed the algorithm for finding the solution of equation (8.129) from the known conditions (8.133), (8.134) and (8.135). Figure 8.1 contains the main steps of this algorithm.

Conclusions

The ISP for the perturbed string equation (8.13) is studied. The results of this ISP are applied to integration of the two-dimensional generalization (8.129) from the Korteweg-de Vries equation. The equation (8.129) is derived from the generalized Lax operator equation (8.128), which is defined by the operators (8.122), (8.124), (8.125), (8.126) and equation (8.132). The problem for equation (8.129) in the space $C(E^2;t)$ with known initial and boundary conditions (8.133) and (8.134), (8.135) is solved by the ISM. Then the solution of this problem is obtained by formula (8.160) expressed through the found solution of the system of time-dependent equations (8.109). The known functions in the fundamental equation (8.109) are kernels of the known scattering operator F, which obeys the time-evolution equation (8.158).

Chapter 9

Connections Between the Inverse Scattering Method and Related Methods

In this chapter, first we introduce preliminarily Fokass methodology for the analysis of IBVPs for NLEEs on the half-line [38], and a Riemann–Hilbert problem on the half-line [51, 53]. Further, the connections between the ISM and the Hirota method are considered. The exact complex-valued solution of the KdV and of the attractive NLS equations in the class of non-scattering potentials are presented, where the inverse scattering problem (ISP) associated with these NLEEs can be solved exactly. It is shown that in the class of non-scattering potentials, if the singular numbers of the associated scattering problem (SP) satisfy some conditions, then the restored non-scattering potential in the associated SP is the solution of the KdV equation (the solution of the attractive NLS equation). The results of constructing explicit solutions of considered KdV and attractive NLS equations are published in the papers [138, 139, 149, 150].

9.1 Fokass methodology for the analysis of IBVPs [38]

A rigorous methodology for analysis of IBVP on the half-line, $0 < x < \infty$, $t > 0$ is applied to the NLS, to the sG in laboratory coordinates and to the KdV equation with dominant surface tension.

Statement of the problem. Let $q(x,t)$ satisfy a NLEE with partial derivatives of order n on the half-line $0 < x < \infty$ and for $0 < t < T$, where T is a positive constant. Let $q(x,t)$ satisfy decaying initial condition at $t = 0$, as well as appropriate boundary conditions at $x = 0$. Assume that this NLEE admits a Lax pair. Then such an IBVP can be analyzed, using the following steps:

Step 1. Assume that the solution $q(x,t)$ of the considered IBVP exists, express $q(x,t)$ through the solution of a Riemann–Hilbert (R–H) problem uniquely defined in terms of the so-called spectral functions denoted by $\{s(k), S(k)\}$. The "jump matrix" of the RH problem is uniquely defined in terms of the 2×2 matrix valued functions $s(k), S(k)$. Express $s(k)$ and $S(k)$ through the

DOI: 10.1201/9781003370543-9

solutions of linear integral equations uniquely defined in terms of the initial condition $q_0(x) \in S(\mathbb{R}^+)$ and boundary values $\{g_l(t)\}_{l=0}^{n-1}$, respectively. The solutions of these linear integral equations denoted by $\mu_j(x,t,k), j = 1,2,3$ are the fundamental eigenfunctions needed for the formulation of a RH problem in the complex k plane. Furthermore, derive the global algebraic relation satisfied by spectral functions $s(k), S(k)$.
Step 2. Given $q_0(x) \in S(\mathbb{R}^+)$, use integral equation for $\mu_3(x,0,k)$ to define $s(k)$. If $S(k)$ is defined in terms of smooth functions $\{g_l(t)\}_{l=0}^{n-1}$ by integral equation for $\mu_2(0,t,k)$, then spectral functions satisfy the global relation. Define $\mu_j(x,t,k), j = 1,2,3$, as the solution of the RH problem formulated in Step 1 and define $q(x,t)$ in terms of $\mu_j(x,t,k)$. Then prove that: a) $q(x,t)$ is defined for all $0 < x < \infty$, $0 < t < T$; b) $q(x,t)$ solves the NLEE; c) $q(x,0) = q_0(x), 0 < x < \infty$; d) $\{\partial_x^l q(0,t) = g_l(t)\}_{l=0}^{n-1}, 0 < t < T$.

The global relation is not only a necessary but it is also a sufficient condition. Thus, given $q_0(x)$ and a subset of the functions $\{g_l(t)\}_{l=0}^{n-1}, 0 < t < T$, the main problem becomes to show that the global relation characterizes the unknown part of the set $\{g_l(t)\}_{l=0}^{n-1}$.
Step 3. Analyse the global condition. a) Identify a class of boundary conditions, for which it is possible to compute explicitly $S(k)$, using only manipulation of the global relation; b) Given $q_0(x) \in S(\mathbb{R}^+)$ and subset of smooth functions $\{g_l(t)\}_{l=0}^{n-1}$ for all $0 < t < T$, prove that the global relation yields the unknown part of the set $\{g_l(t)\}_{l=0}^{n-1}$ for all $0 < t < T$. To achieve the solvability for the unknown part, we make crucial use of analyticity structure of the global relation.

9.2 A Riemann–Hilbert problem

The time-dependence of the scattering matrix $s(k,t)$ for the SP (7.70)–(7.71) in Chapter 7 has been investigated by Habibullin [51, 53]. Due to the linear change dependent of variables (7.89), the system (7.72) is reduced to the form (7.88):

$$\tilde{y}_t(\mu,t) = (-4i\mu\sigma_3 + p_x(0,t)\sigma_1)\,\tilde{y}(\mu,t), \quad \mu = k^3, 0 \le t < \infty, \tag{9.1}$$

where $p_x(0,t)$ is a smooth function.

Habibullin proposed that a certain Riemann–Hilbert (RH) problem with a discontinuity across the cut $\mathrm{Im}\mu = 0, \mu = k^3$, corresponds to system (9.1). The conjugation condition of the proposed RH problem is

$$\psi^-(\mu,t) = \psi^+(\mu,t)\,\Re(\mu,t), \quad \text{for} \quad t > 0, \tag{9.2}$$

where ψ^- and ψ^+ are analytic in the half-plane Im$\mu < 0$ and Im$\mu > 0$, respectively, and normalized at infinity by the condition: $\psi^-(\mu, \infty) = \psi^+(\mu, \infty) = 1$. In addition, the functions $\psi^-(\mu, t)\, e^{-4i\mu t\sigma_3}$ and $\psi^+(\mu, t)\, e^{-4i\mu t\sigma_3}$ are solutions of system (9.1). Therefore, the conjugation matrix $\Re(\mu, t)$ depends on t trivially, i.e., it satisfies the condition:

$$\Re(\mu, t) = e^{-4i\mu t\sigma_3}\Re(\mu, 0)\, e^{4i\mu t\sigma_3}. \tag{9.3}$$

The problem of finding the solutions of the RH problem (9.2) for $t > 0$ is reduced to the linear conjugation problem with the sought functions $\psi^+(\mu, t)$ and $\psi^-(\mu, t)$, that are analytic in the respective domains Im$\mu > 0$ and Im$\mu < 0$, and continuous and non-degenerate in the closures of these domains. The conjugation condition of the RH problem is therefore satisfies on the real axis. The uniqueness of the solution of the RP (9.1) follows from the additional normalization condition at infinity.

Further, the consistency condition of systems (7.70) and (7.71) in the quarter of the plane $x \geq 0, t \geq 0$ is used for finding the connections between the solutions of system (9.1) on μ-spectral plane and solutions of system (7.70) on k-spectral plane at the corner point $(x, t) = (0, 0)$. Thus, the functions $\psi^-(\mu, 0)$ and $\psi^+(\mu, 0)$ are calculated in terms of solutions of system (7.70) at the corner point $(0, 0)$, therefore, $\Re(\mu, 0)$ is constructed by (9.2) from $\psi^-(\mu, 0)$ and $\psi^+(\mu, 0)$, and the conjugation matrix $\Re(\mu, t)$ defined by (9.3) is found. Thus, the search for the functions $\psi^-(\mu, t)$ and $\psi^+(\mu, t)$ for $t > 0$ eventually reduces to solving the RH problem (9.2) with conjugation matrix (9.3). The linear conjugation problem (9.2)–(9.3) is uniquely solvable in [53].

9.3 Hirota's method

One of the areas of interest in the study of non-linear wave propagation is the development of techniques for finding special solutions to the underlying equations. For these equations admitting soliton solutions, Hirota has made many significant contributions [55, 57, 58]. It should be pointed out that these direct methods virtually always work for equations where the ISM is known and sometimes work for equations where the ISM is still unavailable. In practice, these direct methods have often motivated a search for the ISM and sometimes have suggested them. Ablowitz and Segur have shown the basic ideas in this direct method as follows [5]:

(i) Introduce a dependent variable transformation (this may require some ingenuity although there are some standard forms). The transformation should reduce the evolution equation to a so-called bilinear equation, quadratic in the dependent variables. Hirota has developed a novel differential calculus and it is convenient to use it at this stage.

(ii) Introduce a formal perturbation expansion into this bilinear equation. In the case of soliton solutions, this expansion truncates.

(iii) Use mathematical induction to prove that the suggested soliton form is indeed correct.

More specifically, we present the main results from our works [133, 139, 149, 150] about constructing solutions of the KdV and NLS equations by the ISM and Hirota's method. The construction of solutions of the KdV and NLS equations is presented without proving in Subsections 9.3.1–9.3.3 and 9.3.4–9.3.5, respectively.

9.3.1 The scattering problem (SP) associated with the KdV equation

By $L(t)$ we mean the operator generated in the space $L^2[0,\infty)$ by the differential expression [78, 79, 139]:

$$l(t)y = -y_{xx} + p(x,t)y, \quad (x,t) \in [0,\infty) \times (-\infty,\infty), \tag{9.4}$$

with the boundary condition:

$$y(0,t) = 0, \quad t \in (-\infty,\infty), \tag{9.5}$$

where the potential $p(x,t)$ is a complex-valued function satisfying the estimate for all $t \in (-\infty,\infty)$:

$$\int_0^\infty e^{\varepsilon x}|p(x,t)|dx < \infty, \quad \varepsilon > 0. \tag{9.6}$$

The operator $L(t)$ is in general non self-adjoint. We consider the eigenvalue-problem for the Schrödinger equation:

$$-y_{xx} + p(x,t)y = \rho^2 y, \quad (x,t) \in [0,\infty) \times (-\infty,\infty), \tag{9.7}$$

with the asymptotic condition at infinity:

$$\lim_{x\to\infty} y(x,t,\rho)e^{-i\rho x} = 1,$$

where the potential $p(x,t)$ satisfies estimate (9.6). The solution of this eigenvalue problem may be represented in the form of the Jost function [139]:

$$e(x,t,\rho) = e^{i\rho x} + \int_x^\infty K(x,u,t)e^{i\rho u}du, \quad \operatorname{Im}\rho \geqslant 0, t > 0, \tag{9.8}$$

where $K(x,u,t)$ is analog of the kernel of the transformation operator usually used in the scattering problem (SP) [86].

The equation (9.7) processes, besides the solution (9.8), a solution $\tilde{e}(x,t,\rho)$ satisfying the asymptotic condition at infinity:

$$\lim_{x\to\infty} \tilde{e}(x,t,\rho)e^{i\rho x} = 1.$$

The function $S(\rho,t) = \frac{e(t,-\rho)}{e(t,\rho)}$ for $|\operatorname{Im}\rho| < \frac{\varepsilon}{2}$, $e(t,\rho) = e(0,t,\rho)$, is called the scattering function. Since $e(t,\rho)$ is holomorphic for $\operatorname{Im}\rho > -\frac{\varepsilon}{2}$, $e(t,\rho)$ has a finite number of zeros in the domain $\operatorname{Im}\rho \geqslant 0, |\rho| \neq 0$. We call these zeros the singular numbers of the operator $L(t)$. We denote the non-real singular numbers by ρ_j, $\operatorname{Im}\rho_j > 0, j = 1,\ldots,N$, and the real singular number by $\rho_{N+n}, n = 1,\ldots,m$. The multiplicities of these singular numbers are denoted by $m_1,\ldots,m_{N+m}$. The numbers $\lambda_j = \rho_j^2$, $j = 1,\ldots,N$ are eigenvalues of the operator $L(t)$ and this operator does not have other eigenvalues.

We put

$$f_j(x,t) = iRes|_{\rho=\rho_j}\left\{\frac{\tilde{e}(t,\rho)}{e(t,\rho)}e^{i\rho x}\right\}, M_j(x,t) = e^{-i\rho_j x}f_j(x,t),$$
$$(x,t) \in [0,\infty)\times(-\infty,\infty), \operatorname{Im}\rho_j > \varepsilon_0, j = 1,\cdots,N, \qquad (9.9)$$

where $\varepsilon_0 = \min\left\{\frac{\varepsilon}{2}, \operatorname{Im}\rho_1,\ldots,\operatorname{Im}\rho_N\right\}, \tilde{e}(t,\rho) = \tilde{e}(0,t,\rho)$ and $M_j(x,t)$ are polynomials of x of degree $m_j - 1$ with coefficients depending on t and they are called the standardized polynomials of the operator $L(t)$.

The scattering function $S(p,t)$, the non-real singular numbers $\rho_1,\ldots,\rho_N$ and the standardized polynomials $M_1(x,t),\ldots,M_N(x,t)$ are called the scattering data (SD) of the operator $L(t)$.

We introduce in the sense of generalized functions, the inverse Fourier transform of the function $S(\rho,t) - 1$ [78, 79]:

$$f_S^+(x,t) = \frac{1}{2\pi}\int_{+} [S(\rho;t) - 1]e^{i\rho x}d\rho, \quad (x,t) \in [0,\infty)\times(-\infty,\infty), \qquad (9.10)$$

where the contour of integration in (9.10) is any straight line contained in the strip $0 < |\operatorname{Im}\rho| < \varepsilon_0$.

The kernel $K(x,y,t)$ in (9.8) satisfies the fundamental equation in inverse problem [139]:

$$K(x,y,t) = \int_x^\infty K(x,u,t)f(u+y,t)du + f(x+y,t), 0 \leqslant x \leqslant y < \infty, -\infty < t < \infty, \qquad (9.11)$$

where the function $f(x,t)$ is constructed from the given SD as follows:

$$f(x,t) = f_S^+(x,t) - \left\{M_1(x,t)e^{i\rho_1 x} + \cdots + M_N(x,t)e^{i\rho_N x}\right\},$$
$$(x,t) \in [0,\infty)\times(-\infty,\infty). \qquad (9.12)$$

The potential $p(x,t)$ is recovered from the given SD by the formula [139]:

$$p(x,t) = -2\frac{\partial}{\partial x}K(x,x,t), \quad (x,t) \in [0,\infty) \times (-\infty,\infty). \tag{9.13}$$

There exists a remarkable class of potentials, for which the equation (9.11) can be solved exactly [133]. These are non-scattering potentials on the half-line, for which the inverse Fourier transform $f_S^+(x,t)$ defined by (9.10) in the sense of generalized functions is equal to zero. In the class of non-scattering potentials, the formulas (9.10) and (9.12) are

$$\begin{aligned} &f_S^+(x,t) = 0, \quad f(x,t) = -\left\{M_1(x,t)e^{i\rho_1 x} + \cdots + M_N(x,t)e^{i\rho_N x}\right\}, \\ &(x,t) \in [0,\infty) \times (-\infty,\infty). \end{aligned} \tag{9.14}$$

Assume that the potential in equation (9.7) belongs to the class of non-scattering isospectral potentials and so we consider the non-scattering data set:

$$s(t) = \left\{S(\rho;t) \equiv 1 \text{ for } |\text{Im}\rho| < \frac{\varepsilon_0}{2}, \rho_j, \quad \text{Im}\rho_j > \varepsilon_0, M_j(x,t), j = 1,\ldots,N\right\}, \tag{9.15}$$

where ρ_j are the non-real singular numbers with multiplicities m_j and $M_j(x,t)$ are the standardized polynomials of degree $m_j - 1$ with coefficients depending on $t, j = 1,\ldots,N$.

Let $B(x,t)$ denote an $N \times N$ square matrix consisting of the elements:

$$\begin{aligned} &B_{jn}(x,t) = \delta_{jn} + \tilde{M}_{jn}(x,t)e^{i(\rho_j+\rho_n)x}, \quad j,n = 1,\ldots,N, \\ &(x,t) \in [0,\infty) \times (-\infty,\infty), \delta_{jn} = 1 \text{ for } j = n, \delta_{jn} = 0 \text{ for } j \neq n, \end{aligned} \tag{9.16}$$

where

$$\tilde{M}_{jn}(x,t)\, e^{i(\rho_j+\rho_n)x} = \int_x^{\infty} M_j(x+u,t)\, e^{i(\rho_j+\rho_n)u} du, j = 1,\ldots,N. \tag{9.17}$$

Then we obtain the explicit expression for recovering the non-scattering potential $p(x,t)$ from the non-scattering data set (9.15) by formula [139]:

$$p(x,t) = -2\frac{\partial^2}{\partial x^2}\ln\tau(x,t), \quad (x,t) \in [0,\infty) \times (-\infty,\infty), \tag{9.18}$$

where

$$\tau(x,t) = \det B(x,t). \tag{9.19}$$

The time-dependent Schrödinger equation (9.7) is associated with the KdV equation:

$$p_t - 6pp_x + p_{xxx} = 0, \quad (x,t) \in [0,\infty) \times (-\infty,\infty). \tag{9.20}$$

This means that the equation (9.20) is identical to the equation defined by the Lax equation [139]:

$$\dot{L} = [L, A] = LA - AL,$$

where $L = -\frac{d^2}{dx^2} + p, A = 4\frac{\partial^3}{\partial x^3} - 3p_x - 6p\frac{\partial}{\partial x} - \gamma$, γ is some constant.

We denote a class of isospectral non-scattering potentials satisfying estimate (9.6) by $\mathfrak{L}$. The potential $p(x, t)$ defined by (9.18) satisfies the KdV equation (9.20) and belongs to the class $\mathfrak{L}$ [139].

9.3.2 Bilinear equation for the KdV equation

Hirota developed a direct method for finding N-soliton solution of NLEEs [55, 57, 58]. Substituting (9.18) into (9.20), integrating once yields, see [4]:

$$\tau_{xt}\tau - \tau_x\tau_t + \tau\tau_{xxxx} - 4\tau_x\tau_{xxx} + 3\tau_{xx}^2 = 0, (x, t) \in [0, \infty) \times (-\infty, \infty). \quad (9.21)$$

Here, the constant of integration is taken to be zero, because the potential $p(x, t)$ satisfies estimate (9.6). The equation (9.21) is quadratic in τ. Hirota usually refers to this as a bilinear equation for the KdV equation (9.20).

The introduction of the following special operators D is helpful in analysis [4]:

$$D_x^m D_t^n a \cdot b = (\partial_x - \partial_{x'})^m (\partial_t - \partial_{t'})^n a(x, t) b(x', t')|_{\substack{x=x' \\ t=t'}}. \quad (9.22)$$

Using the definition (9.22), the bilinear equation (9.21) can be written in a compact form:

$$\left(D_x D_t + D_x^4\right) \tau \cdot \tau = 0.$$

9.3.3 The degree of standardized polynomials

It is easy to prove by induction the following lemma.

Lemma 9.1. ([149]) *Let $a_j(x), j = 1, \ldots, m$ be polynomials of x and let $\alpha_1, \ldots, \alpha_m$ be different complex numbers. Then the following identity*

$$a_1(x)e^{\alpha_1 x} + a_2(x)e^{\alpha_2 x} + \ldots + a_m(x)e^{\alpha_m x} \equiv 0,$$

holds if and only if $a_j(x) \equiv 0, j = 1, \ldots, m$.

Let the SD (9.15) consist of only one singular number $\rho_1, \mathrm{Im}\rho_1 > 0$ with the multiplicity m_1. In this case, the standardized polynomial M_1 calculated by (9.9) at the singular number ρ_1 is a polynomial of degree $m_1 - 1$. We write formula (9.19) for this case:

$$\tau = 1 + M(x, t)e^{kx}, k = 2i\rho_1, \mathrm{Im}\rho_1 > 0, \quad (9.23)$$

where

$$M(x,t) = e^{-kx} \int\limits_x^\infty M_1(x+u,t)\, e^{ku} du = a_{m_1} x^{m_1 - 1} + \ldots + a_1, \tag{9.24}$$

$a_j, j = 1, \ldots, m_1$ are coefficients depending on t.

Substituting (9.23) into (9.21), then using Lemma 9.1 and (9.23), (9.24), we can prove that the multiplicity m_1 of ρ_1 is equal to 1, therefore the degree of M_1 is zero, and [149]:

$$M = a(t), \quad \tau = 1 + a(t)e^{kx}, \tag{9.25}$$

$$a_t + k^3 a = 0, \quad M = a(t) \equiv Ce^{-k^3 t}, \tag{9.26}$$

where C is a complex number depending on the singular number ρ_1.

Since M in the left-hand side of equality (9.24) is calculated by (9.26), then the polynomial M_1 of zero degree with respect to x in the right-hand side of this equality is defined by:

$$M_1 = -Cke^{-k^3 t}, \quad k \approx 2i\rho_1, \mathrm{Im}\rho_1 > 0. \tag{9.27}$$

Conversely, if M_1 is represented in the form (9.27), then substituting $\tau = 1 + Me^{kx}$ into (9.21), we can verify that τ satisfies the bilinear equation (9.21). The result obtained above can be stated in the following lemma.

Lemma 9.2 ([149]). *If the non-scattering data set* (9.15) *consist of only one singular number* ρ_1*,* $\mathrm{Im}\rho_1 > 0$*, then the dependent variable* τ *defined by* (9.23) *satisfies the bilinear equation* (9.21) *for the KdV equation* (9.20)*, if and only if* M_1 *is a standardized polynomial of zero degree with respect to* x *and evolves according to the time-evolution equation* (9.27).

We consider now the general case, when the non-scattering data set of the form (9.15) consists of N singular number, $N \geqslant 2$. We assume that any singular number $\rho_j = \frac{1}{2i} k_j, 1 \leqslant j \leqslant N, N \geqslant 2$ from the set (9.15) is to be subject to the assumption that for any j, $1 \leqslant j \leqslant N$ and $\alpha \in \{1, 2\}$:

$$\alpha k_j \neq \sum_{\substack{1 \leqslant l \leqslant N \\ l \neq j}} \beta_l k_l \text{ for any } \beta_l \in \{0, 1, 2\} \text{ and } k_l = 2i\rho_l. \tag{9.28}$$

The assumption (9.28) will be referred to as the condition (9.28) of given singular numbers. Note that the non-scattering data set (9.15) with any singular number $\rho_j, 1 \leqslant j \leqslant N, N \geqslant 2$ satisfying condition (9.28) is the subset of the set (9.15).

Theorem 9.1. ([149]) *Let the solution of the KdV equation* (9.20) *be given by formulas of the form* (9.18)–(9.19) *in terms of the non-scattering data set* (9.15) *with given singular numbers* $\rho_j, j = 1, \ldots, N, N \geqslant 2$*, satisfying* *condition* (9.28). *Then standardized polynomials* $M_j(x,t)$, $j=1,\dots,N$ *of the given non-scattering data set* (9.15) *are polynomials of zero degree with respect to variable* x *and are reduced to normalization factors evolving according to the time-evolution equation:*

$$M_j = -C_j k_j e^{-k_j^3 t}, \quad k_j = 2i\rho_j, j = 1,\dots,N, \tag{9.29}$$

where C_j *is a complex number depending on* ρ_j.

The following theorem is considered as the inverse assertion of Theorem 9.1.

Theorem 9.2 ([149]). *Let given singular number* ρ_j *and standardized polynomials* $M_j(x,t), 1 \leqslant j \leqslant N$, *of the non-scattering data set* (9.15) *satisfy the conditions:*

(1) The given singular numbers ρ_j *satisfy condition* (9.28),

(2) The standardized polynomials $M_j(x,t), 1 \leq j \leq N_j$ *are polynomials of zero degree with respect to variable* x *and depend on* t *according to formula* (9.29).

Then

(1) The dependent variable τ_N *defined by formula* (9.19) *constructed from given non-scattering data* (9.15) *satisfying conditions (1) and (2) can be represented in the form:*

$$\tau_N = \det B = \sum_{\{\mu_1,\mu_2,\dots,\mu_N\}} \left\{ \prod_{j=1}^{N} e^{\mu_j} \prod_{1\leqslant l<j\leqslant N} A_{lj}^{\mu_l \mu_j} \right\} \tag{9.30}$$

where the summation is taken over 2^N*-tuples* $\mu = \{\mu_1,\mu_2,\dots,\mu_N\}$, *such that* $\mu_j = 0$ *or 1, refer to each index* j, *in addition:*

$$e_j = C_j e^{k_j x - k_j^3 t}, (x,t) \in [0,\infty) \times (-\infty,\infty), A_{lj} = \left(\frac{k_l - k_j}{k_l + k_j}\right)^2, \mathrm{Re} k_j < 0 \tag{9.31}$$

where C_j *is a complex number,* $j = 1,\dots,N$.

(2) The function $\tau_N(x,t)$ *defined by* (9.30)–(9.31) *is a solution of the bilinear equation* (9.21).

(3) The non-scattering potential $p(x,t)$ *restored from the given non-scattering data by formula*

$$p_N(x,t) = -2\frac{\partial^2}{\partial x^2} \ln \tau_N(x,t) \tag{9.32}$$

is a solution of the KdV equation (9.20).

Remark 9.1. (1) By the ISM and Hirota's method, the solution (9.32) of the KdV equation (9.20) is the complex-valued function constructed from the non-scattering data set (9.15) with given singular numbers satisfying condition (9.28).

(2) By Theorem 6.7 in Chapter 6, the soliton-solution (6.147) of the same KdV equation is the real-valued fuction constructed by the ISM from the non-scattering data (6.149) without any condition imposed on the given singular numbers of set (6.149).

9.3.4 The SP associated with the attractive NLS equation

Under reduction:

$$c_1(x,t) = -\overline{c_2(x,t)} = c(x,t), \tag{9.33}$$

the system of NLEEs (2.1) in Chapter 2 reduces to the attractive NLS equation [133]:

$$ic_t = c_{xx} + 2|c|^2 c, \quad (x,t) \in [0,\infty) \times (-\infty,\infty). \tag{9.34}$$

The system (2.1) can be written as the Lax compatibility condition of two linear eigenvalue equations, i.e., the x-part and the t-part. The x part is the SP generated by the system of first-order ODEs of the form [97]:

$$-iJu_x + Cu = \lambda u, u = (u_1(x,\lambda;t), u_2(x,\lambda;t)), 0 \leqslant x < \infty, -\infty < t < \infty, \tag{9.35}$$

with a potential non-self-adjoint matrix $C = C(x,t) = \begin{pmatrix} 0 & c_1(x,t) \\ c_2(x,t) & 0 \end{pmatrix}$ and with the boundary condition at the origin $x = 0$:

$$u_1(0,\lambda;t) = u_2(0,\lambda;t), -\infty < t < \infty, \tag{9.36}$$

where $J = \begin{pmatrix} -1 & 0 \\ 0 & 1 \end{pmatrix}$, λ is a parameter.

The potentials $c_1(x,t)$ and $c_2(x,t)$ are complex-valued functions satisfying the estimate for any $t \in (-\infty,\infty)$:

$$|c_k(x,t)| \leqslant \tilde{C}e^{-\varepsilon x}, \tilde{C} \text{ is a constant}, \varepsilon > 0, k = 1,2. \tag{9.37}$$

The fundamental system of solutions of system (9.35) is the Jost solutions defined by the vector-functions:

$$e^{(1)}(x,\lambda;t) = \left(e_1^{(1)}(x,\lambda;t), e_2^{(1)}(x,\lambda;t)\right) =$$

$$\left(e^{-i\lambda x} + e^{-i\lambda x}\int_0^\infty H_{11}(x,x+\xi;t)e^{-i\lambda\xi}d\xi, e^{-i\lambda x}\int_0^\infty H_{21}(x,x+\xi;t)e^{-i\lambda\xi}d\xi\right),$$

$$e^{(2)}(x,\lambda;t) = \left(e_1^{(2)}(x,\lambda;t), e_2^{(2)}(x,\lambda;t)\right) =$$

$$\left(e^{i\lambda x}\int_0^\infty H_{12}(x,x+\xi;t)e^{i\lambda\xi}d\xi, e^{i\lambda x} + e^{i\lambda x}\int_0^\infty H_{22}(x,x+\xi;t)e^{i\lambda\xi}d\xi\right),$$

where λ is a real number and $H_{ij}(x,x+\xi;t), i,j = 1,2$ are elements of the kernel matrix $H(x,x+\xi;t)$ of the transformation operator usually used in SPs. In addition, $H_{ij}(x,x+\xi;t)$ satisfies the estimate analogous to the estimate (9.37):

$$|H_{ij}(x,s;t)| \leq \tilde{C}\exp\left\{-\frac{1}{2}\varepsilon(x+s)\right\}, 0 \leqslant x \leqslant s, i,j = 1,2. \tag{9.38}$$

We call

$$S(\lambda;t) = \frac{e_1^{(1)}(0,\lambda;t) - e_2^{(1)}(0,\lambda;t)}{e_2^{(2)}(0,\lambda;t) - e_1^{(2)}(0,\lambda;t)}, \quad |\mathrm{Im}\lambda| < \varepsilon_0,$$

the scattering function for SP (9.35)–(9.36). Due to estimate (9.38), the function $e_2^{(2)}(0,\lambda;t) - e_1^{(2)}(0,\lambda;t)$ is holomorphic for $\mathrm{Im}\lambda > -\frac{\varepsilon}{2}$, and the function $e_1^{(1)}(0,\lambda;t) - e_2^{(1)}(0,\lambda;t)$ is holomorphic for $\mathrm{Im}\lambda < \frac{\varepsilon}{2}$, therefore they have a finite number of zeros: $\lambda_k^+, \mathrm{Im}\lambda_k^+ > 0, k = 1,\ldots,\gamma_1$ and $\lambda_j^-, \mathrm{Im}\lambda_j^- < 0, j = 1,\ldots,\gamma_2$. We denote the multiplicities of these zeros by m_k and n_j, respectively. We introduce in the sense of generalized functions the inverse Fourier transformations of the functions $S(\lambda;t) - 1$ and $S^{-1}(\lambda;t) - 1$:

$$\begin{aligned} f_S(x;t) &= \frac{1}{2\pi}\int_{-\infty+i\eta}^{\infty+i\eta}[S(\lambda;t) - 1]e^{-i\lambda x}d\lambda, \\ g_S(x;t) &= \frac{1}{2\pi}\int_{-\infty-i\eta}^{\infty-i\eta}\left[S^{-1}(\lambda;t) - 1\right]e^{-i\lambda x}d\lambda, \end{aligned} \tag{9.39}$$

where $S(\lambda;t), S^{-1}(\lambda;t)$ are functions of the type (S) in the strip $|\mathrm{Im}\alpha| < \varepsilon_0$, and η is some number such that $0 < \eta < \varepsilon_0$.

The function (9.39) characterize the SP (9.35)–(9.36) on the continuous spectrum. The following functions characterize this SP on the point spectrum:

$$f_k^+(-x) = \frac{1}{2\pi} \int\limits_{c_k^+} \frac{1}{e_2^{(2)}(0,\lambda;t) - e_1^{(2)}(0,\lambda;t)} \Big[A_0 + A_1\left(\lambda - \lambda_k^+\right) + \ldots + \frac{A_{m_k-1}}{(m_k-1)!}\left(\lambda - \lambda_k^+\right)^{m_k-1}\Big] e^{i\lambda x} d\lambda, \quad k = 1, \ldots, \gamma_1, \tag{9.40}$$

$$g_j^-(x) = \frac{1}{2\pi} \int\limits_{c_j^-} \frac{1}{e_1^{(1)}(0,\lambda;t) - e_2^{(1)}(0,\lambda;t)} \Big[B_0 + B_1\left(\lambda - \lambda_j^-\right) + \ldots + \frac{B_{n_j-1}}{(n_j-1)!}\left(\lambda - \lambda_j^-\right)^{n_j-1}\Big] e^{-i\lambda x} d\lambda, \quad j = 1, \ldots, \gamma_2, \tag{9.41}$$

where c_k^+ and c_j^- stand for circles with a sufficiently small radius that avoid the singular points λ_k^+ and λ_j^-, respectively.

By using the definition of a residue [114], from (9.40), (9.41) we obtain:

$$f_k^+(-x;t) = p_k(x;t)\, e^{i\lambda_k^+ x}, \quad k = 1, \ldots, \gamma_1,$$
$$g_j^-(x;t) = q_j(x;t) e^{-i\lambda_j^- x}, \quad j = 1, \ldots, \gamma_2,$$

where $p_k(x;t)$ and $q_j(x;t)$ are standardized polynomials of degree $m_k - 1$ and of $n_j - 1$, respectively.

In the case of reduction (9.33), it follows from the integral equations for the kernels of the transformation operator that [133]:

$$H_{11}(x,s;t) = \overline{H_{22}(x,s;t)}, H_{12}(x,x;t) = -\overline{H_{21}(x,x;t)}, 0 \leqslant x \leqslant s. \tag{9.42}$$

We assume that the SD of the SP (9.35)–(9.36) belong to the class of non-scattering isospectral potentials, and so we consider the non-scattering data set [133]:

$$s(t) = \Big\{S(\lambda,t) \equiv 1 \text{ for } |Im\lambda| < \varepsilon_0, \lambda_j^+, \mathrm{Im}\lambda_j^+ > \varepsilon_0, p_j(x,t), \lambda_j^- = \overline{\lambda_j^+}, \mathrm{Im}\lambda_j^- < -\varepsilon_0, \overline{p_j(x,t)}, j = 1, \ldots, N, \varepsilon_0 > 0\Big\}. \tag{9.43}$$

We construct the sought non-scattering potential from the data set (9.43), so the kernels will be sought in the form:

$$H_{11}(x,u;t) = \sum_{k=1}^{N} H_{11,k}(x;t) e^{i\lambda_k^+ u}, H_{12}(x,u;t) = \sum_{k=1}^{N} H_{12,k}(x;t) e^{-i\lambda_k^- u}. \tag{9.44}$$

Using formulas (9.44), from equations (2.69) and (2.72) in Chapter 2 we obtain the system:

$$H_{11,k}(x;t) - \sum_{j=1}^{N} H_{12,j}(x;t) \int_x^\infty p_k(x+\tau;t)\, e^{i(\lambda_k^+ - \lambda_j^-)\tau} d\tau = 0,$$

$$H_{12,k}(x;t) + \sum_{j=1}^{N} H_{11,j}(x;t) \int_x^\infty p_k(x+\tau;t)\, e^{i(\lambda_j^+ - \lambda_k^-)\tau} d\tau = -\overline{p_k(2x;t)} e^{-i\lambda_k^- x},$$

(9.45)

where $\lambda_k^+ = \overline{\lambda_k^-} \in s(t),\, k = 1, \ldots, N, (x,t) \in [0,\infty) \times (-\infty,\infty)$.

We calculate $H_{12,k}(x;t), k = 1, \ldots, N$ from system (9.45), then with the help of formula (9.44), we find

$$H_{12}(x,x;t) = (\det A)^{-1} \sum_{i=1}^{N} \left(\det A^{(j+N)}\right) e^{\frac{1}{2}k_{j+N}x}, \tag{9.46}$$

where $k_j = 2i\lambda_j^+$, $k_{j+N} = \overline{k_j} = -2i\lambda_j^-$, the matrix A are written in the block form. Namely,

$$A = \begin{pmatrix} I & M \\ \tilde{M} & I \end{pmatrix}, I \text{ is the unit matrix}, \tag{9.47}$$

$$M = \left[-M_{lj}(x,t)\, e^{\frac{1}{2}(k_l + k_{j+N})x}\right]_{l,j=1}^{N}, \tilde{M} = \left[\overline{M_{lj}(x,t)} e^{\frac{1}{2}(k_j + k_{l+N})x}\right]_{l,j=1}^{N},$$

$$M_{lj}(x,t)\, e^{\frac{1}{2}(k_l + k_{j+N})x} = \int_x^\infty p_l(x+\tau,t)\, e^{\frac{1}{2}(k_l + k_{j+N})\tau} d\tau.$$

(9.48)

$A^{(j+N)}$ stands for the matrix obtained from the matrix A substituting the elements in its $(j+N)$-th column by the column of the right-hand side of (9.45), i.e.,

$$\left(0, 0, \ldots, -\overline{p_1(2x,t)} e^{\frac{1}{2}k_{1+N}x}, -\overline{p_2(2x,t)} e^{\frac{1}{2}k_{2+N}x}, \ldots, -\overline{p_N(2x,t)} e^{\frac{1}{2}k_{2N}x}\right)^T$$

for $j = 1, 2, \ldots, N$.

The found kernel (9.46) is related to the potential by the equality:

$$2iH_{12}(x,x;t) = c(x,t). \tag{9.49}$$

From (9.46) and (9.49) it follows that the solution of the attractive NLS equation (9.34) has a complicated structure:

$$c(x,t) = -i\frac{G(x,t)}{F(x,t)}, \tag{9.50}$$

where

$$F(x,t) = \det A, \tag{9.51}$$

$$G(x,t) = -2\sum_{j=1}^{N}\left(\det A^{(j+N)}\right)e^{\frac{1}{2}k_{j+N}x}. \tag{9.52}$$

9.3.5 The representations of F and G

Lemma 9.3 ([150]). *The function F constructed from the non-scattering data set (9.43) by formula (9.51) is real-valued and can be represented in the form:*

$$F(x,t) = \sum_{(\mu_1,\mu_2,\dots,\mu_{2N})} D_1(\mu)a_\mu(x,t)\prod_{j=1}^{2N} e^{\mu_j k_j x}, (x,t) \in [0,\infty)\times(-\infty,\infty) \tag{9.53}$$

where the summation is taken over 2^{2N}-tuples $\mu = (\mu_1,\mu_2,\dots,\mu_{2N})$, such that $\mu_j = 0$ or $\mu_j = 1$, refers to each index j, $a_\mu(x,t)$ is a polynomial of x with coefficients depending on t, and

$$D_1(\mu) = \begin{cases} 1 & when \sum_{j=1}^{N}\mu_j = \sum_{j=1}^{N}\mu_{j+N}, \\ 0 & when \sum_{j=1}^{N}\mu_j \neq \sum_{j=1}^{N}\mu_{j+N}. \end{cases} \tag{9.54}$$

Lemma 9.4 ([150]). *The function G constructed from the non-scattering data set (9.43) by formula (9.52) can be represented in the form:*

$$G(x,t) = (-2)\sum_{(\mu_1,\mu_2,\dots,\mu_{2N})}\left(D_2(\mu)b_\mu(x,t)\prod_{j=1}^{2N} e^{\mu_j k_j x}\right), (x,t) \in [0,\infty)\times(-\infty,\infty) \tag{9.55}$$

where the summation is taken over 2^{2N}-tuples $\mu = (\mu_1,\mu_2,\dots,\mu_{2N})$ such that $\mu_j = 0$ or $\mu_j = 1$, refers to each index j, $b_\mu(x,t)$ is a polynomial of x with coefficients depending on t, and

$$D_2(\mu) = \begin{cases} 1 & when\ 1 + \sum_{j=1}^{N}\mu_j = \sum_{j=1}^{N}\mu_{j+N}, \\ 0 & when\ 1 + \sum_{j=1}^{N}\mu_j \neq \sum_{j=1}^{N}\mu_{j+N}. \end{cases} \tag{9.56}$$

The formula (9.50) is the Hirota's change variable for NLS equation (9.34). Substituting (9.50) into (9.34), using the reality of $F(x,t)$ and the operator D given by (9.22), we lead equation (9.34) to the form:

$$\begin{aligned} &F\left\{iG_tF - iGF_t - G_{xx}F_+2G_xF_x - GF_{xx}\right\} + \\ &G\left\{2F_{xx}F - 2F_x^2 - 2G\overline{G}\right\} = 0. \end{aligned} \tag{9.57}$$

9.3.6 The degree of standardized polynomials and solutions of the NLS equation

Let the non-scattering data set (9.43) consist of one pair of singular number λ_1^+, $\lambda_1^- = \overline{\lambda_1^+}, \mathrm{Im}\lambda_1^+ > 0$, and let n be the multiplicity of the given singular numbers. Then the standardized polynomial p_1 is a polynomial of $n-1$ degree:

$$p(2x,t) = a_{n-1}x^{n-1} + \ldots + a_0,$$

where $a_j = a_j(t), j = 1, \ldots, n-1$.

In this case the functions F and G given by (9.53)–(9.54) and (9.55)–(9.56) are represented in the form:

$$F = 1 + |M_{11}|^2 e^{(k_1+k_2)x}, G = 2\bar{p}_1 e^{k_2 x}, k_1 = 2i\lambda_1^+, k_2 = -2i\lambda_1^-. \tag{9.58}$$

From (9.48) it follows that M_{11} is defined by:

$$M_{11}(x,t)e^{\frac{1}{2}(k_1+k_2)x} = \int_x^\infty p_1(x+\xi)e^{\frac{1}{2}(k_1+k_2)\xi}d\xi. \tag{9.59}$$

Using Lemma 9.1, we can prove the following lemma.

Lemma 9.5. *Let the non-scattering data set* (9.43) *of the SP* (9.35)–(9.36) *be given by the set:*

$$s(t) = \left\{\lambda_1^+, \lambda_1^- = \overline{\lambda_1^+}, \mathrm{Im}\lambda_1^+ > 0, p_1(x,t), \overline{p_1(x,t)}\right\} \tag{9.60}$$

If the functions F and G constructed by the formulas (9.58) *and* (9.59) *from the data set* (9.60) *satisfy equation* (9.57), *then p_1 and $\overline{p_1}$ of the given data set* (9.60) *are standardized polynomials of zero degree with respect to variable x and evolve according to the time-evolution equation:*

$$p_1 = \tilde{C}e^{ik_1^2 t}, \tilde{C} \textit{ is a complex numbers.} \tag{9.61}$$

Substituting (9.61) into (9.59), we have

$$M_{11} = -\frac{2\tilde{C}}{k_1+k_2}e^{ik_1^2 t}. \tag{9.62}$$

Hence, the solution (9.50) constructed from the data set (9.60) for the attractive NLS equation (9.34) is obtain from (9.58) and (9.61)–(9.62):

$$c(x,t) = -2i\frac{\overline{p_1}e^{k_2 x}}{1+|M_{11}|^2 e^{(k_1+k_2)x}}. \tag{9.63}$$

We can prove the inverse assertion of Lemma 9.5, i.e., if p_1 and M_{11} are defined by (9.61) and (9.62), respectively, then $c(x,t)$ constructed by (9.63) satisfies the attractive NLS equation (9.34).

We proceed now to consider the general case when the non-scattering data set (9.43) consists of N pairs of singular numbers. For this purpose we assume that any pair of singular numbers from the set (9.43) $\lambda_j^+ = \frac{1}{2i}k_j$, $\lambda_j^- = -\frac{1}{2i}k_{j+N}$, $j = 1, \ldots, N$ is to be subject to the assumption that for any j, $1 \leqslant j \leqslant N$ and $\alpha \in \{0, 1, 2\}$:

$$\alpha k_j + (\alpha+1)k_{j+N} \neq \sum_{\substack{1 \leqslant l \leqslant 2N \\ l \neq j}} \beta_l k_l \text{ for any } \beta_l \in \{0,1,2,3\}, k_l = 2i\rho_l,$$

$$k_l = 2i\lambda_l^+, k_{l+N} = -2i\lambda_l^-, l = 1, \ldots, N. \tag{9.64}$$

The assumption (9.64) will be referred to as condition (9.64) of given singular number.

Theorem 9.3. *Let the functions $F(x,t)$ and $G(x,t)$ be constructed by formulas* (9.51) *and* (9.52)*, respectively, from the non-scattering data set* (9.43) *with given singular numbers λ_j^+, $\mathrm{Im}\lambda_j^+ > \varepsilon_0$ and $\lambda_j^- = \overline{\lambda_j^+}$, $\mathrm{Im}\lambda_j^- < -\varepsilon_0, \varepsilon_0 > 0, j = 1, \ldots, N$ satisfying condition* (9.64). *If the functions $F(x,t)$ and $G(x,t)$ satisfy equation* (9.57)*, then $p_j(x,t)$, $j = 1, \ldots, N$ of the given non-scattering data set* (9.43) *are standardized polynomials of zero degree with respect to variable x and are reduced to the corresponding factors evolving according to the time-evolution equation:*

$$p_j(x,t) = C_j e^{ik_j^2 t}, -\infty < t < \infty, \tag{9.65}$$

where C_j is a complex number, $j = 1, \ldots, N$.

The following theorem is considered as the inverse assertion of Theorem 9.3.

Theorem 9.4. *Let the non-scattering data set* (9.43) *consist of N pairs of singular numbers such that the given data of this set satisfy the conditions:*

(1) The given singular numbers of the given data set (9.43) *satisfy condition* (9.64).

(2) The standardized polynomials $p_j, j = 1, \ldots, N$ are polynomials of zero degree with respect to variable x and are reduced to the corresponding factors evolving according to the time-evolution equation (9.65).

Let $F(x,t)$ and $G(x,t)$ be functions defined by formulas (9.51) *and* (9.52) *constructed form the non-scattering data set* (9.43) *satisfying conditions (1) and (2). Then*

(1) The function $F(x,t)$ can be represented by the formula:

$$F(x,t) = \sum_{(\mu_1,\mu_2,\ldots,\mu_{2N})} D_1(\mu) \left\{ \prod_{j=1}^{2N} e_j^{\mu_j} \prod_{1 \leqslant l < j \leqslant 2N} \phi_{lj}^{\mu_l \mu_j} \right\}, \tag{9.66}$$

wherein the summation is taken 2^{2N}-tuples $\mu = (\mu_1, \mu_2, \ldots, \mu_{2N})$, such that $\mu_j = 0$ or $\mu_j = 1$, refers to each index j, $D_1(\mu)$ is determined by (9.54). *In* (9.66), *the functions e_j are constructed from the given non-scattering data set* (9.43) *satisfying conditions (1) and (2) by the formulas:*

$$e_j = 2C_j e^{k_j x - \Omega_j t}, \Omega_j = -ik_j^2, \Omega_{j+N} = \bar{\Omega}_j, e_{j+N} = \bar{e}_j, \tag{9.67}$$

C_j is a complex number, $j = 1, \ldots, N$, ϕ_{lj} are constants

$$\phi_{lj} = \phi_{jl} = \begin{cases} (k_l + k_j)^{-2} & \text{for } l \leqslant N < j \text{ or } j \leqslant N < l, \\ (k_l - k_j)^2 & \text{for } l, i \leqslant N \text{ or } l, j > N. \end{cases}$$

(2) *The function $G(x,t)$ can be represented by the formula:*

$$G(x,t) = \sum_{(\mu_1,\mu_2,\ldots,\mu_{2N})} D_2(\mu) \left\{ \prod_{j=1}^{2N} e_j^{\mu_j} \prod_{1 \le l < j \le 2N} \phi_{lj}^{\mu_l \mu_j} \right\}, \tag{9.68}$$

wherein the summation is taken 2^{2N}-tuples $\mu = (\mu_1, \mu_2, \ldots, \mu_{2N})$, such that $\mu_j = 0$ or $\mu_j = 1$, refers to each index j, $D_2(\mu)$ is determined by (9.56).

(3) *The functions $F(x,t)$ and $G(x,t)$ given by* (9.66), (9.67) *and* (9.68) *respectively, satisfy the Hirota's equations:*

$$\begin{aligned} &i(G_t F - GF_t) - (G_{xx}F - 2G_x F_x + GF_{xx}) = 0, \\ &2F_{xx}F - 2F^2 - 2G\bar{G} = 0, \end{aligned}$$

therefore $F(x,t)$ and $G(x,t)$ satisfy equation (9.57).

(4) *The non-scattering potential $c(x,t)$ restored by the formula:*

$$c(x,t) = -i\frac{G(x,t)}{F(x,t)}, \tag{9.69}$$

is the solution of the attractive NLS equation (9.34).

Remark 9.2. (1) By the ISM and Hirota's method, the solution (9.69) of the attractive NLS equation (9.34) is constructed from N pairs of singular numbers satisfying condition (9.64).

(2) By Theorem 2.4 in Chapter 2, the solution (2.111) of the same attractive NLS equation (2.60) is the complex-valued function constructed by formula (2.111) from only one pair of singular numbers of the non-scattering data set (2.108).

Chapter 10

The Bäcklund Transformations Between A Common Solution of Both Linear Equations in The Lax Pair and The Solution of The Associated IBVP for NLEEs on The Half-Line

In Section 10.1 we present the BTs for NLEEs defined in the whole space. In Sections 10.2 and 10.3, we construct the BT between a common solution of both linear equations in the Lax pair and the solution of the associated IBVP for NLEEs defined on the half-line. A close relationship between the BT and the associated ISP is established.

The Bäcklund transformation (BT) was developed in 1883 arising from the pseudospherical surfaces construction. The name of Albert Victor Bäcklund (1845–1922) is associated with the transformation of surfaces that bears his name, the extensions of which have a great impact in soliton theory. The BT is an important tool in soliton theory and integrable systems.

We first mention some works that closely border on the topic of this chapter to be discussed. The ISM is a well-developed mathematical theory that provides means of solving the IVP for a broad class of integrable NLEEs [2]. Lax [69] showed first the criterion for constructing NLEEs. A new form of BTs which gives a unified method of finding BTs for solutions of certain NLEEs has been presented in the work [56], the sine-Gordon equation, the KdV, and mKdV equations have been considered in this work, as examples, The relation between the BT and the ISP for the KdV and mKdV equations has been presented in work [20]. A particular 2×2 system of a couple of Lax equations to describe the KdV, sine-Gordon, and non-linear Schrdinger equations has been considered in [2]. The derivation of BT from the eigenvalue problem for the generalized Zakharov and Shabat system, and vice versa has been investigated in the work [91]. In this work, the interrelation between BT and the ISP was shown. The L-A pair corresponding to the boundary value problem with the boundary condition at the origin $x = 0$ of the KdV equation has been presented in [6]. Applications of BTs in physics and of generalized BTs to continuum mechanics have been presented in [103, 104, 107], respectively.

DOI: 10.1201/9781003370543-10

It is clear that in the study of integrable NLEEs, two important research problems are to find soliton solutions and an inverse scattering problem. The BT is a possible solution to each problem. However, it isn't easy to set up an appropriate ISP, as shown in the works [56, 90]. There is no generally accepted definition of BT [90]. The majority of the above-mentioned problems have been considered in the whole space. Hence, it is required to set up appropriate ISPs associated with IBVPs for integrable NLEEs satisfying the given initial and boundary conditions. So, in the present chapter, we prove that the ISPs considered in previous chapters are set up appropriately for constructing BTs, and there exist interrelations between the considered ISP and the associated IBVP for NLEEs.

To describe, we consider limited cases, where the BTs are obtained from the correspondence relations between scattering data (SD) of scattering problems (SPs) generated by two linear equations in the Lax pair and solution (recovered potentials in these equations) of the associated IBVP or Cauchy problem for NLEEs. We present in this chapter a way of obtaining the BTs between a common solution of both equations in the Lax pair and the solution (recovered potentials) of the associated problem for NLEEs satisfying the given initial and boundary conditions on a half-line, and conversely.

The one-to-one correspondence relations between the SD of the considered SPs generated by both linear equations in the Lax pair and the recovered potentials in these equations are established. These recovered potentials are solutions of the IBVP for NLEEs satisfying the given initial and boundary conditions. Using these proven assertions, we now construct the BTs from the established correspondence relations between the SD and the solution (recovered potential) of the associated IBVP for NLEEs. Two cases may arise:

(1) By using the compatible equalities for two linear equations in the Lax pair, we can prove that the solution of the SP for the first equation in the Lax pair satisfies the second equation in this Lax pair, i.e., this solution is the common solution of the SP for both equations in the Lax pair.

(2) If the solution of the first equation in the Lax pair does not satisfy the second equation in this pair, we must therefore correct this solution. More specifically, we multiply the solution of the first equation from the right by a factor function appropriately chosen such that the resulting factor function satisfies both equations in the Lax pair.

The chapter is based on the papers [29, 90, 151, 152].

10.1 The BTs for NLEEs defined in the whole space

Definition 10.1. *A Bäcklund transformation (BT) is essentially defined as being a system of equations relating the solution of a given equation either to* *another solution of the same equation, or to a solution of another equation (see Ablowitz and Segur [5], p. 154 for a precise definition).*

For simplicity, we restrict our discussion to (system of) partial differential equations. It is convenient to use

$$D(u) = 0 \text{ and } E(v) = 0,$$

to denote partial differential equations. Depending on the context, they may denote the same equations or different ones.

Note that by usual definition, a mapping uniquely defines u from given v, but does not specify either $D(u) = 0$ or $E(v) = 0$. A BT need not define u unique even given v, but it does specify both $D(u) = 0$ and $E(v) = 0$. The distinction between a BT and a mapping is this. Often a BT can be constructed from a mapping by specifying an appropriate evolution equation.

We first consider in Subsection 10.1.1 and 10.1.2 two following known simplest BTs.

10.1.1 The Cauchy–Riemann system

One of the simplest BTs is the Cauchy–Riemann relations:

$$u_x = v_y, \quad u_y = -v_x,$$

which relates the real and imaginary parts u and v of a holomorphic function. This first order partial differential equation has the following properties:

(1) If u and v are solutions of the Cauchy–Riemann equations, then u and v are solutions of the Laplace equation:

$$u_{xx} + u_{yy} = 0, \quad v_{xx} + v_{yy} = 0.$$

(2) Conversely, if $v(x, y) = xy$ is solution of the Laplace equation, then there exits another solution of the Laplace equation $u(x, y) = \frac{1}{2}(x^2 - y^2)$ which solves the Cauchy–Riemann equation together with v:

$$u_x = v_y = x, \quad u_y = -v_x = -y.$$

This example is rather trivial, because all three equations (the equation for u, the equation for v and the BT relating them) are linear.

10.1.2 The Liouville equation

$$u_{xy} = \exp u, \tag{10.1}$$

Let us introduce the wave equation, simple to be solved

$$v_{xy} = 0.$$

The general solution of this wave equation is

$$v(x, y) = a_1(x) + a_2(y),$$

where $a_j(s), j = 1, 2$ are arbitrary uniformly bounded continuous functions from $C(E)$.

Suppose that we have a pair of relations that relate the solution u of the Liouville equation (10.1) and solution v of the wave equation:

$$\begin{aligned} u_x + v_x &= \sqrt{2}\exp\left\{\frac{1}{2}(u - v)\right\}, \\ u_y - v_y &= \sqrt{2}\exp\left\{\frac{1}{2}(u + v)\right\}. \end{aligned} \tag{10.2}$$

From the pair of relations (10.2) and using these relations, we obtain:

$$\begin{aligned} u_{xy} + v_{xy} &= \frac{1}{\sqrt{2}}(u_y - v_y)\exp\left\{\frac{1}{2}(u - v)\right\} \\ &= \frac{1}{\sqrt{2}}\sqrt{2}\exp\left\{\frac{1}{2}(u + v)\right\}\exp\left\{\frac{1}{2}(u - v)\right\} = \exp u, \\ u_{yx} - v_{yx} &= \frac{1}{\sqrt{2}}(u_x + v_x)\exp\left\{\frac{1}{2}(u + v)\right\} \\ &= \frac{1}{\sqrt{2}}\sqrt{2}\exp\left\{\frac{1}{2}(u - v)\right\}\exp\left\{\frac{1}{2}(u + v)\right\} = \exp u. \end{aligned}$$

Hence, by the pair of relations (10.2), using the wave equation we can solve the Liouville equation (10.1). This pair of relations is a BT. By this BT if u is the Liouville equation's solution, then v is the solution of the wave equation. Therefore, the BT can turn a non-linear partial differential equation into a simpler linear partial differential equation.

It is well-known that a large class of NLEEs exists with multisoliton solution structure. Lax [69] showed first the criterion for constructing such NLEEs. Hereafter, we present in the following way of obtaining BTs, which transform a common solution of both equations (the x pair and t-part) in the Lax pair into the solution of the IBVP (or Cauchy problem) for considered NLEEs with given initial and boundary conditions.

10.1.3 The BT for the solution of the two-dimensional generalization from the KdV equation (8.129) in $C(E^2; t)$

The SP for the perturbed string equation (8.13) with the conditions (8.14) and (8.15) is associated with the equation (8.129) in $C(E^2; t)$. By Theorem 8.1, if the profiles of the incident wave $a(s) = (a_1(s), a_2(s))$ are given, then there exists a unique solution of the SP (8.13)–(8.14)–(8.15). This solution is represented in the asymptotic form in terms of profiles of the reflected wave $b(s) = (b_1(s), b_2(s))$. The operator S, that transforms $a(s)$ into $b(s)$ is called the scattering operator for this SP. By Theorem 8.2 the potential function in the equation (8.13) is uniquely restored from the known scattering operator S. The time-dependence of S is derived with the help of the generalized Lax equation (8.121). Then by Theorem 8.3, the solution of the problem for the equation (8.129) with the given initial condition (8.133) and boundary conditions (8.134), (8.135) is found uniquely from the given time-dependent scattering operator. Hence, there exists a one-to-one correspondence relation between the solution of the SP (8.13)–(8.14)–(8.15) and the solution of the problem (8.129), (8.133)–(8.135). This correspondence relation is a BT, which transforms the solution of the SP (8.13)–(8.14)–(8.15) into the solution of the problem (8.129), (8.133)–(8.135), and conversely (see Figure 10.1).

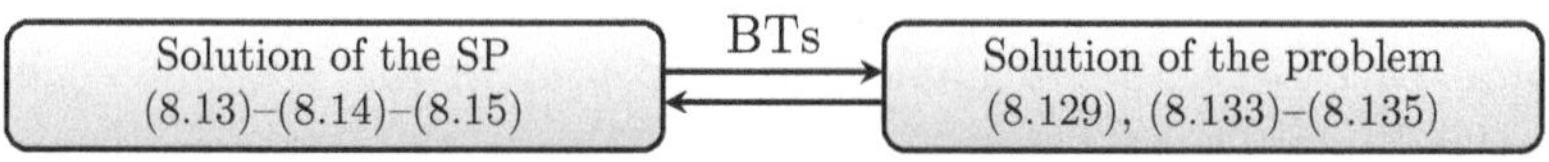

FIGURE 10.1: The BT between the solution of SP (8.13)–(8.14)–(8.15) and the solution of the problem (8.129), (8.133)–(8.135), and conversely.

10.1.4 The BT for the solution of the continual system of non-linear interaction waves (5.213) on the whole line

The SP for the transport equation (5.205) is associated with the continual system of non-linear interaction waves (5.213). The time-dependence of the solution of equation (5.205) is given by the equation (5.207). The potential functions $u(x, y, \tau; t)$ and $v(x, y, \tau; t), -\infty < x < \infty, y, \tau \in (-1, 1), -\infty < t < \infty$, in equations (5.205) and (5.207) are continuous functions possessing the first-order partial continuous derivatives with respect to t and x, and satisfying estimates (5.209). Suppose that the transport equation (5.205) has no generalized eigenvalues, then

(1) By Lemma 5.8 for every given bounded solution of equation (5.105), there exist the bounded functions $A(y, \lambda; t)$ and $B(y, \lambda; t)$: $A(y, \lambda; t) = \lim_{x \to -\infty} \varphi(x, y; t)e^{i\lambda yx}, B(y, \lambda; t) = \lim_{x \to \infty} \varphi(x, y; t)e^{i\lambda yx}$ for a real parameter λ. The solution $\varphi(x, y; t)$ of equation (5.205) is determined uniquely by each of these functions. The one-to-one correspondence relation between the incident wave $A(y, \lambda)$ and the scattered wave $B(y, \lambda)$ defines the transition operator $S(\lambda; t)$: $S(\lambda; t)A(y, \lambda; t) = B(y, \lambda; t)$.

(2) By Lemma 5.17, the operator equations (5.205) and (5.207) are compatible, i.e., $\varphi_{xt} = \varphi_{tx}$.

(3) By Theorem 5.6 the continual system (5.213) is equivalent to the compatibility condition (5.222) for equations (5.205) and (5.207).

There exists the one-to-one correspondence relation between the transition operator $S(\lambda;t)$ for the transport equation (5.205) and the solution of the Cauchy problem for the continual system (5.213) with the preassigned initial condition (5.236).

Let $\varphi = \varphi(x,y;t)$ be a solution of equation (5.205), then this solution satisfies equation (5.207). Indeed, by the assertion (2): $i\varphi_{xt} = (\lambda J\varphi - U\varphi)_t = i\varphi_{tx} = (\lambda T\varphi - V\varphi)_x$.

The function $i\varphi_t = \lambda T\varphi - V\varphi$ is an antiderivative with respect to variable x of function $i\varphi_{tx}$ from the given equability: $i\varphi_{xt} = i\varphi_{tx}$. Hence, the solution φ of equation (5.205) satisfies equation (5.207).

Conversely, if φ is a solution of (5.207), then the function $i\varphi_x = \lambda J\varphi - U\varphi$ is antiderivative with respect to variable t of the function $i\varphi_{tx} = i\varphi_{xt} = (\lambda J\varphi - U\varphi)_t$. Hence, the function φ satisfies equation (5.205).

The following lemma is deduced from the above discussion.

Lemma 10.1. *(1) The function* $\varphi = \varphi(x,y;t)$ *in the operator equations* (5.205) *and* (5.207) *is a common solution of these equations;*

(2) The one-to-one correspondence relation between the common solution of equations (5.205) *and* (5.207) *and the solution of the continual system* (5.213) *is a BT, which transforms the common solution of both equations* (5.205) *and* (5.207) *into the solution of system* (5.213)*, and conversely. See Figure 10.2.*

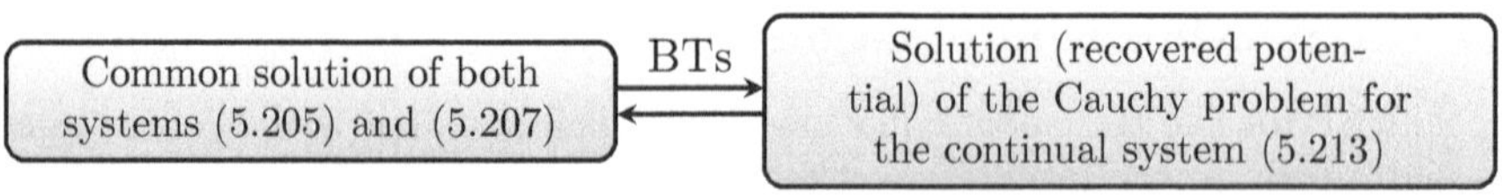

FIGURE 10.2: The BTs between the common solution of the both equations (5.205), (5.207) and the solution of Cauchy problem for the continual system (5.213), and conversely.

10.1.5 A note on the equivalence of BTs and the ISM for a class of non-linear differential equations [29]

In the paper [29], El-Sherbiny and L. Debnath study the relationship between the ISM and the BTs for a class of non-linear partial differential equations:

$$U_t + U_x + a_1 UU_x + a_2 UU_t + a_3 U_{xxx} + a_4 U_{xxt} + a_5 U_{xtt} + a_6 U_{ttt} = 0, \quad (10.3)$$

where $a_i \in R$ $(i = 1, 2, \ldots, 6)$.

In the paper, El-Sherbiny [27] proved that if (a_1/a_2) satisfies the cubic equation

$$a_6 k^3 - a_5 k^2 + a_4 k - a_3 = 0, \quad (10.4)$$

then the class of equations (10.3) satisfying the cubic equations (10.4) has an infinite number of conservation laws. Otherwise, the class has at most three conservation laws, see the papers [29, 27, 28, 30]. In these papers, the following assertions are proved.

1. The class of non-linear differential equations (10.3) satisfying the cubic equations (10.4) has a BT;
2. The fundamental equations of ISM that correspond to the class of equations (10.3) satisfying conditions (10.4) are derived;
3. The BTs of the class of equations (10.3) satisfying conditions (10.4) are equivalent to the fundamental equations of the corresponding ISM.

As noted in [29, 27, 28, 30], an appropriate inverse scattering problem seems to depend on the existence of an infinite number of independent conservation laws for the non-linear partial differential equations. The BTs were originally developed to study the related theories of differential geometry and differential equations and also provide a method of finding the single-soliton or the multi-soliton solutions of some non-linear partial differential equations. The method of solutions of certain classes of non-linear equations is also associated appropriately with appropriate BTs. Indeed, there is a close relationship between the ISM and BTs in the sense that the scattering data and the associated time dependence that constitute an ISM also constitute a BT.

10.2 The BT between a constructed common solution of both equations in the Lax pair and the solution of the associated IBVP for NLEEs on a half line

Since the considered ISPs in the previous chapters are well-solved, then by the unique solvability of the ISP there exists a one-to-one correspondence relation between the scattering data of the SP and the solution (recovered potentials) of the associated IBVP (Cauchy problem) for NLEEs with the given initial and boundary conditions. In Section 10.2, we prove that these ISPs are set up appropriately for constructing BTs.
We propose a Definition of BTs for the considered NLEEs on a half-line.

Definition 10.2. *A one-to-one correspondence relation between the constructed common solution of both equations in the Lax pair of the scattering problem (SP) on a half-line and the solution (recovered potential functions in equations in the Lax par) of the associated IBVP (Cauchy problem) for NLEEs satisfying the given initial and boundary conditions determines a BT, which transforms the common solution into the solution of the associated IBVP (Cauchy problem) for NLEEs, and conversely.*

Note that our proposed definition 10.2 simultaneously satisfies the known definition of BTs for the solution of NLEEs defined in the whole space (see [5, 29]), and the definition of the solution of the boundary problem for a system of two linear equations on the half-line with the boundary condition at $x = 0$.

Denote the first and second operator equations in the Lax pair by U and V equations, respectively. Based on Definition 10.2 and using mentioned results of solving IBVPs (Cauchy problems) for NLEEs on the half-line, we present in this subsection a way of obtaining the BTs from the U-V equations on a half-line.

10.2.1 The BT for the solution of the IBVP for the system of NLEEs (2.5)

The SP for system (2.8) with the boundary condition (2.9) has been solved in Chapter 2. The time-dependence of the solution of system (2.8) is given by system (2.11). The potentials $q_1(x,t), q_2(x,t), 0 \leq x < \infty, -\infty < t < \infty$ in (2.8) and (2.11) are isospectral and continuous functions possessing second partial continues derivatives with respect to x and first partial continuous derivatives with respect to t and satisfying estimate (2.10). Then

(1) By Lemma 2.1, the system (2.8) and (2.11) are compatible, i.e., $u_{xt} = u_{tx}$, and the Lax compatibility condition (2.15) for the pair of systems (2.8) and (2.11) is derived. The matrix equation (2.15) is equivalent to the system of NLEEs (2.5);

(2) By Theorem 2.1 there exists a one-to-one correspondence relation between every scattering data set (2.50) of the SP generated by systems (2.8), (2.11) and the solution (the recovered potential matrix in (2.8), (2.11)) of the IBVP (2.5)–(2.6)–(2.7).

Let $u = (u_1(x,\lambda;t), u_2(x,\lambda;t))$ be a solution of the SP (2.8)–(2.9). Then this solution satisfies equation (2.11). Indeed, by the assertion (1):

$$u_{xt} = (Bu)_t = u_{tx} = (\tilde{V}u)_x, \quad B = i\lambda J + Q,$$
$$J = \begin{pmatrix} -1 & 0 \\ 0 & 1 \end{pmatrix}, Q = \begin{pmatrix} 0 & ic_1(x,t) \\ -ic_2(x,t) & 0 \end{pmatrix} = \begin{pmatrix} 0 & q_1(x,t) \\ q_2(x,t) & 0 \end{pmatrix},$$

where the matrix $\tilde{V}$ is determined by (2.12).

The function $u_t = \tilde{V}u$ is an antiderivative with respect to variable x of the function u_{tx} from the given compatible equality: $u_{xt} = u_{tx}$. Hence, the solution u of equation (2.8) satisfies equation (2.11).

Conversely, if u is a solution of (2.11), then the function $u_x = Bu$ is an antiderivative with respect to variable t of the function $u_{tx} = u_{xt} = (Bu)_t$. Hence, the function u satisfies equation (2.8).

It remains to verify that the boundary condition (2.9) is satisfied. Indeed, by the relations (2.33) and (2.34) between the boundary values of functions $q_j(x,t)$ and their first x-derivatives $q_{jx}(x,t), j = 1,2$ at $x = 0$, from equation (2.11) at $x = 0$, we obtain: $u_{1t}(0,\lambda;t) = u_{2t}(0,\lambda;t) = 0$ for any $t \in (-\infty,\infty)$, therefore $u_1(0,\lambda;t) = u_2(0,\lambda;t) =$ constant for any $t \in (-\infty,\infty)$; i.e., the boundary condition (2.9) is satisfied. The following lemma is deduced from the above discussion.

Lemma 10.2. *(1) The vector function* $u = (u_1(x,\lambda;t), u_2(x,\lambda;t))$ *in systems* (2.8) *and* (2.11) *is the common solution of these systems with boundary condition* (2.9).

(2) Every scattering data set of the type (2.50) *of the SP generated by systems* (2.8), (2.11) *with the recovered potential matrix* (2.56) *and the boundary condition* (2.9) *corresponds to an unique solution of the IBVP* (2.5)–(2.6)–(2.7). *This one-to-one correspondence relation is a BT, which transforms the common solution of both systems* (2.8), (2.11) *into the solution* (2.56) *of system* (2.5) *with the initial and boundary conditions* (2.6) *and* (2.7), *and conversely (see Figure 10.3).*

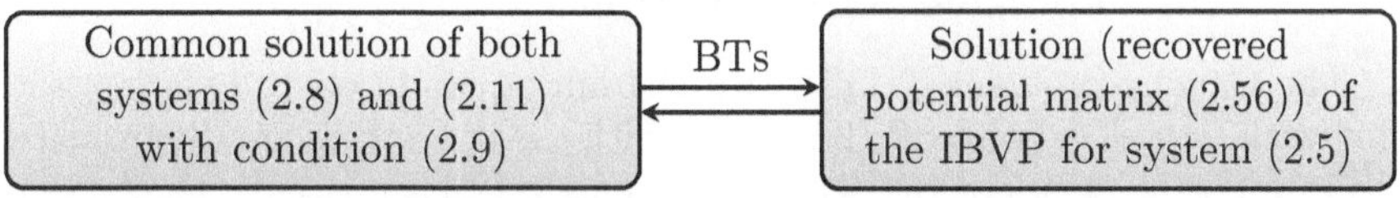

FIGURE 10.3: The BTs between the common solution of both systems (2.8), (2.11) with condition (2.9) and the solution of IBVP (2.5)–(2.6)–(2.7), and conversely.

10.2.2 The BT for the solution of the IBVP for the attractive NLS equation (2.60)

We consider a case of special interest of system (2.5), when the potentials satisfy condition (2.59):

$$c_1(x,t) = -\overline{c_2(x,t)}.$$

Under this condition, the system (2.5) reduces to the attractive NLS equation (2.60). Therefore, the solution of the IBVP (2.60)–(2.61)–(2.62) can be obtained from the solution of the IBVP (2.5)–(2.6)–(2.7). By Theorem 2.2 and Lemma 10.2:

(1) The vector-function $u = (u_1(x,y;t), u_2(x,y;t))$ in systems (2.8) and (2.11) with potentials satisfying condition (2.59) is the common solution of these system with boundary condition (2.9);

(2) Every scattering data set of the type (2.50) of the SP generated by systems (2.8) and (2.11) with the reconstructed potential matrix with the entries defined by (2.65) corresponds to an unique solution of the IBVP (2.60)–(2.61)–(2.62). This one-to-one correspondence relation is a BT, which transforms the common solution into the solution of the IBVP for the attractive NLS equation (2.60), and conversely (see Figure 10.4).

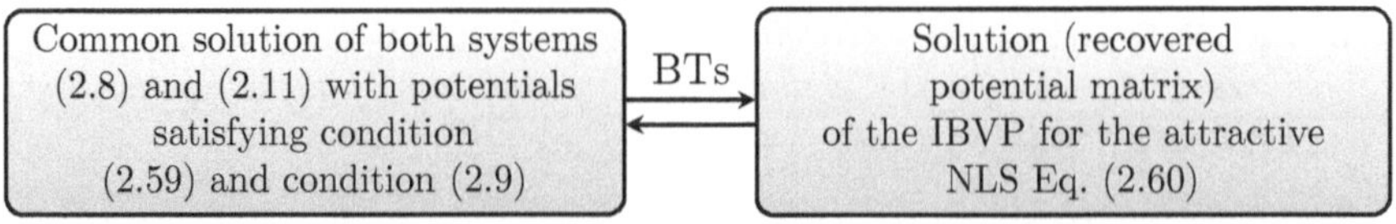

FIGURE 10.4: The BTs between the common solution of both systems (2.8) and (2.11) satisfying condition (2.59) and the solution of IBVP (2.60)–(2.61)–(2.62), and conversely.

10.2.3 The BT for the solution of the Cauchy problem for the repulsive NLS equation (2.113)

The Cauchy problem (2.113) with the initial condition (2.114) is associated with the self-adjoint SP (2.8)–(2.9) with the potential self-adjoint matrix (2.116). The elements of the self-adjoint matrix (2.116) satisfy the condition (2.112), i.e.,

$$c_1(x,t) = \overline{c_2(x,t)}, \quad 0 \le x < \infty, -\infty < t < \infty.$$

Under the reduction (2.112) and estimate (2.115), the SP (2.8)–(2.9) with the potential self-adjoint matrix (2.116) has no discrete spectrum. Then

(1) The vector-function $u = (u_1(x,\lambda;t), u_2(x,\lambda;t))$ in systems (2.8) and (2.11) with the potential self-adjoint matrix (2.116) and condition (2.9) is the common solution of these systems with condition (2.9);

(2) By Theorem 1.2 and Lemma 2.2 there exists the scattering function $S(x,\lambda;t)$, which obeys the evolutionary equations (2.117);

(3) By Theorem 2.5, there exists the one-to-one correspondence relation between the scattering function $S(x,\lambda;t)$ and the reconstructed potential matrix (2.118). By Lemma 10.2 this correspondence relation is a BT, which transforms the common of both systems (2.8) and (2.11) with conditions (2.9) and (2.112) into the solution (recovered potential matrix (2.118)) of the repulsive NLS equation (2.113) of the Cauchy problem (2.113) with the initial condition (2.114), and conversely (see Figure 10.5).

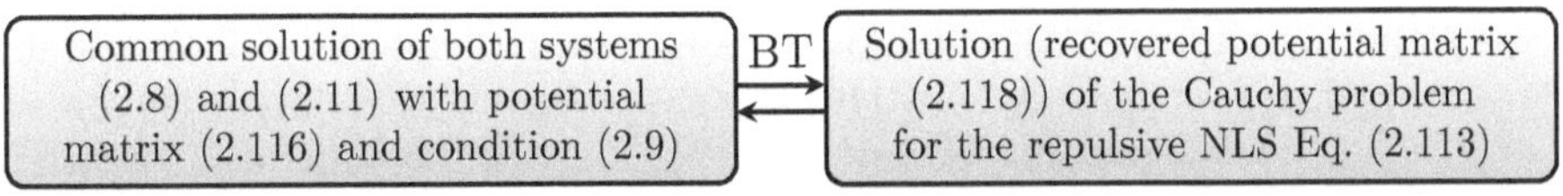

FIGURE 10.5: The BTs between the common solution of both systems (2.8) and (2.11) with conditions (2.9) and (2.112) and the solution of the Cauchy problem (2.113) with the initial condition (2.114).

10.2.4 The BTs for the solutions of the IBVP for the mKdV equations (3.59) and (3.60)

The SP for the operator equation (3.42):

$$u_x = Bu, \quad B = i\lambda J_1 + iQ, \quad 0 \le x < \infty, \quad -\infty < t < \infty,$$
$$J_1 = \begin{pmatrix} 1 & 0 \\ 0 & -1 \end{pmatrix}, Q(x,t) = \begin{pmatrix} 0 & -ic_1(x,t) \\ ic_2(x,t) & 0 \end{pmatrix} = \begin{pmatrix} 0 & q_1(x,t) \\ q_2(x,t) & 0 \end{pmatrix},$$

with the boundary condition (3.43) at the origin $x = 0$:

$$u_1(0, \lambda; t) = u_2(0, \lambda; t),$$

has been solved in Chapter 3. The time-dependence of the solution of this SP is given by system (3.45):

$$u_t = \tilde{V}u,$$

where the matrix $\tilde{V} = \tilde{V}(x,t)$ is determined by (3.46).

The potential functions in systems (3.42) and (3.45) are continuous smooth functions satisfying estimate (3.44). Therefore,

(1) By Lemma 3.12 the systems (3.42) and (3.45) are compatible, i.e., $(u_1(x,\lambda;t), u_2(x,\lambda;t))_{xt} = (u_1(x,\lambda;t), u_2(x,\lambda;t))_{tx}$, and the potential functions $q_1(x,t), q_2(x,t)$ in (3.42) and (3.45) vary with time as the solution of the cubic NLEEs (3.41). The matrix equation for the compatibility condition for systems (3.42) and (3.45) is equivalent to the system of the cubic NLEEs (3.41).

We adopt the procedure for a special case of the relationship (3.58) between the potential functions $q_1(x,t)$ and $q_2(x,t)$ in systems (3.42) and (3.45), a case of special interest, i.e.,

$$q_1(x,t) = q_2(x,t) \text{ or } q_1(x,t) = -q_2(x,t).$$

Under the reduction (3.58), the system of cubic NLEEs (3.41) reduces to the mKdV equation (3.59) or to the mKdV equation (3.60). Due to (3.58), (3.67), (3.68), the relations (3.69), (3.70) between boundary values of $q_1(x,t), q_2(x,t)$ and their first and second x-derivatives evaluated ad $x = 0$ are derived.

(2) By the relations (3.69) and (3.70) from system (3.45) at $x = 0$, we obtain $u_{1t}(0, \lambda; t) = u_{2t}(0, \lambda; t) = 0$ for any $t \in (-\infty, \infty)$, therefore $u_1(0, \lambda; t) = u_2(0, \lambda; t) =$ constant for any $t \in (-\infty, \infty)$, i.e., the boundary condition (3.43) is satisfied.

Let $(u_1(x, \lambda; t), u_2(x, \lambda; t))$ be a solution of the SP (3.42)–(3.43), then by the compatibility condition for systems (3.42) and (3.45):

$$u_{xt} = (Bu)_t = u_{tx} = (\tilde{V}u)_x.$$

The function $u_t = \tilde{V}u$, is an antiderivative with respect to variable x of the function u_{tx} from the given equality $u_{xt} = u_{tx}$. Hence, the solution u of the SP (3.42)–(3.43) satisfies the system (3.45) with the boundary condition (3.43).

Conversely, if u is a solution of the SP (3.45), (3.43), then the function $u_x = Bu$ is an antiderivative with respect to variable to t of the function $u_{xt} = (Bu)_t$. Hence, u is the solution of the SP (3.42)–(3.43) and u is the common solution of both systems (3.42) and (3.45) with the boundary condition (3.43).

(3) By Theorem 3.2, every scattering data set $s_1(t), (s_2(t))$ of the type (3.88) corresponds to a unique solution (reconstructed potential matrix (3.94) ((3.95))) of the IBVP (3.59), (3.61)–(3.62) ((3.60)–(3.61)–(3.62)), respectively.

The following lemma is deduced form the above assertions (1)–(3).

Lemma 10.3. *(1) The vector function* $u = (u_1(x, \lambda; t), u_2(x, \lambda; t))$ *in system* (3.42), (3.45) *is the common solution of these systems with the boundary condition* (3.43)*;*

(2) The one-to-one correspondence relation between the scattering data set $s_1(t)$, $(s_2(t))$ *of the type* (3.88) *and the unique solution (recovered potential matrix* (3.94) ((3.95))*) of the IBVP* (3.59), (3.61)–(3.62) ((3.60)–(3.61)–(3.62)), *respectively is a BT, which transforms the common solution of systems* (3.42), (3.45) *with the boundary condition* (3.43) *and the potential function* $q_1(x,t) = q_2(x,t)$ $(q_1(x,t) = -q_2(x,t))$ *into the solution (recovered potential matrix* (3.94), ((3.95))*) of the mKV equation* (3.59) ((3.60)), *and conversely (see Figure 10.6).*

10.2.5 The Miura's transformations (3.103) and (3.107)

By using the simple substituting

$$\psi_1(x, \lambda; t) = \frac{1}{2}[u_1(x, \lambda; t) + u_2(x, \lambda; t)], \quad \psi_2(x, \lambda; t) = \frac{1}{2i}[u_2(x, \lambda; t) - u_1(x, \lambda; t)],$$

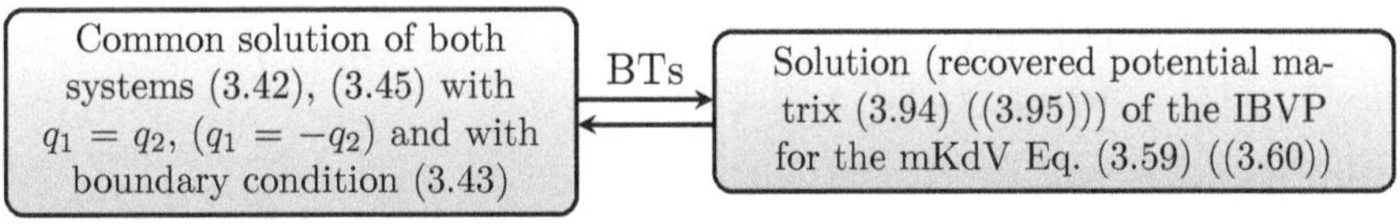

FIGURE 10.6: The BTs between the common solution of both systems (3.42), (3.45) with $q_1 = q_2$ $(q_1 = -q_2)$ and with boundary condition (3.43) and the solution of IBVP for equations (3.59) ((3.60)), and conversely.

we can reduce the system (3.42) to the system (3.99). Further, from this system using the potential matrix (3.86) for the mKdV equation (3.59), we derive equations (3.100) and (3.101). Substituting equation (3.101) into equation (3.100), we obtain the Schrödinger equation (3.102):

$$-\psi_{xx}(x,\lambda;t) + r(x,t)\psi(x,\lambda;t) = \lambda^2\psi(x,\lambda;t), \;\; 0 \le x < \infty, \;\; -\infty < t < \infty,$$

where the potential function $r(x,t)$ is defined by the non-linear Miura's transformation (3.103) [1]:

$$r(x,t) = -c^2(x,t) - ic_x(x,t).$$

Since the solution $(u_1(x,\lambda;t), u_2(x,\lambda;t))$ of system (3.42) satisfies the boundary condition (3.43), then the solution of the Schrödinger equation satisfies the zero boundary condition (3.104) at $x = 0$:

$$\psi(0,\lambda;t) = \frac{1}{2}\left[u_2(0,\lambda;t) - u_1(0,\lambda;t)\right] = 0.$$

As shown in Subsection 3.2.6 of Chapter 3, the SP (3.102), (3.104) is related with the KdV equation (3.98)

$$r_t - 6rr_x + r_{xxx} = 0, \;\; r = r(x,t), \;\; 0 \le x < \infty, \;\; -\infty < t < \infty.$$

By Theorem 3.3, the Miura's transformation (3.103) transforms the solution $c(x,t)$ of the mKdV equation (3.59) into the solution of the above KdV equation (3.98):

If $c_1(x,t) = -c_2(x,t) = -i\mu(x,t)$, where $\mu(x,t)$ is a real-valued function satisfying the mKdV equation (3.60), then the Miura's transformation (3.107):

$$r(x,t) = -(i\mu(x,t))^2 - i(i\mu(x,t))_x = \mu^2(x,t) + \mu_x(x,t),$$

transforms the real solution $\mu(x,t)$ of the mKdV equation (3.60) into the real solution of the KdV equation (3.98).

Hence, the Miura's transformation $r(x,t)$ is the BT, which transforms the solution of the mKdV equation (3.59) ((3.60)) into the solution of the KdV equation (3.98).

10.2.6 The BT for the solution of the Cauchy problem for the cubic NLEEs (3.144)

In this subsection we turn attention to the reduction: $q_1(x,t) = -\overline{q_2(x,t)}$. Under this reduction the system of cubic NLEEs (3.41) reduces to the cubic non-linear equation (3.144):

$$q_t + q_{xxx} - 6\,|q|^2\, q_x = 0, \;\; 0 \le x < \infty, \;\; -\infty < t < \infty.$$

Due to (3.42), the reduction $q_1(x,t) = -\overline{q_2(x,t)} = q(x,t)$ is equivalent to the condition (3.145):

$$q(x,t) = -c_1(x,t) = -\overline{c_2(x,t)}, \;\; 0 \le x < \infty, \;\; -\infty < t < \infty.$$

The Cauchy problem for the cubic NLEEs (3.144) with known initial condition (3.146) is associated with the self-adjoint SP (3.42)–(3.43). In this case, the potential matrix in system (3.42) is self-adjoint matrix with the potential elements satisfying condition (3.145) and estimate (3.147). This SP has no discrete spectrum. Then

(1) By Lemma 10.3, the vector function $u(x,\lambda;t)$ in systems (3.42) and (3.45) with the potential self-adjoint matrix (3.145) and condition (3.43) is the common solution of these systems with condition (3.43);

(2) By Lemma 3.12 and Theorem 3.5, the scattering function $S(x,\lambda;t)$ satisfies the evolutionary equation (3.150). By Theorem 3.6, there exits a one-to-one correspondence relation between the scattering function $S(x,\lambda;t)$ and the reconstructed potential matrix (3.152). This correspondence relation is a BT, which transforms the common solution of both systems (3.42), (3.45) with condition (3.43) and the potential function $q(x,t)$ satisfying condition (3.145) into the solution (recovered potential matrix) of the cubic NLEE (3.144), and conversely (see Figure 10.7).

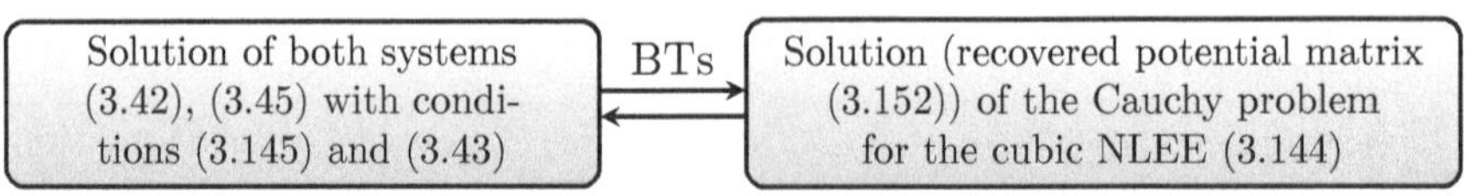

FIGURE 10.7: The BTs between the common solution of both systems (3.42), (3.45) with conditions (3.145) and (3.43) into the solution of Cauchy problem for the cubic NLEE (3.144) with known initial condition (3.146), and conversely.

10.2.7 The BT for the solution of the IBVP for the sine-Gordon equation (4.14)

The Dirichlet IBVP for the sine-Gordon (sG) equation (4.14) is associated with the SP for system (4.18):

$$u_x = U_1 u, \quad u = (u_1(x,\lambda;t), u_2(x,\lambda;t)), \quad 0 \le x < \infty, \quad 0 \le t \le T,$$

with the boundary condition (4.19) at $x = 0$:

$$u_1(0,\lambda;t) = u_2(0,\lambda;t),$$

where $U_1 = i\lambda J_1 + Q$, $J_1 = \begin{pmatrix} 1 & 0 \\ 0 & -1 \end{pmatrix}$,

$$Q = \begin{pmatrix} 0 & -ic_1(x,t) \\ ic_2(x,t) & 0 \end{pmatrix} = \frac{i}{2}\begin{pmatrix} 0 & v_x(x,t) \\ v_x(x,t) & 0 \end{pmatrix}.$$

The potential function $v_x(x,t)$ belongs to the class of real-valued continuous functions possessing continuous partial derivative with respect to t and satisfying estimate (4.17). The time-dependence of eigenfunctions of the SP (4.18)–(4.19) obeys the system (4.21):

$$u_t = V_1 u, \text{ where the matrix } V_1 \text{ is determined by (4.21).}$$

The following assertions are obtained from the results of solving the Dirichlet IBVP for the sG equation (4.84) and Lemma 10.3 for the considered case.

(1) By Lemma 4.1, the system (4.18) and (4.21) are compatible, i.e.,

$$u_{xt} = (U_{1t} + U_1 V_1)u = u_{tx} = (V_{1x} + V_1 U_1)u.$$

and the compatibility condition for the matrix equations (4.18) and (4.21) is derived by the matrix equation (4.33), which is equivalent to the sG equation (4.14).

(2) By Theorem 4.1, the time-dependent scattering data (SD) set (4.75) of the SP (4.18)–(4.19) is constructed. By Theorem 4.2, every SD set of the form (4.75) corresponds a unique solution (recovered potential matrix (4.80)) of the IBVP (4.14)–(4.15)–(4.16).

(3) The potential matrix in system (4.18) is determined by (4.20). It is a special case of the potential matrix in system (3.42), then by Lemma 10.3, the vector function $u = (u_1(x,\lambda;t), u_2(x,\lambda;t))$ is the common solution of both systems (4.18), (4.21) with the potential matrix (4.20) and boundary condition (4.19). The one-to-one correspondence relation between the SD set (4.75) and the solution (recovered potential matrix (4.80)) of the sG equation (4.14) is a BT, which transforms the common solution of both systems (4.18) and (4.21) with condition (4.19) into the solution of the sG equation (4.14), and conversely (see Figure 10.8).

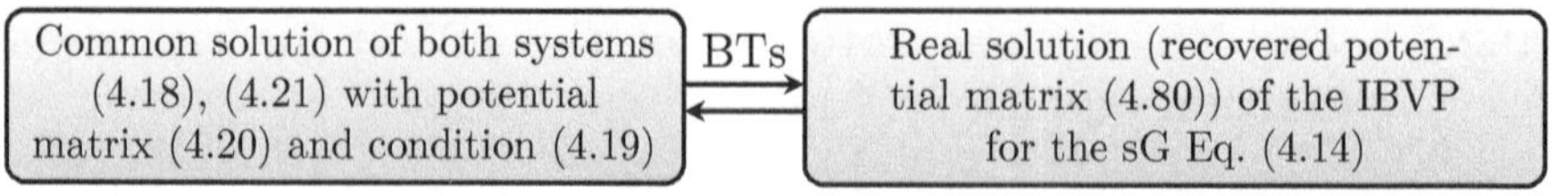

FIGURE 10.8: The BTs between the common solution of both systems (4.18), (4.21) with potential matrix (4.20) and condition (4.19) and the real solution of IBVP for the sG equation (4.14), and conversely.

10.2.8 The BT for the solution of the IBVP for the sinh-Gordon equation (4.84)

The Dirichlet IBVP for the sinh-Gordon (shG) equation (4.84) is associated with the SP for the system (4.88):

$$u_x = U_2 u, \quad u = (u_1(x,\lambda;t), u_2(x,\lambda;t)), \quad 0 \le x < \infty, \quad 0 \le t \le T,$$

with the boundary condition (4.89) at $x = 0$:

$$u_1(0,\lambda;t) = u_2(0,\lambda;t),$$

where $U_2 = i\lambda J_1 + iQ$, $J_1 = \begin{pmatrix} 1 & 0 \\ 0 & -1 \end{pmatrix}$, $Q = Q(x,t) = -J_1 C(x,t)$, $C = C(x,t) = \begin{pmatrix} 0 & c_1(x,t) \\ c_2(x,t) & 0 \end{pmatrix} = \begin{pmatrix} 0 & -\frac{1}{2}v_x(x,t) \\ -\frac{1}{2}v_x(x,t) & 0 \end{pmatrix}$, $Q(x,t) = \begin{pmatrix} 0 & -c_1(x,t) \\ c_2(x,t) & 0 \end{pmatrix} = \frac{1}{2}\begin{pmatrix} 0 & v_x(x,t) \\ -v_x(x,t) & 0 \end{pmatrix}$.

The potential function $v_x(x,t)$ in the matrix belongs to the class of real-valued functions possessing continuous partial derivative with respect to t and satisfying estimate (4.87). The potential matrix C is self-adjoint. Under the reduction of the matrix C and estimate (4.87), the self-adjoint SP (4.88)–(4.89) has no discrete spectrum. The time-dependence of the solution of this SP is given by system (4.95): $u_t = V_2 u$, where the matrix V_2 is given by (4.95).

The following assertions are deduced from the results of solving the IBVP (4.84)–(4.86) and Lemma 10.3 for the considered case.

(1) By Lemma 4.2, the system (4.88) and (4.95) are compatible, i.e.,

$$u_{xt} = (U_{2t} + U_2 V_2)u = u_{tx} = (V_{2x} + V_2 U_2)u,$$

and the compatibility condition for the matrix equations (4.88) and (4.95) is derived by the matrix equation (4.99), which is equivalent to the shG equation (4.84).

(2) By Theorem 4.3, the scattering function $S(x,\lambda;t)$ of the self-adjoint SP (4.88)–(4.89) is defined by the factorization (4.121). The correspondence relation between $S(x,\lambda;t)$ and the potential function $v_x(x,t)$ is one-to-one. Hence, the anti-derivative of $v_x(x,t)$ is the function $v(x,t)$ which

is found uniquely from the initial and boundary conditions (4.85) and (4.86).

(3) The potential matrix C in matrix equations (4.88) and (4.95) is a special case of the potential matrix in system (3.42), then Lemma 10.3. The vector-function $u = (u_1(x, \lambda; t), u_2(x, \lambda; t))$ is the common solution of both system (4.88) and (4.95) with the potential matrix (4.90) and boundary condition (4.89). The one-to-one correspondence relation between the scattering function $S(x, \lambda; t)$ and the recovered potential matrix (4.124) is a BT, which transforms the common solution of both systems (4.88) and (4.95) with condition (4.89) into the solution of the shG equation (4.84), and conversely (see Figure 10.9).

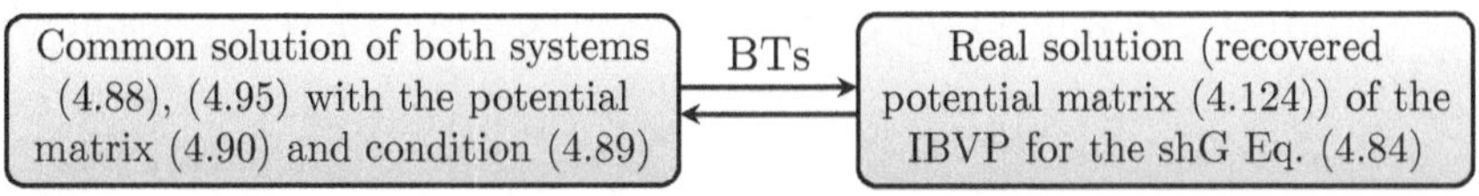

FIGURE 10.9: The BTs between the common solution of both systems (4.88), (4.95) with the potential matrix (4.90) and condition (4.89) and the solution of IBVP for the sG equation (4.84), and conversely.

10.2.9 The BT for the solution of the IBVP for the KdV equation (6.48)

The IBVP for the KdV equation (6.48):

$$p_t - 6pp_x + p_{xxx} = 0, \quad x \geq 0, \quad t \geq 0,$$

with the initial and boundary conditions (6.49) and (6.50) is associated with the Schrödinger operator equation (6.52):

$$Ly = \lambda y, \quad L = -\frac{d^2}{dx^2} + p, \quad \lambda = \rho^2,$$

with the boundary condition (6.53) at $x = 0$: $y(\rho, 0; t) = 0$ for $t \geq 0$.

The time-dependence of the solution of the SP (6.52)–(6.53) is given by the operator equation (6.54):

$$y_t = -Ay, \quad A = 4\frac{d^3}{dx^3} - 3p_x - 6p\frac{d}{dx} - 4i\rho^3.$$

The SP (6.52)–(6.53) has been studied in Chapter 6 in the class of real-valued functions satisfying estimate (6.51) and possessing the third-order partial continuous derivative with respect to x and first-order derivative with respect to t. By Lemma 6.9, the equations (6.52) and (6.54) are compatible, i.e.,

$y_{xxt} = y_{txx}$ and the compatibility condition (6.63) for the equations (6.52) and (6.54) are derived. The equation (6.63) is equivalent to the KdV equation (6.48) in the class of isospectral potentials. We prove that the compatible equality: $y_{xxt} = y_{txx}$ for equations (6.52) and (6.54) produces a common solution of these equations with the boundary condition (6.53). Indeed, let y be the solution of the SP (6.52)–(6.53), then

$$y_{xxt} = y_{txx} = (-Ay)_{xx}.$$

The partial second-order continuous derivative with respect to x of the function $(-Ay)$ is the function y_{txx}, i.e., the function $(-Ay)$ satisfies equation (6.54): $y_t = -Ay$. Then $y_t(\rho, 0; t) = 0$ for any $t \geq 0$, and $y(\rho, 0; t) =$ constant for any $t \geq 0$. Due to condition (6.2): $y(\rho, 0; 0) = 0$, the solution $y(\rho, x; t)$ of equation (6.54) satisfies the boundary condition (6.53). Conversely, if the function y is the solution of the problem for the equation (6.54) with condition (6.53), then

$$y_{txx} = y_{xxt} = ((p(x,t) - L)y)_t$$

The function $(p(x,t)-\lambda)y$ is an anti-derivative with respect to t of the function y_{xxt}. The function $(p(x,t) - \lambda)y$ satisfies equation (6.52): $p(x,t)y - \lambda y = y_{xx}$. Then $y_{xx}(\rho, 0; t) = 0$ for any $t \geq 0$, and $y_x(\rho, 0; t) =$ constant for any $t \geq 0$. Due to condition (6.2): $y(\rho, 0; 0) = 0$, the solution of equation (6.52) satisfies condition (6.53). Hence, the function $y(\rho, x; t)$ is the common solution of both equations (6.52) and (6.54) with condition (6.53).

The Jost solution $e(\rho, x; t)$ of the Schorödinger equation (6.52) with boundary condition at infinity (6.66) is represented in the form (6.67), where the kernel $K(x, \xi; t)$ is real-valued function, satisfying equation (6.11) and estimate (6.6). The properties of the Jost solution $e(\rho, x; t)$ and its BVs $e(\rho; t)$ are listed in Lemma 6.10.

The normalization eigenfunction of the problem (6.52)–(6.53) is defined by (6.72):

$$\Omega(\rho, x; t) = \frac{2i\omega(\rho, x; t)}{e(\rho; t)}, \qquad \text{Im}\rho \geq 0, \quad x \geq 0,$$

where $\omega(\rho, x; t)$ is an entire function of ρ satisfying equation (6.52), with the boundary conditions (6.22) and estimate (6.23).
Due to (6.22) the normalization eigenfunction (6.72) satisfies the boundary conditions at $x = 0$ (6.73) and (6.74).

The normalization eigenfunction $\Omega(k, x; t)$ corresponding to the continuous spectrum of the SP (6.52)–(6.53) is represented in the form (6.80). The time-dependence of $\Omega(k, x; t)$ is derived by the evolutionary equation (6.82). Owing to conditions (6.73) and (6.74), the passage in equation (6.82) to the limit as $x \to 0$ leads to the evolution equation (6.83) for the scattering function $S(k; t)$:

$$S_t(k; t) = 8ik^3 S(k; t) - 4ik \frac{2k^2 + p(0,t)}{e^2(k; t)} \quad \text{for } k \neq 0, \ \ t \geq 0.$$

We put

$$e(k;t) = |e(k;t)|e^{i\delta(k;t)}, \quad |e(k;t)| \neq 0 \text{ for } k \neq 0, \ t > 0.$$

where the scattering phase $\delta(k;t)$ possesses the property: $\delta(k;t) = \delta(-k;t)$.

The BVs $e(\rho;t)$ evaluated at $x = 0$ of the Jost solution give the complete characterization of the spectrum of the operator $L(t)$. The time-dependence of the scattering function is determined by the evolution equation (6.83), which contains the unknown BVs $e(k;t)$. Therefore, the main difficulty in solving the IBVP (6.48)–(6.49)–(6.50) lies in the determination of these unknown BVs in terms of the known conditions (6.49) and (6.50). We overcome the difficulty by considering the Sturm–Liouville scattering problem for equation (6.52) on the interval $(0, b)$, (b is a large positive number), with the boundary conditions (6.73) at $x = 0$ and the zero boundary condition (6.85) at $x = b$:

$$\Omega(k, b;t) = 0 \text{ for } k \neq 0 \text{ and } t \geq 0.$$

We regard the Sturm–Liouville problem for equation (6.52) with boundary conditions (6.73) and (6.85) as a linear problem associated with the IBVP (6.48)–(6.49)–(6.50). Theorem 6.6 proves that the potential in equation (6.52) is uniquely reconstructed from the scattering data of the Sturm-Lioville (6.52), (6.73), (6.85) and is the solution of the IBVP (6.48)–(6.49)–(6.50). In order to prove this theorem, we need to derive the time-dependence of the scattering data. The normalization eigenfunction $\Omega(k, x;t)$ of equation (6.52) with boundary condition (6.53) is represented by the formula (6.80). Suppose that the given eigenfunction $\Omega(k, x;t)$ of equation (6.52) with boundary conditions (6.73) at $x = 0$ satisfies one more zero boundary condition (6.85) at $x = b$. Then the equation (6.85) is determined as the zero boundary condition at $x = b$ of the normalization eigenfunction $\Omega(k, x;t)$ of the Sturm–Liouville problem (6.52), (6.73), (6.85), wherein t enters equation (6.85) as a parameter and $\Omega(k, b;t)$ is the function of k. Therefore, the positive eigenvalues of this spectral Sturm–Liouville problem in essence are the roots of equation (6.85). Since the improper integral in (6.80) is regularly convergent, then the asymptotic form of the normalization eigenfunction (6.80) as $b \to \infty$ can be written as follows:

$$\Omega(k, b;t) = -2ie^{i\delta(k;t)} \sin(kb - \delta(k;t)) + o(1) = 0 \quad (b \to \infty),$$

whence it follows that

$$\sin(kb - \delta(k;t)) = o(1) \ (b \to \infty) \text{ for } k \neq 0, \ t \geq 0.$$

The roots of above equation are determined approximately by the asymptotic equation:

$$kb - \delta(k;t) = \arcsin(o(1)) = j\pi + o(1) \ (b \to \infty) \text{ for } k \neq 0, \ t \geq 0, \quad j = 0, 1, 2, ...$$

It follows from the above asymptotic equation that the scattering phase $\delta(k;t)$ for the Sturm–Liouville problem (6.52), (6.73), (6.85) is approximately time-independent, i.e.,

$$\delta_t(k;t) = o(1) \ \ (b \to \infty) \ \text{ for } \ k \neq 0, \ \ t \geq 0.$$

The asymptotic relation between $\delta(k;t)$ and $\delta(k)$ is deduced from (6.87) and (6.88):

$$\delta(k;t) = \delta(k) + o(1) = kb - j\pi + o(1) \ \ (b \to \infty) \ \text{ for } \ k \neq 0, \ \ t \geq 0, \\ j = 0, 1, 2, ...$$

Owing to (6.89) the scattering function $S(k;t)$ given by (6.78) approximately tends to $S(k) = S(k;0)$ as $b \to \infty$:

$$S(k;t) = e^{-2i\delta(k,t)} = e^{-2i[\delta(k)+o(1)]} = S(k)[1 + o(1)], (b \to \infty) \text{ for } k \neq 0, t \geq 0,$$

where $\delta(k) = kb - j\pi, j = 0, 1, 2, ...$

Thus, the scattering function for the considered Sturm–Liouville problem is approximately time-independent. The positive eigenvalues of this Sturm–Liouville problem are found approximately by (6.87) and (6.89):

$$\lambda_j = k_j^2 = \left(\frac{\delta(k) + j\pi}{b}\right)^2 + \left(\frac{o(1)}{b}\right)^2 \ \ (b \to \infty) \ \text{ for } \ k \neq 0, \ \ j = 0, 1, 2, ...$$

Owing to (6.89) and (6.94), we obtain the asymptotic formulas (6.95) for finding the unknown time-dependent BVs:

$$e(k,t) = |e(k;t)|e^{i\delta(k;t)} = |e(k;t)|e^{i[\delta(k)+o(1)]}, \ \ (b \to \infty),$$

$$|e(k;t)| = \sqrt{1 + \frac{p(0,t)}{2k^2}} + o(1) \ \ (b \to \infty) \ \text{ for } \ k \neq 0, \ \ t \geq 0,$$

subject to $1 + \frac{p(0,t)}{k^2} \geq 0$.

In the formulas (6.95) the scattering phase $\delta(k)$ is calculated from the known initial condition (6.49), therefore the unknown BV $e(k;t)$ is determined by (6.95) in terms of the known initial and boundary conditions (6.49) and (6.50).

By Lemma 6.10 the BV $e(\rho;t)$ has the simple zeros on the imaginary axis at the points: $\rho = i\mu_j, \ \ \mu_j > 0$:

$$e(i\mu_j, 0; t) = e(i\mu_j; t) = 0, \ \ j = 1, ..., N.$$

According to (6.28), the normalization multipliers $m_j(t)$ are calculated by the formula:

$$m_j^2(t) = \frac{2\mu_j}{ie_x(i\mu_j;t)\dot{e}(i\mu_j;t)}, \quad j = 1, ..., N,$$

where $m_j^2(0) = m_j^2$ and m_j^{-2} is defined by the formula (6.28) from the known initial condition (6.49). The functions $\dot{e}(i\mu_j;t)$ and $e_t(i\mu_j;t)$ are determined by the bounded analytic continuation $e(\rho;t)$ into the upper half-plane $\text{Im}\rho > 0$. This continuation is obtained from the calculated BV (6.95).

The scattering function $S(k;t)$ defined by (6.90) is approximately time-independent. This function and the simple zeros $i\mu_j, \mu_j > 0, j = 1, ..., N$, of $e(\rho, t)$, the normalization multipliers $m_j(t)$ characterize the Sturm–Liouville scattering problem (6.52), (6.73), (6.85) on the continuous and discrete spectrum, respectively. Thereby, the time-dependent scattering data set $s(t)$ for this problem is constructed:

$$s(t) = \{S(k)[1+o(1)], (b \to \infty), -\infty < k < \infty; i\mu_j, \mu_j > 0; m_j(t), j = 1, ..., N\},$$

where the quantities of the constructed set $s(t)$ for every $t \geq 0$ satisfy the conditions (I) and (II) enumerated in the Theorem 6.4.

By Theorem 6.6 the data set $s(t)$ of the type (6.102) constructed from the given conditions (6.49) and (6.50) is the scattering data of the Sturm–Liouville SP (6.52), (6.73), (6.85) with the recovered potential (6.108). There exists a one-to-one correspondence between every data set $s(t)$ of the type (6.102) and the recovered potential (6.108). The following lemma is deduced from the above discussion.

Lemma 10.4. *(1) The function* $y = y(\rho, x;t)$ *in systems* (6.52) *and* (6.54) *is the common solution of these systems with the boundary condition* (6.53).

(2) The one-to-one correspondence relation between every data set $s(t)$ *of the type* (6.102) *and the recovered potential* (6.108) *is a BT, which transforms the common solution of both equations* (6.52) *and* (6.54) *with the boundary condition* (6.53) *into the solution (recovered potential* (6.108)*) of the IBVP* (6.48)–(6.49)–(6.50), *and conversely (see Figure 10.10).*

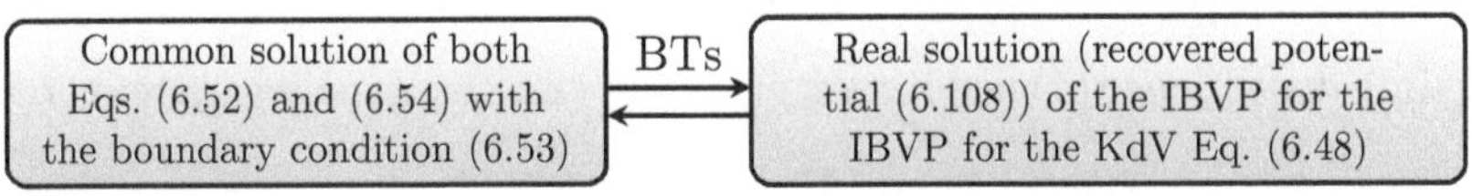

FIGURE 10.10: The BTs between the common solution of both equations (6.52), (6.54) with the boundary condition (6.53) and the solution of IBVP for equation (6.48), and conversely.

10.3 The BT between a common solution of both systems (7.70), (7.71) and the solution of the IBVP for KdV equation (7.1) with a negative dispersive coefficient

The IBVP for the KdV equation (7.1):

$$p_t - p_{xxx} + 6_{pp_x} = 0, \;\; x > 0, \;\; t > 0, \tag{10.5}$$

with the boundary and initial conditions:

$$p_{|x=0} = 0, \;\; p_{xx|x=0} = 0, \tag{10.6}$$

$$p_{|t=0} = p(x), \;\; p(x,t) \to 0, \text{ as } x \to \infty \tag{10.7}$$

are considered in the class S of potential real-valued functions satisfying Condition I and II in Chapter 7.

The boundary condition (10.6) is a special type that agrees with the main integrability criteria.

This subsection aims to construct a BT from both linear equations in the Lax pair and to establish the interrelations between a common solution of both equations and the solution of the associated IBVP for NLEEs on the half-line.

10.3.1 The SP associated with the IBVP for KdV equation (10.5)

The IBVP (10.5)–(10.7) is associated with the SP for system:

$$Y_x = UY, \;\; U = \begin{pmatrix} 0 & 1 \\ p - \lambda & 0 \end{pmatrix}, \; Y = (y_1(k,x,t), y_2(k,x,t)), \;\; 0 \le x < \infty, t > 0, \tag{10.8}$$

with the boundary condition at $x = 0$:

$$(y(k,x,t), y_x(k,x,t)) = (1, -ik) + o(1), \;\; (x \to 0), \quad \lambda = k^2. \tag{10.9}$$

The time-dependence of the solution of system (10.8) is given by system (7.71):

$$Y_t = VY, \tag{10.10}$$

where $V(k,x,t) = \begin{pmatrix} p_x(x,t) & -2p(x,t) - 4k^2 \\ p_{xx}(x,t) - (2p(x,t) + 4k^2)(p(x,t) - k^2) & -p_x(x,t) \end{pmatrix}$.

Putting $y = y_1$ and $y_2 = y_{1x}$, then the function y satisfies the Schrödinger equation (7.68):

$$-y_{xx} + p(x,t)y = k^2 y, \;\; k^2 = \lambda, \;\; 0 \le x < \infty, \;\; t > 0, \tag{10.11}$$

and the time-dependence of the solution y of equation (10.11) is given by the (7.69):

$$y_t = p_x\,(x,t)\,y - 2(p(x,t) + 2\lambda)y_x, \tag{10.12}$$

where the potential function $p(x,t)$ for $t > 0$ belongs to class S of isospectral potentials, therefore the equations (10.11) and (10.12) are compatible, i.e.,

$$y_{2tx} = y_{2xt} \text{ or } y_{1xxt} = y_{1xtx}, \tag{10.13}$$

where the equations (10.11) and (10.12) are equivalent to systems (10.8) and (10.10), respectively.

The compatible equalities (10.13) will be referred to as the Lax compatibility condition for equations (10.11) and (10.12). The KdV equation is identical to the equation generated by the operator $L = -\frac{d^2}{dx^2} + p(x,t)$ and the operator A given by $A = 4\frac{d^3}{dx^3} - 3\left(p\frac{d}{dx} + \frac{d}{dx}p\right)$. Indeed,

$$L_t + (LA - AL) = p_t + 6pp_x - p_{xxx} = 0, \;\; 0 \le x < \infty, \;\; t > 0,$$

where $p = p(x,t)$ is an isospectral potential belonging to the class S of all real-valued functions satisfying condition I and II.

For convenience, we pass from scalar L-A pair of equations (10.11) and (10.12) to the U-V pair of equations (10.8) and (10.10),

The matrix solutions of system (10.8) are constructed from the Jost solution $(e(k,x,t), e_x(k,x,t))$ and analytical solution $(\omega\,(k,x,t)\,, \omega_x(k,x,t))$:
$$E(k,x,t) = \begin{pmatrix} e(k,x,t) & e(-k,x,t) \\ e_x(k,x,t) & e_x(-k,x,t) \end{pmatrix}, \; W(k,x,t) = \begin{pmatrix} \omega(-k,x,t) & \omega(k,x,t) \\ \omega_x(-k,x,t) & \omega_x(k,x,t) \end{pmatrix}$$
with the condition for all real $k = \sqrt{\lambda}$:

$$E = T(k)\begin{pmatrix} e^{ikx} & 0 \\ 0 & e^{-ikx} \end{pmatrix} + o(1), \;\; (x \to \infty), \;\; T(k) = \begin{pmatrix} 1 & 1 \\ ik & -ik \end{pmatrix}, \tag{10.14}$$

$$W = T(k) + o(1), \;\; (x \to 0). \tag{10.15}$$

The Jost solution $e(k,x,t)$ of equation (10.11) satisfies the integral equation:

$$e\,(k,x,t) = e^{ikx} + \int_x^{\infty} \frac{\sin k(s-x)}{k} p(s,t)e(k,s,t)ds, \operatorname{Im} k \ge 0,$$

and can be represented as

$$e\,(k,x,t) = e^{ikx} + \int_x^{\infty} K(x,\xi,t)e^{ik\xi}d\xi, \operatorname{Im} k \ge 0,$$

where the function $K(x,\xi,t)$ is the kernel of the transformation operator and is determined from the integral equation of the Volterra type (7.13).

The analytical solution $\omega(k,x,t)$ of equation (10.11) satisfies the integral equation:

$$\omega\,(k,x,t) = e^{-ikx} + \int_0^{x} \frac{\sin k(x-s)}{k} p^-(s,t)\omega(k,s,t)ds, \operatorname{Im} k \ge 0,$$

and can be represented as

$$\omega(k,x,t)=\begin{cases} e^{-ikx}+\int_{-\infty}^{x} K^{-}(x,\xi,t)e^{-ik\xi}d\xi & \text{for } x>0 \\ e^{-ikx} & \text{for } x<0 \end{cases}$$

where $K^{-}(x,\xi,t)=0$ for $\xi \le x<0$ and $2\frac{d}{dx}K^{-}(x,x,t)=\begin{cases} p^{-}(x) & \text{for } x>0, \\ 0 & \text{for } x\le 0. \end{cases}$

In view of reality of the potential function $p(x,t)$, the kernel function $K(x,\xi,t)$ is real-valued, therefore

$$e(k,x,t)=\overline{e(-k,x,t)},\quad e_x(k,x,t)=\overline{e_x(-k,x,t)},$$
$$\omega(k,x,t)=\overline{\omega(-k,x,t)},\quad \omega_x(k,x,t)=\overline{\omega_x(-k,x,t)}.$$

The solution $e(k,x,t)$ and $\omega(k,x,t)$ of equations (10.11) admit from the real line an analytical continuous into the upper half-plane Im$k>0$. Since the Wronskian of solution do not depend on x, then

$$W\{\omega(-k,x,t),\omega(k,x,t)\}=W\{e(-k,x,t),e(k,x,t)\}=-2ik \text{ for real } k\neq 0.$$

Hence, $(e(-k,x,t),e(k,x,t))$ and $(\omega(-k,x,t),\omega(k,x,t))$ for real $k\neq 0$ constitute the fundamental systems of solution of system (10.8). Therefore, for real $k\neq 0$:

$$\begin{pmatrix} e(k,x,t) & e(-k,x,t) \\ e_x(k,x,t) & e_x(-k,x,t) \end{pmatrix}=\begin{pmatrix} \omega(-k,x,t) & \omega(k,x,t) \\ \omega_x(-k,x,t) & \omega_x(k,x,t) \end{pmatrix}\begin{pmatrix} s_{11}(k,t) & s_{12}(k,t) \\ s_{21}(k,t) & s_{22}(k,t) \end{pmatrix}$$
$$\begin{pmatrix} \omega(-k,x,t) & \omega(k,x,t) \\ \omega_x(-k,x,t) & \omega_x(k,x,t) \end{pmatrix}=\begin{pmatrix} e(k,x,t) & e(-k,x,t) \\ e_x(k,x,t) & e_x(-k,x,t) \end{pmatrix}\begin{pmatrix} s_{11}(-k,t) & -s_{21}(-k,t) \\ -s_{21}(k,t) & s_{11}(k,t) \end{pmatrix} \tag{10.16}$$

The matrix $s(k,t)=[s_{ij}(k,t)],i,j=1,2$, thus defined is called the scattering matrix for the SP for system (10.8) with the boundary condition (10.9). The properties of the entries $s_{ij}(k,t)$ of $s(k,t)$ are clarified in Lemma 7.1.

10.3.2 The time-dependence of eigenfunctions $E(k,x,t)$ and $W(k,x,t)$ of system (10.8)

We derive differential equations governing the time-dependence of the matrix eigenfunctions $E(k,x,t),W(k,x,t)$ of the SP generated by the systems (10.8) and (10.10) with the boundary conditions (10.14) and (10.15), respectively.

Let y be a solution of the Sturm–Liouville equation:

$$Ly=-y_{xx}+py=\lambda y,$$

where $p=p(x,t)$ is an isospectral potential belonging to the class S and λ is isospectral parameter.

Since $L_t = AL - LA$, then $L_t y + L y_t = A(Ly) - L(Ay) + L y_t = \lambda y_t$. Hence,

$$L(y_t - Ay) = \lambda(y_t - Ay),$$

i.e., the function $y_t - Ay$ is a solution of the above Sturm–Liouville equation. Due to the uniqueness of the Jost solution of the Sturm–Liouville equation with condition (10.14), the function $e_t - Ae$ is a solution of equation (10.11), where e is the Jost solution of equation (10.11) with condition (10.14). Therefore, in the class S we obtain:

$$e_t - Ae = 4ik^3 e.$$

Since $Ae = -4\frac{d}{dx}(p - k^2)e + 6pe_x + 3p_x e$, then from the above equality we derive the time-evolution equation for the Jost solution of system (10.11):

$$e_t(k,x,t) = (p_x(x,t) + 4ik^3)e(k,x,t) - (2p(x,t) + 4k^2)e_x(k,x,t),$$
$$e_{xt}(k,x,t) = \left[p_{xx}(x,t) - (2p(x,t) + 4k^2)(p(x,t) - k^2)\right] e(k,x,t) + \left[4ik^3 - p_x(x,t)\right] e_x(k,x,t)$$

The time-dependence of the matrix Jost solution $E(k,x,t)$ is obtained from the above derived equation:

$$E_t(k,x,t) = V(k,x,t)E(k,x,t) + E(k,x,t)4ik^3\sigma_3, \quad \sigma_3 = \begin{pmatrix} 1 & 0 \\ 0 & -1 \end{pmatrix}. \quad (10.17)$$

The time-dependence of the matrix analytical solution $W(k,x,t)$ is derived from the matrix equation:

$$W_t(k,x,t) = V(k,x,t)W(k,x,t) + W(k,x,t)\begin{pmatrix} a(-k,t) & b(k,t) \\ b(-k,t) & a(k,t) \end{pmatrix}, \quad (10.18)$$

where the coefficients $a(k,t)$ and $b(k,t)$ are determined by the boundary condition (10.15). Indeed, using (10.10) and (10.15), from (10.18) we get the system of equations for determining the coefficients $a(k,t)$ and $b(k,t)$:

$$-ik[2p(0,t) + 4k^2] - p_x(0,t) = a(k,t) + b(k,t),$$
$$[2p(0,t) + 4k^2][p(0,t) - k^2] - ikp_x(0,t) - p_{xx}(0,t) = -ik[a(k,t) - b(k,t)].$$

The coefficients $a(k,t)$ and $b(k,t)$ are calculated from the above system:

$$\begin{aligned} a(k,t) &= -4ik^3 + \frac{p_{xx}(0,t) - 2p^2(0,t)}{2ik}, \\ b(k,t) &= -2ik\,p(0,t) - p_x(0,t) + \frac{2p^2(0,t) - p_{xx}(0,t)}{2ik} \end{aligned} \quad (10.19)$$

Due to the boundary condition (10.6), the coefficients $a(k,t)$ and $b(k,t)$ in (10.19) are reduced to the form:

$$a(k,t) = -4ik^3, \quad b(k,t) = -p_x(0,t). \quad (10.20)$$

10.3.3 The system of linear differential equations for $s(k,t)$

Upon differentiating equality (10.16) with respect to t, we have

$$E_t(k,x,t) = W_t(k,x,t)s(k,t) + W(k,x,t)s_t(k,t), \tag{10.21}$$

where the scattering matrix $s(k,t)$ is defined by (10.16) as ratio of two fundamental systems of solutions of system (10.8):

$$s(k,t) = W^{-1}(k,x,t)E(k,x,t),$$

By using the above ratio and (10.17), (10.18), from (10.21) we get

$$W(k,x,t)s_t(k,t) = V(k,x,t)[E(k,x,t) - W(k,x,t)W^{-1}(k,x,t)E(k,x,t)]+$$
$$E(k,x,t)4ik^3\sigma_3 - W(k,x,t)\begin{pmatrix} a(-k,t) & b(k,t) \\ b(-k,t) & a(k,t)\end{pmatrix} W^{-1}(k,x,t)E(k,x,t),$$
$$s_t(k,t) = s(k,t)4ik^3\sigma_3 - \begin{pmatrix} a(-k,t) & b(k,t) \\ b(-k,t) & a(k,t)\end{pmatrix} s(k,t).$$

Substituting (10.19) into the right-hand side of the last equality gives:

$$s_t(k,t) = 4ik^3[s,\sigma_3] + \Big\{p_x(0,t)\sigma_1 + \frac{2p^2(0,t) - p_{xx}(0,t)}{2k}i\sigma_3$$
$$+\frac{p_{xx}(0,t) - 2p^2(0,t) - 4k^2p(0,t)}{2k}\sigma_2\Big\}\, s(k,t)$$

where the Pauli matrices σ_i are determined by:

$$\sigma_1 = \begin{pmatrix} 0 & 1 \\ 1 & 0\end{pmatrix}, \quad \sigma_2 = \begin{pmatrix} 0 & i \\ -i & 0\end{pmatrix}, \quad \sigma_3 = \begin{pmatrix} 1 & 0 \\ 0 & -1\end{pmatrix}.$$

Owing to the boundary condition (10.6), the above system reduces to the system of linear differential equations governing the time-dependence of $s(k,t)$:

$$s_t(k,t) = 4ik^3[s(k,t),\sigma_3] + p_x(0,t)\sigma_1 s(k,t). \tag{10.22}$$

We recall the necessary results of the direct and inverse SP (10.8)–(10.9) for clarifying the matrix evolution equation (10.22) and for constructing a common solution of both systems (10.8) and (10.10).

Remark 10.1. (1) The properties of the entries $s_{ij}(k)$ of the scattering matrix $s(k)$ are clarified in Lemmas 7.1 and 7.2. The refraction and reflection coefficients $s_{11}(k)$ and $s_{21}(k), -\infty < k < \infty$ are required to satisfy the conditions enumerated in Theorem 7.1. Then, the functions $e(k,x)$ and $\omega(k,x)$ constructed from the solutions $K(x,y) \in L_j[x,\infty)$ and $K^-(x,y) \in L_j(-\infty,x], j = 1,2$ of equation (7.45) and (7.46), respectively satisfy the same Schoödinger equation (10.11). This equation is restored precisely from the given right-reflection coefficient $\tilde{R}(k) = -\dfrac{s_{12}(k)}{s_{11}(k)}$;

(2) The boundary conditions (10.6) are satisfied if and only if the system (10.10) along the axis $x = 0$ takes the form (7.72):

$$Y_t = \begin{pmatrix} p_x(0,t) & -4\lambda \\ 4\lambda^2 & -p_x(0,t) \end{pmatrix} Y.$$

Using the linear change of dependent variables given by

$$Y(k,t) = T(k)y(\mu,t), \quad T(k) = \begin{pmatrix} 1 & 1 \\ ik & -ik \end{pmatrix}, \quad \mu = k^3, \ \ 0 \le t < \infty,$$

we reduce system (7.72) to the form (7.92):

$$i\sigma_3 y_t(\mu,t) + C(t)y(\mu,t) = 4\mu y(\mu,t), \ \ y(\mu,t) = (y_1(\mu,t), y_2(\mu,t)), \tag{10.23}$$

with the boundary condition at the origin $t = 0$ (7.93):

$$y_1(\mu,0) = y_2(\mu,0), \tag{10.24}$$

where the potential matrix $C(t)$ is self-adjoint:
$C(t) = \begin{pmatrix} 0 & c_1(t) \\ c_2(t) & 0 \end{pmatrix} = \overline{C(t)}$, $c_1(t) = -ip_x(0,t), c_2(t) = ip_x(0,t)$, the potential $p_x(0,t)$ is a real-valued function.

By Theorem 7.2, the scattering function $S(\mu,t), \mu = k^3$ of the self-adjoint SP (7.92)–(7.93) on the half-line $t \le \tau < \infty$ is constructed from the scattering function $S(\mu) = S(\mu,0)$, which is constructed from entries $s_{ij}(k)$ of the scattering matrix $s(k)$. The potential self-adjoint matrix (7.91) in system (7.92) is recovered uniquely from $S(\mu,t)$, and the unknown object $p_x(0,t)$ in (10.24) is expressed through entries of the potential self-adjoint matrix (7.91).

(3) By Theorem 7.3, the entries $s_{ij}(k,t)$ of the scattering matrix $s(k,t)$ are derived in their sectors on the k-spectral plane (see Figure 7.2). The solution $p(x,t)$ (recovered potential) of the linear equation (7.147) is found from the known right-reflection coefficient $\tilde{R}(k,t)$ and satisfies the KdV equation (7.1) with condition (7.2) and (7.3). The algorithm for finding the solution of the IBVP (7.1)–(7.2)–(7.3) consists of the three successive steps, as shown at the end of Chapter 7 (see Figure 7.3).

10.3.4 A common solution of both systems (10.8) and (10.10)

We construct the matrix function $P(k,x,t)$ from the Jost solution $(e(k,x,t), e_x(k,x,t))$ and the analytical solution $(\omega(k,x,t), \omega_x(k,x,t))$ of system (10.8):

$$P(k,x,t) = \begin{pmatrix} e\,(k,x,t) & \omega\,(k,x,t) \\ e_x\,(k,x,t) & \omega_x\,(k,x,t) \end{pmatrix}, \quad 0 \le x < \infty, t > 0. \tag{10.25}$$

The matrix function $P(k,x,t)$ satisfies only the first system (10.8) for the U-V pair in (10.8) and (10.10). We use the right multiplication by the factor $N(k,t)$ to correct $P(k,x,t)$ such that the corrected matrix function:

$$Q(k,x,t) = P(k,x,t)N(k,t), \tag{10.26}$$

satisfies both systems (10.8) and (10.10). We prove the following lemma for finding an appropriate factor function $N(k,t)$.

Lemma 10.5. *The matrix function $Q(k,x,t)$ given by (10.26) is a common solution of both systems (10.8) and (10.10) in the U-V pair, if the factor function $N(k,t)$ is the solution of the Cauchy problem for the matrix equation:*

$$\frac{d}{dt}N(k,t) = \begin{pmatrix} -4ik^3 & -\frac{p_x(0,t)}{s_{11}(k,t)} \\ 0 & 4ik^3 - \frac{p_x(0,t)s_{21}(k,t)}{s_{11}(k,t)}, \end{pmatrix} N(k,t) \tag{10.27}$$

with the given initial condition:

$$N(k;0) = \begin{pmatrix} 0 & 1 \\ 1 & 0 \end{pmatrix}. \tag{10.28}$$

Proof. The matrix solution $Q(k,x,t)$ determined by (10.26) is required to be a common solution of both systems (10.8) and (10.10). Therefore it satisfies system (10.10):

$$(PN)_t = V(PN). \tag{10.29}$$

Then the time-evolution equation for the matrix factor function N is derived from (10.29):

$$N_t = P^{-1}(VP - P_t)N. \tag{10.30}$$

The t-time evolution equation P_t for the matrix function P is deduced from evolution equations (10.17) and (10.18):

$$\begin{aligned} P_t = \begin{pmatrix} e(k,x,t) & \omega(k,x,t) \\ e_x(k,x,t) & \omega_x(k,x,t) \end{pmatrix}_t &= \{V(k,x,t) + 4ik^3\} \begin{pmatrix} e(k,x,t) \\ e_x(k,x,t) \end{pmatrix} \\ + V(k,x,t)\begin{pmatrix} \omega(k,x,t) \\ \omega_x(k,x,t) \end{pmatrix} &+ \begin{pmatrix} \omega(-k,x,t) & \omega(k,x,t) \\ \omega_x(-k,x,t) & \omega_x(k,x,t) \end{pmatrix} \begin{pmatrix} b(k,t) \\ a(k,t) \end{pmatrix}, \end{aligned} \tag{10.31}$$

where the coefficient $a(k,t)$ and $b(k,t)$ are determined by (10.20).

Using (10.31), from equation (10.30) we derive the t-evolution for $\ln N(k,t)$

$$\begin{aligned} N_tN^{-1} &= P^{-1}(VP - P_t) \\ &= -P^{-1}\begin{pmatrix} 4ik^3e(k,x,t) & \omega(-k,x,t)b(k,t) + \omega(k,x,t)a(k,t) \\ 4ik^3e_x(k,x,t) & \omega_x(-k,x,t)b(k,t) + \omega_x(k,x,t)a(k,t) \end{pmatrix}. \end{aligned} \tag{10.32}$$

Using the properties of the matrix solution $W(k,x,t)$ of the SP (10.8)–(10.9), we calculate:

$$W^{-1}(k,x,t)=\begin{pmatrix}\omega(-k,x,t) & \omega(k,x,t)\\ \omega_x(-k,x,t) & \omega_x(k,x,t)\end{pmatrix}^{-1}$$
$$=\frac{1}{\det W(k,x,t)}\begin{pmatrix}\omega_x(k,x,t) & -\omega(k,x,t)\\ -\omega_x(-k,x,t) & \omega(-k,x,t)\end{pmatrix},$$

where $\det W(k,x,t)=-2ik$.

The entries $s_{ij}(kt)$ of scattering matrix $s(k,t)$ are determined by (10.16):

$$\begin{pmatrix}s_{11}(k,t) & s_{12}(k,t)\\ s_{21}(k,t) & s_{22}(k,t)\end{pmatrix}=\frac{1}{2ik}\begin{pmatrix}\omega_x(k,x,t) & -\omega(k,x,t)\\ -\omega_x(-k,x,t) & \omega(-k,x,t)\end{pmatrix}\begin{pmatrix}e(k,x,t) & e(-k,x,t)\\ e_x(k,x,t) & e_x(-k,x,t)\end{pmatrix}$$
$$=\frac{1}{2ik}\begin{pmatrix}\omega_x(k,x,t)e(k,x,t)-\omega(k,x,t)e_x(k,x,t)\\ -\omega_x(-k,x,t)e(k,x,t)+\omega(-k,x,t)e_x(k,x,t)\end{pmatrix.}$$
$$\begin{matrix}\omega_x(k,x,t)e(-k,x,t)-\omega(k,x,t)e_x(-k,x,t)\\ -\omega_x(-k,x,t)e(-k,x,t)+\omega(-k,x,t)e_x(-k,x,t)\end{matrix}\Bigg).$$

The matrix P^{-1} is defined by:

$$P^{-1}=\frac{1}{\det P}\begin{pmatrix}\omega_x(k,x,t) & -\omega(k,x,t)\\ -e_x(k,x,t) & e(k,x,t)\end{pmatrix},\quad \text{where } \det P=2iks_{11}(k,t).$$

Using the above calculation and the calculated coefficients (10.20), from (10.32) we get:

$$\begin{aligned}&-P^{-1}\begin{pmatrix}4ik^3e(k,x,t)\\ 4ik^3e_x(k,x,t)\end{pmatrix}\\ &=-\frac{4ik^3}{2iks_{11}(k,t)}\begin{pmatrix}\omega_x(k,x,t)e(k,x,t)-\omega(k,x,t)e_x(k,x,t)\\ -e(k,x,t)e_x(k,x,t)+e(k,x,t)e_x(k,x,t)\end{pmatrix}\\ &=-\frac{4ik^3}{2iks_{11}(k,t)}\begin{pmatrix}2iks_{11}(k,t)\\ 0\end{pmatrix}=4ik^3\begin{pmatrix}-1\\ 0\end{pmatrix},\end{aligned}\tag{10.33}$$

$$\begin{aligned}&-P^{-1}\begin{pmatrix}b(k,t)\omega(-k,x,t)+a(k,t)\omega(k,x,t)\\ b(k,t)\omega_x(-k,x,t)+a(k,t)\omega_x(k,x,t)\end{pmatrix}\\ &=-\frac{1}{2iks_{11}(k,t)}\begin{pmatrix}\omega_x(k,x,t) & -\omega(k,x,t)\\ -e_x(k,x,t) & e(k,x,t)\end{pmatrix}\begin{pmatrix}b(k,t)\omega(-k,x,t)+a(k,t)\omega(k,x,t)\\ b(k,t)\omega_x(-k,x,t)+a(k,t)\omega_x(k,x,t)\end{pmatrix}\\ &=-\frac{1}{2iks_{11}(k,t)}\begin{pmatrix}-2ikb(k,t)\\ -2ikb(k,t)s_{21}(k,t)+2ika(k,t)s_{11}(k,t)\end{pmatrix}\\ &=\begin{pmatrix}\frac{b(k,t)}{s_{11}(k,t)}\\ -a(k,t)+b(k,t)\frac{s_{21}(k,t)}{s_{11}(k,t)}\end{pmatrix}=\begin{pmatrix}-\frac{p_x(0,t)}{s_{11}(k,t)}\\ 4ik^3-p_x(0,t)\frac{s_{21}(k,t)}{s_{11}(k,t)}\end{pmatrix}.\end{aligned}\tag{10.34}$$

Substituting (10.33) and (10.34) into the right-hand side of equality (10.32), we obtain the matrix equation (10.27):

$$\begin{pmatrix} n_{11t} & n_{12t} \\ n_{21t} & n_{22t} \end{pmatrix} = \begin{pmatrix} -4ik^3 & -\frac{p_x(0,t)}{s_{11}(k,t)} \\ 0 & 4ik^3 - \frac{p_x(0,t)s_{21}(k,t)}{s_{11}(k,t)} \end{pmatrix} \begin{pmatrix} n_{11}(k,t) & n_{12}(k,t) \\ n_{21}(k,t) & n_{22}(k,t) \end{pmatrix}. \tag{10.35}$$

We find the entries of the matrix solution $N(k,t)$ of the Cauchy problem for equation (10.35) with the known initial condition (10.28) in terms of entries $s_{ij}(k,t)$ of the scattering matrix $s(k,t)$. The time-evolution equation for $s(k,t)$ is given by (10.22), which takes the expansion form:

$$\begin{pmatrix} s_{11t} & s_{12t} \\ s_{21t} & s_{22t} \end{pmatrix} = 4ik^3 \begin{pmatrix} 0 & -2s_{12} \\ 2s_{21} & 0 \end{pmatrix} + p_x(0,t) \begin{pmatrix} s_{21} & s_{22} \\ s_{11} & s_{12} \end{pmatrix}, \tag{10.36}$$

whence it follows that

$$\begin{aligned} s_{11t} &= p_x(0,t)s_{21}, \quad s_{12t} = -8ik^3 s_{12} + p_x(0,t)s_{22}, \\ s_{21t} &= 8ik^3 s_{21} + p_x(0,t)s_{11}, \quad s_{22t} = p_x(0,t)s_{12}, \end{aligned} \tag{10.37}$$

Using (10.37) and the initial condition (10.28), from (10.35) we find the entries $n_{ij}(k,t)$ of the matrix $N(k,t)$:

$$n_{21t}(k,t) = \left\{4ik^3 - \frac{p_x(0,t)s_{21}(k,t)}{s_{11}(k,t)}\right\} n_{21}(k,t),$$

$$n_{21}(k,t) = n_{21}(k,0)\exp\left\{4ik^3 t - \int_0^t \frac{p_x(0,\tau)s_{21}(0,\tau)}{s_{11}(0,\tau)}d\tau\right\}.$$

Since $n_{21}(k,0) = 0$ then $n_{21}(k,t) = 0$ for all $t > 0$.

$$n_{11t}(k,t) = -4ik^3 n_{11}(k,t),\ n_{11}(k,t) = n_{11}(k,0)e^{-4ik^3 t} = e^{-4ik^3 t},$$
$$n_{11}(k,0) = 1,$$

$$n_{22t}(k,t) = \left(4ik^3 - \frac{p_x(0,t)s_{21}(k,t)}{s_{11}(k,t)}\right) n_{22}(k,t) = \left(4ik^3 - \frac{s_{11t}(k,t)}{s_{11}(k,t)}\right) n_{22}(k,t),$$

$$\frac{d}{dt}\ln n_{22}(k,t) + \frac{d}{dt}\ln s_{11}(k,t) = \frac{d}{dt}\ln n_{22}(k,t)s_{11}(k,t) = 4ik^3,$$

$$n_{22}(k,t)s_{11}(k,t) = n_{22}(k,0)s_{11}(k,0)e^{4ik^3 t}, n_{22}(k,0) = 1, n_{22}(k,t) = \frac{s_{11}(k,0)}{s_{11}(k,t)}e^{4ik^3 t},$$

$$n_{12t}(k,t) = -4ik^3 n_{12}(k,t) - \frac{p_x(0,t)}{s_{11}(k,t)} n_{22}(k,t).$$

Putting $n_{12}(k,t) = \tilde{n}_{12}(k,t)e^{-4ik^3t}$, and using the representation of $n_{22}(k,t)$, we have:

$$\tilde{n}_{12t}(k,t) = -\frac{p_x(0,t)}{s_{11}(k,t)}n_{22}(k,t)e^{4ik^3t} = -\frac{p_x(0,t)}{s_{11}(k,t)}\frac{s_{11}(k,0)}{s_{11}(k,t)}e^{8ik^3t}$$
$$= -\frac{p_x(0,t)}{s_{11}^2(k,t)}\left\{s_{11}(k,t)\overline{s_{11}(k,t)} - s_{21}(k,t)\overline{s_{21}(k,t)}\right\}s_{11}(k,0)e^{8ik^3t}$$
$$= \left\{\frac{p_x(0,t)s_{21}(k,t)\overline{s_{21}(k,t)}}{s_{11}^2(k,t)} - \frac{p_x(0,t)\overline{s_{11}(k,t)}}{s_{11}(k,t)}\right\}s_{11}(k,0)e^{8ik^3t}$$

The following relation equalities are deduced from (10.37):

$$p_x(0,t)s_{21}(k,t) = s_{11t}(k,t),$$
$$p_x(0,t)\overline{s_{11}(k,t)} = \overline{s_{21t}(k,t)} + 8ik^3\overline{s_{21}(k,t)}.$$

By the above equalities we get:

$$\tilde{n}_{12t}(k,t) = \left\{\frac{s_{11t}(k,t)\overline{s_{21}(k,t)}}{s_{11}^2(k,t)} - \frac{8ik^3\overline{s_{21}(k,t)} + \overline{s_{21t}(k,t)}}{s_{11}(k,t)}\right\}s_{11}(k,0)e^{8ik^3t}$$
$$= -\frac{d}{dt}\left(\frac{\overline{s_{21}(k,t)}}{s_{11}(k,t)}e^{8ik^3t}\right)s_{11}(k,0),$$
$$\tilde{n}_{12}(k,t) = -\frac{\overline{s_{21}(k,t)}}{s_{11}(k,t)}s_{11}(k,0)e^{8ik^3t} + \text{const}.$$

Since $\tilde{n}_{12}(k,0) = n_{12}(k,0) = 0$, then const $= \overline{s_{21}(k,0)}$, and $n_{12}(k,t) = \tilde{n}_{12}(k,t)e^{-4ik^3t} = \dfrac{\overline{s_{21}(k,0)}s_{11}(k,t)e^{-4ik^3t} - s_{11}(k,0)\overline{s_{21}(k,t)}e^{4ik^3t}}{s_{11}(k,t)}$.

Thus, the entries $n_{ij}(k,t)$ of the matrix solution $N(k,t)$ of the Cauchy problem (10.27)–(10.28) are found:

$$N(k,t) = \begin{pmatrix} e^{-4ik^3t} & \dfrac{\overline{s_{21}(k,0)}s_{11}(k,t)e^{-4ik^3t} - s_{11}(k,0)\overline{s_{21}(k,t)}e^{4ik^3t}}{s_{11}(k,t)} \\ 0 & \dfrac{s_{11}(k,0)}{s_{11}(k,t)}e^{-4ik^3t} \end{pmatrix}, \tag{10.38}$$

where the functions $p_x(0,t)$ and $s_{11}(k,t), t > 0$ are calculated in terms of the given initial and boundary conditions (10.6) and (10.7), as shown in Remark 10.1.

The matrix function $Q(k,x,t)$ constructed by (10.26) satisfies system (10.8) because the factor $N(k,t)$ does not depend on x, and the function $Q(k,x,t)$ satisfies equation (10.8). Moreover, this constructed function $Q(k,x,t)$ satisfies the second system (10.10) in the U-V pair. Indeed, the matrix function $Q(k,x,t)$ satisfies system (10.29), which has the same form of system (10.10) with the same matrix function V governing the time-dependence of solution of system (10.10). Thus, $Q(k,x,t)$ is the common solution of both systems (10.8) and (10.10) in the U-V pair. The lemma is proved. □

The compatible equalities (10.13) are equivalent to the KdV equation (10.5) in the class of isospectral potentials S. As shown in Remark 10.1, by Theorem 7.3 the solution $p(x,t)$ of equation (7.1) is constructed by formula (7.152) expressed through the solution $K(x,y,t)$ of the GLM equation (7.147). By Theorem 7.1 the solution $K(x,y,t)$ of equation (7.147) is the kernel of the transformation operator and is related to the potential by formulas (7.52) and (7.53). Thus, the solution of the KdV equation (10.5) is the recovered potential $p(x,t)$ satisfying the initial and boundary condition (10.6) and (10.7). Because the KdV equation (10.5) is the Lax compatibility condition (10.13) for the pair of the overdetermined equations (10.11) and (10.12), it suffices to use Lemma 10.5 for finding a common solution of both equations (10.8) and (10.10) that are equivalent to equations (10.11) and (10.12), respectively. The following theorem is deduced from the above discussion.

Theorem 10.1. *By Theorems 7.1–7.3 proved by the ISM, there exists a one-to-one correspondence relation between the scattering data of SPs generated by both matrix equations* (10.8) *and* (10.10) *in the U-V pair and the solution $p(x,t)$ (recovered potential) of the IBVP* (10.5)–(10.6)–(10.7). *By Lemma 10.5, there exists the factor function $N(k,t)$ such that the resulting function $Q(x,k,t) = P(x,k,t)N(k,t)$ simultaneously satisfies both matrix equations* (10.8) *and* (10.10). *This correspondence relation is a BT which transforms the constructed common solution $Q(k,x,t)$ of both systems* (10.8), (10.10) *into the solution of the IBVP for the KdV equation* (10.5) *with given initial and boundary conditions* (10.6) *and* (10.7), *and conversely (see Figure 10.11).*

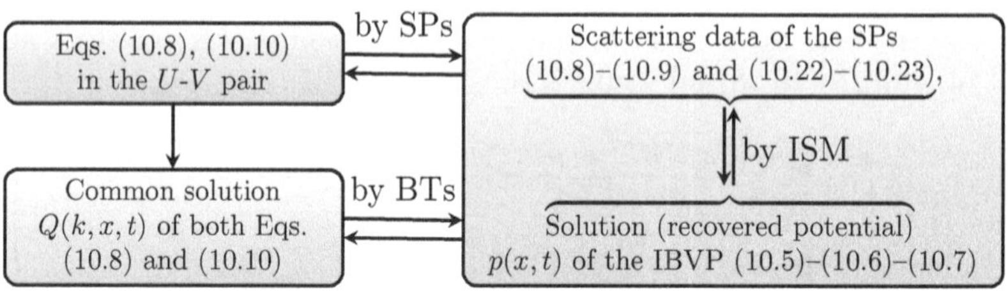

FIGURE 10.11: There exist the transformations by the ISM between the scattering data of SPs (10.8)–(10.9) and (10.23)–(10.24) and the recovered potential $p(x,t)$, and conversely. These transformations are BTs, which transform the common solution $Q(k,x,t)$ of both matrix equations (10.8) and (10.10) into the solution $p(x,t)$ of the IBVP for the KdV equation (10.5) with given initial and boundary conditions (10.6) and (10.7), and conversely.

Note that the Jost solution $e(k,x,t)$ of equation (10.11) is analytic in the closed upper half-plane Im$k \geq 0$. It follows from the entries of the matrix

(10.38) that the constructed common solution $Q(x,k,t)$ of both systems (10.8) and (10.10) is not analytic in the upper half-plane Im$k > 0$, therefore this common $Q(k,x,t)$ is not a solution of the SP for system (10.11).

Remark 10.2. We turn to the relationship between the ISM and the BT for the solution of the IBVP (10.5)–(10.7). The problem (10.5)–(10.7) has infinitely many integrals of motion and higher symmetries, (see [50]). The first three integrals of motion in this problem are given by

$$J_1 = \int_0^\infty p dx, \quad J_2 = \int_0^\infty (p_x^2 + 2p^3) dx, \quad J_3 = \int_0^\infty (p_{xx}^2 - 5p^2 p_{xx} + 5p^4) dx.$$

However, not all the conservation laws of the KdV equation (10.5) are compatible with the IBVP (10.5)–(10.7). For example, the well-known functional $\int_0^\infty p^2 dx$ does not preserve its value.

An infinite set of conservation laws seems to allow proving that the IBVP (10.5)–(10.7) is a completely integrable Hamiltonian system, although this question has not been studied anywhere. Note that it is tough to set up an appropriate ISP which depends seemingly on the existence of an infinite number of independent conservation laws for the evolution equations defined on the half-line.

However, an appropriate inverse scattering problem for differential equations defined in the whole space seems to depend on the existence of an infinite number of independent conservation laws for the non-linear partial differential equations (see for example [29, 27, 28, 30]).

Indeed, the way of obtaining BTs for non-linear differential equations and the relationship between the BTs and associated ISPs for linear differential equations defined in the whole space are different from the way of finding BTs for IBVPs for NLEEs and the relationship between the BTs and the associated ISPs for linear differential equations defined on the half-line, respectively.

Summary

Studying inverse scattering problems (ISPs) on a half-line and applying obtained results of ISPs to solving the considered IBVPs or IVPs for integrable NLEEs are two steps of carrying out the inverse scattering method (ISM).

Step 1. The transformation operators and special Volterra integral representations are utilized to the representation of solutions of direct and inverse scattering problems (ISPs). The unique solvability of the ISPs is proven. The necessary and sufficient conditions for given quantities to be the scattering data (SD) for the scattering problem (SP) generated by differential equations with the known initial and boundary conditions (BCs) are established.

Step 2. The NLEEs can be written as the compatibility of the x-part and t-part of the Lax pair or of generalized Lax equation. The Volterra integral equations for unknown boundary values (BVs) are derived, and solutions of derived integral equations are BVs, that are found in terms of the known initial and boundary conditions. Thus, the time-dependence of SD is derived, then the time-dependent functions in the fundamental equations in the ISP are constructed from the time-dependent SD, therefore, they are known. Hence, the potentials are recovered uniquely in terms of time-dependent SD, and the solution of the NLEEs is expressed uniquely in terms of the found solutions of the ISP. There exists a one-to-one correspondence relation between the time-dependent SD and the found solution of the IBVP or IVP for NLEEs.

Hence, the ISP and associated time-dependence of the scattering data constitute the ISM. Note that instead of *Step 2*, the IBVP for NLEEs can be lead to the associated Riemann–Hilbert problem and then followed by using the unique solvability of the R–H problem (see Sections 9.1, 9.2 of Chapter 9). The ISM is a generalization of the Fourier transformation. The relations between the ISM and related methods, such as Forkas's methodology for analysis of IBVP on the half-line, Riemann-Hilbert problem on the half-line, Hirota's method are presented in Chapter 9. The common solution of both equations in the Lax pair is constructed from these equations. If the solution of the SP for the first equation of the Lax pair does not satisfies the second equation of this pair, then we must therefore correct this solution, as shown by Lemma 10.5. By Theorem 10.1, the BTs are constructed, which transform the constructed common solution into the solution of the IBVP for the KdV equation (10.5) with initial and boundary conditions (10.6) and (10.7), and conversely. Hence, the solutions of NLEEs on the half-line and on the whole line considered in this monograph can be obtained by three different methods: the ISM, the Hirota's method and the BT.

Concluding Remarks

1. Since the considered ISPs are well-solved, then the ISM can be applied for finding solutions of the IBVPs or Cauchy problems for NLEEs. The considered ISPs on the half-line and in the whole space are set up appropriately for constructing BTs for solutions of the associated IBVPs or IVPs for NLEEs.

2. Due to the unique solvability of the associated ISP, there exists a one-to-one correspondence relation between the scattering data of the scattering problem and the solution (recovered potential functions) of the IBVP for NLEEs on the half-line. This one-to-one correspondence relation is a BT that transforms a common solution of both equations in the Lax

pair into the solution of the IBVP for the considered NLEEs on the half-line. Due to the boundary conditions at the origin $x = 0$, the procedure for finding a BT for the IBVP for the considered NLEEs defined on the half-line differs from the one used for obtaining a BT for non-linear equations defined in whole space, (see Remark 10.2).

3. The interrelations between the ISP and the BT are established to become the new powerful unified transformation (UT) for solving IVPs and IBVPs for integrable NLEEs that can be used in different areas of physics and mechanics (see [103,104,107]). The application of the UT is consistent and efficiently embedded in the scheme of the ISP (see Figure 10.11).

Bibliography

[1] M. J. Ablowitz and P. A. Clarkson. *Solitons, nonlinear evolution equations and inverse scattering*, volume 149. London Mathematical Society, Lecture Note Series, 1991.

[2] M. J. Ablowitz, D. J. Kaup, A. C. Newell, and H. Segur. Nonlinear-evolution equations of physical significance. *Physical Review Letters*, 31(2):125–127, jul 1973.

[3] M. J. Ablowitz, D. J. Kaup, A. C. Newell, and H. Segur. The inverse scattering transform Fourier analysis for nonlinear problems. *Studies in Applied Mathematics*, 53(4):249–315, 1974.

[4] M. J. Ablowitz and H. Segur. The inverse scattering transform: Semi-infinite interval. *Journal of Mathematical Physics*, 16(5):1054–1056, 1975.

[5] M. J. Ablowitz and H. Segur. *Solitons and the inverse scattering transform*. SIAM, Studies in Applied Mathematics, Philadelphia, 1981.

[6] V. E. Adler, L. T. Habibullin, and A. B. Shabat. Boundary value problem for the KdV equation on a half-line. *Theoretical and Mathematical Physics*, 110(1):78–90, 1997.

[7] Z. S. Agranovich and V. A. Marchenko. *The inverse problem of scattering theory*. Izd. Khar'kov., Univ. Khar'kov., 1960. (in Russian) English transl., Gordon and Breach, New York, 1963.

[8] V. Ambarzumian. Über eine frage der eigenwerttheorie. *Zeitschrift für Physik A Hadrons and Nuclei*, 53(9):690–695, 1929.

[9] Y. E. Anikonov. *Some methods of investigation of multidimensional inverse problems for differential equations*. Novosiberian: Nauka, 1978.

[10] R. Beals, P. Deift, and C. Tomei. *Direct and inverse scattering on the line*. Number 28. American Mathematical Society, 1988.

[11] J. M. Berezansky. The uniqueness theorem in the inverse problem of spectral analysis for the Schrödinger equation. *American Mathematical Society Translations*, 35:167–235, 1964.

[12] J. M. Berezansky. *Expansions in eigenfunctions of self-adjoint operators.* American Mathematical Society Translations, 17, 1968.

[13] R. F. Bikbaev and V. O. Tarasov. Initial boundary value problem for the nonlinear Schrodinger equation. *Journal of Physics A: Mathematical and General*, 24(11):2507–2516, 1991.

[14] V. A. Blashchak. An analogue of inverse problem in scattering theory for a non-selfadjoint operator, I. *Differentsial'nye Uravneniya*, 4(8):1519–1533, 1968.

[15] V. A. Blashchak. An analogue of inverse problem in scattering theory for a non-selfadjoint operator, II. *Differentsial'nye Uravneniya*, 4(8):1519–1533, 1968.

[16] J. Bona and R. Winther. The Korteweg–de Vries equation, posed in a quarter-plane. *SIAM Journal on Mathematical Analysis*, 14(6):1056–1106, 1983.

[17] G. Borg. Eine umkehrung der Sturm-Liouvilleschen eigenwertaufgabe. *Acta Mathematica*, 78(1):1–96, 1946.

[18] F. Cakoni and D. Colton. *Qualitative methods in inverse scattering theory: an introduction.* Springer-Verlag, Berlin, 2006.

[19] K. Chadan and P. C. Sabatier. *Inverse problems in quantum scattering theory.* Springer-Verlag, Berlin, 2nd edition, 2012.

[20] H.-H. Chen. Relation between Bäcklund transformations and inverse scattering problems. In *Lecture Notes in Mathematics 515, NSF Research Workshop on Contact Transformations*, pages 241–252. Springer-Verlag, Berlin-Heidelberg-New York, 1976.

[21] L. A. Chudov. The inverse Sturm-Liouville problem. *Matematicheskii Sbornik* (in Russian), 25(3):451–454, 1949.

[22] D. Colton and R. Kress. *Inverse acoustic and electromagnetic scattering theory*, volume 93. Applied Mathematical Sciences, Springer-Verlag, Berlin, 2012.

[23] A. Degasperis, S. V. Manakov, and P. M. Santini. On the initial-boundary value problems for soliton equations. *Journal of Experimental and Theoretical Physics Letters*, 74(10):481–485, 2001.

[24] A. Degasperis, S. V. Manakov, and P. M. Santini. Initial-boundary value problems for linear PDEs: the analyticity approach. In *New Trends in Integrability and Partial Solvability*, pages 79–103. Springer-Verlag, Berlin, 2004.

[25] B. A. Dubrovin, V. B. Matveev, and S. P. Novikov. Non-linear equations of Korteweg-de Vries type, finite-zone linear operators, and Abelian varieties. *Russian Mathematical Surveys*, 31(1):59–146, 1976.

[26] W. Eckhaus and A. M. Van Harten. *The inverse scattering transformation and the theory of solitons: An introduction*, volume 50. Mathematics Studies, North-Holland, 1981.

[27] H. M. El-Sherbiny. *The Korteweg-de Vries equation and its homologues: A comparative analysis II.* PhD thesis, The City University, London, 1987.

[28] H. M. El-Sherbiny. Conservation laws of a class of differential equations II. *Applied Mathematics Letters*, 6(4):99–103, 1993.

[29] H. M. El-Sherbiny and L. Debnath. A note on the equivalence of Bcklund transformations and the inverse scattering method for a class of non-linear differential equations. *International Journal of Non-Linear Mechanics*, 31(4):451–457, 1996.

[30] H. M. El-Sherbiny, S. A. Hussein, and K. A. Moustafa. Conservation laws of a class of differential equations I. In *Prof. Assiut. Frist Int. Conf. Math. Stat.*, volume V, page 71, 1990.

[31] L. D. Faddeev. On the relation between S-matrix and potential for the one-dimensional Schrodinger operator. *Doklady Akademii Nauk SSSR*, 121(1):63–66, 1958.

[32] L. D. Faddeev. The inverse problem in the quantum theory of scattering. *Journal of Mathematical Physics*, 4:72–104, 1963. [Uspekhi Matem. Nauk. 14, 57–119 (1959)].

[33] L. D. Faddeev. Factorization of the S-matrix for the multidimensional Schrödinger operator. In *Soviet Physics Doklady SSSR*, volume 11, pages 209–211, 1966.

[34] L. D. Faddeev and L. A. Takhtajan. *Hamiltonian methods in the theory of solitons.* Springer-Verlag, Berlin, 1987.

[35] A. S. Fokas. An initial-boundary value problem for the nonlinear Schrödinger equation. *Physica D: Nonlinear Phenomena*, 35(1-2):167–185, 1989.

[36] A. S. Fokas. A unified transform method for solving linear and certain nonlinear pdes. *Proceedings of the Royal Society of London. Series A: Mathematical, Physical and Engineering Sciences*, 53:1411–1443, 1997.

[37] A. S. Fokas. On the integrability of linear and nonlinear partial differential equations. *Journal of Mathematical Physics*, 41(6):4188–4237, 2000.

[38] A. S. Fokas. Integrable nonlinear evolution equations on the half-line. *Communications in Mathematical Physics*, 230(1):1–39, 2002.

[39] A. S. Fokas and M. J. Ablowitz. The inverse scattering transform for multidimensional (2+1) problems. In K. B. Wolf, editor, *Nonlinear Phenomena. Lecture Notes in Physics*, volume 189, pages 137–183. Springer, Berlin, Heidelberg, 1983.

[40] A. S. Fokas and A. R. Its. An initial-boundary value problem for the sine-Gordon equation in laboratory coordinates. *Theoretical and Mathematical Physics*, 92(3):964–978, 1992.

[41] A. S. Fokas and A. R. Its. The linearization of the initial-boundary value problem of the nonlinear Schrödinger equation. *SIAM Journal on Mathematical Analysis*, 27(3):738–764, 1996.

[42] A. S. Fokas, A. R. Its, and L. Y. Sung. The nonlinear Schrödinger equation on the half-line. *Nonlinearity*, 18(4):1771–1822, 2005.

[43] A. S. Fokas and V. E. Zakharov. *Important developments in soliton theory*. Springer-Verlag, Berlin, 1993.

[44] C. S. Gardner, J. M. Greene, M. D. Kruskal, and R. M. Miura. Method for solving the Korteweg-de Vries equation. *Physical Review Letters*, 19(19):1095–1097, 1967.

[45] C. S. Gardner, J. M. Greene, M. D. Kruskal, and R. M. Miura. Korteweg-de Vries equation and generalizations. VI. Methods for exact solution. *Communications on Pure and Applied Mathematics*, 27(1):97–133, 1974.

[46] C. S. Gardner and G. K. Morikawa. *Similarity in the asymptotic behaviour of collision free hydrodynamic waves and water waves*. Courant Institute of Mathematical and Scientic Research, Report NYO-9082, New York University, New York, 1960.

[47] M. G. Gasymov and B. M. Levitan. The inverse problem for the Dirac system. *Doklady Akademii Nauk SSSR*, 167(5):967–970, 1966.

[48] I. M. Gelfand and B. M. Levitan. On the determination of a differential equation from its spectral function. *American Mathematical Society Translations*, (1):253–304, 1951.

[49] G. M. Gladwell. *Inverse problems in scattering: An introduction*. Springer-Verlag, Berlin, 1993.

[50] B. Grel, M. Grses, and I. Habibullin. Boundary value problems for integrable equations compatible with the symmetry algebra. *Journal of Mathematical Physics*, 36(12):6809–6821, 1995.

[51] I. T. Habibullin. KdV equation on a half-line with the zero boundary condition. *Teoreticheskaya i Matematicheskaya Fizika*, 119(3):397–404, 1999. English Translate: Theoretical and Mathematical Physics, 119:712–718, 1999.

[52] I. T. Habibullin. An initial-boundary value problem on the half-line for the mKdV equation. *Functional Analysis and Its Applications*, 34(1):52–59, 2000.

[53] I. T. Habibullin. Initial boundary value problem for the KdV equation on a semiaxis with homogeneous boundary conditions. *Teoreticheskaya i Matematicheskaya Fizika*, 130(1):31–53, 2002. English Translation: Theoretical and Mathematical Physics, 130(1):25–44, 2001.

[54] I. T. Habibullin and A. N. Vildanov. The KdV equation on a half-line. *arXiv preprint solv-int/9910002*, 2:1–14, 1999.

[55] R. Hirota. Exact solution of the Korteweg-de Vries equation for multiple collisions of solitons. *Physical Review Letters*, 27(18):1192–1194, 1971.

[56] R. Hirota. A new form of Bäcklund transformations and its relation to the inverse scattering problem. *Progress of Theoretical Physics*, 52(5):1498–1512, nov 1974.

[57] R. Hirota. Direct method of finding exact solutions of nonlinear evolution equations. In *Bäcklund Transformations, the Inverse Scattering Method, Solitons, and Their Applications*, pages 40–68. Springer-Verlag, Berlin, 1976.

[58] R. Hirota. Direct methods in soliton theory. In *Solitons*, pages 157–176. Springer-Verlag, Berlin, 1980.

[59] N. S. Iskenderov and M. I. Ismailov. On the inverse scattering transform of a nonlinear evolution equation with 2+1 dimensions related to nonstrict hyperbolic systems. *Nonlinearity*, 25(7):1967–1979, 2012.

[60] S. Kichenassamy and P. J. Olver. Existence and nonexistence of solitary wave solutions to higher-order model evolution equations. *SIAM Journal on Mathematical Analysis*, 23(5):1141–1166, 1992.

[61] A. N. Kolmogorov and S. V. Fomin. *Elements of the theory of functions and functional analysis.* Gostekhizdat, Moscow, 3rd edition, 1972. Translation by L. F. Boron (Volume 1), Graylock, Rochester, NY, 1957; and H. Kemiel and H. Komm (Volume 2), Academic Press, New York, 1961.

[62] V. E. Korepin, N. M. Bogoliubov, and A. G. Izergin. *Quantum inverse scattering method and correlation functions.* Cambridge University Press, United Kingdom, 1993.

[63] M. G. Krein. Solution of the inverse Sturm-Liouville problem. *Dokl. Akad. Nauk SSSR*, 64(1):21–24, 1951. (in Russian).

[64] M. G. Krein. On inverse problems for a nonhomogeneous cord. *Dokl. Akad. Nauk SSSR*, 82(5):669–672, 1952. (in Russian).

[65] M. G. Krein. On the transfer function of a one-dimensional boundary problem of second order. *Dokl. Akad. Nauk SSSR*, 88(3):405–408, 1953. (in Russian).

[66] M. G. Krein. Integral equations on the half-line with a kernel depending on the difference of the arguments. *Uspekhi Matematicheskikh Nauk*, 13(5):3–120, 1958.

[67] M. M. Lavrent'ev, Rezniskaia K. G., and V. G. Iakho. *One-dimensional inverse problems of mathematical physics.* Novosiberian Nauka, 1982.

[68] M. M. Lavrent'ev, V. G Vasil'ev, and V. G. Romanov. *Multidimensional inverse problems for differential equations.* Novosiberian Nauka, 1969.

[69] P. D. Lax. Integrals of nonlinear equations of evolution and solitary waves. *Communications on Pure and Applied Mathematics*, 21(5):467–490, 1968.

[70] J. Leon. Solution of the Dirichlet boundary value problem for the sine-Gordon equation. *Physics Letters A*, 298(4):243–252, 2002.

[71] J. Leon and A. Spire. The Zakharov-Shabat spectral problem on the semi-line: Hilbert formulation and applications. *Journal of Physics A: Mathematical and General*, 34(36):7359–7380, 2001.

[72] N. Levinson. On the uniqueness of the potential in a Schrodinger equation for a given asymptotic phase. *Kgl. Danske Videnskab Selskab. Mat. Fys. Medd.*, 25(1), 1949.

[73] N. Levinson. The inverse Sturm-Liouville problem. *Matematisk Tidsskrift. B*, pages 25–30, 1949.

[74] N. Levinson. Certain relations between phase shifts and scattering potentials. *Physical Review*, 89(4):755–757, 1953.

[75] B. M. Levitan. *Theory of operators of generalized shift.* Nauka, Moscow (in Russian), 2nd edition, 1973.

[76] B. M. Levitan. *Inverse Sturm-Liouville problems.* Nauka, Moscow, 1984. (English Translation VNU Science Press Utiecht, 1987).

[77] B. M. Levitan and M. G. Gasymov. Determination of a differential equation by two of its spectra. *Russian Mathematical Surveys*, 19(2):3–63, 1964.

[78] V. E. Ljance. The inverse problem for a nonselfadjoint operator. *Doklady Akademii Nauk SSSR*, 166(1):30–33, 1966.

[79] V. E. Ljance. An analog of the inverse problem of scattering theory for a nonselfadjoint operator. *Mathematics of the USSR-Sbornik*, 1(4):485–504, 1967.

[80] L. A. Ljusternik and V. J. Sobolev. *Elements of functional analysis.* Nauka, Moscow, 2nd edition, 1965. English Translation: Frederick Ungar Publishing, 1961.

[81] F. G. Maksudov and S. G. Veliev. The inverse scattering problem for the nonselfadjoint Dirac operator on the whole axis. *Doklady Akademii Nauk SSSR*, 225(6):1263–1266, 1975.

[82] T. R. Marchant and N. F. Smyth. Initial-boundary value problems for the Korteweg-de Vries equation. *IMA Journal of Applied Mathematics*, 47(3):247–264, 1991.

[83] T. R. Marchant and N. F. Smyth. The initial boundary problem for the Korteweg-de Vries equation on the negative quarter-plane. In *Proceedings of the Royal Society of London. Series A: Mathematical, Physical and Engineering Sciences*, volume 458, pages 857–871, 2002.

[84] V. A. Marchenko. Certain problems of the theory of one dimensional linear differential operators of the second order I. *Trudy Moskovskogo Matematicheskogo Obshchestva* (in Russian), 1:327–340, 1952.

[85] V. A. Marchenko. On reconstruction of the potential energy from phases of the scattered waves. *Dokl. Akad. Nauk SSSR*, 104(5):695–698, 1955. (in Russian).

[86] V. A. Marchenko. Spectral theory of Sturm-Liouville operators. *Nauk, Kiev*, 1972. English Translation: Sturm-Liouville Operators and Applications, Birkhauser, Basel, 1986.

[87] R. M. Miura. Korteweg-de Vries equation and generalizations. I. A remarkable explicit nonlinear transformation. *Journal of Mathematical Physics*, 9(8):1202–1204, 1968.

[88] R. M. Miura. The Korteweg-de Vries equation - A model equation for nonlinear dispersive waves. *Nonlinear waves,* Cornell University Press, Ithaca, NY, pages 212–234, 1974.

[89] R. M. Miura. The Korteweg-de Vries equation: A survey of results. *SIAM Review*, 18(3):412–459, 1976.

[90] Robert M. Miura. Bäcklund transformations, the inverse scattering method, solitons, and their applications. In *Lecture Notes in Mathematics 515, NSF Research Workshop on Contact Transformations.* Springer-Verlag, Berlin-Heidelberg-New York, 1976.

[91] A. C. Newell. The interrelation between Bäcklund transformations and the inverse scattering transform. In *Lecture Notes in Mathematics 515, NSF Research Workshop on Contact Transformations*, pages 227–240. Springer-Verlag, Berlin-Heidelberg-New York, 1976.

[92] L. P. Nizhnik. The inverse nonstationary scattering problem. *Dokl. Acad. Sci. USSR*, 196(5):1016–1019, 1971.

[93] L. P. Nizhnik. The inverse nonstationary scattering problem. *Naukova Dumka, Kiev*, 1973.

[94] L. P. Nizhnik. Integration of multidimensional nonlinear equations by the method of the inverse problem. *Doklady Akademii Nauk*, 254(2):332–335, 1980.

[95] L. P. Nizhnik. The inverse scattering problems for the hyperbolic equations and their application to non-linear integrable systems. *Reports on Mathematical Physics*, 26(2):261–283, 1988.

[96] L. P. Nizhnik and M. D. Pochinaiko. Integration of the nonlinear two-dimensional spatial Schrodinger equation by the inverse-problem method. *Functional Analysis and Its Applications*, 16(1):66–69, 1982.

[97] L. P. Nizhnik and P. L. Vu. The inverse scattering problem on a half-line with a non-seleconjugate potential matrix. *Ukrainian Mathematical Journal*, 26(4):384–398, 1974.

[98] L. P. Nizhnik and P. L. Vu. Nonstationary inverse scattering problem for a system of second-order equations. *Ukrainian Mathematical Journal*, 34(6):580–585, 1982.

[99] L. P. Nizhnik and P. L. Vu. *The inverse scattering problem for a system of ordinary second-order differential equation on the whole line*. Preprint (in Russian), Institute of Mathematics, National Academy of Sciences of the Ukraine, 1983.

[100] A. R. Osborne. Nonlinear ocean wave and the inverse scattering transform. In *Scattering*, pages 637–666. Elsevier, New York, 2002.

[101] W. Pogorzelski. *Integral equations and their applications*, volume I. Pergamon, New York, 1966.

[102] A. I. Prilepko. Existence of solutions of inverse problems in the potential theory. *Doklady Akademii Nauk SSSR*, 199(1):30–32, 1971.

[103] C. Rogers. On applications of generalized Bäcklund transformations to continuum mechanics. In *Lecture Notes in Mathematics 515, NSF Research Workshop on Contact Transformations*, pages 106–135. Springer-Verlag, Berlin-Heidelberg-New York, 1976.

[104] C. Rogers and W. F. Shadwick. *Bäcklund transformations and their applications.* Academic Press, New York London, 1982.

[105] V. G. Romanov. *Inverse problems for differential equations.* Novosiberian University, 1973.

[106] P. C. Sabatier. Elbow scattering and inverse scattering applications to LKdV and KdV. *Journal of Mathematical Physics*, 41(1):414–436, 2000.

[107] A. C. Scott. The application of Bäcklund transforms to physical problems. In *Lecture Notes in Mathematics 515, NSF Research Workshop on Contact Transformations*, pages 80–105. Springer-Verlag, Berlin-Heidelberg-New York, 1976.

[108] H. Segur. The Korteweg-de Vries equation and water waves. Solutions of the equation. Part 1. *Journal of Fluid Mechanics*, 59(4):721–736, 1973.

[109] A. B. Shabat. Inverse-scattering problem for a system of differential equations. *Functional Analysis and Its Applications*, 9(3):244–247, 1975.

[110] A. B. Shabat. An inverse scattering problem. *Differentsial'nye Uravneniya*, 15(10):1824–1834, 1979.

[111] A. B. Shabat and V. E. Zakharov. Exact theory of two-dimensional self-focusing and one-dimensional self-modulation of waves in nonlinear media. *Soviet Physics JETP*, 34(1):62–69, 1972.

[112] S. L. Sobolev. *Partial differential equations of mathematical physics.* Pergamon Press Ltd, 1964. Translation of Uravnenia matematicheskoi fiziki, 4th edition, Nauka, Moscow, 1966.

[113] L. N. Sretensky. On the uniqueness of the determination of the shape of an attractive body from its outer potential. *Dokl. Akad. Nauk SSSR*, 99(1):21–22, 1954.

[114] A. G. Sveshnikov and A. N. Tikhonov. *The theory of functions of a complex variable.* Nauka, Moscow, 1967. (English Translation: Mir, 1982).

[115] V. S. Vladimirov. Mathematical problems of the uniform-speed theory of transport. *Trudy Mat. Inst. Steklov., Acad. Sci. USSR, Moscow*, 61:3–158, 1961.

[116] V. S. Vladimirov. *Equations of mathematical physics.* Nauka, Moscow, 1967.

[117] P. L. Vu. The scattering problem for a system of Dirac equations on the whole line. In *Selected works: Boundary value problems of Mathematical Physics* (in Russian), pages 323–339. Institute of Mathematicics of National Academy of Sciences of the Ukraine, Kiev, 1971.

[118] P. L. Vu. The inverse scattering problem for a system of Dirac equations of the whole line. *Ukrainian Mathematical Journal*, 24(5):537–544, 1972.

[119] P. L. Vu. The inverse scattering problem for a system of Dirac equations on a half-line. In *Selected works: Linear boundary value problems of Mathematical Physics* (in Russian), pages 174–207. Institute of Mathematics of National Academy of Sciences of the Ukraine, Kiev, 1973.

[120] P. L. Vu. Solving an inverse problem in medium with absorption by the net method. *Selected works of Geophysics* (in Russian), 59:56–62, 1974.

[121] P. L. Vu. The inverse scattering problem for a system of ordinary first-order differential equations. In *Selected works: Spectral theory of operators and of infinitely dimensional analysis* (in Russian), pages 102–146. Institute of Mathematics of National Academy of Sciences of the Ukraine, Kiev, 1976.

[122] P. L. Vu. Nonstationary inverse scattering problem for perturbed string equation on the whole axis. *Ukrainian Mathematical Journal*, 32(5):413–419, 1980.

[123] P. L. Vu. *The direct and inverse scattering problem for perturbed string equation*. Preprint (in Russian), Institute of Mathematics of National Academy of Sciences of the Ukraine, 1981.

[124] P. L. Vu. Inverse scattering problem for the transport equation on the whole line. *Ukrainian Mathematical Journal*, 33(5):482–489, 1981.

[125] P. L. Vu. Non-stationary scattering problem for a system of second-order equations. In *Selected works: The direct and inverse problems of scattering theory* (in Russian), pages 136–144. Institute of Mathematics of National Academy of Sciences of the Ukraine, 1981.

[126] P. L. Vu. *Two-dimensional inverse scattering problems* (in Russian). Summary of Doctoral Dissertation. Institute of Mathematics of National Academy of Sciences of the Ukraine, 1983.

[127] P. L. Vu. Determination of parabolic equation by boundary conditions. *Mathematische Nachrichten*, 127:265–274, 1986.

[128] P. L. Vu. Inverse problems for parabolic equation. *Mathematische Nachrichten*, 141:325–334, 1989.

[129] P. L. Vu. Determination of a diffusion coefficient in the diffusion equation with a source and an initial distribution. *Pubblicazioni dell'Istituto di Analisi Globale e Applicazioni*, (56):11, 1991.

[130] P. L. Vu. Weak Cauchy problems for diffusion equations. *Inverse Problems*, 7(2):299–305, 1991.

[131] P. L. Vu. Weak Cauchy problems for turbolent diffusion equations with a gradient traverse velocity coefficient. *Pubblicazioni dell'Istituto di Analisi Globale e Applicazioni*, (58):14, 1991.

[132] P. L. Vu. Cauchy problems for turbulent diffusion equations with a traverse velocity contribution. *Bollettino dell Unione Matematica Italiana*, 7:893–906, 1993.

[133] P. L. Vu. Cauchy problems for a system of non-linear equations and for the non-linear Schrodinger equation. *Inverse Problems*, 10(2):415–429, 1994.

[134] P. L. Vu. Exact solutions of a system of nonlinear equations and of the Korteweg–de Vries equation. In *Proceedings of the International Workshop on Inverse Problems*, volume 2, pages 196–207, 1995.

[135] P. L. Vu. Initial value problems for a system of nonlinear equations on a half-line. *Acta Applicandae Mathematica*, 40(1):1–23, 1995.

[136] P. L. Vu. Le probleme de Cauchy pour l'equation de Korteweg-de Vries sur le demi-axe. *Vietnam Journal of Mathematics*, 23:199–209, 1995.

[137] P. L. Vu. Soliton solutions of a system of non-linear equations and of the non-linear Schrodinger equations. *Southeast Asian Bulletin of Mathematics*, 19(2):131–140, 1995.

[138] P. L. Vu. Exact solutions of nonlinear equations on a half-line. *Rendiconti dell'Istituto Lombardo*, (A 129):147–182, 1996.

[139] P. L. Vu. Explicit complex-valued solutions of the Korteweg–de Vries equation on the half-line and on the whole-line. *Acta Applicandae Mathematica*, 49(2):107–149, 1997.

[140] P. L. Vu. Some problems for a system of nonlinear equations on a half-line. *Inverse Problems*, 17(6):1889–1906, 2001.

[141] P. L. Vu. The solution of the Korteweg-de Vries equation on the half-line. In *Proceedings of the Conference on Partial Differential Equations and Their Applications*, pages 57–66. National Center for Natural Science and Technology, Hanoi, Institute of Mathematics, 2001.

[142] P. L. Vu. Some problems for cubic nonlinear equations on a half-line. *Acta Applicandae Mathematica*, 84(1):97–120, 2004.

[143] P. L. Vu. The Dirichlet initial-boundary-value problems for sine and sinh-Gordon equations on a half-line. *Inverse Problems*, 21(4):1225–1248, 2005.

[144] P. L. Vu. The initial-boundary value poblem for the Korteweg-de Vries equation on the positive quarter-plane. *Journal of Nonlinear Mathematical Physics*, 14(1):28–43, 2007.

[145] P. L. Vu. Integration of the continual system of nonlinear interaction waves. *Acta Applicandae Mathematicae*, 109(3):765–787, 2010.

[146] P. L. Vu. An initial-boundary value problem for the Korteweg–de Vries equation with dominant surface tension. *Acta Applicandae Mathematicae*, 129(1):41–59, 2014.

[147] P. L. Vu. The description of reflection coefficients of the scattering problems for finding solutions of the Korteweg–de Vries equations. *Journal of Nonlinear Mathematical Physics*, 25(3):399–432, 2018.

[148] P. L. Vu. The inverse scattering problem for the perturbed string equation and its application to integration of the two-dimensional generalization from Korteweg-de Vries equation. *Acta Mathematica Vietnamica*, 45(4):807–831, 2020.

[149] P. L. Vu and N. H. Hoang. On the degree of normalization polynomials of the scattering data for constructing solutions of the Korteweg-de Vries equation. *Southeast Asian Bulletin of Mathematics*, 24:631–641, 2000.

[150] P. L. Vu and N. H. Hoang. Constructing soliton solutions of the nonlinear Schrödinger equation by inverse scattering and Hirota's direct methods. *Vietnam Journal of Mathematics*, 30(2):149–165, 2002.

[151] Pham Loi Vu. *Inverse scattering problems and their application to nonlinear integrable equations.* CRC Press, Boca Raton, FL, USA, 2019.

[152] Pham Loi Vu. The Bäcklund transformations between a common solution of both linear equations in the Lax pair and solution of the associated initial-boundary value problem for nonlinear evolution equations on the half-line. *Submitted*, 2022.

[153] M. Wadati, K. Konno, and Y. H. Ichikawa. A generalization of inverse scattering method. *Journal of the Physical Society of Japan*, 46:1965–1966, 1979.

[154] G. B. Whitham. Nonlinear waves. In *Proceedings of the Royal Society A: Mathematical, Physical and Engineering Sciences*, volume 283, pages 238–261, 1965.

[155] G. B. Whitham. *Linear and nonlinear waves.* John Wiley, New York, 1974.

[156] V. E. Zakharov, S. V. Manakov, S. Novikov, and L. P. Pitaievskii. *Theory of solitons, the inverse method.* Nauka, Moscow, 1980.

Index